STATISTICS
Concepts and Controversies
Eighth Edition

David S. Moore
Purdue University

William I. Notz
The Ohio State University

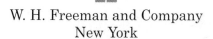

W. H. Freeman and Company
New York

Senior Publisher:	Ruth Baruth
Acquisitions Editor:	Karen Carson
Marketing Manager:	Steve Thomas
Developmental Editors:	Anne Scanlan-Rohrer, Katrina Wilhelm
Senior Media Editor:	Laura Judge
Associate Media Editor:	Catriona Kaplan
Marketing Assistant:	Alissa Nigro
Photo Editor:	Bianca Moscatelli
Photo Researcher:	Deborah Anderson
Cover and Text Designer:	Vicki Tomaselli
Project Editor:	Elizabeth Geller
Illustrations:	Network Graphics
Illustration Coordinator:	Bill Page
Production Coordinator:	Susan Wein
Composition:	MPS Limited
Printing and Binding:	Quad Graphics

Library of Congress Control Number: 2012943500

Instructor Complimentary Copy:

ISBN-13: 978-1-4641-2373-3
ISBN-10: 1-4641-2373-X

Student Edition Paperback (w/CrunchIT/EESEE Access Card):

ISBN-13: 978-1-4641-2566-9
ISBN-10: 1-4641-2566-X

Student Edition Loose leaf (w/CrunchIT/EESEE Access Card):

ISBN-13: 978-1-4641-2565-2
ISBN-10: 1-4641-2565-1

W. H. Freeman and Company
41 Madison Avenue
New York, NY 10010
Houndmills, Basingstoke RG21 6XS, England
www.whfreeman.com

BRIEF CONTENTS

*This material is optional.

CONTENTS

*This material is optional.

*This material is optional.

TO THE TEACHER: Statistics as a Liberal Discipline

tatistics: Concepts and Controversies (SCC) is a book on statistics as a liberal discipline, that is, as part of the general education of "non-mathematical" students. The book grew out of one of the author's experience in developing and teaching a course for freshmen and sophomores from Purdue University's School of Liberal Arts. We are pleased that other teachers have found *SCC* useful for unusually diverse audiences, extending as far as students of philosophy and medicine. This eighth edition is a revision of the text, with several new features. It retains, however, the goals of the original: to present statistics not as a technical tool but as part of the intellectual culture that educated people share.

Statistics among the liberal arts

Statistics has a widespread reputation as the least liberal of subjects. When statistics is praised, it is most often for its usefulness. Health professionals need statistics to read accounts of medical research; managers need statistics because efficient crunching of numbers will find its way to the bottom line; citizens need statistics to understand opinion polls and government statistics such as the unemployment rate and the Consumer Price Index. Because data and chance are omnipresent, our propaganda line goes, everyone will find statistics useful, and perhaps even profitable.

This is true. We would even argue that for most students the conceptual and verbal approach in *SCC* is better preparation for future encounters with statistical studies than the usual methods-oriented introduction. The joint curriculum committee of the American Statistical Association and the Mathematical Association of America recommends that any first course in statistics "emphasize the elements of statistical thinking" and feature "more data and concepts, fewer recipes and derivations." *SCC* does this, with the flavor appropriate to a liberal education: more concepts, more thinking, only simple data, fewer recipes, and no formal derivations.

There is, however, another justification for learning about statistical ideas: statistics belongs among the liberal arts. A liberal education emphasizes fundamental intellectual skills, that is, general methods of inquiry that apply in a wide variety of settings. The traditional liberal arts present such methods: literary and historical studies, the political and social analysis of human societies, the probing of nature by experimental science, the power of abstraction and deduction in mathematics. The case that statistics belongs among the liberal arts rests on the fact that reasoning from uncertain empirical data is a similarly general intellectual method. *Data* and *chance,* the topics of this book, are pervasive aspects of our experience. Though we employ the tools of mathematics to work with data and chance, the mathematics implements ideas that are not strictly mathematical. In fact, psychologists argue convincingly that mastering formal mathematics does little to improve our ability to reason effectively about data and chance in everyday life.

SCC is shaped, as far as the limitations of the authors and the intended readers allow, by the view that statistics is an independent and fundamental intellectual method. The focus is on statistical thinking, on what others might call *quantitative literacy* or *numeracy.*

The nature of this book

There are books on statistical theory and books on statistical methods. This is neither. It is a book on statistical ideas and statistical reasoning and on their relevance to public policy and to the human sciences from medicine to sociology. We have included many elementary graphical and numerical techniques to give flesh to the ideas and muscle to the reasoning. Students learn to think about data by working with data. We have not, however, allowed technique to dominate concepts. Our intention is to teach verbally rather than algebraically, to invite discussion and even argument rather than mere computation, though some computation remains essential. The coverage is considerably broader than one might traditionally cover in a one-term course, as the table of contents reveals. In the spirit of general education, we have preferred breadth to detail.

Despite its informal nature, *SCC* is a textbook. It is organized for systematic study and has abundant exercises, many of which ask students to offer a discussion or make a judgment. Even those admirable individuals who seek pleasure in uncompelled reading should look at the exercises as well as the text. Teachers should be aware that the book is more serious than its low mathematical level suggests. The emphasis on ideas and reasoning asks more of the reader than many recipe-laden methods texts.

New in this edition

This new version of a classic text fits the current teaching environment while continuing to present statistics to "nonmathematical" readers as an aid to clear thinking in personal and professional life. The following new features and enhancements build on *SCC*'s strong pedagogical foundation:

- **Content.** The use of the Internet for surveys is increasing but is controversial. We have therefore included a discussion of Web-based surveys in Chapter 4. To help make the concept of reliability more precise, we introduce variance in Chapter 8. In Chapter 17, we now include a mention of Bayes procedures as part of the discussion of personal probability.

- **Chapter summaries.** The "Statistics in Summary" sections at the end of each chapter now consist of two sections. One, titled "Chapter Specifics," summarizes the material presented in the chapter. The second section, titled "Link It," relates the chapter content to material in previous and upcoming chapters. The goal of this new format is to help students understand how individual chapters relate to each other and to the overall practice of statistics.

- **Examples and exercises.** Over one-third of the examples and exercises are revised to reflect current data and a variety of topics. They cover a wide range of application areas, adding interest and relevance for students. New example and exercise topics include Facebook and grades, low-fat foods and obesity, and texting.

- **Design.** The contemporary design incorporates colorful figures to aid students' understanding of text material. Sleek marginal notes invite students to explore "Statistics in Your World." Exploring the Web exercises are now labeled with a QR code icon, bringing students directly to the Book Companion Site for updated links and sources.

- **EESEE (Electronic Encyclopedia of Statistical Examples and Exercises) case studies.** Developed by The Ohio State University Department of Statistics, these electronic case studies provide students with a wide variety of timely, real examples with real data. EESEE case studies are available via an access code–protected Web site. Access codes are included with new copies of the eighth edition of *SCC*, or subscriptions can be purchased online. Over 40 new case studies have been added, and several are referenced in chapter-opening Case Studies.

- **Applets.** An applet icon signals where related, interactive statistical applets can be found on the Book Companion Site. The applets have been revised, and new applets have been created.

In addition to the new eighth-edition enhancements, *SCC* has retained the successful pedagogical features from previous editions:

- **Case Studies.** Beginning each chapter, Case Studies engage students in real-life scenarios related to the chapter concepts. The **Case Study Evaluated** section at the end of each chapter revisits the chapter-opening Case Study with follow-up questions, asking students to evaluate what they have learned from the chapter and to apply their knowledge to the Case Study.

- **Statistical Controversies.** These boxed features explore controversial topics and relate them to the chapter material. There is follow-up discussion and a proposed resolution to each of these topics in the back of the text, in the Resolving the Controversy section.

- **In the News exercises.** From popular news media outlets, these exercises use current events and cite recent data sources.

- **Now It's Your Turn exercises.** These appear after a worked example, allowing students to test their understanding. Full solutions to these exercises are provided in the back of the text.

- **Exploring the Web exercises.** Found in each chapter, these exercises point students to the Web to investigate topics and think critically about statistical data and concepts. These are now labeled with a QR code directing students to the Book Companion Site for updated links and sources.

Media and Supplements

For Students

STATS**PORTAL**

www.yourstatsportal.com (Access code or online purchase required.) StatsPortal is the digital gateway to *Statistics: Concepts and Controversies*, eighth edition (*SCC* 8e), and is designed to enrich the course and enhance students' study skills through a collection of Web-based tools. StatsPortal integrates a rich suite of diagnostic, assessment, tutorial, and enrichment features, enabling students to master statistics at their own pace. It is organized around three main teaching and learning components:

1. **Interactive e-Book** offers a complete and customizable online version of the text, fully integrated with all the media resources available with *SCC* 8e. The e-Book allows students to quickly search the text, highlight key areas, and add notes about what they're reading. Instructors can customize the e-Book to add, hide, and reorder content, add their own material, and highlight key text for students.

2. **Resources** organizes all the resources for *SCC* 8e into one location for ease of use. These resources include the following:

   - **Statistical Video Series** consisting of StatClips, StatClips Examples, and Statistically Speaking "Snapshots." View animated lecture videos, whiteboard lessons, and documentary-style footage that illustrate key statistical concepts and help students visualize statistics in real-world scenarios.

- **StatTutor Tutorials** offer audio-multimedia tutorials tied directly to the textbook, containing videos, applets, and animations.

- **LEARNING**Curve is a formative quizzing system that offers immediate feedback at the question level to help students master course material.

- **Statistical applets** offer a series of interactive applets to help students master key statistical concepts and work exercises from the text.

- **CrunchIt!**® **statistical software** allows users to analyze data from any Internet location. Designed with the novice user in mind, the software is not only easily accessible but also easy to use. CrunchIt!® offers all the basic statistical routines covered in introductory statistics courses and more.

- **Stat Tutorials** are algorithmically generated quizzing with step-by-step feedback and are easily assignable for homework.

- **Stats@Work Simulations** put students in the role of statistical consultants, helping them better understand statistics interactively within the context of real-life scenarios.

- **EESEE case studies,** developed by The Ohio State University Department of Statistics, teach students to apply their statistical skills by exploring actual case studies using real data.

- **Data sets** are available in ASCII, Excel, TI, Minitab, SPSS (an IBM Company),* and JMP formats.

- **Statistical software manuals** for TI-83/84, Minitab, and Excel provide instruction, examples, and exercises using specific statistical software packages.

- **Tables**

Resources for Instructors Only

- **Instructor's Solutions Manual** includes teaching suggestions, chapter comments, and detailed solutions to all exercises.

- **Test Bank** offers hundreds of multiple-choice questions.

- **Lecture PowerPoint slides** offer a detailed lecture presentation of statistical concepts covered in each chapter of *SCC* 8e.

---

*SPSS was acquired by IBM in October 2009.

- **SolutionMaster** is a Web-based version of the solutions in the Instructor's Solutions Manual. This easy-to-use tool allows instructors to generate a solution file for any set of homework exercises. Solutions can be downloaded in PDF format for convenient printing and posting. For more information or a demonstration, contact your local W. H. Freeman sales representative.

3. **Assignments** organizes assignments and guides instructors through an easy-to-create assignment process providing access to questions from the Test Bank and exercises from the text, including many algorithmic problems. The Assignment Center enables instructors to create their own assignments from a variety of question types for machine-gradable assignments. This powerful assignment manager allows instructors to select their preferred policies in regard to scheduling, maximum attempts, time limitations, feedback, and more!

Companion Web site www.whfreeman.com/scc8e This open-access Web site includes statistical applets, data sets, self-quizzes, updated links for Web exercises, and more.

Special Software Packages Student versions of JMP, Minitab, and SPSS are available on a CD-ROM packaged with the textbook. This software is not sold separately and must be packaged with a text or a manual. Contact your local W. H. Freeman sales representative for information or visit **www.whfreeman.com.**

Video Tool Kit (Access code or online purchase required.) This new Statistical Video Series consists of three types of videos aimed to illustrate key statistical concepts and help students visualize statistics in real-world scenarios:

- **StatClips** lecture videos, created and presented by Alan Dabney, PhD, Texas A&M University, are innovative visual tutorials that illustrate key statistical concepts. In three to five minutes, each StatClips video combines dynamic animation, data sets, and interesting scenarios to help students understand the concepts in an introductory statistics course.

- In **StatClips Examples,** Alan Dabney walks students through step-by-step examples related to the StatClips lecture videos to reinforce the concepts through problem solving.

- **Snapshots** videos are abbreviated, student-friendly versions of the **Statistically Speaking** video series that bring the world of statistics into the classroom. In the same vein as the successful PBS series

Against All Odds: Inside Statistics, Statistically Speaking videos use new and updated documentary footage and interviews that show real people using data analysis to make important decisions in their careers and in their daily lives. From business to medicine, from the environment to understanding the census, Snapshots focus on why statistics is important for students' careers, and how statistics can be a powerful tool to understand their world.

Lab and Activities Manual by Dennis Pearl, The Ohio State University. Available electronically through a custom StatsPortal, this manual provides a variety of projects and exercises to help students develop a fuller appreciation of statistical concepts. It features computer lab and hands-on activities illustrating key concepts in the text, as well as additional end-of-chapter-type problems and activities. Additionally, there are exercises based on the statistical applets and EESEE Case Studies (both accessed through the book's Web site).

Software Manuals covering Minitab, Excel, and TI-83/84 are offered within StatsPortal. These manuals are also available in printed versions through custom publishing.

For Instructors

Instructor's Web site (www.whfreeman.com/scc8e) requires user registration as an instructor and features all of the student Web materials plus:

- Instructor version of **EESEE** (Electronic Encyclopedia of Statistical Examples and Exercises), with solutions to the exercises in the student version.

- **PowerPoint slides** containing all textbook figures and tables.

- **Lecture PowerPoint slides** offering a detailed lecture presentation of statistical concepts covered in each chapter of *SCC* 8e.

Printed Instructor's Solutions Manual and Test Bank includes full solutions to all exercises and hints on teaching from *SCC*. The test bank contains hundreds of multiple-choice questions to generate quizzes and tests for each chapter of the text. ISBN: 1-4292-7760-2 The test bank is also available electronically on CD-ROM (for Windows and Mac), allowing questions to be downloaded, edited, and resequenced. ISBN: 1-4292-7909-5

Enhanced Instructor's Resource CD-ROM allows instructors to **search** and **export** (by key term or chapter) the following material:

- Data sets
- All text images and tables
- Instructor's Solutions Manual and Test Bank files
- PowerPoint files and lecture slides

ISBN: 1-4292-7908-7

Course Management Systems W. H. Freeman and Company provides courses for Blackboard, WebCT (Campus Edition and Vista), Angel, Desire2Learn, Moodle, and Sakai course management systems. These are completely integrated solutions that you can easily customize and adapt to meet your teaching goals and course objectives. Visit **www .macmillanhighered.com/Catalog/other/Coursepack** for more information.

i-clicker

i-clicker is a two-way radio-frequency classroom response solution developed by educators for educators. University of Illinois physicists Tim Stelzer, Gary Gladding, Mats Selen, and Benny Brown created the i-clicker system after using competing classroom response solutions and discovering that they were neither classroom-appropriate nor student-friendly. Each step of i-clicker's development has been informed by teaching and learning. i-clicker is superior to other systems from both a pedagogical and a technical standpoint. To learn more about packaging i-clicker with this textbook, please contact your local W. H. Freeman sales representative or visit **www.iclicker.com.**

Acknowledgments

The staff of W. H. Freeman, especially Mary Louise Byrd, Pamela Bruton, Karen Carson, Elizabeth Geller, Bianca Moscatelli, Anne Scanlan-Rohrer, Vicki Tomaselli, Susan Wein, and Katrina Wilhelm, have done their usual excellent job in editing, designing, and producing the book. We also thank Jackie Miller for carefully checking the accuracy of the manuscript. We are grateful to many colleagues who commented on successive drafts of the manuscript.

John Deely, *Purdue University*
Linda M. Deptola, *Rappahannock Community College*
Robert Floden, *Michigan State University*
Brian Garant, *Prairie State College*
Mark A. Gebert, *University of Kentucky*

Michael Granaas, *University of South Dakota*
Brian Habing, *University of South Carolina*
Leslie Hendrix, *University of South Carolina*
Mic Jackson, *Earlham College*

Stacy Karl, *University of Minnesota*

Yoon G. Kim, *Humboldt State University*

Rose Martinez-Dawson, *Clemson University*

Philip Meyers, *Marymount Manhattan College*

Robert Adam Molnar, *Bellarmine University*

Jacquelyn O'Donohue, *Plymouth State University*

Robert Poulson, *Kansas State University*

Sarah Quesen, *West Virginia University*

Kelly Quinn, *University of Illinois at Chicago*

Mamunur Rashid, *Indiana University–Purdue University Indianapolis*

Donald Richards, *Pennsylvania State University*

Ned Schillow, *Lehigh Carbon Community College*

Dana Sylvan, *Hunter College, The City College of New York*

Dennis Wacker, *Saint Louis University*

Sheila O'Leary Weaver, *University of Vermont*

Andrew Wiesner, *Pennsylvania State University*

Hongling Yang, *University of Texas at El Paso*

Andrew Zieffler, *University of Minnesota*

We are also grateful to those who reviewed previous editions.

Marcus Agustin, *Southern Illinois University, Edwardsville*

Ma. Zenia Agustin, *Southern Illinois University, Edwardsville*

Eric Agyekum, *Malaspina University-College*

Georgiana Baker, *University of South Carolina*

Jennifer Beineke, *Western New England College*

Melody L. Boyd, *Temple University*

Patricia M. Buchanan, *Pennsylvania State University*

Barbara K. Caress, *The City College of New York*

Melissa Cass, *State University of New York at New Paltz*

Joanne Christopherson, *California State University, Fullerton*

Samuel J. Clark, *University of Washington*

Diane Conway, *Bowling Green State University*

Colette Currie, *National Louis University*

Jimmy Doi, *California Polytechnic State University*

John Dugan, *University of Idaho*

Rick L. Edgeman, *University of Idaho*

Christopher J. Ferguson, *Texas A&M International University*

Joseph Gershtenson, *Eastern Kentucky University*

Jane J. Gringauz, *Minneapolis Community and Technical College*

Timothy Grosse, *Jefferson Community College*

Debra Hall, *Indiana University–Purdue University Indianapolis*

Pamela Harman, *California State University, Los Angeles*

Richard John, *University of Southern California*

Patricia A. Kan, *Cleveland State University*

Bonnie Kegan, *University of Maryland, Baltimore*

Rasul A. Khan, *Cleveland State University*

Josh Klugman, *Temple University*

Patrick Lang, *Idaho State University*

Natalie Lochner, *Rollins College*

Ulric Lund, *California Polytechnic State University*

Megan E. Lutz, *Virginia Polytechnic Institute and State University*

Antoinette Marquard, *Cleveland State University*

Jason Martin, *Temple University*

Eric Matsuoka, *Leeward Community College*

James E. Mays, *Virginia Commonwealth University*

Michael McGill, *Virginia Polytechnic Institute and State University*

Alana Northrop, *California State University, Fullerton*

William Rayens, *University of Kentucky*

Jonathan P. Schinohofen, *Bluegrass Community and Technical College*

Majid Shahidi, *Cegep Vanier College*

Louis Soukup, *Bellevue University*

W. Scott Street IV, *Virginia Commonwealth University*

Richard Tardanico, *Florida International University*

Sharon Taylor, *Georgia Southern University*

Agnes Tuska, *California State University, Fresno*

Lewis VanBrackle, *Kennesaw State University*

Gregg G. Van Ryzin, *Rutgers, The State University of New Jersey*

C. K. Venkateswaran, *Baker College*

Nathalie Viau, *Cegep Vanier College*

Gary P. Visco, *University of Vermont*

Elizabeth Walters, *Loyola College in Maryland*

Paul Watson, *Jefferson Community College*

Bethany White, *University of Western Ontario*

George P. Yanev, *University of South Florida*

Yi Yang, *James Madison University*

Jill C. Zimmerman, *Manchester Community College*

APPLICATIONS INDEX

■ Example   ■ Exercise   ■ Project

Statistics is about data. Data are numbers, but they are not "just numbers." *Data are numbers with a context.* The number 10.5, for example, carries no information by itself. But if we hear that a friend's new baby weighed 10.5 pounds at birth, we congratulate her on the healthy size of the child. The context engages our background knowledge and allows us to make judgments. We know that a baby weighing 10.5 pounds is quite large, and that a human baby is unlikely to weigh 10.5 ounces or 10.5 kilograms. The context makes the number informative.

Statistics uses data to gain insight and to draw conclusions. The tools are graphs and calculations, but the tools are guided by ways of thinking that amount to educated common sense. Let's begin our study of statistics with a rapid and informal guide to coping with data and statistical studies in the news media and in the heat of political and social controversy. We will examine the examples introduced in this prelude in more detail later.

Data beat anecdotes

> *Belief is no substitute for arithmetic.*
>
> HENRY SPENCER

An anecdote is a striking story that sticks in our minds exactly because it is striking. Anecdotes humanize an issue, so news reports usually start (and often stop) with anecdotes. But anecdotes are weak ground for making up your mind—they are often misleading exactly because they are striking. Always ask if a claim is backed by data, not just by an appealing personal story.

Does living near power lines cause leukemia in children? The National Cancer Institute spent 5 years and $5 million gathering data on the question. Result: no connection between leukemia and exposure to magnetic fields of the kind produced by power lines. The editorial that accompanied the study report in the *New England Journal of Medicine* thundered, "It is time to stop wasting our research resources" on the question.

Now compare the impact of a television news report of a 5-year, $5 million investigation with that of a televised interview with an articulate mother whose child has leukemia and who happens to live near a power line. In the public mind, the anecdote wins every time. Be skeptical. Data are more reliable than anecdotes because they systematically describe an overall picture rather than focus on a few incidents.

We are tempted to add, "Data beat self-proclaimed experts." The idea of balance held by much of the news industry is to present a quick statement by an "expert" on either side. We never learn that one expert expresses the consensus of an entire field of science, while the other is a quack with a special-interest axe to grind. As a result of the media's taste for conflict, the public now thinks that for every expert there is an equal and opposite expert. If you really care about an issue, try to find out what the data say and how good the data are. Many issues do remain unsettled, but many others are unsettled only in the minds of people who don't care about evidence. You can start by looking at the credentials of the "experts" and at whether the studies they cite have appeared in journals that require careful outside review before they publish a claim.

Where the data come from is important

Figures won't lie but liars will figure.

CHARLES GROSVENOR

Data are numbers, and numbers always seem solid. Some are and some are not. Where the data come from is the single most important fact about any statistical study. When Ann Landers asked readers of her advice column whether they would have children again and 70% of those who replied shouted "No," readers should have just amused themselves with Ann's excerpts from tear-stained letters describing what beasts the writers' children are. Ann Landers was in the entertainment business. Her invitation attracted parents who regretted having their children. Most parents don't regret having children. We know this because opinion polls have asked large numbers of parents, chosen at random to avoid attracting one opinion or another. Opinion polls have their problems, as we will see, but they beat just asking upset people to write in.

Even the most reputable publications have not been immune to bad data. The *Journal of the American Medical Association* once printed an article claiming that pumping refrigerated liquid through tubes in the stomach relieves ulcers. The patients did respond, but only because patients often respond to *any* treatment given with the authority of a trusted doctor. That is, placebos (dummy treatments) work. When a skeptic finally tried a properly controlled study in which some patients got the tube and some got a placebo, the placebo actually did a bit better. "No comparison, no conclusion" is a good starting point for judging medical studies. We would be skeptical about the ongoing interest in "natural remedies," for example. Few of these have passed a comparative trial to show that they are more than just placebos sold in bottles bearing pretty pictures of plants.

Beware the lurking variable

I have enough money to last me the rest of my life, unless I buy something.

JACKIE MASON

You read that crime is higher in counties with gambling casinos. A college teacher says that students who took a course online did better than the students in the classroom. Government reports emphasize that well-educated people earn a lot more than people with less education. Don't jump to conclusions. Ask first, "What is there that they didn't tell me that might explain this?"

Crime is higher in counties with casinos, but it is also higher in urban counties and in poor counties. What kind of counties are casinos in? Did these counties have high crime rates before the casino arrived? The online students did better, but they were older and better prepared than the in-class students. No wonder they did better. Well-educated people do earn a lot. But educated people have (on the average) parents with more education and more money than the parents of poorly educated people have. They grew up in nicer places and went to better schools. These advantages help them get more education and would help them earn more even without that education.

All these studies report a connection between two variables and invite us to conclude that one of these variables influences the other. "Casinos increase crime" and "Stay in school if you want to be rich" are the messages we hear. Perhaps these messages are true. But perhaps much of the connection is explained by other variables lurking in the background, such as the nature of counties that accept casinos and the advantages that highly educated people were born with. Good statistical studies look at lots of background variables. This is tricky, but you can at least find out if it was done.

Variation is everywhere

When the facts change, I change my mind. What do you do, sir?
JOHN MAYNARD KEYNES

If a thermometer under your tongue reads higher than 98.6°F, do you have a fever? Maybe not. People vary in their "normal" temperature. Your own temperature also varies—it is lower around 6 A.M. and higher around 6 P.M. The government announces that the unemployment rate rose a tenth of a percent last month and that new home starts fell by 3%. The stock market promptly jumps (or sinks). Stocks are jumpier than is sensible. The government data come from samples that give good estimates but not the exact truth. Another run of the same samples would give slightly different answers. And economic facts jump around anyway, due to weather, strikes, holidays, and all sorts of other reasons.

Many people join the stock market in overreacting to minor changes in data that are really nothing but background noise. Here is Arthur

Nielsen, head of the country's largest market research firm, describing his experience:

> Too many business people assign equal validity to all numbers printed on paper. They accept numbers as representing Truth and find it difficult to work with the concept of probability. They do not see a number as a kind of shorthand for a range that describes our actual knowledge of the underlying condition.

Variation is everywhere. Individuals vary; repeated measurements on the same individual vary; almost everything varies over time. Ignore the pundits who try to explain the deep reasons behind each day's stock market moves, or who condemn a team's ability and character after a game decided by a last-second shot that did or didn't go in.

Conclusions are not certain

As far as the laws of mathematics refer to reality they are not certain, and as far as they are certain they do not refer to reality.

ALBERT EINSTEIN

Because variation is everywhere, statistical conclusions are not certain. Most women who reach middle age have regular mammograms to detect breast cancer. Do mammograms really reduce the risk of dying of breast cancer? Statistical studies of high quality find that mammograms reduce the risk of death in women aged 50 to 64 years by 26%. That's an average over all women in the age group. Because variation is everywhere, the results are different for different women. Some women who have mammograms every year die of breast cancer, and some who never have mammograms live to 100 and die when they crash their motorcycles.

What the summary study actually said was "mammography reduces the risk of dying of breast cancer by 26 percent (95 percent confidence interval, 17 to 34 percent)." That 26% is, in Arthur Nielsen's words, "shorthand for a range that describes our actual knowledge of the underlying condition." The range is 17% to 34%, and we are 95% confident that the truth lies in that range. We're pretty sure, in other words, but not certain. Once you get beyond news reports, you can look for phrases like "95% confident" and "statistically significant" that tell us that a study did produce findings that, while not certain, are pretty sure.

Data reflect social values

It's easy to lie with statistics. But it is easier to lie without them.

FREDERICK MOSTELLER

Good data do beat anecdotes. Data are more objective than anecdotes or loud arguments about what might happen. Statistics certainly lies on the factual, scientific, rational side of public discourse. Statistical studies deserve more weight than most other evidence about controversial issues. There is, however, no such thing as perfect objectivity. Statistics shares a social context that influences what we decide to measure and how we measure it.

Suicide rates, for example, vary greatly among nations. It appears that much of the difference in the reported rates is due to social attitudes rather than to actual differences in suicide rates. Counts of suicides come from death certificates. The officials who complete the certificates (details vary depending on the state or nation) can choose to look more or less closely at, for example, drownings and falls that lack witnesses. Where suicide is stigmatized, deaths are more often reported as accidents. Countries that are predominantly Catholic have lower reported suicide rates than others, for example. Japanese culture has a tradition of honorable suicide as a response to shame. This tradition leads to better reporting of suicide in Japan because it reduces the stigma attached to suicide. In other nations, changes in social values may lead to higher suicide counts. It is becoming more common to view depression as a medical problem rather than a weakness of character and suicide as a tragic end to the illness rather than a moral flaw. Families and doctors then become more willing to report suicide as the cause of death.

Social values influence data on matters less sensitive than suicide. The percentage of people who are unemployed in the United States is measured each month by the Bureau of Labor Statistics, using a large and very professionally chosen sample of people across the country. But what does it mean to be "unemployed"? It means that you don't have a job even though you want a job and have *actively looked for work in the last four weeks*. If you went four weeks without seeking work, you are not unemployed; you are "out of the labor force." This definition of unemployment reflects the value we attach to working. A different definition might give a very different unemployment rate.

Our point is not that you should mistrust the unemployment rate. The definition of "unemployment" has been stable over time, so that we can see trends. The definition is reasonably consistent across nations, so that we can make international comparisons. The data are produced by professionals free of political interference. The unemployment rate is important and useful information. Our point is that not everything important can be reduced to numbers and that reducing things to numbers is done by people influenced by many pressures, conscious and unconscious.

This isn't a book about the tools of statistics. It is a book about statistical ideas and their impact on everyday life, public policy, and many different fields of study. You will learn some tools, of course. Life will be easier if you have in hand *a calculator with built-in statistical functions.* Specifically, you need a calculator that will find means, standard deviations, and correlations. Look for a calculator that claims to do "two-variable statistics" or mentions "correlation." If you have access to a computer with statistical software, so much the better. On the other hand, you need little formal mathematics. If you can read and use simple equations, you are in good shape. Be warned, however, that you will be asked to think. Thinking exercises the mind more deeply than following mathematical recipes. *Statistics: Concepts and Controversies* presents statistical ideas in four parts:

I. **Data production** describes methods for producing data that can give clear answers to specific questions. Where the data come from really is important—basic concepts about how to select samples and design experiments are the most influential ideas in statistics.

II. **Data analysis** concerns methods and strategies for exploring, organizing, and describing data using graphs and numerical summaries. You can learn to look at data intelligently even with quite simple tools.

III. **Probability** is the language we use to describe chance, variation, and risk. Because variation is everywhere, probabilistic thinking helps separate reality from background noise.

IV. **Statistical inference** moves beyond the data in hand to draw conclusions about some wider universe, taking into account that variation is everywhere and that conclusions are uncertain.

Ultimately, data are used to draw conclusions or make decisions. The process of reasoning from data consists of several steps that yield a case for the validity of the final conclusion. Each part of this book discusses issues that affect the quality of the steps in this process. It is easy to focus on mastering the details in each chapter and lose track of how these details contribute to the overall argument. To help you see how the individual chapters fit into the overall argument, we end each chapter with a section that we call "Link It," which briefly describes how the contents of the chapter fit into the overall reasoning process. You will find this section immediately before the "Statistics in Summary" subsection.

Statistical ideas and tools emerged only slowly from the struggle to work with data. Two centuries ago, astronomers and surveyors faced the problem of combining many observations that, despite the greatest care, did not exactly match. Their efforts to deal with variation in their data produced some of the first statistical tools. As the social sciences emerged in the 19th century, old statistical ideas were transformed and new ones were invented to describe the variation in individuals and societies. The study of heredity and of variable populations in biology brought more advance. The first half of the 20th century gave birth to statistical designs for producing data and to statistical inference based on probability. By midcentury it was clear that a new discipline had been born. As all fields of study place more emphasis on data and increasingly recognize that variability in data is unavoidable, statistics has become a central intellectual method. Every educated person should be acquainted with statistical reasoning. Reading this book will enable you to make that acquaintance.

David S. Moore is Shanti S. Gupta Distinguished Professor of Statistics, Emeritus, at Purdue University and was 1998 president of the American Statistical Association. He received his AB from Princeton University and his PhD from Cornell University, both in mathematics. He has written many research papers in statistical theory and served on the editorial boards of several major journals. Professor Moore is an elected fellow of the American Statistical Association and of the Institute of Mathematical Statistics and an elected member of the International Statistical Institute. He has served as program director for statistics and probability at the National Science Foundation.

Professor Moore has devoted much of his career to the teaching of statistics. He was the content developer for the Annenburg/Corporation for Public Broadcasting college-level telecourse *Against All Odds: Inside Statistics* and for the series of video modules *Statistics: Decisions through Data,* intended to aid the teaching of statistics in schools. He is the author of influential articles on statistical education and of several leading textbooks. Professor Moore has served as president of the International Association for Statistical Education and has received the Mathematical Association of America's national award for distinguished college or university teaching of mathematics.

William I. Notz is Professor of Statistics at The Ohio State University. He received his BS in physics from Johns Hopkins University and his PhD in mathematics from Cornell University. His first academic job was as an assistant professor in the Department of Statistics at Purdue University. While there, he taught the introductory concepts course with Professor Moore, using the first edition of *Statistics: Concepts and Controversies.* As a result of this experience he developed an interest in statistical education. Professor Notz is a coauthor of EESEE (the Electronic Encyclopedia of Statistical Examples and Exercises) and has coauthored several textbooks.

Professor Notz's research interests have focused on experimental design and computer experiments. He is the author of several research papers and of a book on the design and analysis of computer experiments. He is an elected fellow of the American Statistical Association and an elected member of the International Statistical Institute. He has served as the editor of the journals *Technometrics* and *Journal of Statistics Education,* as well as on the editorial boards of several journals. At The Ohio State University, he has served as the Director of the Statistical Consulting Service, as acting chair of the Department of Statistics, and as associate dean in the College of Mathematical and Physical Sciences. He is a winner of The Ohio State University's Alumni Distinguished Teaching Award.

Producing Data

Y ou and your friends are not typical. What you listen to on the radio, for example, is probably not what we listen to. Of course, we and our friends are also not typical. To get a true picture of the country as a whole (or even of college students), we must recognize that the picture may not resemble us or what we see around us. We need *data*. Data from Arbitron (a media research firm) show that the most popular radio formats are adult contemporary (a weekly cumulative audience of 72 million in Autumn 2009) and country (a weekly cumulative audience of 64 million). If you like pop contemporary hit radio (a weekly cumulative audience of 59 million) and we like all news (a weekly cumulative audience of only 15 million), we may have no clue about the tastes of radio audiences as a whole. If we are in the broadcasting business, or even if we are interested in pop culture, we must put our own tastes aside and look at the data.

You can find data in the library or on the Internet (that's where we found the radio format data). But how can we know whether data can be trusted? Good data are as much a human product as wool sweaters and tablet PCs. Sloppily produced data will frustrate you as much as a sloppily made sweater. You examine a sweater before you buy, and you don't buy if it is not well made. Neither should you use data that are not well made. The first part of this book shows how to tell if data are well made.

Where Do Data Come From?

CASE STUDY You can read the newspaper and watch TV news for months without seeing an algebraic formula. No wonder algebra seems unconnected to life. You can't go a day, however, without meeting data and statistical studies. You hear that last month's unemployment rate was 10.0%. A news article says that 70% of people aged 18 to 24 believe that downloading music from the Internet is no different from buying a used CD or recording music borrowed from a friend, as opposed to 36% of people 65 or older. A longer article says that low-income children who received high-quality day care did better on academic tests given years later and were more likely to go to college and hold good jobs than other similar children.

Hemera/Thinkstock/Getty Images

Where do these data come from? Why can we trust them? Or maybe we can't trust them. Good data are the fruit of intelligent human effort. Bad data result from laziness or lack of understanding, or even the desire to mislead others. "Where do the data come from?" is the first question you should ask when someone throws a number at you.

During the 2008 presidential election, "Fork Over Your Vote" was a national project in which people chose who they thought should be the president by tossing a jelly bean into a jar. At the Coffee Scene in Pembroke Pines, Florida, customers got to choose either a blue or a red jelly bean with every purchase. Blueberry jelly beans were votes for Democrat Barack Obama. Cherry-strawberry candies were for McCain.

More than 100 restaurants in 34 states were involved in the project. Votes were counted every Monday. As of October 13, 2008, McCain had 53.5% of the national jelly bean vote. More than 211,000 bean ballots had been cast. The same jelly bean vote in the 2004 election correctly predicted President George W. Bush would win the presidency. Unfortunately for John McCain, the poll incorrectly predicted a victory for McCain in 2008 and it will be interesting to see if the project is repeated in the 2012 election. No information was available in January 2012 when we last checked.

What can we say about data from this poll? By the end of this chapter you will have learned some basic questions to ask about the data from the jelly bean poll. The answers to these questions will help us assess whether the data from the jelly bean poll are good or bad, as we will explore further in Chapter 2. ■

Talking about data: individuals and variables

Statistics is the science of data. We could almost say "the art of data," because good judgment and even good taste along with good math make good statistics. A big part of good judgment lies in deciding what you must measure in order to produce data that will shed light on your concerns. We begin with some vocabulary to describe the raw materials that go into data.

Individuals and variables

Individuals are the objects described by a set of data. Individuals may be people, but they may also be animals or things.

A **variable** is any characteristic of an individual. A variable can take different values for different individuals.

For example, here are the first lines of a professor's data set at the end of a statistics course:

```
NAME                MAJOR    POINTS   GRADE

ADVANI,  SURA       COMM      397      B
BARTON,  DAVID      HIST      323      C
BROWN,   ANNETTE    LIT       446      A
CHIU,    SUN        PSYC      405      B
CORTEZ,  MARIA      PSYC      461      A
```

The *individuals* are students enrolled in the course. In addition to each student's name, there are three *variables*. The first says what major a student has chosen. The second variable gives the student's total points out of 500 for the course, and the third records the grade received.

Statistics deals with numbers, but not all variables are numerical. Some are "categorical" and simply place an individual into one of several groups or categories. Of the three variables in the professor's data set, only total points has numbers as its values. Major and grade are categorical and to do statistics with these variables, we use *counts* or *percentages*. We might give the percentage of students who got an A, for example, or the percentage who are psychology majors.

Bad judgment in choosing variables can lead to data that cost lots of time and money but don't shed light on the world. What constitutes good judgment can be controversial. Here are examples of the challenges in deciding what data to collect.

EXAMPLE 1 Who recycles?

Who takes the trouble to recycle? Researchers spent lots of time and money weighing the stuff put out for recycling in two neighborhoods in a California city; call them Upper Crust and Lower Mid. The *individuals* here are households, because trash and recycling pickup are done for residences, not for people one at a time. The *variable* measured was the weight in pounds of the curbside recycling basket each week.

The Upper Crust households contributed more pounds per week on the average than did the folk in Lower Mid. Can we say that the rich are more serious about recycling? No. Someone noticed that Upper Crust recycling baskets contained lots of heavy glass wine bottles. In Lower Mid, they put out lots of light plastic soda bottles and light metal beer and soda cans. The conclusion: weight tells us little about commitment to recycling.

EXAMPLE 2 What's your race?

The U.S. census asks, "What is this person's race?" for every person in every household. "Race" is a *variable,* and the Census Bureau must say exactly how to measure it. The census form does this by giving a list of races. Years of political squabbling lie behind this list.

How many races shall we list, and what names shall we use for them? Shall we have a category for people of mixed race? Asians wanted more national categories, such as Filipino and Vietnamese, for the growing Asian population. Pacific Islanders wanted to be separated from the larger Asian group. Black leaders did not want a mixed-race category, fearing that many blacks would choose it and so reduce the official count of the black population.

The 2010 census form (see Figure 1.1) ended up with six Asian groups (plus "Other Asian") and three Pacific Island groups (plus "Other Pacific Islander"). There is no "mixed race" group, but you can mark more than one race. That is, people claiming mixed race can count as both, so that the total of the racial group counts in 2010 is larger than the population count. Unable to decide what the proper term for blacks should be, the Census Bureau settled on "Black, African American, or Negro." What about Hispanics? That's a separate question, because Hispanics can be of

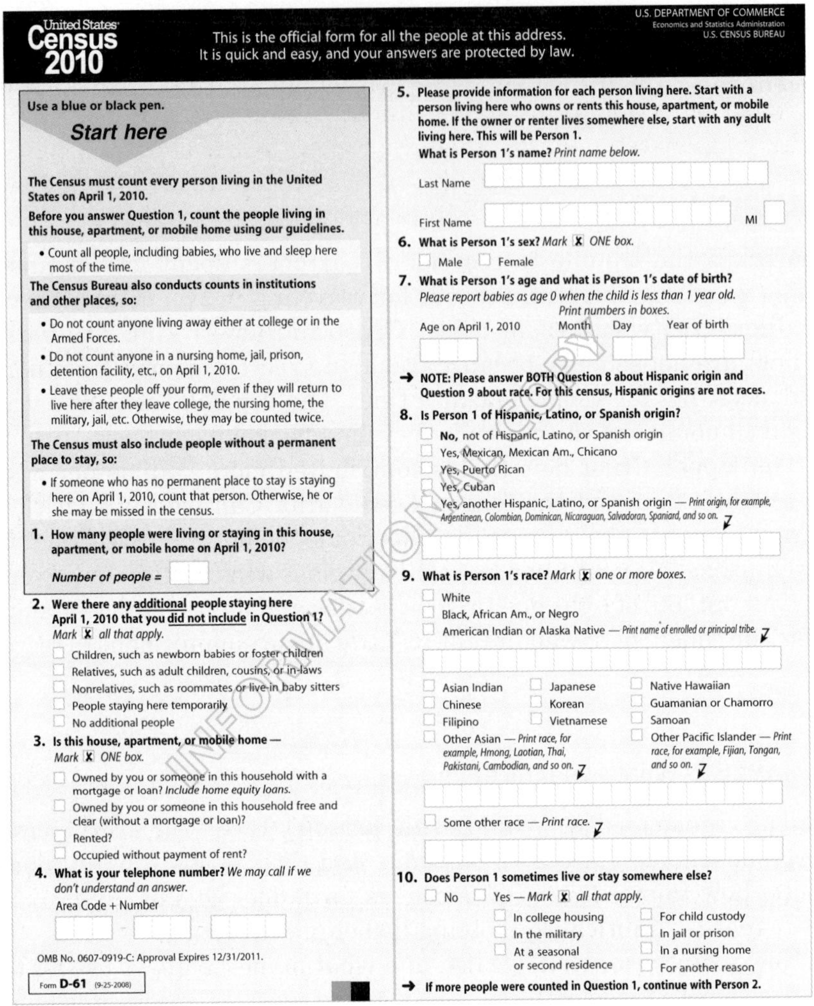

United States Census 2010

This is the official form for all the people at this address.
It is quick and easy, and your answers are protected by law.

U.S. DEPARTMENT OF COMMERCE
Economics and Statistics Administration
U.S. CENSUS BUREAU

Use a blue or black pen.

Start here

The Census must count every person living in the United States on April 1, 2010.

Before you answer Question 1, count the people living in this house, apartment, or mobile home using our guidelines.

- Count all people, including babies, who live and sleep here most of the time.

The Census Bureau also conducts counts in institutions and other places, so:

- Do not count anyone living away either at college or in the Armed Forces.
- Do not count anyone in a nursing home, jail, prison, detention facility, etc., on April 1, 2010.
- Leave these people off your form, even if they will return to live here after they leave college, the nursing home, the military, jail, etc. Otherwise, they may be counted twice.

The Census must also include people without a permanent place to stay, so:

- If someone who has no permanent place to stay is staying here on April 1, 2010, count that person. Otherwise, he or she may be missed in the census.

1. How many people were living or staying in this house, apartment, or mobile home on April 1, 2010?

Number of people = ☐

2. Were there any additional people staying here April 1, 2010 that you did not include in Question 1?
Mark ☒ *all that apply.*

☐ Children, such as newborn babies or foster children
☐ Relatives, such as adult children, cousins, or in-laws
☐ Nonrelatives, such as roommates or live-in baby sitters
☐ People staying here temporarily
☐ No additional people

3. Is this house, apartment, or mobile home —
Mark ☒ *ONE box.*

☐ Owned by you or someone in this household with a mortgage or loan? *Include home equity loans.*
☐ Owned by you or someone in this household free and clear (without a mortgage or loan)?
☐ Rented?
☐ Occupied without payment of rent?

4. What is your telephone number? *We may call if we don't understand an answer.*
Area Code + Number

OMB No. 0607-0919-C: Approval Expires 12/31/2011.

Form **D-61** (9-25-2008)

5. Please provide information for each person living here. Start with a person living here who owns or rents this house, apartment, or mobile home. If the owner or renter lives somewhere else, start with any adult living here. This will be Person 1.
What is Person 1's name? *Print name below.*

Last Name

First Name MI

6. What is Person 1's sex? *Mark* ☒ *ONE box.*
☐ Male ☐ Female

7. What is Person 1's age and what is Person 1's date of birth?
Please report babies as age 0 when the child is less than 1 year old.
Print numbers in boxes.
Age on April 1, 2010 Month Day Year of birth

→ **NOTE: Please answer BOTH Question 8 about Hispanic origin and Question 9 about race. For this census, Hispanic origins are not races.**

8. Is Person 1 of Hispanic, Latino, or Spanish origin?
☐ **No,** not of Hispanic, Latino, or Spanish origin
☐ Yes, Mexican, Mexican Am., Chicano
☐ Yes, Puerto Rican
☐ Yes, Cuban
☐ Yes, another Hispanic, Latino, or Spanish origin — *Print origin, for example, Argentinean, Colombian, Dominican, Nicaraguan, Salvadoran, Spaniard, and so on.*

9. What is Person 1's race? *Mark* ☒ *one or more boxes.*
☐ White
☐ Black, African Am., or Negro
☐ American Indian or Alaska Native — *Print name of enrolled or principal tribe.*

☐ Asian Indian ☐ Japanese ☐ Native Hawaiian
☐ Chinese ☐ Korean ☐ Guamanian or Chamorro
☐ Filipino ☐ Vietnamese ☐ Samoan
☐ Other Asian — *Print race, for example, Hmong, Laotian, Thai, Pakistani, Cambodian, and so on.*
☐ Other Pacific Islander — *Print race, for example, Fijian, Tongan, and so on.*

☐ Some other race — *Print race.*

10. Does Person 1 sometimes live or stay somewhere else?
☐ No ☐ Yes — *Mark* ☒ *all that apply.*
☐ In college housing ☐ For child custody
☐ In the military ☐ In jail or prison
☐ At a seasonal or second residence ☐ In a nursing home
☐ For another reason

→ **If more people were counted in Question 1, continue with Person 2.**

USCENSUSBUREAU

Figure 1.1 The first page of the 2010 census form, mailed to all households in the country. The 2010 census form can be found online at **2010.census.gov/2010census/about/interactive-form.php**.

any race. Again unable to choose a short name that would satisfy everyone, the Census Bureau asked if you are of "Hispanic, Latino, or Spanish origin."

The fight over "race" reminds us that data reflect society. Race is a social idea, not a biological fact. In the census, you say what race you consider yourself to be. Race is a sensitive issue in the United States, so the fight is no surprise and the Census Bureau's diplomacy seems a good compromise.

Observational studies

As Yogi Berra, the former catcher and manager of the New York Yankees who is renowned for his humorous quotes, said, "You can observe a lot by watching." Sometimes all you can do is watch. To learn how chimpanzees in the wild behave, watch. To study how a teacher and young children interact in a schoolroom, watch. It helps if the watcher knows what to look for. The chimpanzee expert may be interested in how males and females interact, in whether some chimps in the troop are dominant, in whether the chimps hunt and eat meat. Indeed, chimps were thought to be vegetarians until Jane Goodall watched them carefully in Gombe National Park, Tanzania. Now it is clear that meat is a natural part of the chimpanzee diet.

At first, the observer may not know what to record. Eventually, patterns seem to emerge and we can decide what variables we want to measure. How often do chimpanzees hunt? Alone or in groups? How large are hunting groups? Males alone, or both males and females? How much of the diet is meat? Observation that is organized and measures clearly defined variables is more convincing than just watching. Here is an example of highly organized (and expensive) observation.

EXAMPLE 3 Do power lines cause leukemia in children?

Electric currents generate magnetic fields. So living with electricity exposes people to magnetic fields. Living near power lines increases exposure to these fields. Really strong fields can disturb living cells in laboratory studies. What about the weaker fields we experience if we live near power lines? Some data suggested that more children in these locations might develop leukemia, a cancer of the blood cells.

We can't do experiments that deliberately expose children to magnetic fields for weeks and months at a time. It's hard to compare cancer rates among children who happen to live in more and less exposed locations because leukemia is quite rare and locations vary a lot in many ways other than magnetic fields. It is easier to start with children who have leukemia and compare them with children who don't. We can look at lots of possible causes—diet, pesticides, drinking water, magnetic fields, and others—to see where children with leukemia differ from those without. Some of these broad studies suggested a closer look at magnetic fields.

A really careful look at magnetic fields took five years and cost $5 million. The researchers compared 638 children who had leukemia and 620 who did not. They went into the homes and actually measured the magnetic fields in the children's bedrooms, in other rooms, and at the front door. They recorded facts about nearby power lines for the family home and also for the mother's residence when she was pregnant. Result: no evidence of more than a chance connection between magnetic fields and childhood leukemia.

"No evidence" that magnetic fields are connected with childhood leukemia doesn't prove that there is no risk. It says only that a very careful study could not find any risk that stands out from the play of chance that distributes leukemia cases across the landscape. In other words, the study could not rule out chance as a plausible explanation for what was observed. Critics continue to argue that the study failed to measure some important variables or that the children studied don't fairly represent all children. Nonetheless, a carefully designed observational study is a great advance over haphazard and sometimes emotional counting of cancer cases.

Response variable and observational study

A **response** is a variable that measures an outcome or result of a study. An **observational study** observes individuals and measures variables of interest but does not intervene in order to influence the responses. The purpose of an observational study is to describe some group or situation.

Sample surveys

You don't have to eat the entire pot of soup to know it needs more salt. That is the idea of sampling: to gain information about the whole by examining only a part. **Sample surveys** are an important kind of observational study. They survey some group of individuals by studying only some of its members, selected not because they are of special interest but because they represent the larger group. Here is the vocabulary we use to discuss sampling.

> **Populations and samples**
>
> The **population** in a statistical study is the entire group of individuals about which we want information.
>
> A **sample** is the part of the population from which we actually collect information and is used to draw conclusions about the whole.

Notice that the *population* is the group we want to study. If we want information about all U.S. college students, that is our population even if students at only one college are available for sampling. To make sense of any sample result, you must know what population the sample represents. Did a preelection poll, for example, ask the opinions of all adults? Or citizens only? Registered voters only? Democrats only? The *sample* consists of the people we actually have information about. If the poll can't contact some of the people it selected, those people aren't in the sample.

The distinction between population and sample is basic to statistics. The following examples illustrate this distinction and also introduce some major uses of sampling. These brief descriptions also indicate the variables measured for each individual in the sample.

EXAMPLE 4 Public opinion polls

Polls such as those conducted by Gallup and many news organizations ask people's opinions on a variety of issues. The *variables* measured are responses to questions about public issues. Though most noticed at election time, these polls are conducted on a regular basis throughout the year. For a typical opinion poll:

Population: U.S. residents 18 years of age and over. Noncitizens and even illegal immigrants are included.

Sample: Between 1000 and 1500 people interviewed by telephone.

EXAMPLE 5 The Current Population Survey

Government economic and social data come from large sample surveys of a nation's individuals, households, or businesses. The monthly Current Population Survey (CPS) is the most important government sample survey in the United States. Many of the *variables* recorded by the CPS concern the employment or unemployment of everyone over 16 years old in a household. The government's monthly unemployment rate comes from the CPS. The CPS also records many other economic and social variables. For the CPS:

Population: The more than 117 million U.S. households. Notice that the individuals are households rather than people or families. A household consists of all people who share the same living quarters, regardless of how they are related to each other.

Sample: About 60,000 households interviewed each month.

EXAMPLE 6 TV ratings

Market research is designed to discover what consumers want and what products they use. One example of market research is the television-rating service of Nielsen Media Research. The Nielsen ratings influence how much advertisers will pay to sponsor a program and whether or not the program stays on the air. For the Nielsen national TV ratings:

Population: The over 114 million U.S. households that have a television set.

Sample: About 25,000 households that agree to use a "people meter" to record the TV viewing of all people in the household.

The *variables* recorded include the number of people in the household and their age and sex, whether the TV set is in use at each time period, and, if so, what program is being watched and who is watching it.

EXAMPLE 7 The General Social Survey

Social science research makes heavy use of sampling. The General Social Survey (GSS), carried out every second year by the National Opinion Research Center at the University of Chicago, is the most important social science sample survey. The *variables* cover the subject's personal and family background, experiences and habits, and attitudes and opinions on subjects from abortion to war.

Population: Adults (aged 18 and over) living in households in the United States. The population does not include adults in institutions such as prisons and college dormitories. It also does not include persons who cannot be interviewed in English.

Sample: About 3000 adults interviewed in person in their homes.

> **NOW IT'S YOUR TURN** | **1.1 Federal funding.** The CNN Opinion Research Corporation conducted a poll on September 1–2, 2010. They asked:
>
> Do you think the federal government should or should not fund research that would use newly created stem cells obtained from human embryos?
>
> The CNN Opinion Research Poll reported that the poll consisted of telephone interviews with 1024 randomly selected adult Americans. What do you think the population is? What is the sample?

Most statistical studies use samples in the broad sense. For example, the 638 children with leukemia in Example 3 are supposed to represent all children with leukemia. We usually reserve the dignified term "sample survey" for studies that use an organized plan to choose a sample that represents some specific population. The children with leukemia were patients at centers that specialize in treating children's cancer. Expert judgment says they are typical of all leukemia patients, even though they come only from special types of hospitals. A sample survey doesn't rely on judgment: it starts with an entire population and uses specific, quantifiable methods to choose a sample to represent the population. Chapters 2, 3, and 4 discuss the art and science of sample surveys.

Census

A sample survey looks at only a part of the population. Why not look at the entire population? A *census* tries to do this.

Census

A **census** is a sample survey that attempts to include the entire population in the sample.

The U.S. Constitution requires a census of the American population every 10 years. A census of so large a population is expensive and takes a long time. Even the federal government, which can afford a census, uses samples such as the Current Population Survey to produce timely data on employment and many other variables. If the government asked every adult in the country about his or her employment, this month's unemployment rate would be available next year rather than next month. In fact, to save

Is a census old-fashioned? The United States has taken a census every 10 years since 1790. Technology marches on, however, and replacements for a national census look promising. Denmark has no census, and France plans to eliminate its census. Denmark has a national register of all its residents, who carry identification cards and change their register entry whenever they move. France will replace its census by a large sample survey that rotates among the nation's regions. The U.S. Census Bureau has a similar idea: the American Community Survey has already started and has eliminated the census "long form."

money, the 2010 census consisted of only ten questions. Five of these were general questions, and five required answers for every person living at the address the form was sent to.

So time and money favor samples over a census. Samples can have other advantages as well. If you are testing fireworks or fuses, the sampled items are destroyed. Moreover, a sample can produce more accurate data than a census. A careful sample of an inventory of spare parts will almost certainly give more accurate results than asking the clerks to count all 500,000 parts in the warehouse. Bored people do not count accurately.

The experience of the Census Bureau reminds us that a census can only *attempt* to sample the entire population. At the time of this writing, results for the 2010 census were not available, but the bureau estimated that the 2000 census missed 0.12% of the American population. These missing persons included an estimated 2.78% of the black population, largely in inner cities. A census is not foolproof, even with the resources of the government behind it. Why take a census at all? The government needs block-by-block population figures to create election districts with equal populations. The main function of the U.S. census is to provide this local information.

"Now eat that banana. The nice statistician is watching us."

Experiments

Our goal in choosing a sample is a picture of the population, disturbed as little as possible by the act of gathering information. All observational studies share the principle "observe but don't disturb." When Jane Goodall first began observing chimpanzees in Tanzania, she set up a feeding station where the chimps could eat bananas. She later said that was a mistake, because it might have changed the apes' behavior.

In *experiments,* on the other hand, we want to change behavior. In doing an experiment, we don't just observe individuals or ask them questions. We actively impose some treatment in order to observe the response. Experiments can answer questions such as "Does aspirin reduce the chance of a heart attack?" and "Do a majority of college students prefer Pepsi to Coke when they taste both without knowing which they are drinking?"

Experiments

An **experiment** deliberately imposes some treatment on individuals in order to observe their responses. The purpose of an experiment is to study whether the treatment causes a change in the response.

EXAMPLE 8   Helping welfare mothers find jobs

The Urban Institute in Washington, DC, reports that most adult welfare recipients are single mothers in their 20s and 30s with one or two children. Observational studies of welfare mothers show that many are able to increase their earnings and leave the welfare system. Some take advantage of voluntary job-training programs to improve their skills. Should participation in job-training and job-search programs be required of all able-bodied welfare mothers? Observational studies of the current system cannot tell us what the effects of such a policy would be. Even if the mothers studied are a properly chosen sample of all welfare recipients, those who seek out training and find jobs may differ in many ways from those who do not. They are observed to have more education, for example, but they may also differ in values and motivation, things that cannot be observed.

To see if a required jobs program will help mothers escape welfare, such a program must actually be tried. Choose two similar groups of mothers when they apply for welfare. Require one group to participate in a job-training program, but do not offer the program to the other group. This is an experiment. Comparing the income and work record of the two groups after several years will show whether requiring training has the desired effect.

NOW IT'S YOUR TURN

1.2 Posting lectures on the class Web site. To determine what students found most helpful, an educational researcher examined student comments from several classes he taught. Mentioned most often was the fact that copies of lectures were posted on the class Web site for students to download. The researcher recommended that instructors post lecture notes on their course Web sites. Was this an observational study or an experiment?

The welfare example illustrates the big advantage of experiments over observational studies: *in principle, experiments can give good evidence for cause and effect*. If we design the experiment properly, we start with two very similar groups of welfare mothers. The *individual* women of course differ from each other in age, education, number of children, and other respects. But the two *groups* resemble each other when we look at the ages, years of education, and number of children for all women in each group. During the experiment, the women's lives differ, but there is only one systematic difference between the two groups: whether or not they are in the jobs program. All live through the same good or bad economic times, the same changes in public attitudes, and so on. If the training group does much better than the untrained group in holding jobs and earning money, we can say that the training program actually causes this happy outcome.

One of the big ideas of statistics is that experiments can give good evidence that a treatment causes a response. A big idea needs a big caution: statistical conclusions hold "on the average" for groups of individuals. They don't tell us much about one individual. If *on the average* the women in the training program earned more than those who were left out, that says that the program achieved its goal. It doesn't say that every woman in such a program will be helped. And a big idea may also raise big questions: if we hope the training will raise earnings, is it ethical to offer it to some women and not to others? Chapters 5 and 6 explain how to design good experiments, and Chapter 7 looks at ethical issues.

STATISTICS IN SUMMARY

Chapter Specifics

- Any statistical study records data about some **individuals** (people, animals, or things) by giving the value of one or more **variables** for each individual.

- Some variables, such as age and income, take numerical values. Others, such as occupation and sex, do not. Be sure the variables in a study really do tell you what you want to know.

- The most important fact about any statistical study is how the data were produced. **Observational studies** try to gather information without disturbing the scene they are observing.

- **Sample surveys** are an important kind of observational study. A sample survey chooses a **sample** from a specific **population** and uses the sample to get information about the entire population.

- A **census** attempts to measure every individual in a population.

- **Experiments** actually do something to individuals in order to see how they respond. The goal of an experiment is usually to learn whether some treatment actually causes a certain response.

 In reasoning from data to a conclusion, we start with the data. Where the data come from is the first step in the argument. The nature and validity of the conclusion are affected by this first step. Two sources of data are observational studies and experiments. Observational studies are best suited for a conclusion that involves describing some group or situation without disturbing the scene we observe. Sample surveys are a type of observational study in which we draw conclusions about a population by observing only a part of the population (the sample). Experiments are best suited for a conclusion that involves determining if a treatment causes a change in a response.

In the next several chapters we discuss these sources of data in more detail. We will see what makes for a good observational study and for a good experiment. And we will see how a bad observational study or experiment undermines the validity of the conclusions we wish to make.

CASE STUDY Use what you have learned in this chapter to answer some basic ques-
EVALUATED tions about the data collected in the jelly bean poll described in the
Case Study that opened the chapter. To participate in the poll, one must go to the store and select a jelly bean from one of the bowls.

1. Is the poll a sample survey, census, or experiment?

2. What is the population of interest?

3. What are the individuals in the poll?

4. For each individual, what variable is measured?

5. Does this variable take numerical values? ∎

CHAPTER 1 EXERCISES

For Exercise 1.1, see page 11; for Exercise 1.2, see page 13.

1.3 Miles per gallon. On the next page is a small part of a data set that describes the fuel economy (in miles per gallon) of 2011 model motor vehicles:

(a) What are the individuals in this data set?

(b) For each individual, what variables are given? Which of these variables take numerical values?

Make and model	Vehicle type	Transmission type	Number of cylinders	City mpg	Highway mpg
⋮					
BMW 328ci	Compact car	Automatic	6	18	28
BMW 335ci	Compact car	Manual	6	19	28
Buick LaCrosse	Midsize car	Automatic	6	17	27
Chevrolet Traverse	Sport utility vehicle (AWD)	Automatic	6	16	23
⋮					

1.4 Athletes' salaries. Here is a small part of a data set that describes Major League Baseball players as of opening day of the 2011 season:

Player	Team	Position	Age	Salary
⋮				
Gomes, Jonny	Reds	Outfield	30	1,750
Beckett, Josh	Red Sox	Pitcher	31	17,000
Sabathia, C. C.	Yankees	Pitcher	31	24,286
Rodriguez, Alex	Yankees	Third base	36	32,000
⋮				

(a) What individuals does this data set describe?

(b) In addition to the player's name, how many variables does the data set contain? Which of these variables take numerical values?

(c) What do you think are the *units* in which each of the numerical variables is expressed? For example, what does it mean to give Josh Beckett's annual salary as 17,000? (*Hint:* The average annual salary of a Major League Baseball player on opening day, 2011, was $3,305,393.)

1.5 Who recycles? In Example 1, weight is not a good measure of the participation of households in different neighborhoods in a city recycling program. What variables would you measure in its place?

1.6 Sampling moms. Pregnant and breast-feeding women should eat at least 12 ounces of fish and seafood per week to ensure their babies' optimal brain development, according to a coalition of top scientists from private groups and federal agencies. A nutritionist wants to know whether pregnant women are eating at least 12 ounces of fish per week. To do so, she obtains a list of the 340 members of a local chain of prenatal fitness clubs and mails a questionnaire to 60 of these women selected at random. Only 21 questionnaires are returned. What is the population in this study? What is the sample from which information is actually obtained? What percentage of the women whom the nutritionist tried to contact responded?

1.7 The death penalty. A press release by the Gallup News Service says that, based on a poll conducted on October 6–9, 2011, it found that 61% of Americans respond "Yes" when asked this question: "Are you in favor of the death penalty for a person convicted of murder?" Toward the end of the article, you read: "These results are based on telephone interviews with a randomly selected national sample of 1005 adults, 18 years and older." What variable did this poll measure? What population do you think Gallup wants information about? What was the sample?

1.8 The political gender gap. There may be a "gender gap" in political party preference in the United States, with women more likely than men to prefer Democratic candidates. A political scientist interviews a large sample of registered voters, both men and women. She asks each voter whether he or she voted for the Democratic or the Republican candidate in the last presidential election. Is this study an experiment? Why or why not? What variables does the study measure?

1.9 What is the population? For each of the following sampling situations, identify the population as exactly as possible. That is, say what kind of individuals the population consists of and say exactly which individuals fall in the population. If the information given is not sufficient, complete the description of the population by making reasonable assumptions about any missing information.

(a) An opinion poll contacts 972 American adults and asks them, "Would you rather have a job working for the government or working for business?"

(b) Video adapter cables have pins that plug into slots in a computer monitor. The adapter will not work if pins are bent or broken. A computer store buys video adapter cables in large lots from a supplier. The store chooses 5 cables from each lot and inspects the pins. If any of the cables have bent or broken pins, the entire lot is sent back.

(c) The American Community Survey contacts 3 million households, including some in every county in the United States. This new Census Bureau survey asks each household questions about their housing, economic, and social status.

1.10 What is the population? For each of the following sampling situations, identify the population as exactly as possible. That is, say what kind of individuals the population consists of and say exactly which individuals fall in the population. If the information given is not sufficient, complete the description of the population in a reasonable way.

(a) A sociologist is interested in determining what proportion of teens believe the drinking age should be lowered to 18 in all U.S. states. She selects a sample of five high schools in a large city and interviews all twelfth-graders in each of the schools.

(b) A medical researcher is interested in the rate of dementia among former NFL football players. From a list of living, former players he selects a sample of 20 and interviews them to determine if signs of dementia are present.

(c) The host of a local radio talk show wonders if people who are actively religious are more likely to trust their neighbors than those who are not. The station receives calls from 51 listeners who voice their opinions.

1.11 Teens' sleep needs. A *Washington Post* article reported on a study about the sleep needs of teenagers. In the study, researchers measured the presence of the sleep-promoting hormone melatonin in teenagers' saliva at different times of the day. They learned that the melatonin levels rise later at night than they do in children and adults and remain at a higher level later in the morning. The teenagers who took part in the study were volunteers. Higher levels of melatonin indicate sleepiness. The researchers recommended that high schools start later in the day to accommodate the sleep needs of teens. Is this study an experiment, a sample survey, or an observational study that is not a sample survey? Explain your answer.

1.12 Power lines and leukemia. The study of power lines and leukemia in Example 3 compared two groups of individuals and measured many variables that might influence differences between the groups. Explain carefully why this study is *not* an experiment.

1.13 Treating prostate disease. A large study used records from Canada's national health care system to compare the effectiveness of two ways to treat prostate disease. The two treatments are traditional surgery and a new method that does not require surgery. The records described many patients whose doctors had chosen one or the other method. The study found that patients treated by the new method were more likely to die within 8 years.

(a) Explain why this is an observational study, not an experiment.

(b) Briefly describe the nature of an experiment to compare the two ways to treat prostate disease.

1.14 Oatmeal and cholesterol. Does eating oatmeal reduce the level of bad cholesterol (LDL)? Here are two ways to study this question.

1. A researcher finds 500 adults over 40 who regularly eat oatmeal or products made from oatmeal. She matches each with a similar adult who does not regularly eat oatmeal or products made from oatmeal. She measures the bad cholesterol (LDL) for each adult and compares both groups.

2. Another researcher finds 1000 adults over 40 who do not regularly eat oatmeal or products made from oatmeal and are willing to participate in a study. She randomly assigns 500 of these to a diet that includes a daily breakfast of

oatmeal. The other 500 continue their usual habits. After 6 months, she compares changes in LDL levels.

(a) Explain why the first is an observational study and the second is an experiment.

(b) Why does the experiment give more useful information about whether oatmeal reduces LDL?

1.15 Alcohol and cancer in women. A *Washington Post* article reported on a study about alcohol consumption and cancer in women. Since 1996, a team of British researchers has been gathering detailed information from 1.28 million women aged 50 to 64. The researchers recorded how much alcohol the women reported consuming when they volunteered for the study and again three years later. The researchers then examined whether there was any link with the 68,775 cancers the women developed over an average of the next seven years. They found that even among women who consumed as little as 10 grams of alcohol a day on average (the equivalent of about one drink), the risk for cancer of the breast, liver, and rectum was elevated.

(a) Is this an experiment? Explain your answer.

(b) We would prefer a sample survey to using women who volunteer for a study. What population does it appear that the researchers were interested in? What variables did they measure?

1.16 The cost of textbooks. A student is interested in determining if the cost of a textbook depends on the number of pages. The student goes to the campus bookstore, picks a textbook rack at random, and records the number of pages and the price of each textbook on the rack. What is the population in this study? What is the sample? What variables does the student measure?

1.17 Choose your study type. What is the best way to answer each of the questions below: an experiment, a sample survey, or an observational study that is not a sample survey? Explain your choices.

(a) Is your school's basketball team called for fewer fouls in home games than in away games?

(b) Are college students satisfied with the quality of recreational facilities available to them?

(c) Do college students who have access to audio recordings of course lectures perform better in the course than those who don't?

1.18 Choose your study purpose. Give an example of a question about college students, their behavior, or their opinions that would best be answered by

(a) a sample survey.

(b) an observational study that is not a sample survey.

(c) an experiment.

EXPLORING THE WEB

1.19 Web-based exercise. All the sample surveys mentioned in Examples 4 to 7 maintain Web sites:

- Gallup Poll (Example 4): **www.gallup.com/**
- Current Population Survey (Example 5): **www.bls.gov/cps/**
- Nielsen Media Research (Example 6): **www.nielsen.com/us/en/ measurement/television-measurement.html**
- General Social Survey (Example 7): **www.norc.org/Research/ Projects/Pages/general-society-survey.aspx**

We recommend the Gallup site for its current poll results and clear explanations of sample survey methods.

Visit the Gallup Poll Web site. Select one of the stories available to nonsubscribers. Identify the population and sample as exactly as possible. How many people were in the sample?

1.20 Web-based exercise. You can find the report of the National Institute of Environmental Health Sciences (NIEHS) on the possible health effects of exposure to power lines at **www.niehs.nih.gov/health/topics/ agents/emf/**. Visit the site. Although the few studies that have been conducted on adult exposures show no evidence of a link between adult cancer and exposure to residential electric and magnetic fields, what does the description given at the site say that NIEHS scientists have concluded? Why?

NOTES AND DATA SOURCES

Page 3 Data on downloading music off the Internet are from "Americans think downloading music for personal use is an innocent act," Harris Poll press release by Robert Leitman, January 28, 2004; available online at **www.harrisinteractive.com/insights/ HarrisVault.aspx?PID=434.**

Page 3 The "Fork Over Your Vote" jelly bean election poll is described in the article "McCain ahead in the jelly bean poll"; available online at **cbs4.com/local/john. mccain.barack.2.844274.html.**

Page 5 Example 1 is suggested by Maxine Pfannkuch and Chris J. Wild, "Statistical thinking and statistical practice: themes gleaned from professional statisticians," *Statistical Science*, 15 (2000), pp. 132–152.

Page 7 Example 3: M. S. Linet et al., "Residential exposure to magnetic fields and acute lymphoblastic leukemia in children," *New England Journal of Medicine*, 337 (1997), pp. 1–7.

Page 11 Exercise 1.1: From "CNN/Opinion Research Poll—September 1–2—Stem Cell Research"; available online at **politicalticker.blogs.cnn.com/2010/09/ 09/cnnopinion-research-poll-september- 1-2-stem-cell-research/.**

Page 12 The estimates of the census under-count come from Professor Eugene Ericksen, "An evaluation of the 2000 census"; available online at **govinfo.library.unt.edu/cmb/cmbp/reports/final_report/fin_sec3_evaluation.pdf.**

Page 17 Exercise 1.7: Frank Newport, "In U.S., support for death penalty falls to 39-year low," *Gallup News Service,* October 13, 2011; available online at **www.gallup.com/poll/150089/Support-Death-Penalty-Falls-Year-Low.aspx.**

Page 17 Exercise 1.9(c): The Web site for the American Community Survey is **www.census.gov/acs/www.**

Page 18 Exercise 1.11: Valerie Strauss, "Schools waking up to teens' unique sleep needs," *Washington Post,* January 10, 2006, p. A08.

Page 19 Exercise 1.15: Rob Stein, "A drink a day raises women's risk of cancer, study indicates," *Washington Post,* February 25, 2009, p. A01.

Samples, Good and Bad

CASE STUDY As discussed in Chapter 1, "Fork Over Your Vote" was a national project conducted during the 2008 presidential election in which people chose who they thought should be the president by tossing a jelly bean into a jar. More than 100 restaurants in 34 states were involved in the project, and votes were counted every Monday. As of October 13, 2008 (the date that the poll was reported on by the CBS station in Pembroke Pines, Florida), McCain had 53.5 % of the national jelly bean vote. The election was still several weeks away, but as of October 13, more than 211,000 bean ballots had been cast. Although it incorrectly predicted a McCain victory in 2008, the same jelly bean vote in the 2004 election correctly predicted President George W. Bush would win the presidency. If the poll is repeated in the 2012 election, what might we expect? Were the 2008 results just bad luck, or is the poll fundamentally flawed? By the end of this chapter you will have learned how to assess whether the data from the jelly bean poll are good or bad. ■

AFP/Getty Images

How to sample badly

For many years in Rapides Parish, Louisiana, only one company had been allowed to provide ambulance service. In 1999, the local paper, the *Town Talk,* asked readers to call in to offer their opinion on whether the company should keep its monopoly. Call-in polls are generally automated: call one telephone number to vote "Yes" and call another number to vote "No." Telephone companies often charge callers to these numbers.

The *Town Talk* got 3763 calls, which suggests unusual interest in ambulance service. Investigation showed that 638 calls came from the ambulance company office or from the homes of its executives. Many more no doubt came from lower-level employees. "We've got employees who are concerned about this situation, their job stability and their families and maybe called more than they should have," said a company vice president. Other sources said employees were told to, as they say in Chicago, "vote early and often."

As the *Town Talk* learned, it is easier to sample badly than to sample well. The paper relied on *voluntary response,* allowing people to call in rather than actively selecting its own sample. The result was *biased*—the sample was overweighted with people favoring the ambulance monopoly. Voluntary response samples attract people who feel strongly about the issue in question. These people, like the employees of the ambulance company, may not fairly represent the opinions of the entire population.

There are other ways to sample badly. Suppose that we sell your company several crates of oranges each week. You examine a sample of oranges from each crate to determine the quality of our oranges. It is easy to inspect a few oranges from the top of each crate, but these oranges may not be representative of the entire crate. Those on the bottom are more often damaged in shipment. If we were less than honest, we might make sure that the rotten oranges are packed on the bottom, with some good ones on top for you to inspect. If you sample from the top, your sample results are again *biased*—the sample oranges are systematically better than the population they are supposed to represent.

Biased sampling methods

The design of a statistical study is **biased** if it systematically favors certain outcomes.

Selection of whichever individuals are easiest to reach is called **convenience sampling.**

A **voluntary response sample** chooses itself by responding to a general appeal. Write-in or call-in opinion polls are examples of voluntary response samples.

Convenience samples and voluntary response samples are often biased.

EXAMPLE 1 Interviewing at the mall

Squeezing the oranges on the top of the crate is one example of convenience sampling. Mall interviews are another. Manufacturers and advertising agencies often use interviews at shopping malls to gather information about the habits of consumers and the effectiveness of ads.

A sample of mall shoppers is fast and cheap. But people contacted at shopping malls are not representative of the entire U.S. population. They are richer, for example, and more likely to be teenagers or retired. Moreover, the interviewers tend to select neat, safe-looking individuals from the stream of customers. Mall samples are biased: they systematically overrepresent some parts of the population (prosperous people, teenagers, and retired people) and underrepresent others. The opinions of such a convenience sample may be very different from those of the population as a whole.

Frank Chmura/Alamy

EXAMPLE 2 Write-in opinion polls

Ann Landers once asked the readers of her advice column, "If you had it to do over again, would you have children?" She received nearly 10,000 responses, almost 70% saying "NO!" Can it be true that 70% of parents regret having children? Not at all. This is a voluntary response sample. People who feel strongly about an issue, particularly people with strong negative feelings, are more likely to take the trouble to respond. Ann Landers's results are strongly biased—the percentage of parents who would not have children again is much higher in her sample than in the population of all parents.

On August 24, 2011, Abigail Van Buren (the niece of Ann Landers) revisited this question in her column "Dear Abby." A reader asked, "Many years ago an advice columnist (your mother?) posed the question to her readers, 'If you had it to do over again, would you still have children?' I'm wondering when the information was collected and what the results of that inquiry were, and if you asked the same question today, what the majority of your readers would answer."

Ms. Van Buren responded, "The results were considered shocking at the time because the majority of responders said they would NOT have children if they had it to do over again. I'm printing your question because it will be interesting to see if feelings have changed over the intervening years."

In October 2011, Ms. Van Buren wrote that this time the majority of respondents would have children again. That is encouraging, but this was again a write-in poll.

"Hey, Pops, what was that letter you sent off to Ann Landers yesterday?"

Write-in and call-in opinion polls are almost sure to lead to strong bias. In fact, only about 15% of the public have ever responded to a call-in poll, and these tend to be the same people who call radio talk shows. That's not a representative sample of the population as a whole.

Simple random samples

In a voluntary response sample, people choose whether to respond. In a convenience sample, the interviewer makes the choice. In both cases, personal choice produces bias. The statistician's remedy is to allow impersonal chance to choose the sample. A sample chosen by chance allows neither favoritism by the sampler nor self-selection by respondents. Choosing a sample by chance attacks bias by giving all individuals an equal chance to be chosen. Rich and poor, young and old, black and white, all have the same chance to be in the sample.

The simplest way to use chance to select a sample is to place names in a hat (the population) and draw out a handful (the sample). This is the idea of *simple random sampling*.

Simple random sample

A **simple random sample (SRS)** of size n consists of n individuals from the population chosen in such a way that every set of n individuals has an equal chance to be the sample actually selected.

An SRS not only gives each individual an equal chance to be chosen (thus avoiding bias in the choice) but also gives every possible sample an equal chance to be chosen. Drawing names from a hat does this. Write 100 names on identical slips of paper and mix them in a hat. This is a population. Now draw 10 slips, one after the other. This is an SRS, because any 10 slips have the same chance as any other 10.

> **NOW IT'S YOUR TURN** | **2.1 Sampling my class.** There are 20 students in my class. They sit in assigned seats, consisting of four rows of 5 students each. I want to take a simple random sample consisting of 4 of the students in my class. To do this I select a single student from each row as follows. I write the numbers 1 to 5 on identical slips of paper. I mix the slips in a hat and draw one at random. I count this number of seats in from the left in the first row and select the student in this seat. For example, if the number selected is 3, I select the third student in from the left in the first row. I replace the slip in the hat, again mix the slips, and draw a new number. The student seated this many seats in from the left in the second row is selected. I repeat this process for the remaining two rows. Every student in the class has a 1-in-5 chance of being selected when I come to their row. Thus, every student has the same chance of being selected. Is the sample a simple random sample? Explain.

Drawing names from a hat makes clear what it means to give each individual and each possible set of n individuals the same chance to be chosen. That's the idea of an SRS. Of course, drawing slips from a hat would be a bit awkward for a sample of the country's 117 million households. In practice, we use computer-generated *random digits* to choose samples. Many statistical software packages have random number generators that generate random digits. Some also allow one to choose an SRS. In Example 4 (page 30) we describe how to use a Web-based tool to choose an SRS.

If you don't use software, you can use a *table of random digits* to choose small samples by hand.

Random digits

A **table of random digits** is a long string of the digits 0, 1, 2, 3, 4, 5, 6, 7, 8, 9 with these two properties:

1. Each entry in the table is equally likely to be any of the 10 digits 0 through 9.

2. The entries are independent of each other. That is, knowledge of one part of the table gives no information about any other part.

Table A at the back of the book is a table of random digits. You can think of Table A as the result of asking an assistant (or a computer) to mix the digits 0 to 9 in a hat, draw one, then replace the digit drawn, mix again, draw a second digit, and so on. The assistant's mixing and drawing save us the work of mixing and drawing when we need to randomize. Table A begins with the digits 19223950340575628713. To make the table easier to read, the digits appear in groups of five and in numbered rows. The groups and rows have no meaning—the table is just a long list of randomly chosen digits. Here's how to use the table to choose an SRS.

EXAMPLE 3 How to choose an SRS

Joan's small accounting firm serves 30 business clients. Joan wants to interview a sample of 5 clients to find ways to improve client satisfaction. To avoid bias, she chooses an SRS of size 5.

Step 1: Label. Give each client a numerical label, using as few digits as possible. Two digits are needed to label 30 clients, so we use labels

01, 02, 03, ..., 28, 29, 30

It is also correct to use labels 00 to 29 or even another choice of 30 two-digit labels. Here is the list of clients, with labels attached, using 01 to 30:

01	A-1 Plumbing	16	JL Appliances
02	Accent Printing	17	Johnson Commodities
03	Action Sport Shop	18	Keiser Construction
04	Anderson Construction	19	Liu's Chinese Restaurant
05	Bailey Trucking	20	MagicTan
06	Balloons, Inc.	21	Peerless Machine
07	Bennett Hardware	22	Photo Arts
08	Best's Camera Shop	23	River City Antiques
09	Blue Print Specialties	24	Riverside Tavern
10	Central Tree Service	25	Rustic Boutique
11	Classic Flowers	26	Satellite Services
12	Computer Answers	27	Scotch Wash
13	Darlene's Dolls	28	Sewer's Center
14	Fleisch Realty	29	Tire Specialties
15	Hernandez Electronics	30	Von's Video Games

Step 2: Table. Enter Table A anywhere and read two-digit groups. Suppose we enter at line 130, which is

69051   64817   87174   09517   84534   06489   87201   97245

The first 10 two-digit groups in this line are

69   05   16   48   17   87   17   40   95   17

Each two-digit group in Table A is equally likely to be any of the 100 possible groups, 00, 01, 02, ..., 99. So two-digit groups choose two-digit labels at random. That's just what we want.

Joan used only labels 01 to 30, so we ignore all other two-digit groups. The first five labels between 01 and 30 that we encounter in the table choose our sample. Of the first 10 labels in line 130, we ignore five because they are too high (over 30). The others are 05, 16, 17, 17, and 17. The clients labeled 05, 16, and 17 go into the sample. Ignore the second and third 17s because that client is already in the sample. Now run your finger across line 130 (and continue to line 131 if needed) until five clients are chosen.

The sample is the clients labeled 05, 16, 17, 20, 19. These are Bailey Trucking, JL Appliances, Johnson Commodities, MagicTan, and Liu's Chinese Restaurant.

Using the table of random digits is much quicker than drawing names from a hat. As Example 3 shows, choosing an SRS has two steps.

Choose an SRS in two steps

Step 1: Label. Assign a numerical label to every individual in the population. Be sure that all labels have the same number of digits if you plan to use a table of random digits.

Step 2: Software or table. Use random digits to select labels at random.

You can assign labels in any convenient manner, such as alphabetical order for names of people. When using a table of random digits, as long as all labels have the same number of digits, all individuals will have the same chance to be chosen. Use the shortest possible labels: one digit for a population of up to 10 members, two digits for 11 to 100 members, three digits for 101 to 1000 members, and so on. As standard practice, we recommend that you begin with label 1 (or 01 or 001, as needed). You can read digits from Table A in any order—across a row, down a column, and so on—because the table has no order. As standard practice, we recommend reading across rows.

Real sample surveys use computer software to choose an SRS, but the software just automates the steps in Example 3. The computer doesn't look in a table of random digits because it can generate them on the spot.

EXAMPLE 4 How to choose an SRS using software

Many commercial statistical software packages allow one to generate an SRS. An example that is available on the Web is the Research Randomizer at **www.randomizer.org**. Click on the link *Randomize Now* and fill in the boxes. You can even ask the Randomizer to arrange your sample in order.

To repeat Example 3, we asked the Randomizer to generate one set of numbers with five numbers per set. We specified the number range as 1 to 30. We requested that each number remain unique and that the numbers be sorted from least to greatest. We asked to view the outputted numbers with the place markers off. See Figure 2.1. After clicking on the "Randomize Now!" button, we obtained the digits 4, 7, 14, 16, and 23.

This time, the sample is the clients labeled 04, 07, 14, 16, 23. These are Anderson Construction, Bennett Hardware, Fleisch Realty, JL Appliances, and River City Antiques.

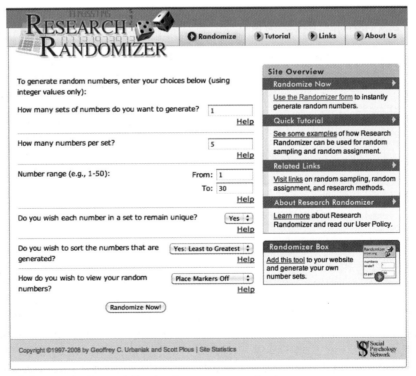

Figure 2.1 Using the Research Randomizer at **www.randomizer.org**.

NOW IT'S YOUR TURN | **2.2 Evaluating teaching assistants.** To assess how its teaching assistants are performing, the statistics department at a large university randomly selects 3 of its teaching assistants each week and sends a faculty member to visit their classes. The current list of 20 teaching assistants is given below. Use either software (for example, the Research Randomizer) or Table A at line 116 to choose 3 to be visited this week. Remember to begin by labeling the teaching assistants from 01 to 20.

01 Dobmeier
02 Joseph
03 Kil
04 Kohlschnidt
05 Koster
06 Landgraf
07 Leatherman
08 Martin
09 Mazzeo
10 Pearl

11 Pflugheisen
12 Sanders
13 Snyder
14 Sonksen
15 Spade
16 Springer
17 Stagner
18 Stettler
19 Tam
20 Tirmenstein

Can you trust a sample?

The *Town Talk*, Ann Landers, and mall interviews produce samples. We can't trust results from these samples, because they are chosen in ways that invite bias. We have more confidence in results from an SRS, because it uses impersonal chance to avoid bias. The first question to ask of any sample is whether it was chosen at random. Opinion polls and other sample surveys carried out by people who know what they are doing use random sampling.

 Statistics in Your World

Golfing at random Random drawings give all the same chance to be chosen, so they offer a fair way to decide who gets a scarce good—like a round of golf. Lots of golfers want to play the famous Old Course at St. Andrews, Scotland. A few can reserve in advance. Most must hope that chance favors them in the daily random drawing for tee times. At the height of the summer season, only 1 in 6 wins the right to pay £150 (about $244) for a round.

EXAMPLE 5 A Gallup Poll

A January 2011 Gallup Poll on mass shootings in America asked the question "Just your opinion, what do you think are the one or two most important things that could be done to prevent mass shootings from occurring in the United States?" The press release reported that the poll found that the most frequent suggestion was stricter gun control laws, with 24% of respondents making this suggestion. Is this actually the most common opinion among Americans? Ask first how Gallup selected its sample. Later in the press release we read this: "Results for this USA Today/Gallup poll are based on telephone interviews conducted Jan. 14–16, 2011, with a random sample of 1,032 adults, aged 18 and older, living in the continental U.S., selected using random-digit-dial sampling."

This is a good start toward gaining our confidence. Gallup tells us what population it has in mind (people at least 18 years old living in the continental United States). We know that the sample from this population was of size 1032 and, most important, that it was chosen at random. There is more to consider in assessing a poll, and we will soon discuss this, but we have at least heard the comforting words "random sample."

STATISTICS IN SUMMARY

Chapter Specifics

- We select a **sample** in order to get information about some **population.**

- How can we choose a sample that fairly represents the population? **Convenience samples** and **voluntary response samples** are common but do not produce trustworthy data. These sampling methods are usually **biased.** That is, they systematically favor some parts of the population over others in choosing the sample.

- The deliberate use of chance in producing data is one of the big ideas of statistics. Random samples use chance to choose a sample, thus avoiding bias due to personal choice.

- The basic type of random sample is the **simple random sample,** which gives all samples of the same size the same chance to be the sample we actually choose.

- To choose an SRS by hand, use a **table of random digits** such as Table A in the back of the book, or use software.

The first step in reasoning from data to a conclusion is obtaining data. In Chapter 1 we discussed sample surveys as one way to collect data in an observational study. The method of selecting the sample in a sample survey affects how well the sample represents the population. Biased sampling methods, such as convenience sampling and voluntary response samples, produce data that can be misleading, resulting in incorrect conclusions. Simple random sampling avoids bias and produces data that give us confidence that the first step in our argument is sound.

In the next chapter we look more closely at what a simple random sample tells us about the population from which it is selected. And in Chapter 4 we discuss some of the problems faced by people who take surveys in the real world.

CASE STUDY To participate in the jelly bean poll described in the Case Study that
EVALUATED opened the chapter, one must make a purchase at one of the participating restaurants. Use what you have learned in this chapter to assess whether the data
collected in such a jelly bean poll are good or bad. Your assessment should be written
so that someone who knows no statistics will understand your reasoning. ■

CHAPTER 2 EXERCISES

For Exercise 2.1, see page 27; for Exercise 2.2, see page 31.

2.3 Letters to the editor. You work for a local newspaper that has recently
reported on a bill that would make it easier to create charter schools in the
state. You report to the editor that 201 letters have been received on the issue,
of which 171 oppose the legislation. "I'm surprised that most of our readers
oppose the bill. I thought it would be quite popular," says the editor. Are you
convinced that a majority of the readers oppose the bill? How would you explain
the statistical issue to the editor?

2.4 Instant opinion. On March 29, 2007, *BusinessWeek* ran an online poll on
their Web site and asked readers the question "Do you think Google is too powerful?" Readers clicked on one of three buttons ("Yes," "No," or "Not sure") to
vote. In all, 1336 (35.9%) said "Yes," 2051 (55.1%) said "No," and 335 (9.0%)
said "Not sure."

(a) What is the sample size for this poll?

(b) At the Web site, *BusinessWeek* includes the following statement about its
online poll. "Note: These are surveys, not scientific polls." Explain why the poll
may give unreliable information.

(c) Just above the poll question was the following statement: "Google's acceler-
ating lead in search and its moves into software and traditional advertising are
sparking a backlash among rivals." How might this statement affect the poll
results?

2.5 More instant opinion. On January 5, 2010, the *Los Angeles Times* ran
an online poll asking readers, "What's your impression of Google's answer to
Apple's iPhone? Does it inspire gadget lust, or is it just a phone?" Readers could
vote by clicking on one of four buttons ("I'm getting one as soon as possible,"
"It's just a phone," "Consumers should hold off until the inevitable glitches are
worked out," or "It's another arm of the fearsome Googlopoly"). Most votes were
recorded by the end of February 2010, and the last time we checked (January 5,
2012), 175 (24%) said, "I'm getting one as soon as possible," 222 (30%) said, "It's
just a phone," 234 (32%) said, "Consumers should hold off until the inevitable
glitches are worked out," and 109 (15%) said, "It's another arm of the fearsome
Googlopoly."

(a) What is the sample size for this poll?

(b) Explain why the poll may give unreliable information.

2.6 Ann Landers takes a sample. Advice columnist Ann Landers once asked her divorced readers whether they regretted their decision to divorce. She received approximately 30,000 responses, about 23,000 of which came from women. Nearly 75% said they were glad they divorced, and most of them said they wished they had done it sooner. Explain why this sample is certainly biased. What is the likely direction of the bias? That is, is 75% probably higher or lower than the truth about the population of all adults who have been divorced?

2.7 We don't like one-way streets. Highway planners decided to make a main street in West Lafayette, Indiana, a one-way street. The *Lafayette Journal and Courier* took a one-day poll by inviting readers to call a telephone number to record their comments. The next day, the paper reported:

> *Journal and Courier readers overwhelmingly prefer two-way traffic flow in West Lafayette's Village area to one-way streets. By nearly a 7-1 margin, callers to the newspaper's Express Yourself opinion line on Wednesday complained about the one-way streets that have been in place since May. Of the 98 comments received, all but 14 said no to one-way.*

(a) What population do you think the newspaper wants information about?
(b) Is the proportion of this population who favor one-way streets almost certainly larger or smaller than the proportion 14/98 in the sample? Why?

2.8 Who won the debate? Following the May 15, 2007 First-in-the-South Republican Presidential Candidates Primary Debate between 10 GOP presidential hopefuls, *FOX News* asked viewers to vote for the winner via text message. More than 40,000 votes were submitted. The results were 29% for former Massachusetts governor Mitt Romney, 25% for Representative Ron Paul (R-Texas), 19% for former New York City mayor Rudy Giuliani, 8% for former Arkansas governor Mike Huckabee, 5% for Representative Duncan Hunter (R-California), 4% for Senator John McCain (R-Arizona), 3% for Representative Tom Tancredo (R-Colorado), 1% for Senator Sam Brownback (R-Kansas), 0% for former Virginia governor Jim Gilmore, and 0% for former Wisconsin governor Tommy Thompson.

(a) What population do you think *FOX News* wanted information about?
(b) At the Web site, *FOX News* included the following statement about its online poll: "The results reflect the opinions of those who chose to participate and do not reflect a scientific sampling of the population." Explain why the poll may give unreliable information.

2.9 Design your own bad sample. Your college wants to gather student opinion about parking for students on campus. It isn't practical to contact all students.
(a) Give an example of a way to choose a sample of students that is poor practice because it depends on voluntary response.
(b) Give an example of a bad way to choose a sample that doesn't use voluntary response.

2.10 A call-in opinion poll. In 2005 the *San Francisco Bay Times* reported on a poll in New Zealand that found that New Zealanders opposed the nation's new gay-inclusive civil-unions law by a 3-1 ratio. This poll was a call-in poll that cost $1 to participate in. The *San Francisco Bay Times* article also reported that a scientific polling organization found that New Zealanders favor the law by a margin of 56.4% to 39.3%. Explain to someone who knows no statistics why the two polls can give such widely differing results and which poll is likely to be more reliable.

2.11 Call-in versus random sample polls. A national survey of TV network news viewers found that 48% said they would believe a phone-in poll of 300,000 persons rather than a random sample of 1000 persons. Of the viewers, 42% said they would believe the random sample poll. Explain to someone who knows no statistics why the opinions of only 1000 randomly chosen respondents are a better guide to what all people think than the opinions of 300,000 callers.

2.12 Choose an SRS. A firm wants to understand the attitudes of its minority managers toward its system for assessing management performance. Below is a list of all the firm's managers who are members of minority groups. Use Table A at line 134 to choose 6 to be interviewed in detail about the performance appraisal system.

Berliner	Hans	Liu	Rumsey
Browne	Herbei	MacEachern	Santner
Calder	Holloman	Miller	Shi
Craigmile	Hsu	Nagaraja	Stasny
Cressie	Kaizar	Notz	Turkmen
Critchlow	Kubatko	Ozturk	Verducci
Dean	Lee	Pearl	Wolfe
Goel	Lin	Peruggia	Xu

2.13 Choose an SRS. Your class in ancient Ugaritic religion is poorly taught and the class members have decided to complain to the dean. The class decides to choose 4 of its members at random to carry the complaint. The class list appears below. Choose an SRS of 4 using the table of random digits, beginning at line 139.

01 Ashmead	10 Katzfuss	19 Sgambellone
02 Blake	11 Lee	20 Sullivan
03 Crowther	12 McFarland	21 Svenson
04 Darneider	13 Nelson	22 Thompson
05 Draguljic	14 Pressler	23 Wang
06 Draper	15 Quan	24 Wennersten
07 Fox	16 Rettiganti	25 Winner
08 Fry	17 Schlessman	26 Xu
09 Jelsing	18 Schuetter	27 Zhang

2.14 An election day sample. You want to choose an SRS of 20 of a city's 480 voting precincts for special voting-fraud surveillance on election day.

(a) Explain clearly how you would label the 480 precincts. How many digits make up each of your labels? What is the greatest number of precincts you could label using this number of digits?

(b) Use Table A to choose the SRS, and list the labels of the precincts you selected. Enter Table A at line 107.

2.15 Is this an SRS? A university has 30,000 undergraduate and 10,000 graduate students. A survey of student opinion concerning health care benefits for domestic partners of students selects 300 of the 30,000 undergraduate students at random and then separately selects 100 of the 10,000 graduate students at random. The 400 students chosen make up the sample.

(a) Explain why this sampling method gives each student an equal chance to be chosen.

(b) Nonetheless, this is not an SRS. Why not?

2.16 How much do students pay for rent? A university's housing and residence office wants to know how much students pay per month for rent in off-campus housing. The university does not have enough on-campus housing for students, and this information will be used in a brochure about student housing. The population contains 12,304 students who live in off-campus housing and have not yet graduated. The university will send a questionnaire to an SRS of 200 of these students, drawn from an alphabetized list.

(a) Describe how you would label the students in order to select the sample.

(b) Use Table A, beginning at line 125, to select the first 5 students in the sample.

2.17 Apartment living. You are planning a report on apartment living in a college town. You decide to select three apartment complexes at random for in-depth interviews with residents. Use Table A, starting at line 112, to select a simple random sample of three of the following apartment complexes.

Albany Commons	Gaslight Village	Oak Run
Apple Run	Georgetowne	Old Nantucket
Bexley Court	Golf Pointe	Parliament Ridge
Brooks Edge	Hickory Mill	Pheasant Run
Canterbury Way	Highview Place	Ravine Bluff
Chablis Villas	Indian Creek	Rocky Creek
Cherryblossom Way	Jefferson Commons	Scioto Commons
Dublin Plaza	Kenbrook Village	Stratford East
English Village	Lawn Manor	Timbercreek
Fairway Lakes	Little Brook Place	Walnut Knolls
Forest Creek	Marble Cliff	Woodland Trace
Forest Park	Morse Glen	York Terrace

2.18 How do random digits behave? Which of the following statements are true of a table of random digits, and which are false? Explain your answers.

(a) Each pair of digits has chance 1/100 of being 33.

(b) There are exactly four 4s in each row of 40 digits.

(c) The digits 99999 can never appear as a group, because this pattern is not random.

2.19 Text in your vote. During the television broadcast on the Big Ten Network of a 2010 basketball game at Ohio State between Ohio State and Penn State, the announcers asked the following question: "Which player has meant the most to his team this year: Talor Battle of Penn State, Evan Turner of Ohio State, or other?" Viewers were asked to text in their vote. Later in the program, the results were announced. Evan Turner received 72% of the vote; Talor Battle, 26%; and "other," 2%. Explain why this opinion poll is almost certainly biased.

2.20 More randomization. Most sample surveys call residential telephone numbers at random. They do not, however, always ask their questions of the person who picks up the phone. Instead, they ask about the adults who live in the residence and choose one at random to be in the sample. Why is this a good idea?

2.21 Racial profiling and traffic stops. The Denver Police Department wants to know if Hispanic residents of Denver believe that the police use racial profiling when making traffic stops. A sociologist prepares several questions about the police. The police department chooses an SRS of 200 mailing addresses in predominantly Hispanic neighborhoods and sends a uniformed Hispanic police officer to each address to ask the questions of an adult living there.

(a) What are the population and the sample?

(b) Why are the results likely to be biased even though the sample is an SRS?

2.22 Random selection? Choosing at random is a "fair" way to decide who gets some scarce good, in the sense that everyone has the same chance to win. But random choice isn't always a good idea—sometimes we don't want to treat everyone the same, because some people have a better claim. In each of the following situations, would you support choosing at random? Give your reasons in each case.

(a) The basketball arena has 4000 student seats, and 7000 students want tickets. Shall we choose 4000 of the 7000 at random?

(b) The list of people waiting for liver transplants is much larger than the number of available livers. Shall we let impersonal chance decide who gets a transplant?

(c) During the Vietnam War, young men were chosen for army service at random, by a "draft lottery." Is this the best way to decide who goes and who stays home?

EXPLORING THE WEB

2.23 Web-based exercise. There are several voluntary response polls available on the Internet. Visit **www.misterpoll.com** and examine several of the current polls. What are the sample sizes in these polls? Who can vote? Is it possible to vote more than once? Do you think you can trust the results of the polls on **www.misterpoll.com**? Why?

2.24 Web-based exercise. To see how software speeds up choosing an SRS, go to the Research Randomizer at **www.randomizer.org**. Click on *Randomize Now* and fill in the boxes. Now use the Research Randomizer to do Exercises 2.12, 2.13, and 2.17.

2.25 Web-based exercise. Random.org maintains an online tool that allows you to generate random numbers and a tool that allows you to rearrange any list in random order. The tools, with instructions, can be found at **www.random.org/integers/** (for generating random numbers) and at **www.random.org/lists/** for rearranging a list in random order.

(a) How could you use these tools to generate an SRS?
(b) What does the Web site say about its method for generating randomness?

NOTES AND DATA SOURCES

Page 23 Case Study. The "Fork Over Your Vote" jelly bean election poll is described in the article "McCain ahead in the jelly bean poll"; originally found online at **cbs4.com/local/john.mccain.barack.2.844274.html.** As far as we have been able to tell, the poll was never again mentioned after this article appeared.

Page 23 "Acadian ambulance officials, workers flood call-in poll," *Baton Rouge Advocate,* January 22, 1999.

Page 31 Example 5: "Americans link gun laws, mental health to mass shootings," Gallup News Service press release by Frank Newport, January 24, 2011; available online at **www.gallup.com/poll/145757/Americans-Link-Gun-Laws-Mental-Health-Mass-Shootings.aspx.**

Page 33 Exercise 2.4: The link to the *BusinessWeek* poll is **www.businessweek.com/magazine/content/07_15/b4029010.htm,** and the associated news story can be found online at **www.businessweek.com/magazine/content/07_15/b4029001.htm.**

Page 33 Exercise 2.5: The link to the *Los Angeles Times* poll is **opinion.latimes.com/opinionla/2010/01/poll-will-googles-nexus-one-kill-the-iphone.html.**

Page 34 Exercise 2.8: The link to the *FOX News* poll is **www.foxnews.com/story/0,2933,272493,00.html.**

Page 35 Exercise 2.10: "Clarification," *San Francisco Bay Times,* November 15, 2005.

What Do Samples Tell Us?

CASE STUDY Same-sex marriages are controversial. Many people oppose them on the basis of religious convictions. Some believe that same-sex marriages undermine the traditional family and the institution of marriage. Those who support same-sex marriages see it as a matter of equal rights. In May 2011, a Gallup Poll asked, "Do you think marriages between same-sex couples should or should not be recognized by the law as valid, with the same rights as traditional marriages?" For the first time since 2004, when Gallup began asking this question annually, a majority (53%) responded, "should be recognized by the law as valid."

This topic became a major news issue, capturing national attention, in February 2004. Several cities, most notably San Francisco, had begun to perform same-sex marriages, although such ceremonies were against state law. President Bush made the declaration, "Today I call upon the Congress to promptly pass, and to send to the states for ratification, an amendment to our Constitution defining and protecting marriage as a union of a man and woman as husband and wife." How strong was the support for such an amendment? A Gallup Poll conducted from July 2003 to February 2004 asked the following question: "Would you favor or oppose a constitutional amendment that would define marriage as being between a man and a woman, thus barring marriages between gay and lesbian couples?" The Gallup Poll found that "such an amendment is supported by a very slim majority of Americans, 51%, with 45% opposed." Can we trust this conclusion?

Reading further, we find that Gallup talked with 2527 randomly selected adults to reach these conclusions. We're happy Gallup chooses at random—we wouldn't get unbiased information about support for such a constitutional amendment by asking people attending a same-sex marriage in San Francisco. However, the Census Bureau said that there were about 220 million adults in the United States in 2004. How can 2527 people, even a random sample of 2527 people, tell us about the opinions of 220 million people?

Is 51% in favor evidence that, in fact, the majority of Americans favor such an amendment? Does the May 2011 poll, which was based on a random sample of 1018 adults, suggest that today that the majority of Americans would be *against* such an amendment? By the end of this chapter you will learn the answers to these questions. ∎

From sample to population

Gallup's 2004 finding that "such an amendment is supported by a very slim majority of Americans, 51%," makes a claim about the population of 220 million adults. But Gallup doesn't know the truth about this population. The poll contacted 2527 people and found that 51% of them said they supported such an amendment. Because the sample of 2527 people was chosen at random, it's reasonable to think that they represent the entire population pretty well. So Gallup turns the *fact* that 51% of the *sample* supported such an amendment into an *estimate* that about 51% of *all adults* support such an amendment. That's a basic move in statistics: use a fact about a sample to estimate the truth about the whole population. To think about such moves, we must be clear whether a number describes a sample or a population. Here is the vocabulary we use:

Parameters and statistics

A **parameter** is a number that describes the **population.** A parameter is a fixed number, but in practice we don't know the actual value of this number.

A **statistic** is a number that describes a **sample.** The value of a statistic is known when we have taken a sample, but it can change from sample to sample. We often use a statistic to estimate an unknown parameter.

So parameter is to population as statistic is to sample. Want to estimate an unknown parameter? Choose a sample from the population and use a sample statistic as your estimate. That's what Gallup did.

EXAMPLE 1 Do you favor a constitutional amendment?

The proportion of all adults who favor a constitutional amendment that would define marriage as being between a man and a woman is a *parameter* describing the population of 220 million adults. Call it p, for "proportion." Alas, we do not know the numerical value of p. To estimate p, Gallup took a sample of 2527 adults. The proportion of the sample who favor such an amendment is a *statistic*. Call it $\hat{p}$, read as "p-hat." It happens that 1289 of this sample of size 2527 said that they favor an amendment, so for this sample,

$$\hat{p} = \frac{1289}{2527} = 0.51 \quad \text{(that is, 51%)}$$

Because all adults had the same chance to be among the chosen 2527, it seems reasonable to use the statistic $\hat{p} = 0.51$ as an estimate of the unknown parameter p. It's a *fact* that 51% of the sample favored an amendment—we know because we asked them. We don't know what percentage of all adults favor an amendment, but we *estimate* that about 51% do.

Sampling variability

If Gallup took a second random sample of 2527 adults, the new sample would have different people in it. It is almost certain that there would not be exactly 1289 favorable responses. That is, the value of the statistic $\hat{p}$ will *vary* from sample to sample. Could it happen that one random sample finds that 51% of adults favor an amendment and a second random sample finds that only 37% favor one? Random samples eliminate *bias* from the act of choosing a sample, but they can still be wrong because of the variability that results when we choose at random. If the variation when we take repeated samples from the same population is too great, we can't trust the results of any one sample.

We are saved by the second great advantage of random samples. The first advantage is that choosing at random eliminates favoritism. That is, random sampling attacks bias. The second advantage is that if we took lots of random samples of the same size from the same population, the variation from sample to sample would follow a predictable pattern. This predictable pattern shows that results of bigger samples are less variable than the results of smaller samples.

EXAMPLE 2 Lots and lots of samples

Here's another big idea of statistics: to see how trustworthy one sample is likely to be, ask what would happen if we took many samples from the same population. Let's try it and see. Suppose that in fact (unknown to Gallup), exactly 50% of all adults favor a constitutional amendment that would define marriage as being between a man and a woman. That is, the truth about the population is that $p = 0.5$. What if Gallup used the sample proportion $\hat{p}$ from an SRS of size 100 to estimate the unknown value of the population proportion p?

Figure 3.1 illustrates the process of choosing many samples and finding $\hat{p}$ for each one. In the first sample, 56 of the 100 people favored the amendment, so $\hat{p} = 56/100 = 0.56$. Only 36 in the next sample favored the amendment, so for that sample $\hat{p} = 0.36$. Choose 1000 samples and make a plot of the 1000 values of $\hat{p}$ like the graph (called a histogram) at the right of Figure 3.1. The different values of the sample proportion $\hat{p}$ run along the horizontal axis. The height of each bar shows how many of our 1000 samples gave the group of values on the horizontal axis covered by the bar. For example, in Figure 3.1 the bar covering the values between 0.40 and 0.42 has a height of slightly over 50. Thus, over 50 of our 1000 samples had values between 0.40 and 0.42.

Of course, Gallup interviewed 2527 people, not just 100. Figure 3.2 shows the results of 1000 SRSs, each of size 2527, drawn from a population in which the true sample proportion is $p = 0.5$. Figures 3.1 and 3.2 are drawn on the same scale. Comparing them shows what happens when we increase the size of our samples from 100 to 2527.

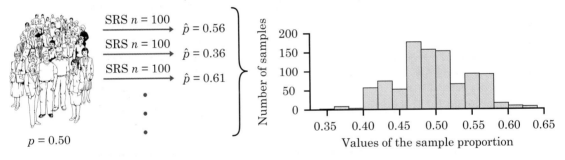

Figure 3.1 The results of many SRSs have a regular pattern. Here, we draw 1000 SRSs of size 100 from the same population. The population proportion is $p = 0.5$. The sample proportions vary from sample to sample, but their values center at the truth about the population.

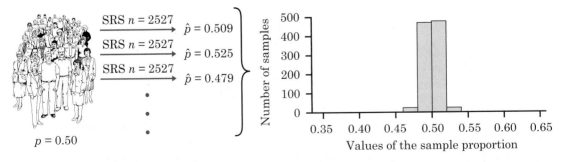

Figure 3.2 Draw 1000 SRSs of size 2527 from the same population as in Figure 3.1. The 1000 values of the sample proportion are much less spread out than was the case for smaller samples.

Look carefully at Figures 3.1 and 3.2. We flow from the population, to many samples from the population, to the many values of $\hat{p}$ from these many samples. Gather these values together and study the histograms that display them.

- In both cases, the values of the sample proportion $\hat{p}$ vary from sample to sample, but the values are centered at 0.5. Recall that $p = 0.5$ is the true population parameter. Some samples have a $\hat{p}$ less than 0.5 and some greater, but there is no tendency to be always low or always high. That is, $\hat{p}$ has no **bias** as an estimator of p. This is true for both large and small samples.

- The values of $\hat{p}$ from samples of size 100 are much more spread out than the values from samples of size 2527. In fact, 95% of our 1000 samples of size 2527 have a $\hat{p}$ lying between 0.4805 and 0.5195. That's within 0.0195 on either side of the population truth 0.5. Our samples of size 100, on the other hand, spread the middle 95% of their values between 0.40 and 0.60. That goes out 0.1 from the truth, about five times as far as the larger samples. So larger random samples have less **variability** than smaller samples.

The result is that we can rely on a sample of size 2527 to almost always give an estimate $\hat{p}$ that is close to the truth about the population. Figure 3.2 illustrates this fact for just one value of the population proportion, but it is true for any population proportion. Samples of size 100, on the other hand, might give an estimate of 40% or 60% when the truth is 50%.

Thinking about Figures 3.1 and 3.2 helps us restate the idea of bias when we use a statistic like $\hat{p}$ to estimate a parameter like p. It also reminds us that variability matters as much as bias.

Two types of error in estimation

Bias is consistent, repeated deviation of the sample statistic from the population parameter in the same direction when we take many samples.

Variability describes how spread out the values of the sample statistic are when we take many samples. Large variability means that the result of sampling is not repeatable.

A good sampling method has both small bias and small variability.

We can think of the true value of the population parameter as the bull's-eye on a target, and of the sample statistic as an arrow fired at the bull's-eye. Bias and variability describe what happens when an archer fires many arrows at the target. *Bias* means that the aim is off, and the arrows

Figure 3.3 Bias and variability in shooting arrows at a target. Bias means the archer systematically misses in the same direction. Variability means that the arrows are scattered.

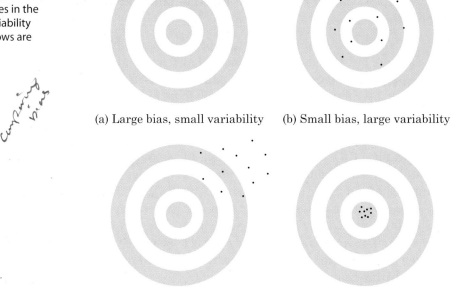

(a) Large bias, small variability   (b) Small bias, large variability

(c) Large bias, large variability   (d) Small bias, small variability

land consistently off the bull's-eye in the same direction. The sample values do not center about the population value. Large *variability* means that repeated shots are widely scattered on the target. Repeated samples do not give similar results but differ widely among themselves. Figure 3.3 shows this target illustration of the two types of error.

Notice that small variability (repeated shots are close together) can accompany large bias (the arrows are consistently away from the bull's-eye in one direction). And small bias (the arrows center on the bull's-eye) can accompany large variability (repeated shots are widely scattered). A good sampling scheme, like a good archer, must have both small bias and small variability. Here's how we do this:

Managing bias and variability

To reduce bias, use random sampling. When we start with a list of the entire population, simple random sampling produces *unbiased* estimates: the values of a statistic computed from an SRS neither consistently overestimate nor consistently underestimate the value of the population parameter.

To reduce the variability of an SRS, use a larger sample. You can make the variability as small as you want by taking a large enough sample.

In practice, Gallup takes only one sample. We don't know how close to the truth an estimate from this one sample is, because we don't know what the truth about the population is. But *large random samples almost always give an estimate that is close to the truth.* Looking at the pattern of many samples shows how much we can trust the result of one sample.

Margin of error and all that

The "margin of error" that sample surveys announce translates sampling variability of the kind pictured in Figures 3.1 and 3.2 into a statement of how much confidence we can have in the results of a survey. Let's start with the kind of language we hear so often in the news.

What margin of error means

"Margin of error plus or minus two percentage points" is shorthand for this statement:

If we took many samples using the same method we used to get this one sample, 95% of the samples would give a result within plus or minus 2 percentage points of the truth about the population.

Take this step-by-step. A sample chosen at random will usually not estimate the truth about the population exactly. We need a margin of error to tell us how close our estimate comes to the truth. But we can't be *certain* that the truth differs from the estimate by no more than the margin of error. Although 95% of all samples come this close to the truth, 5% miss by more than the margin of error. We don't know the truth about the population, so we don't know if our sample is one of the 95% that hit or one of the 5% that miss. We say we are **95% confident** that the truth lies within the margin of error.

EXAMPLE 3 Understanding the news

Here's what the TV news announcer says: "A new Gallup Poll finds that a slim majority of 51% of American adults favor a constitutional amendment that would define marriage as being between a man and a woman, thus barring marriages between gay and lesbian couples. The margin of error for the poll was 2 percentage points." Plus or minus 2% starting at 51% is 49% to 53%. Most people think Gallup claims that the truth about the entire population lies in that range.

This is what Gallup actually said: "For results based on a sample of this size, one can say with 95% confidence that the error attributable to sampling and other random effects could be plus or minus 2 percentage points for adults." That is, Gallup tells us that the margin of error includes the truth about the entire population for only 95% of all its samples. "95% confidence" is shorthand for that. The news report left out the "95% confidence."

Finding the margin of error exactly is a job for statisticians. You can, however, use a simple formula to get a rough idea of the size of a sample survey's margin of error. The reasoning behind this formula is discussed in Chapter 21.

A quick method for the margin of error

Use the sample proportion $\hat{p}$ from a simple random sample of size n to estimate an unknown population proportion p. The margin of error for 95% confidence is roughly equal to $1/\sqrt{n}$.

EXAMPLE 4 What is the margin of error?

The Gallup Poll in Example 1 interviewed 2527 people. The margin of error for 95% confidence will be about

$$\frac{1}{\sqrt{2527}} = \frac{1}{50.27} = 0.020 \quad \text{(that is, 2.0\%)}$$

Gallup announced a margin of error of 2% and our quick method agrees with this. In general, however, our quick method can disagree a bit with Gallup's for two reasons. First, polls usually round their announced margin of error to the nearest whole percent to keep their press releases simple. Second, our rough formula works for an SRS. We will see in the next chapter that most national samples are more complicated than an SRS in ways that tend to slightly increase the margin of error.

Our quick method also reveals an important fact about how margins of error behave. Because the sample size n appears in the denominator of the fraction, larger samples have smaller margins of error. We knew that. Because the formula uses the square root of the sample size, however, *to cut the margin of error in half, we must use a sample four times as large.*

EXAMPLE 5 Margin of error and sample size

In Example 2 we compared the results of taking many SRSs of size $n = 100$ and many SRSs of size $n = 2527$ from the same population. We found that the spread of the middle 95% of the sample results was about five times larger for the smaller samples.

Our quick formula estimates the margin of error for SRSs of size 2527 to be about 2.0%. The margin of error for SRSs of size 100 is about

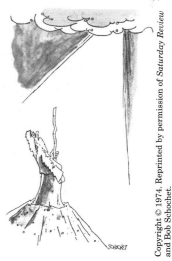

"Seventy-three percent are in favor of one through five, forty-one percent find six unfair, thirteen percent are opposed to seven, sixty-two percent applauded eight, thirty-seven percent …"

$$\frac{1}{\sqrt{100}} = \frac{1}{10} = 0.1 \quad \text{(that is, 10\%)}$$

Because 2527 is roughly 25 times 100 and the square root of 25 is 5, the margin of error is about five times larger for samples of 100 people than for samples of 2527 people.

NOW IT'S YOUR TURN | **3.1 Taxing the rich.** In April 2011, the Gallup Poll asked a random sample of 1077 adults, "People feel differently about how far a government should go. Here is a phrase which some people believe in and some don't. Do you think our government should or should not redistribute wealth by heavy taxes on the rich?" It found that 47% said our government should redistribute wealth by heavy taxes on the rich. What is the margin of error for 95% confidence?

Confidence statements

Here is Gallup's conclusion about Bush's proposal for a constitutional amendment in short form: "The poll found that a very slim majority of Americans, 51%, favor such an amendment. We are 95% confident that the truth about all adults is within plus or minus 2 percentage points of this sample result." Here is an even shorter form: "We are 95% confident that between 49% and 53% of all adults favor such an amendment." These are *confidence statements*.

> ### Confidence statements
>
> A **confidence statement** has two parts: a **margin of error** and a **level of confidence.** The margin of error says how close the sample statistic lies to the population parameter. The level of confidence says what percentage of all possible samples satisfy the margin of error.

A confidence statement is a fact about what happens in all possible samples and is used to say how much we can trust the result of one sample. The phrase "95% confidence" means "We used a sampling method that gives a result this close to the truth 95% of the time." Here are some hints for interpreting confidence statements:

- *The conclusion of a confidence statement always applies to the population, not to the sample.* We know exactly the opinions of the 2527 people in the sample, because Gallup interviewed them. The confidence statement uses the sample result to say something about the population of all adults.

- *Our conclusion about the population is never completely certain.* Gallup's sample *might* be one of the 5% that miss by more than 2 percentage points.

- *A sample survey can choose to use a confidence level other than 95%.* We pay for higher confidence with a larger margin of error. For the same sample, a 99% confidence statement requires a larger margin of error than 95% confidence. If you are content with 90% confidence, you get in return a smaller margin of error. Remember that our quick method gives the margin of error only for 95% confidence.

- *It is usual to report the margin of error for 95% confidence.* If a news report gives a margin of error but leaves out the confidence level, it's pretty safe to assume 95% confidence.

- *Want a smaller margin of error with the same confidence? Take a larger sample.* Remember that larger samples have less variability. You can get as small a margin of error as you want and still have high confidence by paying for a large enough sample.

 The telemarketer's pause People who do sample surveys hate telemarketing. We all get so many unwanted sales pitches by phone that many people hang up before learning that the caller is conducting a survey rather than selling vinyl siding. Here's a tip. Both sample surveys and telemarketers dial telephone numbers at random. Telemarketers automatically dial many numbers, and their sellers come on the line only after you pick up the phone. Once you know this, the telltale "telemarketer's pause" gives you a chance to hang up before the seller arrives. Sample surveys have a live interviewer on the line when you answer.

EXAMPLE 6 2008 election polls

In 2008, shortly before the presidential election, SurveyUSA, a polling organization, asked voters in several states who they would vote for. In Florida they asked a random sample of 691 likely voters, and 50% said they would vote for Barack Obama and 47% said John McCain. SurveyUSA reported the margin of error to be plus or minus 3.8%. In Georgia they sampled 547 likely voters, and 43% said they would vote for Obama and 51% said McCain. The margin of error was reported to be plus or minus 4.3%.

There you have it: the sample of likely voters in Georgia was smaller, so the margin of error for conclusions about voters in Georgia is wider. We are 95% confident that between 38.7% (that's 43% minus 4.3%) and 47.3% (that's 43% plus 4.3%) of likely voters in Georgia would vote for Obama. Note that the actual 2008 election results for Georgia were 47% for Obama, which is within the margin of error.

NOW IT'S YOUR TURN | **3.2 Taxing the rich.** In April 2011, the Gallup Poll asked a random sample of 1077 adults, "People feel differently about how far a government should go. Here is a phrase which some people believe in and some don't. Do you think our government should or should not redistribute wealth by heavy taxes on the rich?" It found that 47% said our government should redistribute wealth by heavy taxes on the rich. Suppose that the sample size had been 4000 rather than 1077. Find the margin of error for 95% confidence in this case. How does it compare with the margin of error for a sample of size 1077?

Sampling from large populations

Gallup's sample of 2527 adults was only 1 out of every 82,700 adults in the United States. Does it matter whether 2527 is 1-in-100 individuals in the population or 1-in-82,700?

Population size doesn't matter

The variability of a statistic from a random sample is essentially unaffected by the size of the population as long as the population is at least 100 times larger than the sample.

STATISTICAL CONTROVERSIES

Should Election Polls Be Banned?

MCT/Newscom

Preelection polls tell us that Senator So-and-So is the choice of 58% of Ohio voters. The media love these polls. Statisticians don't love them, because elections often don't go as forecasted even when the polls use all the right statistical methods. Many people who respond to the polls change their minds before the election. Others say they are undecided. Still others say which candidate they favor but won't bother to vote when the election arrives. Election forecasting is one of the less satisfactory uses of sample surveys because we must ask people now how they will vote in the future.

Exit polls, which interview voters as they leave the voting place, don't have these problems. The people in the sample have just voted. A *good* exit poll, based on a national sample of election precincts, can often call a presidential election correctly long before the polls close. But, as was the case in the 2004 presidential election, exit polls can also fail to call the results correctly. These facts sharpen the debate over the political effects of election forecasts.

Can you think of good arguments *against* public preelection polls? Think about how these polls might influence voter turnout. What about arguments against exit polls? Remember that voting ends in the East several hours before it ends in the West.

What are arguments *for* preelection polls? Consider freedom of speech, for example. What about arguments for exit polls?

For some thought-provoking articles on polls, especially in light of the exit poll failures in the 2004 presidential election, see the following Web sites.

www.washingtonpost.com/wp-dyn/ articles/A47000-2004Nov12_2.html

www.washingtonpost.com/wp-dyn/ articles/A64906-2004Nov20.html

www.edisonresearch.com/exit_ poll_faq.php

http://thehill.com/opinion/ columnists/dick-morris/4723-those- faulty-exit-polls-were-sabotage

You may also go to **www.google.com** and search using key words "exit poll failures in the 2004 presidential election."

Why does the size of the population have little influence on the behavior of statistics from random samples? Imagine sampling harvested corn by thrusting a scoop into a lot of corn kernels. The scoop doesn't know whether it is surrounded by a large sack of corn or by an entire truckload. As long as the corn is well mixed (so that the scoop selects a random sample) and the scoop is a small fraction of the total, the variability of the result depends only on the size of the scoop.

This is good news for national sample surveys like the Gallup Poll. A random sample of size 1000 or 2500 has small variability because the sample size is large. But remember that even a very large voluntary response sample or convenience sample is worthless because of bias. Taking a larger sample doesn't fix bias.

However, the fact that the variability of a sample statistic depends on the size of the sample and not on the size of the population is bad news for anyone planning a sample survey in a university or a small city. For example, it takes just as large an SRS to estimate the proportion of Ohio State University undergraduate students who call themselves political conservatives as to estimate with the same margin of error the proportion of all adult U.S. residents who are conservatives. We can't use a smaller SRS at Ohio State just because there are 49,000 Ohio State undergraduate students and over 232 million adults in the United States in 2009.

STATISTICS IN SUMMARY

Chapter Specifics

- The purpose of sampling is to use a sample to gain information about a population. We often use a sample **statistic** to estimate the value of a population **parameter.**

- This chapter has one big idea: to describe how trustworthy a sample is, ask, "What would happen if we took a large number of samples from the same population?" If almost all samples would give a result close to the truth, we can trust our one sample even though we can't be certain that it is close to the truth.

- In planning a sample survey, first aim for small **bias** by using random sampling and avoiding bad sampling methods such as voluntary response. Next, choose a large enough random sample to reduce the **variability** of the result. Using a large random sample guarantees that almost all samples will give accurate results.

- To say how accurate our conclusions about the population are, make a **confidence statement.** News reports often mention only the **margin of error.** Most often this margin of error is for **95% confidence.** That is, if we chose many samples, the truth about the population would be within the margin of error 95% of the time.

- We can estimate the margin of error for 95% confidence based on a simple random sample of size n by the formula $1/\sqrt{n}$. As this formula suggests, only the size of the sample, not the size of the population, matters. This is true as long as the population is much larger (at least 100 times larger) than the sample.

In Chapter 1 we introduced sample surveys as an important kind of observational study. In Chapter 2 we discussed both good and bad methods for taking a sample survey. Simple random sampling was introduced as a method that deliberately uses chance to produce unbiased data. This deliberate use of chance to produce data is one of the big ideas of statistics.

In this chapter we looked more carefully at how sample information is used to gain information about the population from which it is selected. The big idea is to ask what would happen if we used our method for selecting a sample to take many samples from the same population. If almost all would give results that are close to the truth, then we have a basis for trusting our sample.

In practice, how easy is it to take a simple random sample? What problems do we encounter when we attempt to take samples in the real world? This is the topic of the next chapter.

CASE STUDY   In the Case Study at the beginning of the chapter, 51% of those surveyed in 2004 said that they favored a constitutional amendment that would define marriage as being between a man and a woman, thus barring marriages between gay and lesbian couples. The Gallup Poll stated that such an amendment was supported by a slim majority (51%) of Americans. Is 51% in favor evidence that, in fact, the majority of Americans in 2004 favored such an amendment? Do the results of the 2011 Gallup Poll indicate that today the majority (53%) of Americans would be opposed to such an amendment? Use what you have learned in this chapter to answer this question. Your answer should be written so that someone who knows no statistics will understand your reasoning. ■

CHAPTER 3 EXERCISES

For Exercise 3.1, see page 47; for Exercise 3.2, see page 49.

3.3 The boldface number in the next paragraph is the value of either a **parameter** or a **statistic**. State which it is.

The Bureau of Labor Statistics announces that last month it interviewed all members of the labor force in a sample of 60,000 households; **9.7%** of the people interviewed were unemployed.

3.4 Each boldface number in the next paragraph is the value of either a **parameter** or a **statistic**. In each case, state which it is.

A carload lot of ball bearings has an average diameter of **2.503** centimeters (cm). This is within the specifications for acceptance of the lot by the purchaser. The inspector happens to inspect 100 bearings from the lot with an average diameter of **2.515** cm. This is outside the specified limits, so the lot is mistakenly rejected.

3.5 Each boldface number in the next paragraph is the value of either a **parameter** or a **statistic**. In each case, state which it is.

Voter registration records show that **15.4%** of all voters in Philadelphia are registered as Republicans. However, a radio talk show host in Philadelphia found that of 20 local residents who called the show recently, **60%** were registered Republicans.

3.6 Each boldface number in the next paragraph is the value of either a **parameter** or a **statistic**. In each case, state which it is.

A national polling organization uses a random digit dialing device to dial residential landline phone numbers in the United States. Of the first 100 numbers dialed, **32** are unlisted numbers. This is not surprising, because **34%** of all residential phones in the United States, including cell-phone-only and many unlisted landline phone households, are not covered by current landline telephone number sampling methods.

3.7 A sampling experiment. Figures 3.1 and 3.2 (page 42) show how the sample proportion $\hat{p}$ behaves when we take many samples from the same population. You can follow the steps in this process on a small scale.

Figure 3.4 (next page) represents a small population. Each circle represents an adult. The white circles are people who favor a constitutional amendment that would define marriage as being between a man and a woman, and the colored circles are people who are opposed. You can check that 50 of the 100 circles are white, so in this population the proportion who favor an amendment is $p = 50/100 = 0.5$.

(a) The circles are labeled 00, 01, ... , 99. Use line 101 of Table A to draw an SRS of size 4. What is the proportion $\hat{p}$ of the people in your sample who favor a constitutional amendment?

(b) Take 9 more SRSs of size 4 (10 in all), using lines 102 to 110 of Table A, a different line for each sample. You now have 10 values of the sample proportion $\hat{p}$. Write down the 10 values you should now have of the sample proportion $\hat{p}$.

(c) Because your samples have only 4 people, the only values $\hat{p}$ can take are 0/4, 1/4, 2/4, 3/4, and 4/4. That is, $\hat{p}$ is always 0, 0.25, 0.5, 0.75, or 1. Mark these numbers on a line and make a histogram of your 10 results by putting a bar above each number to show how many samples had that outcome.

(d) Taking samples of size 4 from a population of size 100 is not a practical setting, but let's look at your results anyway. How many of your 10 samples estimated the population proportion $p = 0.5$ exactly correctly? Is the true value 0.5 in the center of your sample values? Explain why 0.5 would be in the center of the sample values if you took a large number of samples.

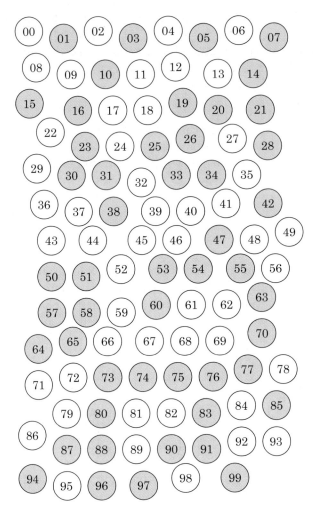

Figure 3.4 A population of 100 individuals for Exercise 3.7. Some individuals (white circles) favor a constitutional amendment, and the others do not.

3.8 A sampling experiment. Let us illustrate sampling variability in a small sample from a small population. Ten of the 25 club members listed below are female. Their names are marked with asterisks in the list. The club chooses 5 members at random to receive free trips to the national convention.

Alonso	Darwin	Herrnstein	Myrdal	Vogt*
Binet*	Epstein	Jimenez*	Perez*	Went
Blumenbach	Ferri	Luo	Spencer*	Wilson
Chase*	Gonzales*	Moll*	Thomson	Yerkes
Chen*	Gupta	Morales*	Toulmin	Zimmer

(a) Draw 20 SRSs of size 5, using a different part of Table A each time. Record the number of females in each of your samples. Make a histogram like that in Figure 3.1 to display your results. What is the average number of females in your 20 samples?

(b) Do you think the club members should suspect discrimination if none of the 5 tickets go to women?

3.9 Canada's national health care. The Ministry of Health in the Canadian province of Ontario wants to know whether the national health care system is achieving its goals in the province. Much information about health care comes from patient records, but that source doesn't allow us to compare people who use health services with those who don't. So the Ministry of Health conducted the Ontario Health Survey, which interviewed a random sample of 61,239 people who live in the province of Ontario.

(a) What is the population for this sample survey? What is the sample?

(b) The survey found that 76% of males and 86% of females in the sample had visited a general practitioner at least once in the past year. Do you think these estimates are close to the truth about the entire population? Why?

3.10 Bigger samples, please. Explain in your own words the advantages of bigger random samples in a sample survey.

3.11 Sampling variability. In thinking about Gallup's sample of size 2527, we asked, "Could it happen that one random sample finds that 51% of adults favor an amendment and a second random sample finds that only 37% favor one?" Look at Figure 3.2, which shows the results of 1000 samples of this size when the population truth is $p = 0.5$, or 50%. Would you be surprised if a sample from this population gave 51%? Would you be surprised if a sample gave 37%?

3.12 Health care satisfaction. A November 2011 Gallup Poll of 1012 adults found that 607 are satisfied with the total cost they pay for their health care. The announced margin of error is ±4 percentage points. The announced confidence level is 95%.

(a) What is the value of the sample proportion $\hat{p}$ who say they are satisfied with the total cost they pay for their health care? Explain in words what the population parameter p is in this setting.

(b) Make a confidence statement about the parameter p.

3.13 Bias and variability. Figure 3.5 (next page) shows the behavior of a sample statistic in many samples in four situations. These graphs are like those in Figures 3.1 and 3.2. That is, the heights of the bars show how often the sample statistic took various values in many samples from the same population. The true value of the population parameter is marked on each graph. Label each of the graphs in Figure 3.5 as showing high or low bias and as showing high or low variability.

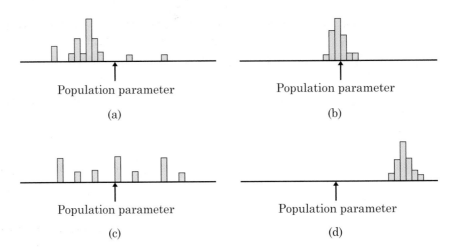

Figure 3.5 Take many samples from the same population and make a histogram of the values taken by a sample statistic. Here are the results for four different sampling methods, for Exercise 3.13.

3.14 Is a larger sample size always better? In February 2004, *USA Today* conducted an online poll. Visitors to their Web site were asked the following question: "Should the U.S. pass a constitutional amendment banning gay marriage?" Visitors could vote by clicking a button. The results as of 3:30 P.M. on February 25 were that 68.61% voted "No" and 31.39% voted "Yes." A total of 63,046 votes had been recorded. Using our quick method, we find that the margin of error for 95% confidence for a sample of this size is roughly equal to four-tenths of one percentage point. Is it correct to say that, based on this *USA Today* online poll, we are 95% confident that 68.61% ±0.4% of American adults are opposed to having the United States pass a constitutional amendment banning gay marriage? Explain your answer. Be careful not to confuse your personal opinion with the statistical issues.

3.15 Predict the election. Just before a presidential election, a national opinion poll increases the size of its weekly sample from the usual 1000 people to 4000 people. Does the larger random sample reduce the bias of the poll result? Does it reduce the variability of the result?

3.16 Take a bigger sample. A management student is planning a project on student attitudes toward part-time work while attending college. She develops a questionnaire and plans to ask 25 randomly selected students to fill it out. Her faculty adviser approves the questionnaire but suggests that the sample size be increased to at least 100 students. Why is the larger sample helpful? Back up your statement by using the quick method to estimate the margin of error for samples of size 25 and for samples of size 100.

3.17 Sampling in the states. An agency of the federal government plans to take an SRS of residents in each state to estimate the proportion of owners of real estate in each state's population. The populations of the states range from about 544,000 people in Wyoming to about 37.0 million in California.

(a) Will the variability of the sample proportion change from state to state if an SRS of size 2000 is taken in each state? Explain your answer.

(b) Will the variability of the sample proportion change from state to state if an SRS of 1/10 of 1% (0.001) of the state's population is taken in each state? Explain your answer.

3.18 Polling women. A *New York Times* Poll on women's issues interviewed 1025 women randomly selected from the United States, excluding Alaska and Hawaii. The poll found that 47% of the women said they do not get enough time for themselves.

(a) The poll announced a margin of error of ±3 percentage points for 95% confidence in its conclusions. Make a 95% confidence statement about the percentage of all adult women who think they do not get enough time for themselves.

(b) Explain to someone who knows no statistics why we can't just say that 47% of all adult women do not get enough time for themselves.

(c) Then explain clearly what "95% confidence" means.

3.19 The Zogby Poll. Here is the language used by the Zogby Poll to explain the accuracy of a recent poll: "The margin of error is plus or minus 1.2 percentage points. Zogby International's sampling and weighting procedures also have been validated through its political polling: more than 95% of the firm's polls have come within 1% of actual election-day outcomes." What does Zogby mean by "95% of the firm's polls have come within 1% of actual election-day outcomes"?

3.20 Polling men and women. The sample survey described in Exercise 3.18 interviewed 472 randomly selected men as well as 1025 women. The poll announced a margin of error of ±3 percentage points for 95% confidence in conclusions about women. The margin of error for results concerning men was ±5 percentage points. Why is this larger than the margin of error for women?

3.21 Explaining confidence. A student reads that we are 95% confident that the average score of young men on the quantitative part of the National Assessment of Educational Progress is 267.8 to 276.2. Asked to explain the meaning of this statement, the student says, "95% of all young men have scores between 267.8 and 276.2." Is the student right? Explain your answer.

3.22 The death penalty. In October 2011, the Gallup Poll asked a sample of 1005 adults, "Are you in favor of the death penalty for a person convicted of murder?" The proportion who said they were in favor was 61%.

(a) How many of the 1005 people interviewed said they were in favor of the death penalty for a person convicted of murder?

(b) Gallup says that the margin of error for this poll is ±4 percentage points. Explain to someone who knows no statistics what "margin of error ±4 percentage points" means.

3.23 Find the margin of error. Example 6 (page 49) tells us that a SurveyUSA Poll asked 547 likely voters in Georgia which presidential candidate they would vote for; 51% said they would vote for John McCain. Use the quick method to estimate the margin of error for conclusions about all likely voters in Georgia. How does your result compare with SurveyUSA's margin of error given in Example 6?

3.24 Find the margin of error. Exercise 3.22 concerns a Gallup Poll sample of 1005 people. Use the quick method to estimate the margin of error for statements about the population of all adults. Is your result close to the ±4% margin of error announced by Gallup?

3.25 Find the margin of error. Exercise 3.9 (page 55) describes a sample survey of 61,239 adults living in Ontario. Estimate the margin of error for conclusions having 95% confidence about the entire adult population of Ontario.

3.26 Belief in God. A Gallup Poll conducted in May 2011 reports that 92% of a sample of 509 adults said "Yes" when asked "Do you believe in God?"

(a) Use the quick method to estimate the margin of error for an SRS of this size.

(b) Assuming that this was a random sample, make a confidence statement about the percentage of all adults who believe in God.

3.27 Abortion. A 2011 Harris Poll of 2362 adults found that 1110 favored permitting abortion under "some [but not all] circumstances," a decrease of 6 percentage points from 2009. Make a confidence statement about the percentage of all adults who favored permitting abortion under "some [but not all] circumstances," at the time the poll was taken. (Assume that this is an SRS, and use the quick method to find the margin of error.)

3.28 Moral uncertainty versus statistical uncertainty. In Exercise 3.27 and in the Case Study we examined polls involving controversial moral issues. In both polls, national opinion was divided, suggesting that there is considerable "moral uncertainty" regarding both issues. What was the margin of error (the "statistical uncertainty") in both polls? Is it possible for issues with a high degree of moral uncertainty to have very little statistical uncertainty? Discuss.

3.29 Smaller margin of error. Exercise 3.22 describes an opinion poll that interviewed 1005 people. Suppose that you want a margin of error half as large as the one you found in that exercise. How many people must you plan to interview?

3.30 Satisfying Congress. Exercise 3.12 (page 55) describes a sample survey of 1012 adults, with margin of error ±4% for 95% confidence.

(a) A member of Congress thinks that 95% confidence is not enough. He wants to be 99% confident. How would the margin of error for 99% confidence based on the same sample compare with the margin of error for 95% confidence?

(b) Another member of Congress is satisfied with 95% confidence, but she wants a smaller margin of error than ±4 percentage points. How can we get a smaller margin of error, still with 95% confidence?

3.31 The Current Population Survey. Though opinion polls usually make 95% confidence statements, some sample surveys use other confidence levels. The monthly unemployment rate, for example, is based on the Current Population Survey of about 60,000 households. The margin of error in the unemployment rate is announced as about two-tenths of one percentage point with 90% confidence. Would the margin of error for 95% confidence be smaller or larger? Why?

3.32 Honesty and Wall Street? In April 2011, the Harris Poll asked a random sample of 1010 adults if they agreed with the following statement: "Most people on Wall Street would be willing to break the law if they believed they could make a lot of money and get away with it." It found that 677 agreed with the statement. Write a short report of this finding, as if you were writing for a newspaper. Be sure to include a margin of error. Be careful not to confuse your personal opinion with the statistical findings.

3.33 Who is to blame? A February 2009 poll conducted by the Marist College Institute for Public Opinion in Poughkeepsie, New York, asked a random sample of 2071 U.S. adults who or what was responsible for a company's failure or success. Of those surveyed, 70% attributed a company's failure or success to the decisions of its top executives. The poll asked the same question of a random sample of 110 business executives. Among executives, 88% said that top executives were responsible for a company's success or failure.

Marist reported that the margin of error for one of these results was ±9 percentage points and for the other it was ±2.5 percentage points. Which result had the margin of error of ±9 percentage points? Explain your answer.

3.34 Simulation. Random digits can be used to *simulate* the results of random sampling. Suppose that you are drawing simple random samples of size 25 from a large number of college students and that 20% of the students are unemployed during the summer. To simulate this SRS, let 25 consecutive digits in Table A stand for the 25 students in your sample. The digits 0 and 1 stand for unemployed students, and other digits stand for employed students. This is an accurate imitation of the SRS because 0 and 1 make up 20% of the 10 equally likely digits.

Simulate the results of 50 samples by counting the number of 0s and 1s in the first 25 entries in each of the 50 rows of Table A. Make a histogram like that in Figure 3.1 to display the results of your 50 samples. Is the truth about the population (20% unemployed, or 5 in a sample of 25) near the center of your graph? What are the smallest and largest counts of unemployed students that you obtained in your 50 samples? What percentage of your samples had either 4, 5, or 6 unemployed?

 EXPLORING THE WEB

3.35 Web-based exercise. Go to **www.gallup.com** and read the Top Story. Click on the *More...* link. What does the article tell you about the margin of error for 95% confidence?

3.36 Web-based exercise. Read the Gallup Organization's own explanation of why surprisingly small samples can give trustworthy results about large populations at **http://media.gallup.com/PDF/FAQ/ HowArePolls.pdf**. What does the article say about interpreting 95% confidence?

NOTES AND DATA SOURCES

Page 39 The Gallup Organization, "For first time, majority of Americans favor legal gay marriage"; available online at **www.gallup.com.**

Page 39 The Gallup Organization, "Constitutional amendment defining marriage lacks 'supermajority' support," Social Issues and Policy detailed report; available online at **www.gallup.com.**

Page 49 Example 6: This example comes from results posted at the Web site **www.surveyusa.com/electionpolls.aspx.**

Page 52 Exercise 3.3: News releases about unemployment can be found online at **www.bls.gov.**

Page 55 Exercise 3.9: Warren McIsaac and Vivek Goel, "Is access to physician services in Ontario equitable?" Institute for Clinical Evaluative Sciences in Ontario, October 18, 1993.

Page 56 Exercise 3.14: "Bush backs gay-marriage ban," *USATODAY.com,* February 24, 2004; available online at **www.usatoday .com/news/washington/2004-02-24-bush-marriage_x.htm.**

Page 57 Exercise 3.18: *New York Times*, August 21, 1989.

Page 57 Exercise 3.19: See **www.zogby.com.**

Page 58 Exercise 3.26: See **www.gallup .com/poll/147887/Americans-Continue-Believe-God.aspx.**

Page 58 Exercise 3.27: See **www .harrisinteractive.com/NewsRoom/ PressReleases/tabid/446/mid/1506/ articleId/841/ctl/ReadCustom% 20Default/Default.aspx.**

Page 59 Exercise 3.32: See **www .harrisinteractive.com/NewsRoom/ HarrisPolls/tabid/447/ctl/ReadCustom% 20Default/mid/1508/ArticleId/783/ Default.aspx.**

Sample Surveys in the Real World

CASE STUDY An opinion poll talks to 1000 people chosen at random, announces its results, and announces a margin of error. Should we be happy? Perhaps not. Many polls don't tell the whole truth about their samples. The Pew Research Center for the People and the Press imitated the methods of the better opinion polls and did tell the whole truth. Here it is.

Tetra Images/SuperStock

Most polls are taken by telephone, dialing numbers at random to get a random sample of households. After eliminating fax and business numbers, Pew had to call 2879 residential numbers to get their sample of 1000 people. Here's the breakdown:

Never answered phone	938
Answered but refused	678
Not eligible: no person aged	
18 or older, or language barrier	221
Incomplete interview	42
Complete interview	1000
Total called	2879

Out of 2879 good residential phone numbers, 33% never answered. Of those who answered, 35% refused to talk. The overall rate of nonresponse (people who never answered, refused, or would not complete the interview) was 1658 out of 2879, or 58%. Pew called every number five times over a five-day period, at different times of day and on different days of the week. Many polls call only once, and it is usual to find that over half of those who answer refuse to talk. Although Pew did obtain the desired sample of 1000 people, can we trust the results of this poll? By the end of this chapter you will learn how to answer this question. ∎

How sample surveys go wrong

Random sampling eliminates bias in choosing a sample and allows control of variability. So once we see the magic words "randomly selected" and "margin of error," do we know we have trustworthy information before us? It certainly beats voluntary response, but not always by as much as we might hope. Sampling in the real world is more complex and less reliable than choosing an SRS from a list of names in a textbook exercise. Confidence statements do not reflect all the sources of error that are present in practical sampling.

Errors in sampling

Sampling errors are errors caused by the act of taking a sample. They cause sample results to be different from the results of a census.

Random sampling error is the deviation between the sample statistic and the population parameter caused by chance in selecting a random sample. The margin of error in a confidence statement includes *only* random sampling error.

Nonsampling errors are errors not related to the act of selecting a sample from the population. They can be present even in a census.

Most sample surveys are afflicted by errors other than random sampling errors. These errors can introduce bias that makes a confidence statement meaningless. Good sampling technique includes the art of reducing all sources of error. Part of this art is the science of statistics, with its random samples and confidence statements. In practice, however, good statistics isn't all there is to good sampling. Let's look at sources of errors in sample surveys and at how samplers combat them.

Sampling errors

Random sampling error is one kind of sampling error. The margin of error tells us how serious random sampling error is, and we can control it by choosing the size of our random sample. Another source of sampling error is the use of *bad sampling methods,* such as voluntary response. We can avoid bad methods. Other sampling errors are not so easy to handle. Sampling begins with a list of individuals from which we will draw our sample. This list is called the **sampling frame.** Ideally, the sampling frame should list every individual in the population. Because a list of the entire population is rarely available, most samples suffer from some degree of *undercoverage.*

Undercoverage

Undercoverage occurs when some groups in the population are left out of the process of choosing the sample.

If the sampling frame leaves out certain classes of people, even random samples from that frame will be biased. Using telephone directories as the frame for a telephone survey, for example, would miss everyone with an unlisted telephone number. More than half the households in many large cities have unlisted numbers, so massive undercoverage and bias against urban areas would result. In fact, telephone surveys use random digit dialing equipment that dials telephone numbers in selected regions at random. In effect, the sampling frame contains all residential telephone numbers.

Opting out of modern society was tiring at times, but Ted felt that increasing the undercoverage of opinion polls made it all worthwhile.

EXAMPLE 1   We undercover

Most opinion polls can't afford to even attempt full coverage of the population of all adult residents of the United States. The interviews are done by telephone, thus missing the 2% of households without phones. Only *households* are contacted, so that students in dormitories, prison inmates, and most members of the armed forces are left out. So are the homeless and people staying in shelters. Because calls to Alaska and Hawaii are expensive, most polls restrict their samples to the contiguous states. Many polls interview only in English, which leaves some immigrant households out of their samples.

The kinds of undercoverage found in most sample surveys are most likely to leave out people who are young or poor or who move often. Nonetheless, random digit dialing comes close to producing a random sample of households with phones outside Alaska and Hawaii. Sampling errors in careful sample surveys are usually quite small. The real problems start when someone picks up (or doesn't pick up) the phone. Now nonsampling errors take over.

Nonsampling errors

Nonsampling errors are those that can plague even a census. They include **processing errors**—mistakes in mechanical tasks such as doing arithmetic or entering responses into a computer. The spread of computers has made processing errors less common than in the past.

EXAMPLE 2 Computer-assisted interviewing

The days of the interviewer with a clipboard are past. Contemporary interviewers carry a laptop computer for face-to-face interviews or watch a computer screen as they carry out a telephone interview. Computer software manages the interview. The interviewer reads questions from the computer screen and uses the keyboard to enter the responses. The computer skips irrelevant items—once a respondent says that she has no children, further questions about her children never appear. The computer can check that answers to related questions are consistent with each other. It can even present questions in random order to avoid any bias due to always asking questions in the same order.

Computer software also manages the record keeping. It keeps records of who has responded and prepares a file of data from the responses. The tedious process of transferring responses from paper to computer, once a source of processing errors, has disappeared. The computer even schedules the calls in telephone surveys, taking account of the respondent's time zone and honoring appointments made by people who were willing to respond but did not have time when first called.

Another type of nonsampling error is **response error,** which occurs when a subject gives an incorrect response. A subject may lie about her age or income or about whether she has used illegal drugs. She may remember incorrectly when asked how many packs of cigarettes she smoked last week. A subject who does not understand a question may guess at an answer rather than appear ignorant. Questions that ask subjects about their behavior during a fixed time period are notoriously prone to response errors due to faulty memory. For example, the National Health Survey asks people how many times they have visited a doctor in the past year. Checking their responses against health records found that they failed to remember 60% of their visits to a doctor. A survey that asks about sensitive issues can also expect response errors, as the next example illustrates.

EXAMPLE 3 The effect of race

In 1989, New York City elected its first black mayor and the state of Virginia elected its first black governor. In both cases, samples of voters interviewed as they left their polling places predicted larger margins of victory than the official vote counts. The polling organizations were certain that some voters lied when interviewed because they felt uncomfortable admitting that they had voted against the black candidate. This phenomenon is known as "social desirability bias" and "the Bradley effect," after Tom Bradley, the former black mayor of Los Angeles who lost the 1982 California gubernatorial election despite leading in final-day preelection polls.

This effect attracted media attention during the 2008 presidential election. A few weeks before the election, polls predicted a victory, possibly a big one, for Barack Obama. Even so, Democrats worried that these polls might be overly optimistic because of the Bradley effect. In this case their fears were unfounded, but some political scientists claimed to detect the Bradley effect in polls predicting outcomes in primary races between Barack Obama and Hilary Clinton (for example, in the New Hampshire primary, polls predicted an 8 percentage point Obama victory but Clinton won by 3 percentage points).

Technology and attention to detail can minimize processing errors. Skilled interviewers greatly reduce response errors, especially in face-to-face interviews. There is no simple cure, however, for the most serious kind of nonsampling error, *nonresponse*.

Nonresponse

Nonresponse is the failure to obtain data from an individual selected for a sample. Most nonresponse happens because some subjects can't be contacted or because some subjects who are contacted refuse to cooperate.

"You can call, you can send email, you can stand at the door all day. The answer is still NO!"

Nonresponse is the most serious problem facing sample surveys. People are increasingly reluctant to answer questions, particularly over the phone. The rise of telemarketing, answering machines, and caller ID drives down response to telephone surveys. Gated communities and buildings guarded by doormen prevent face-to-face interviews. Nonresponse can bias sample survey results because different groups have different rates of nonresponse. Refusals are higher in large cities and among the elderly, for example. Bias due to nonresponse can easily overwhelm the random sampling error described by a survey's margin of error.

EXAMPLE 4 How bad is nonresponse?

The Current Population Survey (CPS) has the best response rate of any poll we know: only about 7% or 8% of the households in the CPS sample don't respond. People are more likely to respond to a government survey such as the CPS, and the CPS contacts its sample in person before doing later interviews by phone.

The General Social Survey (Example 7 in Chapter 1) also contacts its sample in person, and it is run by a university. Despite these advantages, its recent surveys have a 29% rate of nonresponse.

What about polls done by the media and by market research and opinion-polling firms? We often don't know their rates of nonresponse, because they won't say. That itself is a bad sign. The Pew Poll we looked at in the Case Study suggests how bad things are. Pew got 1221 responses (of whom 1000 were in the population they targeted) and 1658 who were never at home, refused, or would not finish the interview. That's a nonresponse rate of 1658 out of 2879, or 58%. The Pew researchers were more thorough than many pollsters. Insiders say that nonresponse often reaches 75% or 80% of an opinion poll's original sample.

Sample surveyors know some tricks to reduce nonresponse. Carefully trained interviewers can keep people on the line if they answer at all. Calling back over longer time periods helps. So do letters sent in advance. Letters and many callbacks slow down the survey, so opinion polls that want fast answers to satisfy the media don't use them. Even the most careful surveys find that nonresponse is a problem that no amount of expertise can fully overcome. That makes this reminder even more important:

What the margin of error doesn't say

The announced margin of error for a sample survey covers only random sampling error. Undercoverage, nonresponse, and other practical difficulties can cause large bias that is not covered by the margin of error.

Careful sample surveys warn us about the other kinds of error. Gallup, for example, says, "In addition to sampling error, question wording and practical difficulties in conducting surveys can introduce error or bias into the findings of public opinion polls." How true it is.

Does nonresponse make many sample surveys useless? Maybe not. We began this chapter with an account of a "standard" telephone survey done by the Pew Research Center. The Pew researchers also carried out a "rigorous" survey, with letters sent in advance, unlimited calls over eight weeks, letters by priority mail to people who refused, and so on. All this drove the rate of nonresponse down to 30%, compared with 58% for the standard survey. Pew then compared the answers to the same questions from the two surveys. The two samples were quite similar in age, sex, and race, though the rigorous sample was a bit more prosperous. The two samples also held similar opinions on all issues except one: race. People who at first refused to respond were less sympathetic toward the plights of blacks and other minorities than those who were willing to respond when contacted the first time. Overall, it appears that standard polls give reasonably accurate results. But, as in Example 3, race is again an exception.

Wording questions

A final influence on the results of a sample survey is the exact wording of questions. It is surprisingly difficult to word questions so that they are completely clear. A survey that asked about "ownership of stock" found that most Texas ranchers owned stock, though probably not the kind traded on the New York Stock Exchange.

 He started it! A study of deaths in bar fights showed that in 90% of the cases, the person who died started the fight. You shouldn't believe this. If you killed someone in a fight, what would you say when the police ask you who started the fight? After all, dead men tell no tales. Now that's nonresponse.

EXAMPLE 5 Words make a big difference

A February 2009 *USA Today*/Gallup Poll asked respondents about the federal government's involvement in the nation's banks, but the sample was divided in half, and each half was asked the question differently. One half were asked whether they favored or opposed "the federal government temporarily taking over major U.S. banks in danger of failing in an attempt to stabilize them." The other half of the sample were asked if they favored or opposed "the federal government temporarily nationalizing major U.S. banks in danger of failing in an attempt to stabilize them." The first question found 54% in favor. The second version of the question found 37% in favor.

It seems that "nationalizing" is a negative word. Small changes in how a question is worded can make a big difference in the response.

Doonesbury

The wording of questions always influences the answers. If the questions are slanted to favor one response over others, we have another source of nonsampling error. A favorite trick is to ask if the subject favors some policy as a means to a desirable end: "Do you favor banning private ownership of handguns in order to reduce the rate of violent crime?" and "Do you favor imposing the death penalty in order to reduce the rate of violent crime?" are loaded questions that draw positive responses from people who are worried about crime. Here is an example of the influence of a slanted question.

EXAMPLE 6 Campaign finance

The financing of political campaigns is a perennial issue. Here are two opinion poll questions on this topic:

Should laws be passed to eliminate all possibilities of special interests giving huge sums of money to candidates?

Should laws be passed to prohibit interest groups from contributing to campaigns, or do groups have a right to contribute to the candidate they support?

The first question was posed by Ross Perot, the third-party candidate in the 1992 presidential election. It drew a 99% "Yes" response in a mail-in poll. We know that voluntary response polls are worthless, so the Yankelovich Clancy Shulman survey firm asked the same question of a nationwide random sample. The result: 80% "Yes." Perot's question almost demands a "Yes" answer, so Yankelovich wrote the second question as a more neutral way of presenting the issue. Only 40% of a nationwide random sample wanted to prohibit contributions when asked this question.

NOW IT'S YOUR TURN | **4.1 Should we recycle?** Is the following question slanted toward a desired response? If so, how?

In view of escalating environmental degradation and incipient resource depletion, would you favor economic incentives for recycling of resource-intensive consumer goods?

How to live with nonsampling errors

Nonsampling errors, especially nonresponse, are always with us. What should a careful sample survey do about this? First, **substitute other households** for the nonresponders. Because nonresponse is higher in cities, replacing nonresponders with other households in the same neighborhood may reduce bias. Once the data are in, all professional surveys use statistical methods to **weight the responses** in an attempt to correct sources of bias. If many urban households did not respond, the survey gives more weight to those that did respond. If too many women are in the sample, the survey gives more weight to the men. Here, for example, is part of a statement in the *New York Times* describing one of its sample surveys:

The results have been weighted to take account of household size and number of telephone lines into the residence and to adjust for variations in the sample relating to geographic region, sex, race, age and education.

The goal is to get results "as if" the sample matched the population in age, sex, place of residence, and other variables.

The practice of weighting creates job opportunities for statisticians. It also means that the results announced by a sample survey are rarely as simple as they seem to be. Gallup announces that it interviewed 1523 adults and found that 57% of them bought a lottery ticket in the last 12 months. It would seem that because 57% of 1523 is 868, Gallup found that 868 people in its sample had played the lottery. Not so. Gallup no doubt used some quite fancy statistics to weight the actual responses: 57% is Gallup's best estimate of what it would have found in the absence of nonresponse. Weighting does help correct bias. It usually also increases variability. The announced margin of error must take this into account, creating more work for statisticians.

Sample design in the real world

The basic idea of sampling is straightforward: take an SRS from the population and use a statistic from your sample to estimate a parameter of the population. We now know that the sample statistic is altered behind the scenes to partly correct for nonresponse. The statisticians also have their hands on our beloved SRS. In the real world, most sample surveys use more complex designs.

EXAMPLE 7 The Current Population Survey

The population that the Current Population Survey (CPS) is interested in consists of all households in the United States (including Alaska and Hawaii). The sample is **chosen in stages.** The Census Bureau divides the nation into 2007 geographic areas called Primary Sampling Units (PSUs). These are generally groups of neighboring counties. At the first stage, 754 PSUs are chosen. This isn't an SRS. If all PSUs had the same chance to be chosen, the sample might miss Chicago and Los Angeles. So 428 highly populated PSUs are automatically in the sample. The other 1579 are divided into 326 groups, called **strata,** by combining PSUs that are similar in various ways. One PSU is chosen at random to represent each stratum.

Each of the 754 PSUs in the first-stage sample is divided into census blocks (smaller geographic areas). The blocks are also grouped into strata, based on such things as housing types and minority population. The households in each block are arranged in order of their location and divided into groups, called **clusters,** of about four households each. The final sample consists of samples of clusters (not of individual households) from each stratum of blocks. Interviewers go to all households in the chosen clusters. The samples of clusters within each stratum of blocks are also not SRSs. To be sure that the clusters spread out geographically, the sample starts at a random cluster and then takes, for example, every 10th cluster in the list.

The design of the CPS illustrates several ideas that are common in real-world samples that use face-to-face interviews. Taking the sample **in several stages** with **clusters** at the final stage saves travel time for interviewers by grouping the sample households first in PSUs and then in clusters. The most important refinement mentioned in Example 7 is *stratified sampling.*

Stratified sample

To choose a **stratified random sample:**

Step 1. Divide the sampling frame into distinct groups of individuals, called **strata.** Choose the strata according to any special interest you have in certain groups within the population or because the individuals in each stratum resemble each other.

Step 2. Take a separate SRS in each stratum and combine these to make up the complete sample.

We must of course choose the strata using facts about the population that are known before we take the sample. You might group a university's students into undergraduate and graduate students or into those who live on campus and those who commute. Stratified samples have some advantages over an SRS. First, by taking a separate SRS in each stratum, we can set sample sizes to allow separate conclusions about each stratum. Second, a stratified sample usually has a smaller margin of error than an SRS of the same size. The reason is that the individuals in each stratum are more alike than the population as a whole, so working stratum-by-stratum eliminates some variability in the sample.

It may surprise you that stratified samples can violate one of the most appealing properties of the SRS—stratified samples need not give all individuals in the population the same chance to be chosen. Some strata may be deliberately overrepresented in the sample.

EXAMPLE 8 Stratifying a sample of students

A large university has 30,000 students, of whom 3000 are graduate students. An SRS of 500 students gives every student the same chance to be in the sample. That chance is

$$\frac{500}{30,000} = \frac{1}{60}$$

We expect an SRS of 500 to contain only about 50 grad students—because grad students make up 10% of the population, we expect them to make up about 10% of an SRS. A sample of size 50 isn't large enough to estimate grad student opinion with reasonable accuracy. We might prefer a stratified random sample of 200 grad students and 300 undergraduates.

You know how to select such a stratified sample. Label the graduate students 0001 to 3000 and use Table A to select an SRS of 200. Then label the undergraduates 00001 to 27,000 and use Table A a second time to select an SRS of 300 of them. These two SRSs together form the stratified sample.

In the stratified sample, each grad student has chance

$$\frac{200}{3000} = \frac{1}{15}$$

to be chosen. Each of the undergraduates has a smaller chance,

$$\frac{300}{27,000} = \frac{1}{90}$$

Because we have two SRSs, it is easy to estimate opinions in the two groups separately. The quick method (page 46) tells us that the margin of error for a sample proportion will be about

$$\frac{1}{\sqrt{200}} = 0.071 \quad \text{(that is, 7.1\%)}$$

for grad students and about

$$\frac{1}{\sqrt{300}} = 0.058 \quad \text{(that is, 5.8\%)}$$

for undergraduates.

Because the sample in Example 8 deliberately overrepresents graduate students, the final analysis must adjust for this to get unbiased estimates of overall student opinion. Remember that our quick method works only for an SRS. In fact, a professional analysis would also take account of the fact that the population contains "only" 30,000 individuals—more job opportunities for statisticians.

NOW IT'S YOUR TURN

4.2 A stratified sample The statistics department at Stochastic University has 5 faculty and 10 undergraduate majors. Use Table A, starting at line 111, to choose a stratified sample of 1 faculty member and 1 student to attend a reception being held by the university president.

EXAMPLE 9 The woes of telephone samples

In principle, it would seem that a telephone survey that dials numbers at random could be based on an SRS. Telephone surveys have little need for clustering. Stratifying can still reduce variability, however, and so telephone surveys often take samples in two stages: a stratified sample of telephone number prefixes (area code plus first three digits) followed by individual numbers (last four digits) dialed at random in each prefix.

New York, New York
New York City, they say, is bigger, richer, faster, ruder. Maybe there's something to that. The sample survey firm Zogby International says that as a national average it takes 5 telephone calls to reach a live person. When calling to New York, it takes 12 calls. Survey firms assign their best interviewers to make calls to New York and often pay them bonuses to cope with the stress.

The real problem with an SRS of telephone numbers is that too few numbers lead to households. Blame technology. Fax machines, modems, and cell phones demand new phone numbers. Between 1988 and 2008, the number of households in the United States grew by 29%, but the number of possible residential phone numbers grew by more than 120%. Some analysts believe that in the near future we may have to increase the number of digits for telephone numbers from 10 (including the area code) to 12. This will further exacerbate this problem. Telephone surveys now use "list-assisted samples" that check electronic telephone directories to eliminate prefixes that have no listed numbers before random sampling begins. Fewer calls are wasted, but anyone living where all numbers are unlisted is missed. Prefixes with no listed numbers are therefore separately sampled (stratification again), perhaps with a smaller sample size than if included in the list-assisted sample, to fill the gap.

The proliferation of cell phones has created additional problems for telephone samples. Random digit dialing using a machine is not allowed for cell phone numbers. Phone numbers assigned to cell phones are determined by the location of the cell phone company providing the service and need not coincide with the actual residence of the user. This makes it difficult to implement sophisticated methods of sampling such as stratified sampling by geographic location.

It may be that the woes of telephone sampling prompted the Gallup Organization, in recent years, to drop the phrase "random sampling" from the description of their survey methods at the end of most of their polls. This presumably prevents misinterpreting the results as coming from simple random samples. In their detailed description of survey methods used for the Gallup World Poll and the Gallup Well-Being Index (available online at the Gallup Web site), the samples are described as involving random sampling.

STATISTICAL CONTROVERSIES

The Harris Poll Online

Blend Images/Hill Street Studios/Getty Images

The Harris Poll Online has created an online research panel of over 6 million volunteers. According to the Harris Poll Online Web site, the "panel consists of a diverse cross-section of people residing in the United States, as well as in over 200 countries around the world," and "this multimillion member panel consists of potential respondents who have been recruited through online, telephone, mail, and in-person approaches to increase population coverage and enhance representativeness." One can join the panel at **www .harrispollonline.com/rewards.asp.**

When the Harris Poll Online conducts a survey, this panel serves as the sampling frame. A probability sample is selected from it, and statistical methods are used to weight the responses. In particular, the Harris Poll Online uses propensity score weighting, a proprietary Harris Interactive technique, which is also applied (when appropriate) to adjust for respondents' likelihood of being online. They claim that "this procedure provides added assurance of accuracy and representativeness."

Are you convinced that the Harris Poll Online provides accurate information about well-defined populations such as all American adults? Why or why not?

For more information about the Harris Poll Online and Web surveys in general, see **www.harrispollonline.com/ question.asp**; **www.harrisinteractive .com/partner/methodology.asp**; and M. P. Crouper, "Web surveys: a review of issues and approaches," *Public Opinion Quarterly,* 64 (Winter 2000), pp. 464–494, available online at **www.jstor.org/sici? sici=0033-362X%28200024%2964% 3A4%3C464%3ARWSARO%3E2.0 .CO%3B2-M.**

The challenge of Internet surveys

The Internet is having a profound effect on many things people do, and this includes surveys. Using the Internet to conduct "Web surveys" is becoming increasingly popular. Web surveys have several advantages over more traditional survey methods. It is possible to collect large amounts of survey data at lower costs than traditional methods allow. Anyone can put survey questions on dedicated sites offering free services; thus, large-scale data collection is available to almost every person with access to the Internet. Furthermore, Web surveys allow one to deliver multimedia survey content to respondents, opening up new realms of survey possibilities that would be extremely difficult to implement using traditional methods. Some argue that eventually Web surveys will replace traditional survey methods.

Although Web surveys are easy to do, they are not easy to do well. The reasons include many of the issues we have discussed in this chapter. Three major problems are voluntary response, undercoverage, and nonresponse. Voluntary response appears in several forms. Some Web surveys invite visitors to a particular Web site to participate in a poll. **Misterpoll.com** is one such example. Visitors to this site can participate in several ongoing polls, create their own poll, and respond multiple times to the same poll. Other Web surveys solicit participation through announcements in newsgroups, email invitations, and banner ads on high-traffic sites. An example is a series of 10 polls conducted by Georgia Tech University's Graphic, Visualization, and Usability Center (GVU) in the 1990s.

Although **misterpoll.com** indicates that the surveys on the site are primarily intended for entertainment, the GVU polls appear to claim some measure of legitimacy. The Web site **www.cc.gatech.edu/gvu/user_surveys/** states that the information from these surveys "is valued as an independent, objective view of developing Web demographics."

A third and more sophisticated example of voluntary response occurs when the polling organization creates what it believes to be a representative panel consisting of volunteers and uses panel members as a sampling frame. A random sample is selected from this panel, and those selected are invited to participate in the poll. A very sophisticated version of this approach is used by the Harris Poll Online.

Web surveys, such as the Harris Poll Online, in which a random sample is selected from a well-defined sampling frame are reasonable when the sampling frame clearly represents some larger population or when interest is only in the members of the sampling frame. Examples of this include Web surveys that use systematic sampling to select every nth visitor to a site and the target population is narrowly defined as visitors to the site. Other examples are some Web surveys on college campuses. All students may be assigned email addresses and have Internet access. A list of these email addresses serves as the sampling frame, and a random sample is selected from this list. If the population of interest is all students at this particular college, these surveys can potentially yield very good results. Here is an example of this type of Web survey.

EXAMPLE 10 Doctors and placebos

A placebo is a "dummy" treatment, such as a salt pill, that has no direct effect on a patient but may bring about a response because patients expect it to. Do academic physicians who maintain private practices sometimes give their patients placebos? A Web survey of doctors in internal medicine departments at Chicago-area medical schools was possible because almost all doctors had listed email addresses.

An email was sent to each doctor explaining the purpose of the study, promising anonymity, and providing a Web link for response. Result: 45% of respondents said they sometimes use placebos in their clinical practice.

Several other Web survey methods have been employed to eliminate problems arising from voluntary response. One is to use the Web as one of many alternative ways to participate in the survey. The Bureau of Labor Statistics and the Census Bureau have used this method. Another method is to select random samples from panels, but instead of relying on volunteers to form the panels, members are recruited using random sampling (for example, random digit dialing). Telephone interviews can be used to collect background information, identify those with Internet access, and recruit eligible persons to the panel. If the target population is current users of the Internet, this method should also potentially yield reliable results. The Pew Research Center has employed this method.

Perhaps the most ambitious approach, and one that attempts to obtain a random sample from a more general population, is the following. Take a random sample from the population of interest. Provide all those selected with the necessary equipment and tools to participate in subsequent Web surveys. This methodology is similar in spirit to that used for the Nielsen TV ratings. It was employed by one company, InterSurvey, several years ago, although InterSurvey is no longer in business.

Several challenges remain for those who employ Web surveys. Even though Internet and email use is growing (according to the 2009 *Statistical Abstract of the United States,* as of 2009, 79% of American adults aged 18 and older have Internet access at home or work, and 71% have Internet access at home), there is still the problem of undercoverage if Web surveys are used to draw conclusions about all American adults aged 18 and older. Weighting responses to correct for possible biases does not solve the problem, because studies indicate that Internet users differ in many ways that traditional methods of weighting do not account for.

In addition, even if 100% of Americans had Internet access, there is no list of Internet users that we can use as a sampling frame, nor is there anything comparable to random digit dialing that can be used to draw random samples from the collection of all Internet users.

Finally, Web surveys often have very high rates of nonresponse. Methods that are used in phone and mail surveys to improve response rates can help, but they make Web surveys more expensive and difficult, offsetting some of their advantages.

Probability samples

It's clear from Examples 7, 8, 9, and 10 that designing samples is a business for experts. Even most statisticians don't qualify. We won't worry about such details. The big idea is that good sample designs use chance to select individuals from the population. That is, all good samples are *probability samples*.

Probability sample

A **probability sample** is a sample chosen by chance. We must know what samples are possible and what chance, or probability, each possible sample has. Some probability samples, such as stratified samples, don't allow all possible samples from the population and may not give an equal chance to all the samples they do allow.

A stratified sample of 300 undergraduate students and 200 graduate students, for example, allows only samples with exactly that makeup. An SRS would allow any 500 students. Both are probability samples. We need only know that estimates from any probability sample share the nice properties of estimates from an SRS. Confidence statements can be made without bias and have smaller margins of error as the size of the sample increases. Nonprobability samples such as voluntary response samples do not share these advantages and cannot give trustworthy information about a population. Now that we know that most nationwide samples are more complicated than an SRS, we will usually go back to acting as if good samples were SRSs. That keeps the big idea and hides the messy details.

Questions to ask before you believe a poll

Opinion polls and other sample surveys can produce accurate and useful information if the pollster uses good statistical techniques and also works hard at preparing a sampling frame, wording questions, and reducing non-response. Many surveys, however, especially those designed to influence public opinion rather than just record it, do not produce accurate or useful information. Here are some questions to ask before you pay much attention to poll results.

- **Who carried out the survey?** Even a political party should hire a professional sample survey firm whose reputation demands that they follow good survey practices.

- **What was the population?** That is, whose opinions were being sought?

- **How was the sample selected?** Look for mention of random sampling.

- **How large was the sample?** Even better, find out both the sample size and the **margin of error** within which the results of 95% of all samples drawn as this one was would fall.

- **What was the response rate?** That is, what percentage of the original subjects actually provided information?

- **How were the subjects contacted?** By telephone? Mail? Face-to-face interview?

- **When was the survey conducted?** Was it just after some event that might have influenced opinion?

- **What were the exact questions asked?**

Academic survey centers and government statistical offices answer these questions when they announce the results of a sample survey. National opinion polls usually don't announce their response rate (which is often low) but do give us the other information. Editors and newscasters have the bad habit of cutting out these dull facts and reporting only the sample results. Many sample surveys by interest groups and local newspapers and TV stations don't answer these questions because their polling methods are in fact unreliable. If a politician, an advertiser, or your local TV station announces the results of a poll without complete information, be skeptical.

STATISTICS IN SUMMARY

Chapter Specifics

- Sampling in the real world is complex. Even professional sample surveys don't give exactly correct information about the population.

- There are many potential sources of error in sampling. The margin of error announced by a sample survey covers only **random sampling error,** the variation due to chance in choosing a random sample.

- Other types of error are in addition to the margin of error and can't be directly measured. **Sampling errors** come from the act of choosing a sample. Random sampling error and **undercoverage** are common types of sampling error. Undercoverage occurs when some members of the population are left out of the **sampling frame,** the list from which the sample is actually chosen.

- The most serious errors in most careful surveys, however, are **nonsampling errors.** These have nothing to do with choosing a sample—they are present even in a census.

- The single biggest problem for sample surveys is **nonresponse:** subjects can't be contacted or refuse to answer.

- Mistakes in handling the data (**processing errors**) and incorrect answers by respondents (**response errors**) are other examples of nonsampling errors.

- Finally, the exact **wording of questions** has a big influence on the answers.

- People who design sample surveys use statistical techniques that help correct nonsampling errors, and they also use **probability samples** more complex than simple random samples, such as **stratified samples.**

- You can assess the quality of a sample survey quite well by just looking at the basics: use of random samples, sample size and margin of error, the rate of nonresponse, and the wording of the questions.

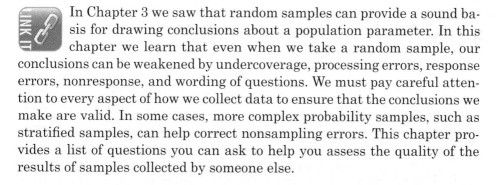 In Chapter 3 we saw that random samples can provide a sound basis for drawing conclusions about a population parameter. In this chapter we learn that even when we take a random sample, our conclusions can be weakened by undercoverage, processing errors, response errors, nonresponse, and wording of questions. We must pay careful attention to every aspect of how we collect data to ensure that the conclusions we make are valid. In some cases, more complex probability samples, such as stratified samples, can help correct nonsampling errors. This chapter provides a list of questions you can ask to help you assess the quality of the results of samples collected by someone else.

CASE STUDY   Use what you have learned in this chapter to evaluate the Case Study
EVALUATED   that opened the chapter. In particular, do the following.

1. Answer the questions given in the section "Questions to Ask before You Believe a Poll" on pages 77–78.

2. Are the results of the Pew Poll useless? You may want to refer to the discussion on pages 74–76. ▪

CHAPTER 4 EXERCISES

For Exercise 4.1, see page 69; for Exercise 4.2, see page 72.

4.3 What kind of error? Which of the following are sources of *sampling error* and which are sources of *nonsampling error?* Explain your answers.

(a) The subject lies about past drug use.

(b) A typing error is made in recording the data.

(c) Data are gathered by asking people to go to a Web site and answer questions online.

4.4 What kind of error? Each of the following is a source of error in a sample survey. Label each as *sampling error* or *nonsampling error*, and explain your answers.

(a) The telephone directory is used as a sampling frame.

(b) The subject cannot be contacted in five calls.

(c) Interviewers choose people on the street to interview.

4.5 Not in the margin of error. According to a February 2008 *USA Today/Gallup* Poll, 43% of Americans identify themselves as baseball fans. That is low by recent standards, as an average of 49% of Americans have said they were fans of the sport since Gallup started tracking this measure in 1993. The high point came in 1998, when Sammy Sosa and Mark McGwire pursued (and ultimately surpassed) Roger Maris's single-season home run record, at which time 56% of Americans considered themselves baseball fans. The Gallup press release says:

For results based on this sample, one can say with 95% confidence that the maximum error attributable to sampling and other random effects is ±5 percentage points.

Give one example of a source of error in the poll result that is *not* included in this margin of error.

4.6 College parents. An online survey of college parents was conducted during February and March 2007. Email was sent to 41,000 parents who were listed in either the College Parents of America database or the Student Advantage database. Parents were invited to participate in the online survey. Out of those invited, 1727 completed the online survey. The survey protected the anonymity of those participating in the survey but did not allow more than one response from an individual IP address.

One of the survey results was that 33% of mothers communicate at least once a day with their child while at school.

(a) What was the *response rate* for this survey? (The response rate is the percentage of the planned sample—that is, those invited to participate—who responded.)

(b) Use the quick method (page 46) to estimate the margin of error for a random sample of size 1727.

(c) Do you think that the margin of error is a good measure of the accuracy of the survey's results? Explain your answer.

4.7 Polling customers. An online store chooses an SRS of 100 customers from its list of all people who have bought something from the store in the last year. It asks those selected how satisfied they are with the store's Web site. If it selected two SRSs of 100 customers at the same time, the two samples would give somewhat different results. Is this variation a source of sampling error or of nonsampling error? Would the survey's announced margin of error take this source of error into account?

4.8 Ring-no-answer. A common form of nonresponse in telephone surveys is "ring-no-answer." That is, a call is made to an active number but no one answers. The Italian National Statistical Institute looked at nonresponse to a government survey of households in Italy during the periods January 1 to Easter and July 1 to August 31. All calls were made between 7 and 10 P.M., but 21.4% gave "ring-no-answer" in one period versus 41.5% "ring-no-answer" in the other period. Which period do you think had the higher rate of no answers? Why? Explain why a high rate of nonresponse makes sample results less reliable.

4.9 The environment and the economy. Here are two opinion poll questions asked in December 2009 about protecting the environment versus protecting the economy.

Often there are trade-offs or sacrifices people must make in deciding what is important to them. Generally speaking, when a trade-off has to be made, which is more important to you: stimulating the economy or protecting the environment?

Which worries you more: that the U.S. will NOT take the actions necessary to prevent the catastrophic effects of global warming because of fears those actions would harm the economy, or that the U.S. WILL take actions to protect against global warming and those actions will cripple the U.S. economy?

In response to the first question, 61% said stimulating the economy was more important. But only 46% of those asked the second question said they were afraid that the U.S. will take actions to protect against global warming and that those actions will cripple the U.S. economy. Why do you think the second wording discouraged more people from expressing more concern about the economy than about the environment?

4.10 Amending the Constitution. You are writing an opinion poll question about a proposed amendment to the Constitution. You can ask if people are in favor of "changing the Constitution" or "adding to the Constitution" by approving the amendment.

(a) Why do you think the responses to these two questions will produce different percentages in favor?

(b) One of these choices of wording will produce a much higher percentage in favor. Which one? Why?

4.11 Paying taxes. In April 2011, a Gallup Poll asked two questions about the amount one pays in federal income taxes. Here are the two questions:

> *Do you regard the income tax which you will have to pay this year as fair?*
>
> *Do you consider the amount of federal income tax you have to pay as too high, about right, or too low?*

One of these questions drew 57% saying the amount was fair or about right; the other, only 43%. Which wording produced the higher percentage? Why?

4.12 Wording survey questions. Comment on each of the following as a potential sample survey question. Is the question clear? Is it slanted toward a desired response? (Survey questions on issues that one might regard as inflammatory are often prone to slanted wording.)

(a) Which of the following best represents your opinion on gun control?

    1. The government should take away our guns.

    2. We have the right to keep and bear arms.

(b) In light of skyrocketing gasoline prices, we should consider opening up a very small amount of Alaskan wilderness for oil exploration as a way of reducing our dependence on foreign oil. Do you agree or disagree?

(c) Do you think that the excessive restrictions placed on U.S. law enforcement agencies hampered their ability to detect the 9/11 terrorist plot before it occurred?

4.13 Bad survey questions. Write your own examples of bad sample survey questions.

(a) Write a biased question designed to get one answer rather than another.

(b) Write a question that is confusing, so that it is hard to answer.

4.14 Appraising a poll. In December 2009 the *Wall Street Journal* published an article on President Obama's approval rating. The article also noted that just 22% of Americans approved of the job Congress was doing. News articles tend to be brief in describing sample surveys. Here is part of the *Wall Street Journal's* description of this poll:

> *The Wall Street Journal/NBC News poll was based on nationwide telephone interviews of 1008 adults, including a sample of 104 interviews with people who only use a cell phone. It was conducted December 11–14 by the polling organizations of Peter D. Hart and Bill McInturff.*
>
>   *The sample was drawn in the following manner: 350 geographic points were randomly selected proportionate to the population of each region and,*

within each region, by size of place. These individuals were selected by a method that gave all telephone numbers, listed and unlisted, an equal chance of being included. The cell phone sample was drawn from a list of cell phone users nationally.

Pages 77–78 lists several "questions to ask" about an opinion poll. What answers does the *Wall Street Journal* give to each of these questions?

4.15 Appraising a poll. A *New York Times* article on attitudes about the economy and joblessness discussed the results of a sample survey that found, for example, that 48% of unemployed adults in the survey "have experienced emotional or mental health issues, like depression or anxiety." Here is part of the *Times*'s statement "How the Poll Was Conducted":

The latest New York Times/CBS News Poll is based on telephone interviews conducted Dec. 5 through Dec. 10 with 1650 adults throughout the United States. . . .

The sample of land-line telephone exchanges called was randomly selected by a computer from a complete list of more than 69,000 active residential exchanges across the country.

Within each exchange, random digits were added to form a complete telephone number, thus permitting access to listed and unlisted numbers alike. Within each household, one adult was designated by a random procedure to be the respondent for the survey.

To increase coverage, this land-line sample was supplemented by respondents reached through random dialing of cell phone numbers. The two samples were then combined.

Interviewers tried several times to reach every phone number in the survey, calling back unanswered numbers on different days at different times in the day and evening.

Pages 77–78 lists several "questions to ask" about an opinion poll. What answers does the *Times* give to each of these questions?

4.16 Closed versus open questions. Two basic types of questions are closed questions and open questions. A closed question asks the subject for one or more of a fixed set of responses. An open question allows the subject to answer in his or her own words; the interviewer writes down the responses and classifies them later. An example of an open question is

What do you believe about the afterlife?

An example of a closed question is

What do you believe about the afterlife? Do you believe
a. there is an afterlife and entrance depends only on your actions?
b. there is an afterlife and entrance depends only on your beliefs?

c. there is an afterlife and everyone lives there forever?

d. there is no afterlife?

e. I don't know.

What are the advantages and disadvantages of open and closed questions?

4.17 Telling the truth? Many subjects don't give honest answers to questions about activities that are illegal or sensitive in some other way. One study divided a large group of white adults into thirds at random. All were asked if they had ever used cocaine. The first group was interviewed by telephone: 21% said "Yes." In the group visited at home by an interviewer, 25% said "Yes." The final group was interviewed at home but answered the question on an anonymous form that they sealed in an envelope. Of this group, 28% said they had used cocaine.

(a) Which result do you think is closest to the truth? Why?

(b) Give two other examples of behavior you think would be underreported in a telephone survey.

4.18 Did you vote? When the Current Population Survey asked the adults in its sample of 60,000 households if they voted in the 2008 presidential election, 63.6% said they had. The margin of error was less than 0.5%. In fact, only 61.6% of the adult population voted in that election. Why do you think the CPS result missed by four times the margin of error?

4.19 A party poll. At a party there are 20 students over age 21 and 40 students under age 21. You choose at random 2 of those over 21 and separately choose at random 4 of those under 21 to interview about attitudes toward alcohol. You have given every student at the party the same chance to be interviewed: what is that chance? Why is your sample not an SRS?

4.20 A stratified sample. A club has 20 student members and 10 faculty members. The students are

Barrett	Dobmeier	Josey	Leatherman	Schuetter
Blake	Douglas	Kohlschmidt	Miller	Svenson
Bayer	Fox	Koster	Nelson	Thompson
Conroy	Fry	Landgraf	Quan	Winner

The faculty members are

Berliner	Dean	Hans	Ozturk	Rumsey
Craigmile	Fligner	Miller	Pearl	Turkmen

The club can send 3 students and 2 faculty members to a convention. It decides to choose those who will go by random selection.

(a) Use Table A to choose a stratified random sample of 3 students and 2 faculty members.

(b) What is the chance that the student named Conroy is chosen? What is the chance that faculty member Dean is chosen?

4.21 A stratified sample. A university has 2000 male and 800 female faculty members. The equal opportunity employment officer wants to poll the opinions of a random sample of faculty members. In order to give adequate attention to female faculty opinion, he decides to choose a stratified random sample of 200 males and 200 females. He has alphabetized lists of female and male faculty members.

(a) Explain how you would assign labels and use random digits to choose the desired sample. Enter Table A at line 122 and give the first 5 females and the first 5 males in your sample.

(b) What is the chance that any one of the 2000 males will be in your sample? What is the chance that any one of the 800 females will be in your sample?

4.22 Sampling by accountants. Accountants use stratified samples during audits to verify a company's records of such things as accounts receivable. The stratification is based on the dollar amount of the item and often includes 100% sampling of the largest items. One company reports 5000 accounts receivable. Of these, 100 are in amounts over $50,000; 500 are in amounts between $1000 and $50,000; and the remaining 4400 are in amounts under $1000. Using these groups as strata, you decide to verify all of the largest accounts and to sample 5% of the midsize accounts and 1% of the small accounts. How would you label the two strata from which you will sample? Use Table A, starting at line 115, to select *only the first 5* accounts from each of these strata.

4.23 A sampling paradox? Example 8 compares two SRSs, of a university's undergraduate and graduate students. The sample of undergraduates contains a smaller fraction of the population, 1 out of 90, versus 1 out of 15 for graduate students. Yet sampling 1 out of 90 undergraduates gives a smaller margin of error than sampling 1 out of 15 graduate students. Explain to someone who knows no statistics why this happens.

4.24 Appraising a poll. Exercise 4.14 (page 82) gives part of the description of a sample survey from the *Wall Street Journal*. It appears that the sample was taken in several stages. Why can we say this? The first stage no doubt used a stratified sample, though the *Journal* does not say this. Explain why it would be bad practice to use an SRS from a large number of "geographic points" across the country rather than a stratified sample of such points.

4.25 Multistage sampling. An article in the journal *Science* looks at differences in attitudes toward genetically modified foods between Europe and the United States. This calls for sample surveys. The European survey chose a sample of 1000 adults in each of 17 European countries. Here's part of the description: "The Eurobarometer survey is a multistage, random-probability face-to-face sample survey."

(a) What does "multistage" mean?

(b) You can see that the first stage was stratified. What were the strata?

(c) What does "random-probability sample" mean?

4.26 Online courses in high schools? What do adults believe about requiring online courses in high schools? Are opinions different in urban, suburban, and rural areas? To find out, researchers wanted to ask adults this question:

It has become common for education courses after high school to be taken online. In your opinion, should public high schools in your community require every student to take at least one course online while in high school?

Because most people live in heavily populated urban and suburban areas, an SRS might contain few rural adults. Moreover, it is too expensive to choose people at random from a large region. We should start by choosing school districts rather than people. Describe a suitable sample design for this study, and explain the reasoning behind your choice of design.

4.27 Systematic random samples. The last stage of the Current Population Survey (Example 7, page 70) uses a **systematic random sample.** An example will illustrate the idea of a systematic sample. Suppose that we must choose 4 rooms out of the 100 rooms in a dormitory. Because $100/4 = 25$, we can think of the list of 100 rooms as four lists of 25 rooms each. Choose 1 of the first 25 rooms at random, using Table A. The sample will contain this room and the rooms 25, 50, and 75 places down the list from it. If 13 is chosen, for example, then the systematic random sample consists of the rooms numbered 13, 38, 63, and 88. Use Table A to choose a systematic random sample of 5 rooms from a list of 200. Enter the table at line 120.

4.28 Systematic isn't simple. Exercise 4.27 describes a systematic random sample. Like an SRS, a systematic sample gives all individuals the same chance to be chosen. Explain why this is true, then explain carefully why a systematic sample is nonetheless *not* an SRS.

4.29 Planning a survey of students. The student government plans to ask a random sample of students their opinions about on-campus parking. The university provides a list of the 20,000 enrolled students to serve as a sampling frame.

(a) How would you choose an SRS of 200 students?

(b) How would you choose a systematic sample of 200 students? (See Exercise 4.27 to learn about systematic samples.)

(c) The list shows whether students live on campus (8000 students) or off campus (12,000 students). How would you choose a stratified sample of 50 on-campus students and 150 off-campus students?

4.30 Sampling students. You want to investigate the attitudes of students at your school toward the school's policy on extra fees for lab courses. You have a grant that will pay the costs of contacting about 500 students.

(a) Specify the exact population for your study. For example, will you include part-time students?

(b) Describe your sample design. For example, will you use a stratified sample with student majors as strata?

(c) Briefly discuss the practical difficulties that you anticipate. For example, how will you contact the students in your sample?

4.31 Mall interviews. Example 1 in Chapter 2 (page 24) describes mall interviewing. This is an example of a convenience sample. Why do mall interviews not produce probability samples?

4.32 Partial-birth abortion? Here are three opinion poll questions on the same issue, with the poll results:

> *As you may know, the Supreme Court recently upheld a law that makes the procedure commonly known as a partial birth abortion illegal. Do you favor or oppose this ruling by the Supreme Court? Result: 53% favor; 34% oppose.*

> *As you may know, the Supreme Court recently upheld a law that makes the procedure commonly known as a partial birth abortion illegal. A partial birth abortion is a procedure performed in the late-term of pregnancy, when in some cases the baby is old enough to survive on its own outside the womb. The court's ruling outlaws using this procedure, and does not make an exception for the health of the mother. Do you favor or oppose this ruling by the Supreme Court? Result: 47% favor; 43% oppose.*

> *Now I would like to ask your opinion about a specific abortion procedure known as a "late-term" abortion or "partial-birth" abortion, which is sometimes performed on women during the last few months of pregnancy. Do you think that the government should make this procedure illegal, or do you think that the procedure should be legal? Result: 66% illegal; 28% legal.*

Using this example, discuss the difficulty of using responses to opinion polls to understand public opinion.

EXPLORING THE WEB

4.33 Web-based exercise. The Web site for the Pew Research Center for the People and the Press is **www.people-press.org.** Go to the Web site and read the featured survey. Pages 77–78 lists several "questions to ask" about an opinion poll. What answers does the Pew Research Center for the People and the Press give to each of these questions? You may find the links at the end of the featured survey helpful for finding answers to some of these questions.

4.34 Web-based exercise. The Web site for the American Association for Public Opinion Research discusses several issues about polls. This information can be found at **www.aapor.org/Poll_andamp_Survey_FAQs.htm.** Click on the link *Questions to Ask When Writing about Polls* for suggestions about how to determine if a poll is good or bad. Click on the link *Bad Samples* for some examples of flawed samples.

4.35 Web-based exercise. The Harris Poll no longer provides a margin of error for their polls. The Harris Poll Web site is **www.harrisinteractive.com/Insights/HarrisVault.aspx?PID=434.** Visit the site, read the Methodology section at the end of a recent poll, and write a brief report about why the Harris Poll no longer provides a margin of error.

4.36 Web-based exercise. Visit the Gallup Web site at **www.gallup.com/Home.aspx** and look at a recent poll that includes, at the end, a description of the survey methods. Are the samples described as coming from random sampling?

NOTES AND DATA SOURCES

Page 64 For more detail on the limits of memory in surveys, see N. M. Bradburn, L. J. Rips, and S. K. Shevell, "Answering autobiographical questions: the impact of memory and inference on surveys," *Science,* 236 (1987), pp. 157–161.

Page 65 Example 3: For more on the effect of race, see Gregory Flemming and Kimberly Parker, "Race and reluctant respondents: possible consequences of non-response for pre-election surveys," Pew Research Center for the People and the Press, 1997; available online at **www.people-press.org.** For con-

cerns about the Bradley effect in the 2008 presidential election, go to **www.politico.com/news/stories/0508/10397.html** and **www.nytimes.com/2008/10/12/weekinreview/12zernike.html.**

Page 65 For more detail on nonsampling errors, along with references, see P. E. Converse and M. W. Traugott, "Assessing the accuracy of polls and surveys," *Science,* 234 (1986), pp. 1094–1098.

Page 66 Example 4: The nonresponse rate for the CPS comes from *BLS Handbook* *of Methods,* Chapter 1, "Labor Force Data Derived from the Current Population Survey," found on the Bureau of Labor Statistics Web site at **www.bls.gov/cps/cps_over .htm#methodology.** The General Social Survey reports its response rate on its Web site, **www3.norc.org/GSS/GSS+Facts.htm.** Don Van Natta, Jr. claims that "pollsters say response rates have fallen as low as 20 percent in some recent polls"; see his article "Polling's 'dirty little secret': no response," *New York Times,* November 11, 1999.

Page 67 Example 5: This example comes from a report "Do Americans back nationalization? Depends how you ask," posted by Sarah Dutton online at **www.cbsnews .com/blogs/2009/03/13/politics/ politicalhotsheet/entry4864121.shtml.**

Page 68 Example 6: D. Goleman, "Pollsters enlist psychologists in quest for unbiased results," *New York Times,* September 7, 1993.

Page 69 The quotation on weighting is from Adam Clymer and Janet Elder, "Poll finds greater confidence in Democrats," *New York Times,* November 10, 1999.

Page 70 Example 7: The most recent account of the design of the CPS is the *BLS Handbook of Methods*; available online at **www.bls.gov/opub/hom/.** The account in Example 7 omits many complications, such as the need to separately sample "group quarters" like college dormitories.

Page 73 A detailed description of the methods Gallup uses in its World Poll can be found online at **www.gallup.com/consulting/ worldpoll/108079/Methodological-Design.aspx.** A detailed description of the methods Gallup uses in its Well-Being Index can be found online at **www.well-beingindex.com/methodology.asp.**

Page 74 For more information on Web surveys see M. P. Crouper, "Web surveys: a review of issues and approaches," *Public Opinion Quarterly,* 64 (Winter 2000), pp. 464–494.

Page 75 Example 10: Rachel Sherman and John Hickner, "Academic physicians use placebos in clinical practice and believe in the mind-body connection," *Journal of General Internal Medicine,* 23 (2008), pp. 7–10.

Page 80 Exercise 4.5: The quotation, from the Gallup Web site, **www.gallup.com/ poll/Topics.aspx,** is typical of Gallup polls.

Page 80 Exercise 4.6: Details of the survey can be found online at **www .collegeparents.org/sites/default/files/ current-parent-survey-summary.pdf.**

Page 81 Exercise 4.8: Giuliana Coccia, "An overview of non-response in Italian telephone surveys," *Proceedings of the 99th Session of the International Statistical Institute, 1993,* Book 3, pp. 271–272.

Page 81 Exercise 4.9: The first question is from an CBS News/*New York Times* Poll, December 4–8, 2009. The second question is from a *USA Today*/Gallup Poll, December 11–13, 2009.

Page 82 Exercise 4.11: The questions are from a Gallup Poll, April 7–11, 2011; available online at **www.gallup.com/poll/1714/ Taxes.aspx.**

Page 82 Exercise 4.14: Peter Wallsten, "Democrats' blues grow deeper in new poll," *Wall Street Journal,* December 17, 2009; available online at **online.wsj.com/article/ SB126100346902694549.html?mod=WSJ_ hpp_MIDDLENexttoWhatsNewsTop.**

Page 83 Exercise 4.15: Michael Luo and Megan Thee-Brenan, "Poll reveals trauma of joblessness in U.S.," *New York Times,*

December 14, 2009; available online at **www.nytimes.com/2009/12/15/us/15poll .html?_r=1.**

Page 84 Exercise 4.17: D. Goleman, "Pollsters enlist psychologists in quest for unbiased results," *New York Times,* September 7, 1993.

Page 86 Exercise 4.25: From the online "Supplementary Material" for G. Gaskell et al., "Worlds apart? The reception of genetically modified foods in Europe and the U.S.," *Science,* 285 (1999), pp. 384–387.

Page 87 Exercise 4.32: The first question is from an NBC News/*Wall Street Journal* Poll conducted by the polling organizations of Peter Hart (D) and Robert Teeter (R), November 8–10, 2003. The second question is from a CNN/*USA Today*/Gallup Poll, October 24–26, 2003. The third question is from a *Los Angeles Times* Poll, January 30–February 2, 2003. Available online at **www.pollingreport.com/abortion.htm.**

Experiments, Good and Bad

CASE STUDY Reports about climate change appear frequently in the media. Climate scientists warn us that major changes will occur in the coming years. For example, scientists predict that the changing climate will probably bring more rain to California, but they don't know whether the additional rain will come during the winter wet season or extend into the long dry season in spring and summer. Is it possible to investigate the effects of possible future changes in climate now?

Researchers at the University of California at Berkeley carried out an experiment to study the effects of more rain in either season. They randomly assigned plots of open grassland to 3 treatments. One treatment was to add water equal to 20% of annual rainfall during January to March (winter). A second treatment was to add water equal to 20% of annual rainfall during April to June (spring). The third treatment was to add no water beyond normal rainfall. Eighteen circular plots of area 70 square meters were used for this study, with six plots used for each treatment. One variable the researchers measured was total plant biomass, in grams per square meter, produced in a plot over a year. Total plant biomass for the three treatments was compared to assess the effect of increased rainfall.

Is this a good study? By the end of this chapter, you will be able to determine the strengths and weaknesses of a study such as this. ▪

Talking about experiments

Observational studies passively collect data. We observe, record, or measure, but we don't interfere. Experiments actively produce. Experimenters intentionally intervene by imposing some treatment in order to see what happens. All experiments and many observational studies are interested in the effect that one variable has on another variable. Here is the vocabulary we use to distinguish the variable that acts from the variable that is acted upon.

The vocabulary of experiments

A **response variable** is a variable that measures an outcome or result of a study.

    An **explanatory variable** is a variable that we think explains or causes changes in the response variable.

    The individuals studied in an experiment are often called **subjects.**

    A **treatment** is any specific experimental condition applied to the subjects. If an experiment has several explanatory variables, a treatment is a combination of specific values of these variables.

EXAMPLE 1   Learning on the Web

An optimistic account of learning online reports a study at Nova Southeastern University, Fort Lauderdale, Florida. The authors of the study claim that students taking undergraduate courses online were "equal in learning" to students taking the same courses in class. Replacing college classes with Web sites saves colleges money, so this study seems to suggest we should all move online.

    College students are the *subjects* in this study. The *explanatory variable* considered in the study is the setting for learning (in class or online). The *response variable* is a student's score on a test at the end of the course. Other variables were also measured in the study, including the score on a test on the course material before the courses started. Although this was not used as an explanatory variable in the study, prior knowledge of the course material might affect the response, and the authors wished to make sure this was not the case.

EXAMPLE 2   The effects of abstinence-only sex education

Whether children should be taught about sex in schools and, if so, how have been controversial issues for many years. A study at the University of Pennsylvania followed 662 black children attending four middle schools in cities in the northeastern United States and tracked their behavior for three years. The children were randomly assigned to one of four different classes: an eight-hour program that taught safe sex; a twenty-hour program that taught abstinence and birth control; an eight-hour program that encouraged abstinence; and a program that simply focused on other ways to be healthy, including eating right and exercising. The result was that the eight-hour program that encouraged abstinence worked better than the other three treatments in terms of the percentage who started having sex during a 24-month follow-up period.

The University of Pennsylvania study is an experiment in which the *subjects* are the 662 children. The experiment compares four *treatments*. The *explanatory variable* is the treatment a child received. Several *response variables* were measured. The primary one was a self-report of ever having sexual intercourse over a 24-month follow-up period.

You will often see explanatory variables called *independent variables* and response variables called *dependent variables*. The idea is that the response variables depend on the explanatory variables. We avoid using these older terms, partly because "independent" has other and very different meanings in statistics.

How to experiment badly

Do students who take a course via the Web learn as well as those who take the same course in a traditional classroom? The best way to find out is to assign some students to the classroom and others to the Web. That's an experiment. The Nova Southeastern University study was not an experiment, because it imposed no treatment on the student subjects. Students chose for themselves whether to enroll in a classroom or an online version of a course. The study simply measured their learning. It turns out that the students who chose the online course were very different from the classroom students. For example, their average score on a test on the course material given before the courses started was 40.70, against only 27.64 for the classroom students. It's hard to compare in-class versus online learning when the online students have a big head start. The effect of online versus in-class instruction is hopelessly mixed up with influences lurking in the background. Figure 5.1 shows the mixed-up influences in picture form.

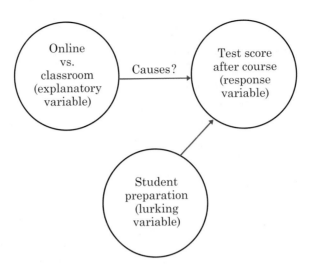

Figure 5.1 Confounding in the Nova Southeastern University study. The influence of course setting (the explanatory variable) cannot be distinguished from the influence of student preparation (a lurking variable).

Lurking variables

A **lurking variable** is a variable that has an important effect on the relationship among the variables in a study but is not one of the explanatory variables studied.

Two variables are **confounded** when their effects on a response variable cannot be distinguished from each other. The confounded variables may be either explanatory variables or lurking variables.

In the Nova Southeastern University study, student preparation (a lurking variable) is confounded with the explanatory variable. The study report claims that the two groups did equally well on the final test. We can't say how much of the online group's performance is due to their head start. That a group that started with a big advantage did no better than the more poorly prepared classroom students is not very impressive evidence of the wonders of Web-based instruction. Here is another example, one in which a second experiment was proposed to untangle the confounding.

EXAMPLE 3 Pig whipworms and the need for further study

Crohn's disease is a chronic inflammatory bowel disease. An experiment reported in *Gut,* a British medical journal, claimed that a drink

containing thousands of pig whipworm eggs was effective in reducing
abdominal pain, bleeding, and diar-
rhea associated with the disease.

Experiments that study the effectiveness of medical treatments on
actual patients are called **clinical
trials.** The clinical trial that suggested that a drink made from pig
whipworm eggs might be effective in
relieving the symptoms of Crohn's
disease had a "one-track" design—
that is, only a single treatment was
applied:

"I want to make one thing perfectly clear, Mr. Smith.
The medication I prescribe *will* cure that run-down
feeling."

Impose treatment $\longrightarrow$ **Measure response**

Pig whipworms $\longrightarrow$ **Reduced symptoms?**

The patients did report reduced symptoms, but we can't say that the
pig whipworm treatment caused the reduced symptoms. It might be just
the **placebo effect.** A **placebo** is a dummy treatment with no active
ingredients. Many patients respond favorably to *any* treatment, even a
placebo. This response to a dummy treatment is the placebo effect. Perhaps the placebo effect is in our minds, based on trust in the doctor and
expectations of a cure. Perhaps it is just a name for the fact that many
patients improve for no visible reason. The one-track design of the experiment meant that the placebo effect was confounded with any effect the
pig whipworm drink might have.

The researchers recognized this and urged further study with a better-
designed experiment. Such an experiment might involve dividing subjects
with Crohn's disease into two groups. One group would be treated with
the pig whipworm drink as before. The other would receive a placebo.
Subjects in both groups would not know which treatment they were receiving. Nor would the physicians recording the symptoms of the subjects
know which treatment a subject received, so that their diagnosis would
not be influenced by such knowledge. An experiment in which neither
subjects nor physicians recording the symptons know which treatment
was received is called **double-blind.**

Both observational studies and one-track experiments often yield useless data because of confounding with lurking variables. It is hard to avoid
confounding when only observation is possible. Experiments offer better

possibilities, as the pig whipworm experiment shows. This experiment could be designed to include a group of subjects who receive only a placebo. This would allow us to see whether the treatment being tested does better than a placebo and so has more than the placebo effect going for it. Effective medical treatments pass the placebo test.

Randomized comparative experiments

The first goal in designing an experiment is to ensure that it will show us the effect of the explanatory variables on the response variables. Confounding often prevents one-track experiments from doing this. The remedy is to *compare* two or more treatments. When confounding variables affect all subjects equally, any systematic differences in the responses of subjects receiving different treatments can be attributed to the treatments rather than to the confounding variables. This is the idea behind the use of a placebo. All subjects are exposed to the placebo effect because all receive some treatment. Here is an example of a new medical treatment that passes the placebo test in a direct comparison.

EXAMPLE 4 Sickle-cell anemia

Sickle-cell anemia is an inherited disorder of the red blood cells that in the United States affects mostly blacks. It can cause severe pain and many complications. The National Institutes of Health carried out a clinical trial of the drug hydroxyurea for treatment of sickle-cell anemia. The subjects were 299 adult patients who had had at least three episodes of pain from sickle-cell anemia in the previous year. An episode of pain was defined to be a visit to a medical facility that lasted more than four hours for acute sickling-related pain. The measurement of the length of the visit included all time spent after registration at the medical facility, including the time spent waiting to see a physician.

Simply giving hydroxyurea to all 299 subjects would confound the effect of the medication with the placebo effect and other lurking variables such as the effect of knowing that you are a subject in an experiment. Instead, approximately half of the subjects received hydroxyurea, and the other half received a placebo that looked and tasted the same. All subjects were treated exactly the same (same schedule of medical checkups, for example) except for the content of the medicine they took. Lurking variables therefore affected both groups equally and should not have caused any differences between their average responses.

The two groups of subjects must be similar in all respects before they start taking the medication. Just as in sampling, the best way to avoid

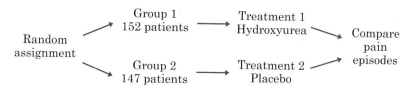

Figure 5.2 The design of a randomized comparative experiment to compare hydroxyurea with a placebo for treating sickle-cell anemia, for Example 4.

bias in choosing which subjects get hydroxyurea is to allow impersonal chance to make the choice. A simple random sample of 152 of the subjects formed the hydroxyurea group; the remaining 147 subjects made up the placebo group. Figure 5.2 outlines the experimental design.

The experiment was stopped ahead of schedule because the hydroxyurea group had many fewer pain episodes than the placebo group. This was compelling evidence that hydroxyurea is an effective treatment for sickle-cell anemia, good news for those who suffer from this serious illness.

Figure 5.2 illustrates the simplest **randomized comparative experiment,** one that compares just two treatments. The diagram outlines the essential information about the design: random assignment to groups; one group for each treatment; the number of subjects in each group (it is generally best to keep the groups similar in size); what treatment each group gets; and the response variable we compare. Random assignment of subjects to groups uses some of the techniques discussed in Chapter 2 for choosing a simple random sample. Label the 299 subjects 001 to 299, then read three-digit groups from the table of random digits (Table A) until you have chosen the 152 subjects for Group 1. The remaining 147 subjects form Group 2.

The placebo group in Example 4 is called a **control group** because comparing the treatment and control groups allows us to control the effects of lurking variables. A control group need not receive a dummy treatment such as a placebo. In Example 2, the students who were randomly assigned to the program that simply focused on other ways to be healthy, including eating right and exercising, were considered to be a control group. Clinical trials often compare a new treatment for a medical condition, not with a placebo, but with a treatment that is already on the market. Patients who are randomly assigned to the existing treatment form the control group. To compare more than two treatments, we can randomly assign the available experimental subjects to as many groups as there are treatments. Here is an example with three groups.

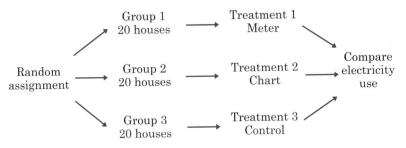

Figure 5.3 The design of a randomized comparative experiment to compare three programs to reduce electricity use by households, for Example 5.

EXAMPLE 5 Conserving energy

Many utility companies have introduced programs to encourage energy conservation among their customers. An electric company considers placing electronic meters in households to show what the cost would be if the electricity use at that moment continued for a month. Will meters reduce electricity use? Would cheaper methods work almost as well? The company decides to design an experiment.

One cheaper approach is to give customers a chart and information about monitoring their electricity use. The experiment compares these two approaches (meter, chart) and also a control. The control group of customers receives information about energy conservation but no help in monitoring electricity use. The response variable is total electricity used in a year. The company finds 60 single-family residences in the same city willing to participate, so it assigns 20 residences at random to each of the 3 treatments. Figure 5.3 outlines the design.

To carry out the random assignment, label the 60 households 01 to 60. Enter Table A to select an SRS of 20 to receive the meters. Continue in Table A, selecting 20 more to receive charts. The remaining 20 form the control group.

NOW IT'S YOUR TURN

5.1 Exercise and heart attacks. Does regular exercise reduce the risk of a heart attack? To answer this question a researcher finds 4000 men over 40 who have not had heart attacks and are willing to participate in a study. She assigns 2000 of the men to a regular program of supervised exercise. The other 2000 continue their usual habits. The researcher follows both groups for 5 years. Outline the design of this study using a diagram like Figures 5.2 and 5.3.

The logic of experimental design

The randomized comparative experiment is one of the most important ideas in statistics. It is designed to allow us to draw cause-and-effect conclusions. Be sure you understand the logic:

- Randomization produces groups of subjects that should be similar, on average, in all respects before we apply the treatments.

- Comparative design exposes all groups to similar conditions, other than the treatments they receive. This ensures that any additional lurking variables operate equally on all groups and, on average, groups differ only in the treatments they receive.

- Therefore, differences in the response variable must be due to the effects of the treatments.

We use chance to choose the groups in order to eliminate any systematic bias in assigning the subjects to groups. In the sickle-cell study, for example, a doctor might subconsciously assign the most seriously ill patients to the hydroxyurea group, hoping that the untested drug will help them. That would bias the experiment against hydroxyurea. Choosing an SRS of the subjects to be Group 1 gives everyone the same chance to be in either group. We expect the two groups to be similar in all respects—age, seriousness of illness, smoker or not, and so on. Chance tends to assign equal numbers of smokers to both groups, for example, even if we don't know which subjects are smokers.

What about the effects of lurking variables not addressed by randomization—for example, those that arise after subjects have been randomly assigned to groups? The placebo effect is such a lurking variable. Its effect occurs only after the treatments are administered to subjects. If the groups are treated at different times of the year, so that some groups are treated during flu season and others not, higher exposure of some groups to the flu could be a lurking variable. In a comparative design, we try to ensure that these lurking variables operate similarly on all groups. All groups receive some treatment in order to ensure they are equally exposed to the placebo effect. All groups receive treatment at the same time, so all experience the same exposure to the flu.

It should not surprise you to learn that medical researchers adopted randomized comparative experiments only slowly—many doctors think they can tell "just by watching" whether a new therapy helps their patients. Not so. There are many examples of medical treatments that became popular on the basis of one-track experiments and were shown to be worth no more than a placebo when some skeptic tried a randomized comparative experiment. One search of the medical literature looked for therapies studied both by proper comparative trials and by trials with "historical controls." A study with historical controls compares the results of a new treatment, not with a control group, but with how well similar patients have done in the past. Of the 56 therapies studied, 44 came out winners with respect to historical controls. But only 10 passed the placebo test in proper randomized comparative experiments. Expert judgment is too optimistic even when aided by comparison with past patients. At present, the law requires that new drugs be shown to be both safe and effective by randomized comparative trials. There is no such requirement for other medical treatments, such as surgery. A Google search of "comparisons with historical controls" found recent studies for other medical treatments that have used historical controls.

There is one important caution about randomized experiments. Like random samples, they are subject to the laws of chance. Just as an SRS of voters might by bad luck choose people nearly all of whom have the same political party preference, a random assignment of subjects might by bad luck put nearly all the smokers in one group. We know that if we choose *large* random samples, it is very likely that the sample will match the population well. In the same way, if we use *many* experimental subjects, it is very likely that random assignment will produce groups that match closely. More subjects means that there is less chance variation among the treatment groups and less chance variation in the outcomes of the experiment. "Use enough subjects" joins "compare two or more treatments" and "randomize" as a basic principle of statistical design of experiments.

Principles of experimental design

The basic principles of statistical design of experiments are:

1. **Control** the effects of lurking variables on the response by ensuring all subjects are affected similarly by these lurking variables. Then simply compare two or more treatments.

2. **Randomize**—use impersonal chance to assign subjects to treatments so treatment groups are similar, on average.

3. **Use enough subjects** in each group to reduce chance variation in the results.

Statistical significance

The presence of chance variation requires us to look more closely at the logic of randomized comparative experiments. We cannot say that *any* difference in the average number of pain episodes between the hydroxyurea group and the control group must be due to the effect of the drug. Even if both treatments are the same, there will always be some chance differences between the individuals in the control group and those in the treatment group. Randomization eliminates just the systematic differences between the groups.

Statistical significance

An observed effect of a size that would rarely occur by chance is called **statistically significant.**

The difference between the average number of pain episodes for subjects in the hydroxyurea group and the average for the control group was "highly statistically significant." That means that a difference of this size would almost never happen just by chance. We do indeed have strong evidence that hydroxyurea beats a placebo in helping sickle-cell disease sufferers. You will often see the phrase "statistically significant" in reports of investigations in many fields of study. It tells you that the investigators found good "statistical" evidence for the effect they were seeking.

Of course, the actual results of an experiment are more important than the seal of approval given by statistical significance. The treatment group in the sickle-cell experiment had an average of 2.5 pain episodes per year, against 4.5 per year in the control group. That's a big enough difference to be important to people with the disease. A difference of 2.5 versus 2.8 would be much less interesting even if it were statistically significant.

How large an observed effect must be in order to be regarded as statistically significant depends on the number of subjects involved. A relatively small effect, one that might not be regarded as practically important, can be statistically significant if the size of the study is large. Thus, in the sickle-cell experiment, an average of 2.50 pain episodes per year versus 2.51 per year in the control group could be statistically significant if the number of subjects involved is sufficiently large. For a very large number of subjects, the average number of pain episodes per year should be almost the same if differences are due only to chance. It is also true that a very large effect may not be statistically significant. If the number of subjects in an experiment is small, it may be possible to observe large effects simply by chance. We will discuss these issues more fully in Parts III and IV.

Thus, in assessing statistical significance it is helpful to know the magnitude of the observed effect and the number of subjects. Perhaps a better term than "statistically significant" might be "statistically dissimilar."

How to live with observational studies

Does regular church attendance lengthen people's lives? Do doctors discriminate against women in treating heart disease? Does talking on a cell phone while driving increase the risk of having an accident? These are cause-and-effect questions, so we reach for our favorite tool, the randomized comparative experiment. Sorry. We can't randomly assign people to attend church or not, because going to religious services is an expression of beliefs or their absence. We can't use random digits to assign heart disease patients to be men or women. We are reluctant to require drivers to use cell phones in traffic, because talking while driving may be risky.

The best data we have about these and many other cause-and-effect questions come from observational studies. We know that observation is a weak second best to experiment, but good observational studies are far from worthless. What makes a good observational study?

First, good studies are **comparative** even when they are not experiments. We compare random samples of people who do and who don't attend religious services regularly. We compare how doctors treat men and women patients. We might compare drivers talking on cell phones with the *same* drivers when they are not on the phone. We can often combine comparison with **matching** in creating a control group. To see the effects of taking a painkiller during pregnancy, we compare women who did so with women who did not. From a large pool of women who did not take the drug, we select individuals who match the drug group in age, education, number of children, and other lurking variables. We now have two groups that are similar in all these ways, so that these lurking variables should not affect our comparison of the groups. However, if other important lurking variables, not measurable or not thought of, are present, they will affect the comparison, and confounding will still be present.

Matching does not entirely eliminate confounding. People who attend church or synagogue or mosque take better care of themselves than nonattenders. They are less likely to smoke, more likely to exercise, and less likely to be overweight. Although matching can reduce some of these differences, direct comparison of ages at death of attenders and nonattenders would still confound any effect of religion with the effects of healthy living. A good comparative study **measures and adjusts for confounding variables.** If we measure weight, smoking, and exercise, there are statistical techniques that reduce the effects of these variables on length of life so that (we hope) only the effect of religion itself remains.

EXAMPLE 6 Living longer through religion

One of the better studies of the effect of regular attendance at religious services gathered data from a random sample of 3617 adults. Random sampling is a good start. The researchers then measured lots of variables, not just the explanatory variable (religious activities) and the response variable (length of life). A news article said:

Churchgoers were more likely to be nonsmokers, physically active, and at their right weight. But even after health behaviors were taken into account, those not attending religious services regularly still were about 25% more likely to have died.

That "taken into account" means that the final results were adjusted for differences between the two groups. Adjustment reduced the advantage of religion but still left a large benefit.

EXAMPLE 7 Sex bias in treating heart disease?

Doctors are less likely to give aggressive treatment to women with symptoms of heart disease than to men with similar symptoms. Is this because doctors are sexist? Not necessarily. Women tend to develop heart problems much later than men, so that female heart patients are older and often have other health problems. That might explain why doctors proceed more cautiously in treating them.

This is a case for a comparative study with statistical adjustments for the effects of confounding variables. There have been several such studies, and they produce conflicting results. Some show, in the words of one doctor, "When men and women are otherwise the same and the only difference is gender, you find that treatments are very similar." Other studies find that women are undertreated even after adjusting for differences between the female and male subjects.

As Example 7 suggests, statistical adjustment is tricky. Randomization creates groups that are similar in *all* variables known and unknown. Matching and adjustment, on the other hand, can't work with variables the researchers didn't think to measure. Even if you believe that the researchers thought of everything, you should be a bit skeptical about statistical adjustment. There's lots of room for cheating in deciding which variables to adjust for. And the "adjusted" conclusion is really something like this:

If female heart disease patients were younger and healthier than they really are, and if male patients were older and less healthy than they really are, then the two groups would get the same medical care.

This may be the best we can get, and we should thank statistics for making such wisdom possible. But we end up longing for the clarity of a good experiment.

STATISTICS IN SUMMARY

Chapter Specifics

- Statistical studies often try to show that changing one variable (the **explanatory variable**) causes changes in another variable (the **response variable**).

- In an **experiment,** we actually set the explanatory variables ourselves rather than just observe them.

- Observational studies and one-track experiments that simply apply a single treatment often fail to produce useful data because **confounding** with **lurking variables** makes it impossible to say what the effect of the treatment was.

- In a **randomized comparative experiment** we compare two or more treatments, use chance to decide which subjects get each treatment, and use enough subjects so that the effects of chance are small.

- Comparing two or more treatments **controls** lurking variables affecting all subjects, such as the **placebo effect**, because they act on all the treatment groups.

- Differences among the effects of the treatments so large that they would rarely happen just by chance are called **statistically significant.**

- Observational studies of cause-and-effect questions are more impressive if they **compare matched groups** and measure as many lurking variables as possible to allow **statistical adjustment.**

 In Chapter 1 we saw that experiments are best suited for drawing conclusions about whether a treatment causes a change in a response. In this chapter we learn that only well-designed experiments, in particular randomized comparative experiments, provide a sound basis for such conclusions. Statistically significant differences among the effects of treatments are the best available evidence that changing the explanatory variable really *causes* changes in the response.

When it is not possible to do an experiment, observational studies that measure as many lurking variables as possible and make statistical adjustments for their effects are sometimes used to answer cause-and-effect questions. However, they remain a weak second best to well-designed experiments.

CASE STUDY   Use what you have learned in this chapter to evaluate the Case Study
EVALUATED   that opened the chapter. Start by reviewing the information on page
91. Then answer each of the following questions in complete sentences. Be sure to communicate clearly enough for any of your classmates to understand what you are saying.

First, here are the results of the study. After one season, the biomass of plants in the
plot receiving additional spring rain was approximately twice that in plots receiving the
other treatments. This difference was statistically significant.

1. Is this study an experiment or an observational study?

2. Explain what the phrase "statistically significant" means.

3. What advantage is gained by randomly assigning the plots to the treatments? ▪

CHAPTER 5 EXERCISES

For Exercise 5.1, see page 98.

5.2 Exhaust is bad for your heart. A *CNET News* article reported that the
artery walls of people living within 100 meters of a highway thicken more than
twice as fast as the average person's. Researchers used ultrasound to measure
the carotid artery wall thickness of 1483 people living near freeways in the Los
Angeles area. The artery wall thickness among those living within 100 meters
of a highway increased by 5.5 micrometers (roughly 1/20th the thickness of a
human hair) each year during the three-year study, which is more than twice
the progression observed in participants who did not live within this distance
of a highway.

(a) What are the explanatory and response variables?

(b) Explain carefully why this study is not an experiment.

(c) Explain why confounding prevents us from concluding that living near a
highway is bad for your heart because it causes increased thickness in the
carotid artery wall.

5.3 Decline in SAT Math scores. A *New York Times* article reported that
average SAT Math scores for the high school class of 2007 dropped 3 points
compared with scores in 2006. Officials of the College Board, the nonprofit organization that administers the SAT, suggested that increased numbers of students taking the SAT had contributed to the decline in scores. "The larger the
population you get that takes the exam, it obviously knocks down the scores,"
said Gaston Caperton, the president of the College Board. Is this conclusion
the result of an experiment? Why or why not? What are the explanatory and
response variables?

5.4 Weight-loss surgery and longer life. An article in the *Washington Post*
reported that, according to two large studies, obese people are significantly less

likely to die prematurely if they undergo stomach surgery to lose weight. But people choose whether to have stomach surgery. Explain why this fact makes any conclusion about cause and effect untrustworthy. Use the language of lurking variables and confounding in your explanation, and draw a picture like Figure 5.1 to illustrate it.

5.5 Is obesity contagious? A study closely followed a large social network of 12,067 people for 32 years, from 1971 until 2003. The researchers found that when a person gains weight, close friends tend to gain weight, too. The researchers reported that obesity can spread from person to person, much like a virus.

Explain why the fact that, when a person gains weight, close friends also tend to gain weight does not necessarily mean that weight gains in a person cause weight gains in close friends. In particular, identify some lurking variables whose effect on weight gain may be confounded with the effect of weight gains in close friends. Draw a picture like Figure 5.1 (page 94) to illustrate your explanation.

5.6 Aspirin and heart attacks. Can aspirin help prevent heart attacks? The Physicians' Health Study, a large medical experiment involving 22,000 male physicians, attempted to answer this question. One group of about 11,000 physicians took an aspirin every second day, while the rest took a placebo. After several years the study found that subjects in the aspirin group had significantly fewer heart attacks than subjects in the placebo group.

(a) Identify the experimental subjects, the explanatory variable and the values it can take, and the response variable.

(b) Use a diagram to outline the design of the Physicians' Health Study. (When you outline the design of an experiment, be sure to indicate the size of the treatment groups and the response variable. The diagrams in Figures 5.2 (page 97) and 5.3 (page 98) are models.)

(c) What do you think the term "significantly" means in "significantly fewer heart attacks"?

5.7 Computer simulations in teaching. Are computer simulations as effective as direct observation in learning science concepts? If so, computer simulations can provide a less time-consuming alternative to direct observations in science classes. Researchers at two universities plan to study this issue. A group of 20 astronomy majors from each university is selected to take part in the study. Half of the students will use computer simulations to learn about moon phases. The other half will directly observe the moon. All students will then be evaluated to determine their conceptual understanding of moon phases. For simplicity, the researchers decide that all the students at one university will use the computer simulation to learn about moon phases and all the students at the other university will learn about moon phases by directly observing the moon. Why is this a bad idea?

5.8 Neighborhood's effect on grades. To study the effect of neighborhood on academic performance, one thousand families were given federal housing vouchers to move out of their low-income neighborhoods. No improvement in the academic performance of the children in the families was found one year after the move.

Explain clearly why the lack of improvement in academic performance after one year does not necessarily mean that neighborhood does not affect academic performance. In particular, identify some lurking variables whose effect on academic performance may be confounded with the effect of neighborhood. Use a picture like Figure 5.1 (page 94) to illustrate your explanation.

5.9 Computer simulations in teaching, continued.

(a) Outline a better design than that of Exercise 5.7 for an experiment to compare the two treatments (computer simulation or direct observation) that students received for learning about moon phases. What do you suggest as a response variable? (When you outline the design of an experiment, be sure to indicate the size of the treatment groups and the response variable. The diagrams in Figures 5.2 (page 97) and 5.3 (page 98) are models.)

(b) Use Table A, starting at line 119, to do the randomization your design requires.

5.10 Learning on the Web. The discussion following Example 1 (page 92) notes that the Nova Southeastern University study does not tell us much about Web versus classroom learning because the students who chose the Web version were much better prepared. Describe the design of an experiment to get better information.

5.11 Do antioxidants prevent cancer? People who eat lots of fruits and vegetables have lower rates of colon cancer than those who eat little of these foods. Fruits and vegetables are rich in "antioxidants" such as vitamins A, C, and E. Will taking antioxidants help prevent colon cancer? A clinical trial studied this question with 864 people who were at risk for colon cancer. The subjects were divided into four groups: daily beta-carotene, daily vitamins C and E, all three vitamins every day, and daily placebo. After four years, the researchers were surprised to find no significant difference in colon cancer among the groups.

(a) What are the explanatory and response variables in this experiment?

(b) Outline the design of the experiment. (The diagrams in Figures 5.2 (page 97) and 5.3 (page 98) are models.)

(c) Assign labels to the 864 subjects and use Table A, starting at line 118, to choose the *first* 5 subjects for the beta-carotene group.

(d) What does "no significant difference" mean in describing the outcome of the study?

(e) Suggest some lurking variables that could explain why people who eat lots of fruits and vegetables have lower rates of colon cancer. The results of the experiment suggest that these variables, rather than the antioxidants, may be responsible for the observed benefits of fruits and vegetables.

5.12 Conserving energy. Example 5 (page 98) describes an experiment to learn whether providing households with electronic meters or with charts will reduce their electricity consumption. An executive of the electric company objects to including a control group. He says, "It would be cheaper to just compare electricity use last year [before the meter or chart was provided] with consumption in the same period this year. If households use less electricity this year, the meter or chart must be working." Explain clearly why this design is inferior to that in Example 5.

5.13 Improving Chicago's schools. The National Science Foundation (NSF) paid for "systemic initiatives" to help cities reform their public education systems in ways that should help students learn better. Does this program work? The initiative in Chicago focused on improving the teaching of mathematics in high schools. The average scores of students on a standard test of math skills were higher after two years of the program in 51 out of 60 high schools in the city. Leaders of NSF said this was evidence that the Chicago program was succeeding. Critics said this doesn't say anything about the effect of the systemic initiative. Are these critics correct? Explain.

5.14 Table saw blades. A manufacturer of table saw blades is interested in determining whether a narrower blade will cause less burning of wood when cutting very hard woods, such as maple. To answer this question, engineers obtain 20 similar one-inch-thick hard maple boards. Half are sawed using the new, narrow-style blade, and the other half are sawed using the standard-width blade. All cuts are done at the same feed rate (the rate at which the board is pushed against the blade to make the cut). The engineers then measure the amount of burning (rated on a scale of 1 to 10, with 10 being the worst) on each board.

(a) The individuals studied in this experiment are not people. What are they?

(b) What is the explanatory variable, and what values does it take?

(c) What is the response variable?

5.15 Reducing health care spending. Will people spend less on health care if their health insurance requires them to pay some part of the cost themselves? An experiment on this issue asked if the percentage of medical costs that is paid by health insurance has an effect both on the amount of medical care that people use and on their health. The treatments were four insurance plans. Each plan paid all medical costs above a ceiling. Below the ceiling, the plans paid 100%, 75%, 50%, or 0% of costs incurred.

(a) Outline the design of a randomized comparative experiment suitable for this study.

(b) Briefly describe the practical and ethical difficulties that might arise in such an experiment.

5.16 Table saw blades. Consider again the table saw blade experiment of Exercise 5.14.

(a) Use a diagram to describe a randomized comparative experimental design for this experiment.

(b) Use Table A, starting at line 120, to do the randomization required by your design.

5.17 Treating drunk drivers. Once a person has been convicted of drunk driving, one purpose of court-mandated treatment or punishment is to prevent future offenses of the same kind. Suggest three different treatments that a court might require. Then outline the design of an experiment to compare their effectiveness. Be sure to specify the response variables you will measure.

5.18 Statistical significance. A randomized comparative experiment examines whether the usual care of patients with chronic heart failure plus aerobic exercise training improves health status compared with the usual care alone. The researchers conclude that usual care plus exercise training confers modest but statistically significant improvements in self-reported health status compared with usual care without training. Explain what "statistically significant" means in the context of this experiment, as if you were speaking to a patient who knows no statistics.

5.19 Statistical significance. A study, mandated by Congress when it passed No Child Left Behind in 2002, evaluated 15 reading and math software products used by 9424 students in 132 schools across the country during the 2004–2005 school year. It is the largest study that has compared students who received the technology with those who did not, as measured by their scores on standardized tests. There were no statistically significant differences between students who used software and those who did not. Explain the meaning of "no statistically significant differences" in plain language.

5.20 Memantine and Alzheimer's disease. Some medical researchers suspect that the drug memantine improves the cognition of patients with moderate to severe Alzheimer's disease. You have available 50 people with moderate to severe Alzheimer's disease who are willing to serve as subjects:

Albright	Han	Lewis	Samara	Tompkins
Ashmead	Hotait	Li	Sanders	Townsend
Asihiro	Hu	Lim	Schneider	Turkmen
Bai	Josey	Madaeni	Smith	Wang
Bayer	Jung	Martin	Stagner	Westra
Biller	Khalaf	Patton	Stettler	Williams
Chen	Koster	Penzenik	Tan	Winner
Critchlow	Landgraf	Powell	Tang	Yontz
Davis	Lathrop	Ren	Thomas	Yulovitch
Dobmeier	Lefevre	Rodriguez	Tirmenstein	Zhang

(a) Outline an appropriate design for the experiment, taking the placebo effect into account.

(b) The names of the subjects appear above. If you have access to statistical software, use it to carry out the randomization required by your design. Otherwise, use Table A, beginning at line 131, to do the randomization required by your design. List the subjects to whom you will give the drug.

5.21 Treating prostate disease. A large study used records from Canada's national health care system to compare the effectiveness of two ways to treat prostate disease. The two treatments are traditional surgery and a new method that does not require surgery. The records described many patients whose doctors had chosen one or the other method. The study found that patients treated by the new method were significantly more likely to die within 8 years.

(a) Further study of the data showed that this conclusion was wrong. The extra deaths among patients treated with the new method could be explained by lurking variables. What lurking variables might be confounded with a doctor's choice of surgical or nonsurgical treatment? For example, why might a doctor avoid assigning a patient to surgery?

(b) You have 300 prostate patients who are willing to serve as subjects in an experiment to compare the two methods. Use a diagram to outline the design of a randomized comparative experiment.

5.22 Prayer and meditation. You read in a magazine that "nonphysical treatments such as meditation and prayer have been shown to be effective in controlled scientific studies for such ailments as high blood pressure, insomnia, ulcers, and asthma." Explain in simple language what the article means by "controlled scientific studies" and why such studies might show that meditation and prayer are effective treatments for some medical problems.

5.23 Exercise and bone loss. Does regular exercise reduce bone loss in post-menopausal women? Here are two ways to study this question. Explain clearly why the second design will produce more trustworthy data.

1. A researcher finds 1000 postmenopausal women who exercise regularly. She matches each with a similar postmenopausal woman who does not exercise regularly, and she follows both groups for 5 years.

2. Another researcher finds 2000 postmenopausal women who are willing to participate in a study. She assigns 1000 of the women to a regular program of supervised exercise. The other 1000 continue their usual habits. The researcher follows both groups for 5 years.

5.24 Safety of anesthetics. The death rates of surgical patients differ for operations in which different anesthetics are used. An observational study found these death rates for four anesthetics:

Anesthetic:	Halothane	Pentothal	Cyclopropane	Ether
Death rate:	1.7%	1.7%	3.4%	1.9%

This is *not* good evidence that cyclopropane is more dangerous than the other anesthetics. Suggest some lurking variables that may be confounded with the choice of anesthetic in surgery and that could explain the different death rates.

5.25 Randomization at work. To demonstrate how randomization reduces confounding, consider the following situation. A nutrition experimenter intends to compare the weight gain of prematurely born infants fed Diet A with those fed Diet B. To do this, she will feed each diet to 10 prematurely born infants whose parents have enrolled them in the study. She has available 10 baby girls and 10 baby boys. The researcher is concerned that baby boys may respond more favorably to the diets, so that if all the baby boys were fed diet A the experiment would be biased in favor of Diet A.

(a) Label the infants 00, 01, ... , 19. Use Table A to assign 10 infants to Diet A. Or, if you have access to statistical software, use it to assign 10 infants to Diet A. Do this four times, using different parts of the table (or different runs of your software), and write down the four groups assigned to Diet A.

(b) The infants labeled 10, 11, 12, 13, 14, 15, 16, 17, 18, and 19 are the 10 baby boys. How many of these infants were in each of the four Diet A groups that you generated? What was the average number of baby boys assigned to Diet A?

EXPLORING THE WEB

5.26 Exploring the Web. Go to the *New England Journal of Medicine* Web site (**http://content.nejm.org**) and find the article "A Randomized, Controlled Trial of Financial Incentives for Smoking Cessation" by Volpp et al. in the February 12, 2009, issue. Was this a comparative study? Was randomization used? How many subjects took part? Were the results statistically significant? (If your institution does not have a subscription to the *New England Journal of Medicine,* you can find an abstract of the article at **www.ncbi.nlm.nih.gov/pubmed/19213683.**)

You can find the latest medical research in the *Journal of the American Medical Association* (**www.jama.ama-assn.org**) and the *New England Journal of Medicine* (**http://content.nejm.org**). Many of the articles describe randomized comparative experiments, and even more use the language of statistical significance.

NOTES AND DATA SOURCES

Page 91 Case Study: K. B. Suttle, Meredith A. Thomsen, and Mary E. Power, "Species interactions reverse grassland responses to changing climate," *Science,* 315 (2007), pp. 640–642.

Page 92 Example 1: Allan H. Schulman and Randi L. Sims, "Learning in an online format versus an in-class format: an experimental study," *T.H.E. Journal,* June 1999, pp. 54–56.

Page 92 Example 2: John B. Jemmott III et al., "Efficacy of a theory-based abstinence-only intervention over 24 months," *Archives of Pediatrics and Adolescent Medicine,* 164 (2010), pp. 152–159.

Page 94 Example 3: R. W. Summers et al., "*Trichuris suis* therapy in Crohn's disease," *Gut,* 54 (2005), pp. 87–90.

Page 96 Example 4: Samuel Charache et al., "Effects of hydroxyurea on the frequency of painful crises in sickle cell anemia," *New England Journal of Medicine,* 332 (1995), pp. 1317–1322.

Page 100 H. Sacks, T. C. Chalmers, and H. Smith, Jr., "Randomized versus historical controls for clinical trials," *American Journal of Medicine,* 72 (1982), pp. 233–240.

Page 102 Example 6: Marilyn Ellis, "Attending church found factor in longer life," *USA Today,* August 9, 1999.

Page 103 Example 7: Dr. Daniel B. Mark, in Associated Press, "Age, not bias, may explain differences in treatment," *New York Times,* April 26, 1994. Dr. Mark was commenting on Daniel B. Mark et al., "Absence of sex bias in the referral of patients for cardiac catheterization," *New England Journal of Medicine,* 330 (1994), pp. 1101–1106. See the correspondence from D. Douglas Miller and Leslee Shaw, "Sex bias in the care of patients with cardiovascular disease," *New England Journal of Medicine,* 331 (1994), p. 883, for comments on a study with opposing results.

Page 105 Exercise 5.2: E. A. Moore, "Highway to hell: Exhaust is bad for your heart," *CNET News,* February 9, 2010; available online at **news.cnet.com/8301-27083_3-10450159-247.html?tag=mncol.**

Page 105 Exercise 5.3: A. Finder, "Math and reading SAT scores drop," *New York Times,* August 28, 2007.

Page 105 Exercise 5.4: R. Stein, "Weight-loss surgery tied to a longer life," *Washington Post,* August 23, 2007.

Page 106 Exercise 5.6: Information about the Physicians' Health Study is available online at **http://phs.bwh.harvard.edu/phs1.htm.**

Page 107 Exercise 5.11: G. Kolata, "New study finds vitamins are not cancer preventers," *New York Times,* July 21, 1994. For the details, look in the *Journal of the American Medical Association* for the same date.

Page 108 Exercise 5.13: Letter to the editor by Stan Metzenberg, *Science,* 286 (1999), p. 2083.

Page 108 Exercise 5.15 is based on Christopher Anderson, "Measuring what works in health care," *Science,* 263 (1994), pp. 1080–1082.

Page 109 Exercise 5.19: A. R. Paley, "Software's benefits on tests in doubt," *Washington Post,* April 5, 2007.

Page 110 Exercise 5.24: L. E. Moses and F. Mosteller, "Safety of anesthetics," in J. M. Tanur et al. (eds.), *Statistics: A Guide to the Unknown,* 3rd edition, Wadsworth, 1989, pp. 15–24.

Experiments in the Real World

CASE STUDY Is caffeine dependence real? Researchers at the Johns Hopkins University School of Medicine wanted to determine if some individuals develop a serious addiction called caffeine dependence syndrome. Eleven volunteers were recruited who were diagnosed as caffeine dependent. For a two-day period these volunteers were given a capsule that either contained their daily amount of caffeine or a fake substance. Over another two-day period, at least one week after the first, the contents of the capsules were switched. Whether the subjects first received the capsule containing caffeine or the capsule with the fake substance was determined by randomization. The subjects' diets were restricted during the study periods. All products with caffeine were prohibited, but to divert the subjects' attention from caffeine, products containing ingredients such as artificial sweeteners were also prohibited. Questionnaires assessing depression, mood, and the presence of certain physical symptoms were administered at the end of each two-day period. The subjects also completed a tapping task in which they were instructed to press a button 200 times as fast as they could. Finally, subjects were interviewed by a researcher, who did not know what was in the capsules the subjects had taken, to find other evidence of functional impairment. The EESEE story "Is Caffeine Dependence Real?" contains more information about this study.

Is this a good study? By the end of this chapter, you will be able to determine the strengths and weaknesses of a study such as this. ■

Equal treatment for all

Probability samples are a big idea, but sampling in practice has difficulties that just using random samples doesn't solve. Randomized comparative experiments are also a big idea, but they don't solve all the difficulties of experimenting. A sampler must know exactly what information she wants and must compose questions that extract that information from her sample. An experimenter must know exactly what treatments and responses he wants information about, and he must construct the apparatus needed to apply the treatments and measure the responses. This is what psychologists or medical researchers or engineers mean when they talk about "designing an experiment." We are concerned with the *statistical* side of designing experiments, ideas that apply to experiments in psychology, medicine, engineering, and other areas as well. Even at this general level, you should understand the practical problems that can prevent an experiment from producing useful data.

The logic of a randomized comparative experiment assumes that all the subjects are treated alike except for the treatments that the experiment is designed to compare. Any other unequal treatment can cause bias. Treating subjects exactly alike is hard to do.

EXAMPLE 1 Rats and rabbits

Rats and rabbits that are specially bred to be uniform in their inherited characteristics are the subjects in many experiments. However, animals, like people, can be quite sensitive to how they are treated. Here are two amusing examples of how unequal treatment can create bias.

Does a new breakfast cereal provide good nutrition? To find out, compare the weight gains of young rats fed the new product and rats fed a standard diet. The rats are randomly assigned to diets and are housed in large racks of cages. It turns out that rats in upper cages grow a bit faster than rats in bottom cages. If the experimenters put rats fed the new product at the top and those fed the standard diet below, the experiment is biased in favor of the new product. Solution: assign the rats to cages at random.

Another study looked at the effects of human affection on the cholesterol level of rabbits. All the rabbit subjects ate the same diet. Some (chosen at random) were regularly removed from their cages to have their furry heads scratched by friendly people. The rabbits who received affection had lower cholesterol. So affection for some but not other rabbits could bias an experiment in which the rabbits' cholesterol level is a response variable.

Double-blind experiments

Placebos work. That bare fact means that medical studies must take special care to show that a new treatment is not just a placebo. Part of equal treatment for all is to be sure that the placebo effect operates on all subjects.

EXAMPLE 2 The powerful placebo

Want to help balding men keep their hair? Give them a placebo—one study found that 42% of balding men maintained or increased the amount of hair on their heads when they took a placebo. Another study told 13 people who were very sensitive to poison ivy that the stuff being rubbed on one arm was poison ivy. It was a placebo, but all 13 broke out in a rash. The stuff rubbed on the other arm really was poison ivy, but the subjects were told it was harmless—and only 2 of the 13 developed a rash.

When the ailment is vague and psychological, like depression, some experts think that about three-quarters of the effect of the most widely used drugs is just the placebo effect. Others disagree (see Exercise 6.26). The strength of the placebo effect in medical treatments is hard to pin down because it depends on the exact environment. How enthusiastic the doctor is seems to matter a lot. But "placebos work" is a good place to start when you think about planning medical experiments.

The strength of the placebo effect is a strong argument for randomized comparative experiments. In the baldness study, 42% of the placebo group kept or increased their hair, but 86% of the men getting a new drug to fight baldness did so. The drug beats the placebo, so it has something besides the placebo effect going for it. Of course, the placebo effect is still part of the reason this and other treatments work.

Because the placebo effect is so strong, it would be foolish to tell subjects in a medical experiment whether they are receiving a new drug or a placebo. Knowing that they are getting "just a placebo" might weaken the placebo effect and bias the experiment in favor of the other treatments. It is also foolish to tell doctors and other medical personnel what treatment each subject is receiving. If they know that a subject is getting "just a placebo," they may expect less than if they know the subject is receiving a promising experimental drug. Doctors' expectations change how they interact with patients and even the way they diagnose a patient's condition. Whenever possible, experiments with human subjects should be *double-blind*.

"Dr. Burns, are you sure this is what the statisticians call a double-blind experiment?"

> **Double-blind experiments**
>
> In a **double-blind experiment,** neither the subjects nor the people who work with them know which treatment each subject is receiving.

Until the study ends and the results are in, only the study's statistician knows for sure. Reports in medical journals regularly begin with words like these, from a study of a flu vaccine given as a nose spray: "This study was a randomized, double-blind, placebo-controlled trial. Participants were enrolled from 13 sites across the continental United States between mid-September and mid-November 1997." Doctors are supposed to know what "randomized," "double-blind," and "placebo-controlled" mean. Now you also know.

Refusals, nonadherers, and dropouts

Sample surveys suffer from nonresponse due to failure to contact some people selected for the sample and the refusal of others to participate. Experiments with human subjects suffer from similar problems.

EXAMPLE 3 Minorities in clinical trials

Refusal to participate is a serious problem for medical experiments on treatments for major diseases such as cancer. As in the case of samples, bias can result if those who refuse are systematically different from those who cooperate.

Clinical trials are medical experiments involving human subjects. Minorities, women, the poor, and the elderly have long been underrepresented in clinical trials. In many cases, they weren't asked. The law now requires representation of women and minorities, and data show that most clinical trials now have fair representation. But refusals remain a problem. Minorities, especially blacks, are more likely to refuse to participate. The government's Office of Minority Health says, "Though recent studies have shown that African Americans have increasingly positive attitudes toward cancer medical research, several studies corroborate that they are still cynical about clinical trials. A major impediment for lack of participation is a lack of trust in the medical establishment." Some remedies for lack of trust are complete and clear information about the experiment, insurance coverage for experimental treatments, participation of black researchers, and cooperation with doctors and health organizations in black communities.

Subjects who participate but don't follow the experimental treatment, called **nonadherers,** can also cause bias. AIDS patients who participate in trials of a new drug sometimes take other treatments on their own, for example. In addition, some AIDS subjects have their medication tested and drop out or add other medications if they were not assigned to the new drug. This may bias the trial against the new drug.

Experiments that continue over an extended period of time also suffer **dropouts,** subjects who begin the experiment but do not complete it. If the reasons for dropping out are unrelated to the experimental treatments, no harm is done other than reducing the number of subjects. If subjects drop out because of their reaction to one of the treatments, bias can result.

EXAMPLE 4   Dropouts in a medical study

The League to Mess Up Experiments meets...

Orlistat is a drug that may help reduce obesity by preventing absorption of fat from the foods we eat. As usual, the drug was compared with a placebo in a double-blind randomized trial. Here's what happened.

The subjects were 1187 obese individuals. They were given a placebo for four weeks, and the subjects who wouldn't take a pill regularly were dropped. This addressed the problem of nonadherers. There were 892 subjects left. These subjects were randomly assigned to orlistat or a placebo, along with a weight-loss diet. After a year devoted to losing weight, 576 subjects were still participating. On the average, the orlistat group lost 3.15 kilograms (about 7 pounds) more than the placebo group. The study continued for another year, now emphasizing main-

"Agent B, you will scratch the heads of the lab rabbits. Agent Q, you will join a clinical trial and not take your pills. Agent K, you will sign up for an experiment, then drop out just before the end."

taining the weight loss from the first year. At the end of the second year, 403 subjects were left. That's only 45% of the 892 who were randomized. Orlistat again beat the placebo, reducing the weight regained by an average of 2.25 kilograms (about 5 pounds).

Can we trust the results when so many subjects dropped out? The overall dropout rates were similar in the two groups: 54% of the subjects taking orlistat and 57% of those in the placebo group dropped out. Were dropouts related to the treatments? Placebo subjects in weight-loss experiments often drop out because they aren't losing weight. This would bias the study against orlistat because the subjects in the placebo group at the end may be those who could lose weight just by following a diet. The researchers looked carefully at the data available for subjects who dropped out. Dropouts from both groups had lost less weight than those who stayed, but careful statistical study suggested that there was little bias. Perhaps so, but the results aren't as clean as our first look at experiments promised.

Can we generalize?

A well-designed experiment tells us that changes in the explanatory variable cause changes in the response variable. More exactly, it tells us that this happened for specific subjects in the specific environment of this specific experiment. No doubt we had grander things in mind. We want to proclaim that our new method of teaching math does better for high school students in general or that our new drug beats a placebo for some broad class of patients. Can we generalize our conclusions from our little group of subjects to a wider population?

The first step is to be sure that our findings are *statistically significant,* that they are too strong to often occur just by chance. That's important, but it's a technical detail that the study's statistician can reassure us about. The serious threat is that the treatments, the subjects, or the environment of our experiment may not be realistic. Let's look at some examples.

EXAMPLE 5 Studying frustration

A psychologist wants to study the effects of failure and frustration on the relationships among members of a work team. She forms a team of students, brings them to the psychology laboratory, and has them play a game that requires teamwork. The game is rigged so that they lose regularly. The psychologist observes the students through a one-way window and notes the changes in their behavior during an evening of game playing.

Playing a game in a laboratory for small stakes, knowing that the session will soon be over, is a long way from working for months developing a new product that never works right and is finally abandoned by your company. Does the behavior of the students in the lab tell us much about the behavior of the team whose product failed?

In Example 5, the subjects (students who know they are subjects in an experiment), the treatment (a rigged game), and the environment (the psychology lab) are all unrealistic if the psychologist's goal is to reach conclusions about the effects of frustration on teamwork in the workplace. Psychologists do their best to devise realistic experiments for studying human behavior, but lack of realism limits the ability to generalize beyond the environment and subjects in their study, and hence the usefulness of experiments in this area.

EXAMPLE 6 The effects of day care

Should the government provide day care for low-income, preschool children? If day care helps these children stay in school and hold good jobs later in life, the government would save money by paying less welfare and collecting more taxes, so even those who are concerned only about the cost to the government might support day care programs. The Carolina Abecedarian Project (the name suggests learning the ABCs) has followed a group of children since 1972.

The Abecedarian Project is an experiment involving 111 people who in 1972 were healthy but low-income black infants in Chapel Hill, North Carolina. All the infants received nutritional supplements and help from social workers. Approximately half, chosen at random, were also placed in an intensive preschool program. The experiment compares these two treatments. Many response variables were recorded over more than 30 years, including academic test scores, college attendance, and employment.

This long and expensive experiment does show that intensive day care has substantial benefits in later life. The day care in the study was intensive indeed—lots of highly qualified staff, lots of parent participation, and detailed activities starting at a very young age, all costing about $11,000 per year for each child. It's unlikely that society will decide to offer such care to all low-income children, so the level of care in this experiment is somewhat unrealistic. The unanswered question is a big one: how good must day care be to really help children succeed in life?

EXAMPLE 7 Are subjects treated too well?

Surely medical experiments are realistic? After all, the subjects are real patients in real hospitals really being treated for real illnesses.

Even here, there are some questions. Patients participating in medical trials get better medical care than most other patients, even if they are in the placebo group. Their doctors are specialists doing research on their specific ailment. They are watched more carefully than other patients. They are more likely to take their pills regularly because they are constantly reminded to do so. Providing "equal treatment for all" except for the experimental and control therapies translates into "provide the best possible medical care for all." The result: ordinary patients may not do as well as the clinical trial subjects when the new therapy comes into general use. It's likely that a therapy that beats a placebo in a clinical trial will beat it in ordinary medical care, but "cure rates" or other measures of success estimated from the trial may be optimistic.

Meta-analysis A single study of an important issue is rarely decisive. We often find several studies in different settings, with different designs, and of different quality. Can we combine their results to get an overall conclusion? That is the idea of "meta-analysis." Of course, differences among the studies prevent us from just lumping them together. Statisticians have more sophisticated ways of combining the results. Meta-analysis has been applied to issues ranging from the effect of secondhand smoke to whether coaching improves SAT scores.

When experiments are not fully realistic, statistical analysis of the experimental data cannot tell us how far the results will generalize. Experimenters generalizing from students in a lab to workers in the real world must argue based on their understanding of how people function, not based just on the data. It is even harder to generalize from rats in a lab to people in the real world. This is one reason why a single experiment is rarely completely convincing, despite the compelling logic of experimental design. The true scope of a new finding must usually be explored by a number of experiments in various settings.

A convincing case that an experiment is sufficiently realistic to produce useful information is based not on statistics but on the experimenter's knowledge of the subject matter of the experiment. The attention to detail required to avoid hidden bias also rests on subject-matter knowledge. Good experiments combine statistical principles with understanding of a specific field of study.

Experimental design in the real world

The experimental designs we have met all have the same pattern: divide the subjects at random into as many groups as there are treatments, then apply each treatment to one of the groups. These are *completely randomized* designs.

Completely randomized design

In a **completely randomized** experimental design, all the experimental subjects are allocated at random among all the treatments.

What is more, our examples to this point have had only a single explanatory variable (for example, drug versus placebo, classroom versus Web instruction). A completely randomized design can have any number of explanatory variables. Here is an example with two.

EXAMPLE 8   Can low-fat food labels lead to obesity?

Tetra Images/Getty Images

What are the effects of low-fat food labels on food consumption? Do people eat more of a snack food when the food is labeled as low-fat? The answer may depend both on whether the snack food is labeled low-fat and whether the label includes serving-size information. An experiment investigated this question using university staff, graduate students, and undergraduate students at a large university as subjects. Over 10 late-afternoon sessions, all subjects viewed episodes of a 60-minute, made-for-television program in a theater on campus and were asked to rate the episodes. They were also told that because it was late in the afternoon, they would be given a cold 24-ounce bottle of water and a bag of granola from a respected campus restaurant called The Spice Box. They were told to enjoy as much or as little of it as they wanted. Each participant received 640 calories (160 grams) of granola in ziplock bags that were labeled with an attractive 3.25 × 4 inch color label. Depending on the condition randomly assigned to the subjects, the bags ware labeled either "Regular Rocky Mountain Granola" or "Low-Fat Rocky Mountain Granola." Below this, the label indicated "Contains 1 Serving" or "Contains 2 Servings," or it provided no serving-size information. As participants left the theater, they were asked how many serving sizes they believed their package contained. Out of sight of the participants, the researchers also weighed each granola bag. Participants' statements about serving size and the actual weights of the granola bags are the response variables.

This experiment has two explanatory variables: fat content, with 2 levels, and serving size, with 3 levels. The 6 combinations of 1 level of each variable form 6 treatments. Figure 6.1 shows the layout of the treatments.

Variable B
Serving size

		No information	1 serving	2 servings
Variable A Fat content	Regular	Treatment 1	Treatment 2	Treatment 3
	Low-fat	Treatment 4	Treatment 5	Treatment 6

Figure 6.1 The treatments in the experiment of Example 8. Combinations of two explanatory variables form 6 treatments.

> **NOW IT'S YOUR TURN**
>
> **6.1 Tasty cakes.** A food company is preparing to market a new cake mix. It is important that the taste of the cake not be changed by small variations in baking time or temperature. In an experiment, cakes made from the mix are baked at 300°F, 320°F, or 340°F and either for 1 hour or for 1 hour and 15 minutes. Ten cakes are baked at each combination of temperature and time. A panel of tasters scores each cake for texture and taste.
>
> What are the explanatory variables and the response variables for this experiment?
>
> Make a diagram like Figure 6.1 to describe the treatments. How many treatments are there? How many cakes are needed?

Experimenters often want to study the combined effects of several variables simultaneously. The interaction of several factors can produce effects that could not be predicted from looking at the effect of each factor alone. Perhaps longer commercials increase interest in a product, and more commercials also increase interest, but if we both make a commercial longer and show it more often, viewers get annoyed and their interest in the product drops. The experiment in Example 8 will help us find out.

Matched pairs and block designs

Completely randomized designs are the simplest statistical designs for experiments. They illustrate clearly the principles of control and randomization. However, completely randomized designs are often inferior to more elaborate statistical designs. In particular, matching the subjects in various ways can produce more precise results than simple randomization.

One common design that combines matching with randomization is the **matched pairs design.** A matched pairs design compares just two treatments. Choose pairs of subjects that are as closely matched as possible. Assign one of the treatments to each subject in a pair by tossing a coin or reading odd and even digits from Table A. Sometimes each "pair" in a matched pairs design consists of just one subject, who gets both treatments one after the other. Each subject serves as his or her own control. The *order* of the treatments can influence the subject's response, so we randomize the order for each subject, again by a coin toss.

EXAMPLE 9 Coke versus Pepsi

Pepsi wanted to demonstrate that Coke drinkers prefer Pepsi when they taste both colas blind. The subjects, all people who said they were Coke drinkers, tasted both colas from glasses without brand markings and said which they liked better. This is a matched pairs design in which each subject compares the two colas. Because responses may depend on which cola is tasted first, the order of tasting should be chosen at random for each subject.

When more than half the Coke drinkers chose Pepsi, Coke claimed that the experiment was biased. The Pepsi glasses were marked M and Coke glasses were marked Q. Aha, said Coke, the results could just mean that people like the letter M better than the letter Q. The matched pairs design is OK, but a more careful experiment would avoid any distinction other than Coke versus Pepsi.

Matched pairs designs use the principles of comparison of treatments and randomization. However, the randomization is not complete—we do not randomly assign all the subjects at once to the two treatments. Instead, we randomize only within each matched pair. This allows matching to reduce the effect of variation among the subjects. Matched pairs are an example of *block designs*.

Block design

A **block** is a group of experimental subjects that are known before the experiment to be similar in some way that is expected to affect the response to the treatments. In a **block design,** the random assignment of subjects to treatments is carried out separately within each block.

A block design combines the idea of creating equivalent treatment groups by matching with the principle of forming treatment groups at random. Blocks are another form of *control*. They control the effects of some outside variables by bringing those variables into the experiment to form the blocks. Here are some typical examples of block designs.

EXAMPLE 10 Men, women, and advertising

Women and men respond differently to advertising. An experiment to compare the effectiveness of three television commercials for the same product will want to look separately at the reactions of men and women, as well as assess the overall response to the ads.

A completely randomized design considers all subjects, both men and women, as a single pool. The randomization assigns subjects to three

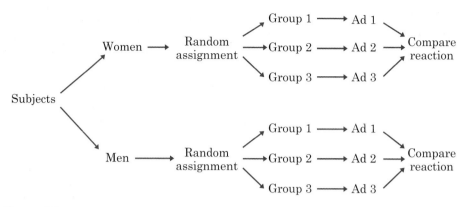

Figure 6.2 A block design to compare the effectiveness of three TV advertisements, for Example 10. Female and male subjects form two blocks.

treatment groups without regard to their sex. This ignores the differences between men and women. A better design considers women and men separately. Randomly assign the women to three groups, one to view each commercial. Then separately assign the men at random to three groups. Figure 6.2 outlines this improved design.

EXAMPLE 11 Comparing welfare systems

A social policy experiment will assess the effect on family income of several proposed new welfare systems and compare them with the present welfare system. Because the future income of a family is strongly related to its present income, the families who agree to participate are divided into blocks of similar income levels. The families in each block are then allocated at random among the welfare systems.

A block is a group of subjects formed before an experiment starts. We reserve the word "treatment" for a condition that we impose on the subjects. We don't speak of 6 treatments in Example 10 even though we can compare the responses of 6 groups of subjects formed by the 2 blocks (men, women) and the 3 commercials. Block designs are similar to stratified samples, which we discussed in Chapter 4. Blocks and strata both group similar individuals together. We use two different names only because the idea developed separately for sampling and experiments. The advantages of block designs are the same as the advantages of stratified samples. Blocks allow us to draw separate conclusions about each block—for example, about men and women in the advertising study in Example 10. Blocking also allows more precise overall conclusions, because the systematic differences between men and women can be removed when we study the overall effects of the 3 commercials. The idea of blocking is an important additional principle of statistical design of experiments. A wise experimenter will form blocks based on the most important unavoidable sources of variability among the experimental subjects. Randomization will then average out the effects of the remaining variation and allow an unbiased comparison of the treatments.

STATISTICAL CONTROVERSIES

Is It or Isn't It a Placebo?

Cordelia Molloy/Photo Researchers

Natural supplements are big business: creatine and amino acid supplements to enhance athletic performance; green tea extract to boost the immune system; yohimbe bark to help your sex life; grapefruit extract and apple cider vinegar to support weight loss; white kidney bean extract to block carbs. Store shelves and Web sites are filled with exotic substances claiming to improve your health.

A therapy that has not been compared with a placebo in a randomized experiment may itself be just a placebo. In the United States, the law requires that new prescription drugs and new medical devices show their safety and effectiveness in randomized trials.

What about those "natural remedies"? The law allows makers of herbs, vitamins, and dietary supplements to claim without any evidence that they are safe and will help "natural conditions." They can't claim to treat "diseases." Of course, the boundary between natural conditions and diseases is vague. Without any evidence whatsoever, we can claim that Dr. Moore's Old Indiana Extract promotes healthy hearts. But without clinical trials and an okay by the Food and Drug Administration (FDA), we can't claim that it reduces the risk of heart disease. No doubt lots of folks will think that "promotes healthy hearts" means the same thing as "reduces the risk of heart disease" when they see our advertisements. We also don't have to worry about what dose of Old Indiana Extract our pills contain or about what dose might actually be toxic.

Should the FDA require natural remedies to meet the same standards as prescription drugs? What does your statistical training tell you about claims not backed up by well-designed experiments? What about the fact that sometimes these natural remedies have real effects? Should that be sufficient for requiring FDA approval on natural remedies?

> **NOW IT'S YOUR TURN**
>
> **6.2 Multiple-choice exams.** A researcher was interested in whether the order of the answers to multiple-choice questions affects exam scores. He made three versions of an exam. Each version had the same questions and the same set of answers, but the order of the possible answers was different for each version. The three versions were given to students in each of two classes having different instructors. Each class had an enrollment of 75 students. The researcher was concerned that scores might also depend on instructor, so instructor (class) was treated as a blocking variable. Use a diagram to outline a block design for this experiment. Use Figure 6.2 (page 124) as a model.

Like the design of samples, the design of complex experiments is a job for experts. Now that we have seen a bit of what is involved, for the remainder of the text we will usually assume that most experiments were completely randomized.

STATISTICS IN SUMMARY

Chapter Specifics

- Because the **placebo effect** is strong, **clinical trials** and other experiments with human subjects should be **double-blind** whenever this is possible.

- The double-blind method helps achieve a basic requirement of comparative experiments: **equal treatment for all subjects** except for the actual treatments the experiment is comparing.

- The most common weakness in experiments is that we can't **generalize** the conclusions widely. Some experiments apply unrealistic treatments, some use subjects from some special group such as college students, and all are performed at some specific place and time. We want to see similar experiments at other places and times confirm important findings.

- Many experiments use designs that are more complex than the basic **completely randomized design,** which divides all the subjects among all the treatments in one randomization. **Matched pairs designs** compare two treatments by giving one to each of a pair of similar subjects or by giving both to the same subject in random order. **Block designs** form blocks of similar subjects and assign treatments at random separately in each block.

- The big ideas of **randomization, control, and adequate numbers of subjects** remain the keys to convincing experiments.

In Chapter 5 we learned that well-designed randomized comparative experiments provide a sound basis for determining if a treatment causes changes in a response. In the real world, simple randomized comparative experiments don't solve all the difficulties of experimenting. The placebo effect and researchers' expectations can introduce biases that undermine our conclusions. Just as samples suffer from nonresponse, experiments suffer from uncooperative subjects. Some subjects refuse to participate; others drop out before the experiment is complete; others don't follow instructions, as when some subjects in a drug trial don't take their pills. More complex designs and techniques, some of which are discussed in this chapter, are used to overcome real-world difficulties. We must pay careful attention to every aspect of an experiment to ensure that the conclusions we make are valid. And when reading about the results of experiments, you should use the ideas provided in this chapter to assess the quality of the conclusions.

CASE STUDY Use what you have learned in this chapter to evaluate the Case Study **EVALUATED** that opened the chapter. Start by reviewing the information on page 113. You can also read the EESEE story "Is Caffeine Dependence Real?" for additional information. Then answer each of the following questions in complete sentences. Be sure to communicate clearly enough for any of your classmates to understand what you are saying.

First, here are the results of the study. The number of subjects who showed withdrawal symptoms during the period in which they took capsules that did not contain caffeine and the magnitude of their symptoms were considered statistically significant.

1. Explain what the phrase "statistically significant" means.

2. Explain why the researchers gave subjects capsules with a fake substance rather than just having them take nothing during one of the periods.

3. What advantage is gained by having subjects take both a capsule with caffeine and a capsule with a fake substance rather than having some of the subjects just take a capsule with caffeine and the remaining subjects just take a capsule with a fake substance? ▪

CHAPTER 6 EXERCISES

For Exercise 6.1, see page 122; for Exercise 6.2, see page 126.

6.3 Magic mushrooms. A *Washington Post* article reported that psilocybin, the active ingredient of "magic mushrooms," promoted a mystical experience in two-thirds of people who took it for the first time, according to a study published in the online journal *Psychopharmacology*. The authors of the article stated that their "double-blind study evaluated the acute and longer-term psychological effects of a high dose of psilocybin relative to a comparison compound administered under comfortable, supportive conditions." Explain to someone who knows no statistics what the term "double-blind" means here.

6.4 Do antidepressants help? A researcher studied the effect of an antidepressant on depression. He randomly assigned subjects with moderate levels of depression to two groups. One group received the antidepressant and the other a placebo. Subjects were blinded with respect to the treatment they received. After four weeks, the researcher interviewed all subjects and rated the change in their symptoms based on the comments of subjects during the interview. Critics said that the results were suspect because the ratings were not blind. Explain what this means and how lack of blindness could bias the reported results.

6.5 Treating acne. An article in a medical journal reports an experiment to see if pulsed laser dye therapy is effective in treating acne. The article describes the experiment as a "randomized, controlled, single-blinded split-face clinical trial of a volunteer sample of 40 patients aged 13 years or older with facial acne conducted at an academic referral center from August 2002 to September 2003." A split-face clinical trial is one in which one side of the face is treated and one side is not. What do you think "single-blinded" means here? Why isn't a double-blind experiment possible?

6.6 Bright bike lights. Will requiring bicyclists to use bright, high-intensity xenon lights mounted on the front and rear of the bike reduce accidents with cars by making bikes more visible?

(a) Briefly discuss the design of an experiment to help answer this question. In particular, what response variables will you examine?

(b) Suppose your experiment demonstrates that using high-intensity xenon lights reduces accidents. What concerns might you have about whether your experimental results will reduce accidents with cars if all bicyclists are required to use such lights? (*Hint:* To help you answer this question, consider the following example. A 1980 report by the Highway Traffic Safety Administration found that adding a center brake light to cars reduced rear-end collisions by as much as 50%. These findings were the result of a randomized comparative experiment. As a result, center brake lights have been required on all cars sold since 1986. Ten years later, the Insurance Institute found only a 5% reduction in rear-end collisions. Apparently, when the study was originally carried out, center brake lights were unusual and caught the eye of following drivers. By 1996, center brake lights were common and no longer captured attention.)

6.7 Red wine and a long and happy life. A *New York Times* article reported that according to a study conducted by researchers at Harvard Medical School and the National Institute of Aging, a substance in red wine could extend life. In the study mice fed a high-fat diet and a large daily dose of resveratrol, a substance found in the skin of grapes and in red wine, were found to live as long as mice on a healthy diet without resveratrol. The article stated that the mice receiving resveratrol "had all the pleasures of gluttony but paid none of the price." The article also reported that the mice were fed doses of resveratrol that would require humans to drink 10 to 20 bottles of red wine a day to get a similar dose. Briefly discuss the questions that arise in using this experiment to decide the benefits of resveratrol for humans.

6.8 Blood-chilling and strokes. A *Science News* article reported a study of the effect of cooling the blood of stroke patients on the extent of recovery ninety days after the stroke. Researchers randomly assigned 58 severe-stroke patients to receive either tPA (the standard treatment for stroke) or tPA plus blood-chilling. Regulators overseeing the study required a one-hour delay from the point at which tPA was given before cooling could be started. The researchers found no significant difference in the effects of the two treatments on recovery. Researchers also noted that the recovery rate for both groups was worse than the average seen in stroke patients nationwide, but were not concerned. Why were they unconcerned?

6.9 Beating sunburn with broccoli. Some recent studies suggest that compounds in broccoli may be helpful in combating the effects of overexposure to ultraviolet radiation. Based on these studies we hope to show that a cream consisting of a broccoli extract reduces sunburn pain. Sixty patients suffering from severe sunburn and needing pain relief are available. We will apply the cream to the sunburn of each patient and ask them an hour later, "About what percent of pain relief did you experience?"

(a) Why should we not simply apply the cream to all 60 patients and record the responses?

(b) Outline the design of an experiment to compare the cream's effectiveness with that of an over-the-counter product for sunburn relief and of a placebo.

(c) Should patients be told which remedy they are receiving? How might this knowledge affect their reactions?

(d) If patients are not told which treatment they are receiving, the experiment is single-blind. Should this experiment be double-blind also? Explain.

6.10 Testing a natural remedy. The statistical controversy presented in this chapter discusses issues surrounding the efficacy of natural remedies. The National Institutes of Health at last began sponsoring proper clinical trials of some natural remedies. In one study at Duke University, 330 patients with mild depression were enrolled in a trial to compare Saint John's wort with a placebo and with Zoloft, a common prescription drug for depression. The Beck Depression Inventory is a common instrument that rates the severity of depression on a 0 to 3 scale.

(a) What would you use as the response variable to measure change in depression after treatment?

(b) Outline the design of a completely randomized clinical trial for this study.

(c) What other precautions would you take in this trial?

6.11 The placebo effect. A survey of physicians found that some doctors give a placebo to a patient who complains of pain for which the physician can find no cause. If the patient's pain improves, these doctors conclude that it had no physical basis. The medical school researchers who conducted the survey claimed that these doctors do not understand the placebo effect. Why?

6.12 The best painkiller for children. A *Washington Post* article reported a study comparing the effectiveness of three common painkillers for children. Three hundred children, aged 6 to 17, were randomly assigned to three groups. Group A received a standard dose of ibuprofen. Group B received a standard dose of acetaminophen. Group C received a standard dose of codeine. The youngsters rated their pain on a 100-point scale before and after taking the medicine.

(a) Outline the design of this experiment. You do not need to do the randomization that your design requires.

(b) You read that "the children and physicians were blinded" during the study. What does this mean?

(c) You also read that there was a significantly greater decrease in pain ratings for Group A than for Groups B and C, but there was no significant difference in the decrease of pain ratings for Groups B and C. What does this mean? What does this finding lead you to conclude about the use of ibuprofen as a painkiller?

6.13 Flu shots. A *New York Times* article reported a study that investigated whether giving flu shots to schoolchildren protects a whole community from the disease. Researchers in Canada recruited 49 remote Hutterite farming colonies in western Canada for the study. In 25 of the colonies, all children aged 3 to 15 received flu shots in late 2008; in the 24 other colonies, they received a placebo. Which colonies received flu shots and which received the placebo was determined by randomization, and the colonies did not know whether they received the flu shots or the placebo. The researchers recorded the percentage of all children and adults in each colony who had laboratory-confirmed flu over the ensuing winter and spring.

(a) Outline the design of this experiment. You do not need to do the randomization that your design requires.

(b) You read that the placebo was actually the hepatitis A vaccine and that "hepatitis was not studied, but to keep the investigators from knowing which colonies received flu vaccine, they had to offer placebo shots, and hepatitis shots do some good while sterile water injections do not." In addition, the article mentions that the colonies were studied "without the investigators being subconsciously biased by knowing which received the placebo." Why was it important that investigators not be subconsciously biased by knowing which received the placebo?

(c) You also read that by June 2009, more than 10% of all the adults and children in colonies that received the placebo had had laboratory-confirmed seasonal flu. Less than 5% of those in the colonies that received flu shots had. This difference was statistically significant. Explain to someone who knows no statistics what "statistically significant" means in this context.

6.14 Comparing corn varieties. New varieties of corn with altered amino acid content may have higher nutritional value than standard corn, which is low in the amino acid lysine. An experiment compares two new varieties, called opaque-2 and floury-2, with normal corn. The researchers mix corn-soybean meal diets using each type of corn at each of three protein levels: 12% protein, 16% protein, and 20% protein. They feed each diet to 10 one-day-old male chicks and record their weight gains after 21 days. The weight gain of the chicks is a measure of the nutritional value of their diet.

(a) What are the individuals and the response variable in this experiment?

(b) How many explanatory variables are there? How many treatments? Use a diagram like Figure 6.1 (page 121) to describe the treatments. How many experimental individuals does the experiment require?

(c) Use a diagram to describe a completely randomized design for this experiment. (Don't actually do the randomization.)

6.15 Price change and fairness. A marketing researcher wishes to study what factors affect the perceived fairness of a change in the price of an item from its advertised price. In particular, does the type of change in price (an increase or decrease) and the source of the information about the change affect the perceived fairness? In an experiment, 20 subjects interested in purchasing a new rug are recruited. They are told that the price of a rug in a certain store was advertised at $500. Subjects are sent, one at a time, to the store, where they learn that the price has changed. Five subjects are told by a store clerk that the price has increased to $550. Five subjects learn that the price has increased to $550 from the price tag on the rug. Five subjects are told by a store clerk that the price has decreased to $450. Five subjects learn that the price has decreased to $450 from the price tag on the rug. After learning about the change in price, each subject is asked to rate the fairness of the change on a 10-point scale with 1 = "very unfair" to 10 = "very fair."

(a) What are the explanatory variables and the response variables for this experiment?

(b) Make a diagram like Figure 6.1 (page 121) to describe the treatments. How many treatments are there?

(c) Explain why it is a bad idea to have the first five subjects learn from a store clerk that the price has increased to $550, the next five learn that the price has increased to \$550 from the price tag on the rug, and so on. Instead, the order in which subjects are sent to the store and which scenario they will encounter (type of change and source of information about the change) are determined randomly.

6.16 Comparing hand strength. Is the right hand generally stronger than the left in right-handed people? You can crudely measure hand strength by placing a bathroom scale on a shelf with the end protruding, then squeezing the scale between the thumb below and the four fingers above. The reading of the scale shows the force exerted. Describe the design of a matched pairs experiment to compare the strength of the right and left hands, using 10 right-handed people as subjects. (You need not actually do the randomization.)

6.17 Does charting help investors? Some investors believe that charts of past trends in the prices of securities can help predict future prices. Most economists disagree. In an experiment to examine the effects of using charts, business students trade (hypothetically) a foreign currency at computer screens. There are 20 student subjects available, named for convenience A, B, C, ..., T. Their goal is to make as much money as possible, and the best performances are rewarded with small prizes. The student traders have the price history of the foreign currency in dollars in their computers. They may or may not also have software that highlights trends. Describe two designs for this experiment: a completely randomized design and a matched pairs design in which each student serves as his or her own control. In both cases, carry out the randomization required by the design.

6.18 Wine tasting. A researcher is interested in whether people who enjoy fine wine can actually tell the difference between a highly rated expensive wine and a good inexpensive wine. She selects several bottles of a highly rated cabernet sauvignon (a red wine) costing over \$100 dollars per bottle and several bottles of a good cabernet sauvignon costing \$12. Subjects are volunteers who identify themselves as enjoying fine wine. Subjects will taste a glass of wine without knowing whether it is the expensive or inexpensive wine. This is the treatment. They will then be asked to estimate how much the wine cost. This is the response variable.

(a) Outline a completely randomized design to compare the cost estimates for the two treatments. Twenty subjects are available.

(b) Because individuals may differ greatly in their tasting ability, the wide variation in individual skills may hide the systematic effect of the differences in the wines unless there are many subjects in each group. Describe in detail the design of a matched pairs experiment in which each subject serves as his or her own control.

6.19 Comparing cancer treatments. The progress of a type of cancer differs in women and men. A clinical experiment to compare 4 therapies for this cancer therefore treats sex as a blocking variable.

(a) You have 500 male and 300 female patients who are willing to serve as subjects. Use a diagram to outline a block design for this experiment. Figure 6.2 (page 124) is a model.

(b) What are the advantages of a block design over a completely randomized design using these 800 subjects? What are the advantages of a block design over a completely randomized design using 800 male subjects?

6.20 Comparing weight-loss treatments. Twenty overweight females have agreed to participate in a study of the effectiveness of 4 weight-loss treatments: A, B, C, and D. The researcher first calculates how overweight each subject is by comparing the subject's actual weight with her "ideal" weight. The subjects and their excess weights in pounds are

Alexander	21	Franken	25	Murray	34	Schumer	42
Barrasso	34	Hatch	33	Nelson	28	Specter	33
Bayh	30	Kerry	28	Pryor	30	Tester	35
Collins	25	Leahy	32	Reed	30	Webb	29
Dodd	24	McCain	39	Sanders	27	Wyden	35

The response variable is the weight lost after 8 weeks of treatment. Because a subject's excess weight will influence the response, a block design is appropriate.

(a) Arrange the subjects in order of increasing excess weight. Form 5 blocks of 4 subjects each by grouping the 4 least overweight, then the next 4, and so on.

(b) Use Table A (or statistical software) to randomly assign the 4 subjects in each block to the 4 weight-loss treatments. Be sure to explain exactly how you used the table.

6.21 In the corn field. An agronomist (a specialist in crop production and soil chemistry) wants to compare the yield of 4 corn varieties. The field in which the experiment will be carried out increases in fertility from north to south. The agronomist therefore divides the field into 20 plots of equal size, arranged in 5 east-west rows of 4 plots each, and employs a block design with the rows of plots as the blocks.

(a) Draw a sketch of the field, divided into 20 plots. Label the rows Block 1 to Block 5.

(b) Do the randomization required by the block design. That is, randomly assign the 4 corn varieties A, B, C, and D to the 4 plots in each block. Mark on your sketch which variety is planted in each plot.

6.22 Speeding the mail? Is the number of days a letter takes to reach another city affected by the time of day it is mailed and whether the standard 5-digit zip code or the more informative 9-digit zip code is used? Describe briefly the design of an experiment with two explanatory variables to investigate this question. Be sure to specify the treatments exactly and to tell how you will handle lurking variables such as the day of the week on which the letter is mailed.

6.23 Burger King versus McDonald's. Do consumers prefer the taste of a double cheeseburger from Burger King or from McDonald's in a blind test in which neither cheeseburger is identified? Describe briefly the design of a matched pairs experiment to investigate this question.

6.24 What do you want to know? The previous two exercises illustrate the use of statistically designed experiments to answer questions that arise in everyday life. Select a question of interest to you that an experiment might answer and briefly discuss the design of an appropriate experiment.

6.25 Doctors and nurses. Nurse-practitioners are nurses with advanced qualifications who often act much like primary-care physicians. An experiment assigned 1316 patients who had no regular source of medical care to either a doctor (510 patients) or a nurse-practitioner (806 patients). All the patients had been diagnosed with either asthma, diabetes, or high blood pressure before being assigned. The response variables included measures of the patients' health and of their satisfaction with their medical care after 6 months.

(a) Is the diagnosis (asthma, etc.) a treatment variable or a block? Why?

(b) Is the type of care (nurse or doctor) a treatment variable or a block? Why?

EXPLORING THE WEB

6.26 Web-based exercise. Not all researchers agree that placebos have a powerful clinical effect. Go to the *New England Journal of Medicine* Web site (**http://content.nejm.org**) and find the article "Is the Placebo Powerless? An Analysis of Clinical Trials Comparing Placebo with No Treatment" by Hrobjartsson and Gotzsche in the May 24, 2001, issue. What conclusions do the authors reach? Also, in the same issue, read the editorial "The Powerful Placebo and the Wizard of Oz" by Bailar. Does Bailar agree with the findings of Hrobjartsson and Gotzsche?

6.27 Web-based exercise. You can find a (very) skeptical look at health claims not backed by proper clinical trials at the "Quackwatch" site, **www.quackwatch.org**. Visit the site and prepare a one-paragraph summary of one of the articles.

NOTES AND DATA SOURCES

Page 114 Example 1: For the study on rats, see E. Street and M. B. Carroll, "Preliminary evaluation of a new food product," in J. M. Tanur et al. (eds.), *Statistics: A Guide to the Unknown,* 3rd edition, Wadsworth, 1989, pp. 161–169.

Page 115 Example 2: The placebo effect examples are from Sandra Blakeslee, "Placebos prove so powerful even experts are surprised," *New York Times,* October 13, 1998. The "three-quarters" estimate is cited by Martin Enserink, "Can the placebo be the cure?," *Science*, 284 (1999), pp. 238–240. An extended treatment is Anne Harrington (ed.), *The Placebo Effect: An Interdisciplinary Exploration*, Harvard University Press, 1997.

Page 116 The flu trial quotation is from Kristin L. Nichol et al., "Effectiveness of live, attenuated intranasal influenza virus vaccine in healthy, working adults," *Journal of the American Medical Association*, 282 (1999), pp. 137–144.

Page 116 Example 3: "Cancer clinical trials: barriers to African American participation," *Closing the Gap*, newsletter of the Office of Minority Health, December 1997–January 1998.

Page 117 Example 4: Michael H. Davidson et al., "Weight control and risk factor reduction in obese subjects treated for 2 years with orlistat: a randomized controlled trial," *Journal of the American Medical Association*, 281 (1999), pp. 235–242.

Page 119 Example 6: Details of the Carolina Abecedarian Project, including references to published work, can be found online at **www.fpg.unc.edu/~abc/**.

Page 121 Example 8: This is a simpler version of an experiment described in Brian Wansik and Pierre Chandon, "Can 'low-fat' nutrition labels lead to obesity?," *Journal of Marketing Research*, 43 (November 2006), pp. 605–617.

Page 122 Example 9: "Advertising: the cola war," *Newsweek*, August 30, 1976, p. 67.

Page 127 Exercise 6.3: David Brown, "Drug's mystical properties confirmed,"

Washington Post, July 11, 2006. Look online at **www.springer.com/biomed/pharmaceutical+science/journal/213** in the journal *Psychopharmacology*, 187, No. 3 (August 2006), pp. 263–268, for the details.

Page 128 Exercise 6.5: Jeffrey S. Orringer et al., "Treatment of acne vulgaris with a pulsed laser dye," *Journal of the American Medical Association*, 291 (2004), pp. 2834–2839.

Page 128 Exercise 6.7: Nicholas Wade, "Substance in red wine could extend life, study says," *New York Times*, November 1, 2006.

Page 129 Exercise 6.8: Nathan Seppa, "Blood-chilling device could save stroke victims from brain damage," *Science News*, February 26, 2010.

Page 129 Exercise 6.10: Hypericum Depression Trial Study Group, "Effect of *Hypericum perforatum* (St. John's wort) in major depressive disorder: a randomized, controlled trial," *Journal of the American Medical Association*, 287 (2002), pp. 1807–1814.

Page 130 Exercise 6.12: Lindsey Tanner, "Study says ibuprofen is best painkiller for children," *Washington Post*, March 5, 2007, p. A09.

Page 130 Exercise 6.13: Donald G. McNeil, Jr., "Flu shots in children can help community," *New York Times*, March 10, 2010, p. A16.

Page 134 Exercise 6.25: Mary O. Mundinger et al., "Primary care outcomes in patients treated by nurse practitioners or physicians," *Journal of the American Medical Association*, 238 (2000), pp. 59–68.

Data Ethics

CASE STUDY Does alcohol increase our perception of the attractiveness of members of the opposite sex? Researchers at the University of Bristol in England attempted to answer this question. Forty-two male and forty-two female students at the University of Bristol were recruited to take part in a study. Students were randomly assigned to receive either a strong alcoholic drink (vodka, tonic water, and lime cordial) or a placebo (tonic water and lime cordial). They were given 15 minutes to consume their drink, after which they were asked to rate the facial attractiveness of 20 male and female faces. The researchers compared the ratings of those receiving the alcoholic drink with those receiving the placebo.

Many students would like to be participants in such an experiment. But is having subjects consume a strong alcoholic drink so that their judgment is impaired ethical? By the end of this chapter you will have learned the principles that will help you answer this question. ▪

First principles

The production and use of data, like all human endeavors, raise ethical questions. We won't discuss the telemarketer who begins a telephone sales pitch with "I'm conducting a survey," when the goal is to sell you something rather than collect useful information. Such deception is clearly unethical. It enrages legitimate survey organizations, which find the public less willing to talk with them. Neither will we discuss those few researchers who, in the pursuit of professional advancement, publish fake data. There is no ethical question here—faking data to advance your career is just wrong. It will end your career when uncovered. But just how honest must researchers be about real, unfaked data? Here is an example that suggests the answer is "More honest than they often are."

EXAMPLE 1 Missing details

Papers reporting scientific research are supposed to be short, with no extra baggage. Brevity can allow the researchers to avoid complete honesty about their data. Did they choose their subjects in a biased way? Did they report data on only some of their subjects? Did they try several statistical analyses and report only the ones that supported what the researchers hoped to find? The statistician John Bailar screened more than 4000 medical papers in more than a decade as consultant to the *New England Journal of Medicine.* He says, "When it came to the statistical review, it was often clear that critical information was lacking, and the gaps nearly always had the practical effect of making the authors' conclusions look stronger than they should have." The situation is no doubt worse in fields that screen published work less carefully.

"That's the gist of what I want to say. Now get me some statistics to base it on."

The most complex issues of data ethics arise when we collect data from people. The ethical difficulties are more severe for experiments that impose some treatment on people than for sample surveys that simply gather information. Trials of new medical treatments, for example, can do harm as well as good to their subjects. Here are some basic standards of data ethics that must be obeyed by any study that gathers data from human subjects, whether sample survey or experiment.

Basic data ethics

The organization that carries out the study must have an **institutional review board** that reviews all planned studies in advance in order to protect the subjects from possible harm.

All individuals who are subjects in a study must give their **informed consent** before data are collected.

All individual data must be kept **confidential.** Only statistical summaries for groups of subjects may be made public.

The law requires that studies funded by the federal government obey these principles. But neither the law nor the consensus of experts is completely clear about the details of their application.

Institutional review boards

The purpose of an institutional review board (often abbreviated IRB) is not to decide whether a proposed study will produce valuable information or whether it is statistically sound. The board's purpose is, in the words of one university's board, "to protect the rights and welfare of human subjects (including patients) recruited to participate in research activities." The board reviews the plan of the study and can require changes. It reviews the consent form to ensure that subjects are informed about the nature of the study and about any potential risks. Once research begins, the board monitors its progress at least once a year.

The most pressing issue concerning institutional review boards is whether their workload has become so large that their effectiveness in protecting subjects drops. When the government temporarily stopped human-subject research at Duke University Medical Center in 1999 due to inadequate protection of subjects, more than 2000 studies at Duke were in progress. That's a lot of review work. There are shorter review procedures for projects that involve only minimal risks to subjects, such as most sample surveys. When a board is overloaded, there is a temptation to put more proposals in the minimal-risk category to speed the work.

Informed consent

Both words in the phrase "informed consent" are important, and both can be controversial. Subjects must be *informed* in advance about the nature of a study and any risk of harm it may bring. In the case of a sample survey, physical harm is not possible. The subjects should be told what kinds of questions the survey will ask and about how much of their time it will take. Experimenters must tell subjects the nature and purpose of the study and outline possible risks. Subjects must then *consent,* usually in writing.

EXAMPLE 2 Who can consent?

Are there some subjects who can't give informed consent? It was once common, for example, to test new vaccines on prison inmates who gave their consent in return for good-behavior credit. Now we worry that prisoners are not really free to refuse, and the law forbids medical experiments in prisons.

Children can't give fully informed consent, so the usual procedure is to ask their parents. A study of new ways to teach reading is about to start at a local elementary school, so the study team sends consent forms home to parents. Many parents don't return the forms. Can their children take part in the study because the parents did not say "No," or should we allow only children whose parents returned the form and said "Yes"?

What about research into new medical treatments for people with mental disorders? What about studies of new ways to help emergency room patients who may be unconscious or have suffered a stroke? In most cases, there is no time even to get the consent of the family. Does the principle of informed consent bar realistic trials of new treatments for unconscious patients?

These are questions without clear answers. Reasonable people differ strongly on all of them. There is nothing simple about informed consent.

NOW IT'S YOUR TURN   **7.1 Informed consent?** A 72-year-old man with multiple sclerosis is hospitalized. His doctor feels he may need to be placed on a feeding tube soon to ensure adequate nourishment. He asks the patient about this in the morning and the patient agrees. However, in the evening (before the tube has been placed), the patient becomes disoriented and seems confused about his decision to have the feeding tube placed. He tells the doctor he doesn't want it in. The doctor revisits the question in the morning, when the patient is again lucid. Unable to recall his state of mind from the previous evening, the patient again agrees to the procedure. Do you believe the patient has given informed consent to the procedure?

The difficulties of informed consent do not vanish even for capable subjects. Some researchers, especially in medical trials, regard consent as a barrier to getting patients to participate in research. They may not explain all possible risks; they may not point out that there are other therapies that might be better than those being studied; they may be too optimistic when talking with patients even when the consent form has all the right details. On the other hand, mentioning every possible risk leads to very long consent forms that really are barriers. "They are like rental car contracts," one lawyer said. Some subjects don't read forms that run five or six printed pages. Others are frightened by the large number of possible (but unlikely) disasters that might happen and so refuse to participate. Of course, unlikely disasters sometimes happen. When they do, lawsuits follow and the consent forms become yet longer and more detailed.

Confidentiality

Ethical problems do not disappear once a study has been cleared by the review board, has obtained consent from its subjects, and has actually collected data about the subjects. It is important to protect the subjects' privacy by keeping all data about individuals confidential. The report of an opinion poll may say what percentage of the 1500 respondents felt that legal immigration should be reduced. It may not report what *you* said about this or any other issue.

Confidentiality is not the same as **anonymity.** Anonymity means that subjects are anonymous—their names are not known even to the director of the study. It is not possible to determine which subject produced which data. Anonymity is rare in statistical studies. Even where anonymity is possible (mainly in surveys conducted by mail), it prevents any follow-up to improve nonresponse or inform subjects of results.

Any breach of confidentiality is a serious violation of data ethics. The best practice is to separate the identity of the subjects from the rest of the data at once. Sample surveys, for example, use the identification only to check on who did

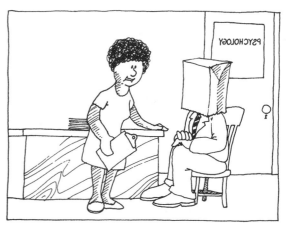

"I realize the participants in this study are to be anonymous, but you're going to have to expose your eyes."

or did not respond. In an era of advanced technology, however, it is no longer enough to be sure that each individual set of data protects people's privacy. The government, for example, maintains a vast amount of information about citizens in many separate databases—census responses, tax returns, Social Security information, data from surveys such as the Current Population Survey, and so on. Many of these databases can be searched by computers for statistical studies. A clever computer search of several databases might be able, by combining information, to identify you and learn a great deal about you even if your name and other identification have been removed from the data available for search. Privacy and confidentiality of data are hot issues among statisticians in the computer age.

EXAMPLE 3 Use of government databases

Citizens are required to give information to the government. Think of tax returns and Social Security contributions. The government needs these data for administrative purposes—to see if we paid the right amount of tax and how large a Social Security benefit we are owed when we retire. Some people feel that individuals should be able to forbid any other use of their data, even with all identification removed. This would prevent using government records to study, say, the ages, incomes, and household sizes of Social Security recipients. Such a study could well be vital to debates on reforming Social Security.

NOW IT'S YOUR TURN

7.2 Anonymous or confidential? A Web site for a medical clinic offers information about HIV testing that they provide. The Web site explains that appointments are not necessary—you can just drop in. When you arrive, you are given a number code that will match you to your test results. A blood sample is taken by pricking your finger. You must return, in person, in two weeks to obtain results. Does this practice offer anonymity or confidentiality?

Statisticians, honest and dishonest Developed nations rely on government statisticians to produce honest data. We trust the monthly unemployment rate, for example, to guide both public and private decisions. Honesty can't be taken for granted, however. In 1998, the Russian government arrested the top statisticians in the State Committee for Statistics. They were accused of taking bribes to fudge data to help companies avoid taxes. "It means that we know nothing about the performance of Russian companies," said one newspaper editor.

Clinical trials

Clinical trials are experiments that study the effectiveness of medical treatments on actual patients. Medical treatments can harm as well as heal, so clinical trials spotlight the ethical problems of experiments with human subjects. Here are the starting points for a discussion:

- Randomized comparative experiments are the only way to see the true effects of new treatments. Without them, risky treatments that are no better than placebos will become common.

- Clinical trials produce great benefits, but most of these benefits go to future patients. The trials also pose risks, and these risks are borne by the subjects of the trial. So we must balance future benefits against present risks.

- Both medical ethics and international human rights standards say that "the interests of the subject must always prevail over the interests of science and society."

The quoted words are from the 1964 Helsinki Declaration of the World Medical Association, the most respected international standard. The most outrageous examples of unethical experiments are those that ignore the interests of the subjects.

EXAMPLE 4 The Tuskegee syphilis study

In the 1930s, syphilis was common among black men in the rural South, a group that had almost no access to medical care. The Public Health Service recruited 399 poor black sharecroppers with syphilis and 201 others without the disease in order to observe how syphilis progressed when no treatment was given. Beginning in 1943, penicillin became available to treat syphilis. However, the study subjects were not treated, even after penicillin became a standard treatment for syphilis. In fact, the Public Health Service tried to prevent any treatment until word leaked out and forced an end to the study in 1972.

The Tuskegee study is an extreme example of investigators following their own interests and ignoring the well-being of their subjects. A 1996 review said, "It has come to symbolize racism in medicine, ethical misconduct in human research, paternalism by physicians, and government abuse of vulnerable people." In 1997, President Clinton formally apologized to the surviving participants in a White House ceremony.

The Tuskegee study helps explain the lack of trust that lies behind the reluctance of many blacks to take part in clinical trials.

Because "the interests of the subject must always prevail," medical treatments can be tested in clinical trials only when there is reason to hope that they will help the patients who are subjects in the trials. Future benefits alone aren't enough to justify any experiment with human subjects. Of course, if there is already strong evidence that a treatment works and is safe, it is unethical *not* to give it. Here are the words of Dr. Charles Hennekens of the Harvard Medical School, who directed the large clinical trial that showed that aspirin reduces the risk of heart attacks in men:

There's a delicate balance between when to do or not do a randomized trial. On the one hand, there must be sufficient belief in the agent's potential to justify exposing half the subjects to it. On the other hand, there must be sufficient doubt about its efficacy to justify withholding it from the other half of subjects who might be assigned to placebos.

Why is it ethical to give a control group of patients a placebo? Well, we know that placebos often work. Patients on placebos often show real improvement. What is more, placebos have no harmful side effects. So in the state of balanced doubt described by Dr. Hennekens, the placebo group may be getting a better treatment than the drug group. If we *knew* which treatment was better, we would give it to everyone. When we don't know, it is ethical to try both and compare them. Here are some harder questions about placebos, with arguments on both sides.

STATISTICAL CONTROVERSIES

Hope for Sale?

We have pointed to the ethical problems of experiments with human subjects, clinical trials in particular. *Not* doing proper experiments can also pose problems. Here is an example. Women with advanced breast cancer will eventually die. A promising but untried treatment appears.

The promising treatment is "bone marrow transplant" (BMT for short). The idea of BMT is to harvest a patient's bone marrow cells, blast the cancer with very high doses of drugs, then return the harvested cells to keep the drugs from killing the patient. BMT has become popular, but it is painful, expensive, and dangerous.

New anticancer drugs are first available through clinical trials, but there is no constraint on therapies such as BMT. When small, uncontrolled trials seemed to show success, BMT became widely available. The economics of medicine had a lot to do with this. The early leaders in offering BMT were for-profit hospitals that advertise heavily to attract patients. Others soon jumped in. The *New York Times* reported: "Every entity offering the experimental procedure tried a different sales pitch. Some promoted the prestige of their institutions, others the convenience of their locations, others their caring attitudes and patient support." The profits for hospitals and doctors are high.

Should we have waited for controlled clinical trials to show that the treatment works, or was it right to make it available immediately? What do you think? What are some of the issues one should consider?

EXAMPLE 5 Placebo controls?

You are testing a new drug. Is it ethical to give a placebo to a control group if an effective drug already exists?

Yes: The placebo gives a true baseline for the effectiveness of the new drug. There are three groups: new drug, best existing drug, and placebo. Every clinical trial is a bit different, and not even genuinely effective treatments work in every setting. The placebo control helps us see if the study is flawed so that even the best existing drug does not beat the placebo. Sometimes the placebo wins, so the doubt about the efficacy of the new and the existing drugs is justified. Placebo controls are ethical except for life-threatening conditions.

No: It isn't ethical to deliberately give patients an inferior treatment. We don't know whether the new drug is better than the existing drug, so it is ethical to give both in order to find out. If past trials showed that the existing drug is better than a placebo, it is no longer right to give patients a placebo. After all, the existing drug includes the placebo effect. A placebo group is ethical only if the existing drug is an older one that did not undergo proper clinical trials or doesn't work well or is dangerous.

EXAMPLE 6 Sham surgery

"Randomized, double-blind, placebo-controlled trials are the gold standard for evaluating new interventions and are routinely used to assess new medical therapies." So says an article in the *New England Journal of Medicine* that discusses the treatment of Parkinson's disease. The article isn't about the new treatment, which offers hope of reducing the tremors and lack of control brought on by the disease, but about the ethics of studying the treatment.

The law requires well-designed experiments to show that new drugs work and are safe. Not so with surgery—only about 7% of studies of surgery use randomized comparisons. Surgeons think their operations succeed, but innovators always think their innovations work. Even if the patients are helped, the placebo effect may deserve most of the credit. So we don't really know whether many common surgeries are worth the risk they carry. To find out, do a proper experiment. That includes a "sham surgery" to serve as a placebo. In the case of Parkinson's disease, the promising treatment involves surgery to implant new cells. The placebo subjects get the same surgery, but the cells are not implanted.

Placebos work. Patients on placebos often show improvement and their inclusion produces a better experiment. As more doctors recognize this fact, more begin to ask, "If we accept a placebo in drug trials, why don't we accept it in surgery trials?" This is a very controversial question. Here are two arguments about whether placebos should be used in surgery trials.

Yes: Most surgeries have not been tested in comparative experiments, and some are no doubt just placebos. Unlike placebo pills, these surgeries carry risks. Comparing real surgeries to placebo surgeries can eliminate thousands of unnecessary operations and save many lives. The placebo surgery can be made quite safe. For example, placebo patients can be given a safe drug that removes their memory of the operation rather than a more risky anesthetic required for the more serious real surgery. Subjects are told that they are in a placebo-controlled trial and they agree to take part. Placebo-controlled trials of surgery are ethical (except for life-threatening conditions) if the risk to the placebo group is small and there is informed consent.

No: Placebo surgery, unlike placebo drugs, always carries some risk, such as postoperative infection. Remember that "the interests of the subject must always prevail." Even great future benefits can't justify risks to subjects today unless those subjects receive some benefit. We might give a patient a placebo drug as a medical therapy, because placebos work and are not risky. No doctor would do a sham surgery as ordinary therapy, because there is some risk. If we would not use it in medical practice, it isn't ethical to use it in a clinical trial.

"I'm doing a little study on the effects of emotional stress. Now, just take the axe from my assistant."

Behavioral and social science experiments

When we move from medicine to the behavioral and social sciences, the direct risks to experimental subjects are less acute, but so are the possible benefits to the subjects. Consider, for example, the experiments conducted by psychologists in their study of human behavior.

EXAMPLE 7 Keep out of my space

Psychologists observe that people have a "personal space" and get annoyed if others come too close to them. We don't like strangers to sit at our table in a coffee shop if other tables are available, and we see people move apart in elevators if there is room to do so. Americans tend to require more personal space than people in most other cultures. Can violations of personal space have physical, as well as emotional, effects?

Investigators set up shop in a men's public restroom. They blocked off urinals to force men walking in to use either a urinal next to an experimenter (treatment group) or a urinal separated from the experimenter (control group). Another experimenter, using a periscope from a toilet stall, measured how long the subject took to start urinating and how long he kept at it.

This personal space experiment illustrates the difficulties facing those who plan and review behavioral studies.

- There is no risk of harm to the subjects, although they would certainly object to being watched through a periscope. What should we protect subjects from when physical harm is unlikely? Possible emotional harm? Undignified situations? Invasion of privacy?

- What about informed consent? The subjects in Example 7 did not even know they were participating in an experiment. Many behavioral experiments rely on hiding the true purpose of the study. The subjects would change their behavior if told in advance what the investigators were looking for. Subjects are asked to consent on the basis of vague information. They receive full information only after the experiment.

The "Ethical Principles" of the American Psychological Association require consent unless a study merely observes behavior in a public place. They allow deception only when it is necessary to the study, does not hide information that might influence a subject's willingness to participate, and is explained to subjects as soon as possible. The personal space study of Example 7 (from the 1970s) does not meet current ethical standards.

We see that the basic requirement for informed consent is understood differently in medicine and psychology. Here is an example of another setting with yet another interpretation of what is ethical. The subjects get no information and give no consent. They don't even know that an experiment may be sending them to jail for the night.

EXAMPLE 8 Domestic violence

How should police respond to domestic violence calls? In the past, the usual practice was to remove the offender and order him to stay out of the household overnight. Police were reluctant to make arrests because the victims rarely pressed charges. Women's groups argued that arresting offenders would help prevent future violence even if no charges were filed. Is there evidence that arrest will reduce future offenses? That's a question that experiments have tried to answer.

A typical domestic violence experiment compares two treatments: arrest the suspect and hold him overnight or warn the suspect and release him. When police officers reach the scene of a domestic violence call, they calm the participants and investigate. Weapons or death threats require an arrest. If the facts permit an arrest but do not require it, an officer radios headquarters for instructions. The person on duty opens the next envelope in a file prepared in advance by a statistician. The envelopes contain the treatments in random order. The police either arrest the suspect or warn and release him, depending on the contents of the envelope. The researchers then monitor police records and visit the victim to see if the domestic violence reoccurs.

    The first such experiment appeared to show that arresting domestic violence suspects does reduce their future violent behavior. As a result of this evidence, arrest has become the common police response to domestic violence.

    The domestic violence experiments shed light on an important issue of public policy. Because there is no informed consent, the ethical rules that govern clinical trials and most social science studies would forbid these experiments. They were cleared by review boards because, in the words of one domestic violence researcher, "These people became subjects by committing acts that allow the police to arrest them. You don't need consent to arrest someone."

STATISTICS IN SUMMARY

Chapter Specifics

- Data ethics begin with some principles that go beyond just being honest. Studies with human subjects must be screened in advance by an **institutional review board.**

- All subjects must give their **informed consent** before taking part.

- All information about individual subjects must be kept **confidential.**

 The production and use of data to make decisions, like all human endeavors, raise ethical questions. In real-world applications of statistics, these must be addressed as part of the process of reasoning from data to a conclusion. The principles discussed in this chapter are a good start in addressing these questions, but many ethical debates remain, especially in the area of experiments with humans. Many of the debates concern the right balance between the welfare of the subjects and the future benefits of the experiment. Remember that randomized comparative experiments can answer questions that can't be answered without them. Also remember that "the interests of the subject must always prevail over the interests of science and society."

CASE STUDY Use what you have learned in this chapter to evaluate the Case Study
EVALUATED that opened the chapter. In particular, do the following:

1. Based on the principles discussed in this chapter, would you consider the experiment to be ethical? Explain.

2. Federal regulations say that "minimal risk" means that the risks are no greater than "those ordinarily encountered in daily life or during the performance of routine physical or psychological examinations or tests." Do you think this study qualifies as "minimal risk"? Explain. ▪

CHAPTER 7 EXERCISES

Most of the exercises in this chapter pose issues for discussion. There are no right or wrong answers, but there are more and less thoughtful answers.

For Exercise 7.1, see page 140; for Exercise 7.2, see page 142.

7.3 Minimal risk? You are a member of your college's institutional review board. You must decide whether several research proposals qualify for less rigorous review because they involve only minimal risk to subjects. Federal regulations say that "minimal risk" means the risks are no greater than "those ordinarily encountered in daily life or during the performance of routine physical or psychological examinations or tests." That's vague. Which of these do you think qualifies as "minimal risk"? Explain your reasoning.

(a) Take hair and nail clippings in a nondisfiguring manner. Yes

(b) Draw a drop of blood by pricking a finger in order to measure blood sugar. No *Diabetics okay*

(c) Draw blood from the arm for a full set of blood tests.

(d) Insert a tube that remains in the arm so that blood can be drawn regularly.

(e) Take extra specimens from a subject who is undergoing an invasive clinical procedure such as a bronchoscopy (a procedure in which a physician views the inside of the airways for diagnostic and therapeutic purposes using an instrument that is inserted into the airways, usually through the nose or mouth).

7.4 Who serves on the review board? Government regulations require that institutional review boards consist of at least five people, including at least one scientist, one nonscientist, and one person from outside the institution. Most boards are larger, but many contain just one outsider.
(a) Why should review boards contain people who are not scientists?
(b) Do you think that one outside member is enough? How would you choose that member? (For example, would you prefer a medical doctor? A member of the clergy? An activist for patients' rights?)

7.5 Institutional review boards. If your college or university has an institutional review board that screens all studies that use human subjects, get a copy of the document that describes this board (you can probably find it online). At larger institutions you may find multiple institutional review boards—for example, separate boards for medical studies and for studies in the social sciences.

(a) According to this document, what are the duties of the board?

(b) How are members of the board chosen? How many members are not scientists? How many members are not employees of the college? Do these members have some special expertise, or are they simply members of the "general public"?

7.6 Informed consent. A researcher suspects that people with ultraliberal political beliefs tend to be more prone to depression. She prepares a questionnaire that measures depression and that also asks many political questions. Write a description of the purpose of this research to be read by subjects in order to obtain their informed consent. You must balance the conflicting goals of not deceiving the subjects as to what the questionnaire will tell about them and of not biasing the sample by scaring off people with ultraliberal political views.

7.7 Is consent needed? In which of the circumstances below would you allow collecting personal information without the subjects' consent? Why?

(a) A government agency takes a random sample of income tax returns to obtain information on the marital status and average income of people who identify themselves as clergy. Only the marital status and income are recorded from the returns, not the names.

(b) A social psychologist attends public meetings of a religious group to study the behavior patterns of members.

(c) A social psychologist pretends to be converted to membership in a religious group and attends private meetings to study the behavior patterns of members.

7.8 Students as subjects. Students taking Psychology 001 are required to serve as experimental subjects. Students in Psychology 002 are not required to serve, but they are given extra credit if they do so. Students in Psychology 003 are required either to sign up as subjects or to write a term paper. Serving as an experimental subject may be educational, but current ethical standards frown on using "dependent subjects" such as prisoners or charity medical patients. Students are certainly somewhat dependent on their teachers. Do you object to any of these course policies? If so, which ones, and why?

7.9 How common is HIV infection? Researchers from Yale University, working with medical teams in Tanzania, wanted to know how common infection with the AIDS virus is among pregnant women in that African country. To do this, they planned to test blood samples drawn from pregnant women.

Yale's institutional review board insisted that the researchers get the informed consent of each woman and tell her the results of the test. This is the usual procedure in developed nations. The Tanzanian government did not want to tell the women why blood was drawn or tell them the test results. The government feared panic if many people turned out to have an incurable disease for which the country's medical system could not provide care. The study was canceled. Do you think that Yale was right to apply its usual standards for protecting subjects? Explain your answer.

7.10 Anonymous or confidential? One of the most important nongovernment surveys in the United States is the General Social Survey (see Example 7 in Chapter 1, page 10). The GSS regularly monitors public opinion on a wide variety of political and social issues. Interviews are conducted in person in the subject's home. Are a subject's responses to GSS questions anonymous, confidential, or both? Explain your answer. You may wish to visit the GSS Web site at **www.norc.org/projects/General+Social+Survey.htm.**

7.11 Anonymous or confidential? The University of Wisconsin at Madison, like many universities, offers free screening for HIV, the virus that causes AIDS. The announcement at the University Health Services Web site says that for persons who seek testing one option is the following. "A code is used instead of a name. The person tested receives a copy of the report for their own information, but only the code identifies the report as theirs. The report does not go into a medical record." Does this practice offer anonymity or just confidentiality?

7.12 Anonymous or confidential? A Web site is looking for volunteers for a research study involving methicillin-resistant *Staphylococcus aureus* (MRSA), a bacterial infection that is highly resistant to some antibiotics. The Web site contains the following information about the study. "The nonprofit organization the Alliance for the Prudent Use of Antibiotics is looking for individuals who have or have had MRSA to fill out an anonymous survey and provide suggestions on how to improve treatment. The survey will help us to find out more about the concerns of people affected by MRSA and should take about 25 minutes to complete." Following the announcement is a Web link that takes you to the questionnaire. Does this study really provide anonymity or just confidentiality? Explain your answer.

7.13 Not really anonymous. Some common practices may appear to offer anonymity while actually delivering only confidentiality. Market researchers often use mail surveys that do not ask the respondent's identity but contain hidden codes on the questionnaire that identify the respondent. A false claim of anonymity is clearly unethical. If only confidentiality is promised, is it also unethical to say nothing about the identifying code, perhaps causing respondents to believe their replies are anonymous?

7.14 Human biological materials. Long ago, doctors drew a blood specimen from you as part of treating minor anemia. Unknown to you, the sample was stored. Now researchers plan to use stored samples from you and many other people to look for genetic factors that may influence anemia. It is no longer possible to ask your consent because you are no longer alive. Modern technology can read your entire genetic makeup from the blood sample.

(a) Do you think it violates the principle of informed consent to use your blood sample if your name is on it but you were not told that it might be saved and studied later?

(b) Suppose that your identity is not attached. The blood sample is known only to come from (say) "a 20-year-old white female being treated for anemia." Is it now okay to use the sample for research?

(c) Perhaps we should use biological materials such as blood samples only from patients who have agreed to allow the material to be stored for later use in research. It isn't possible to say in advance what kind of research, so this falls short of the usual standard for informed consent. Is this practice nonetheless acceptable, given complete confidentiality and the fact that using the sample can't physically harm the patient?

7.15 Equal treatment. Researchers on depression proposed to investigate the effect of supplemental therapy and counseling on the quality of life of adults with depression. Eligible patients on the rolls of a large medical clinic were to be randomly assigned to treatment and control groups. The treatment group would be offered dental care, vision testing, transportation, and other services not available without charge to the control group. The review board felt that providing these services to some but not other persons in the same institution raised ethical questions. Do you agree? Explain your answer.

7.16 Sham surgery? Clinical trials like the Parkinson's disease study mentioned in Example 6 are becoming more common. One medical researcher says, "This is just the beginning. Tomorrow, if you have a new procedure, you will have to do a double-blind placebo trial." Example 6 (page 145) outlines the arguments for and against testing surgery just as drugs are tested. When would you allow sham surgery in a clinical trial of a new surgery?

7.17 AIDS clinical trials. Now that effective treatments for AIDS are at last available, is it ethical to test treatments that may be less effective? Combinations of several powerful drugs reduce the level of HIV in the blood and at least delay illness and death from AIDS. But effectiveness depends on how damaged the patient's immune system is and what drugs he or she has previously taken. There are strong side effects, and patients must be able to take more than a dozen pills on time every day. Because AIDS is often fatal and the combination therapy works, we might argue that it isn't ethical to test any new treatment for AIDS that might possibly be less effective. But that might prevent discovery of better treatments. This is a strong example of the conflict between doing the best we know for patients now and finding better treatments for other patients in the future. How can we ethically test new drugs for AIDS?

7.18 AIDS trials in Africa. Effective drugs for treating AIDS are very expensive, so most African nations cannot afford to give them to large numbers of people. Yet AIDS is more common in parts of Africa than anywhere else. A few clinical trials are looking at ways to prevent pregnant mothers infected with

HIV from passing the infection to their unborn children, a major source of HIV infections in Africa. Some people say these trials are unethical because they do not give effective AIDS drugs to their subjects, as would be required in rich nations. Others reply that the trials are looking for treatments that can work in the real world in Africa and that they promise benefits at least to the children of their subjects. What do you think?

7.19 AIDS trials in Africa. One of the most important goals of AIDS research is to find a vaccine that will protect against HIV. Because AIDS is so common in parts of Africa, that is the easiest place to test a vaccine. It is likely, however, that a vaccine would be so expensive that it could not (at least at first) be widely used in Africa. Is it ethical to test in Africa if the benefits go mainly to rich countries? The treatment group of subjects would get the vaccine, and the placebo group would later be given the vaccine if it proved effective. So the actual subjects would benefit—it is the future benefits that would go elsewhere. What do you think? Explain your answer.

7.20 Opinion polls. The congressional campaigns are in full swing, and the candidates have hired polling organizations to take regular polls to find out what the voters think about the issues. What information should the pollsters be required to give out?

(a) What does the standard of informed consent, as discussed in this chapter, require the pollsters to tell potential respondents?

(b) The standards accepted by polling organizations also require giving respondents the name and address of the organization that carries out the poll. Why do you think this is required?

(c) The polling organization usually has a professional name such as "Samples Incorporated," so respondents don't know that the poll is being paid for by a political party or candidate. Would revealing the sponsor to respondents bias the poll? Should the sponsor always be announced whenever poll results are made public?

7.21 A right to know? Some people think that the law should require that all political poll results be made public. Otherwise, the possessors of poll results can use the information to their own advantage. They can act on the information, release only selected parts of it, or time the release for best effect. A candidate's organization replies that they are paying for the poll in order to gain information for their own use, not to amuse the public. Do you favor requiring complete disclosure of political poll results? What about other private surveys, such as market research surveys of consumer tastes?

7.22 Telling the government. The 2010 census was a short-form-only census. The decennial long form was eliminated. The American Community Survey (ACS) replaced the long form in 2010 and will collect long-form-type information throughout the decade rather than only once every 10 years. The 2010 ACS asked detailed questions, for example:

Does this house, apartment, or mobile home have a) hot and cold piped water?; b) a flush toilet?; c) a bathtub or shower?; d) a sink or faucet?; e) a stove or range?; f) a refrigerator?; and g) telephone service from which you can both make and receive calls? Include cell phones.

The form also asked your income in dollars, broken down by source, and whether any "physical, mental, or emotional condition" causes you difficulty in "concentrating, remembering, or making decisions."

Give brief arguments for and against the use of the ACS form: the government has legitimate uses for such information, but the questions seem to invade people's privacy.

7.23 Charging for data? Data produced by the government are often available free or at low cost to private users. For example, satellite weather data produced by the U.S. National Weather Service are available free to TV stations for their weather reports and to anyone on the Web. *Opinion 1: Government data should be available to everyone at minimal cost.* European governments, on the other hand, charge TV stations for weather data. *Opinion 2: The satellites are expensive, and the TV stations are making a profit from their weather services, so they should share the cost.* Which opinion do you support, and why?

7.24 Surveys of youth. The Centers for Disease Control and Prevention, in a survey of teenagers, asked the subjects if they had ever had sexual intercourse. Males who said "Yes" were then asked, "That very first time that you had sexual intercourse with a female, how old were you?" and "Please tell me the name or initials of your first sexual partner so that I can refer to her during the interview." Should consent of parents be required to ask minors about sex, drugs, and other such issues, or is consent of the minors themselves enough? Give reasons for your opinion.

7.25 Deceiving subjects. Students sign up to be subjects in a psychology experiment. When they arrive, they are placed in a room and assigned a task. During the task, the subject hears a loud thud from an adjacent room and then a piercing cry for help. Some subjects are placed in a room by themselves. Others are placed in a room with "confederates" who have been instructed by the researcher to look up upon hearing the cry, then return to their task. The treatments being compared are whether the subject is alone in the room or in the room with confederates. Will the subject ignore the cry for help?

The students had agreed to take part in an unspecified study, and the true nature of the experiment is explained to them afterward. Do you think this study is ethically okay?

7.26 Tempting subjects. A psychologist conducts the following experiment: he measures the attitude of subjects toward cheating, then has them take a mathematics skills exam in which the subjects are tempted to cheat. Subjects are told that high scores will receive a $100.00 gift certificate and that the purpose of the experiment is to see if rewards affect performance. The exam is computer-based and multiple choice. Subjects are left alone in a room with a computer on which the exam is available and are told that they are to click on the answer they believe is correct. However, when subjects click on an answer a small pop-up window appears with the correct answer indicated. When the pop-up window is closed, it is possible to change the answer selected. The computer records—unknown to the subjects—whether or not they change their answers after closing the pop-up window. After completing the exam, attitude toward cheating is retested.

Subjects who cheat tend to change their attitudes to find cheating more acceptable. Those who resist the temptation to cheat tend to condemn cheating more strongly on the second test of attitude. These results confirm the psychologist's theory.

This experiment tempts subjects to cheat. The subjects are led to believe that they can cheat secretly when in fact they are observed. Is this experiment ethically objectionable? Explain your position.

7.27 Decency and public money. Congress has often objected to spending public money on projects that seem "indecent" or "in bad taste." The arts are most often affected, but science can also be a target. Congress once refused to fund an experiment to study the effect of marijuana on sexual response. The journal *Science* reported:

Dr. Harris B. Rubin and his colleagues at the Southern Illinois Medical School proposed to exhibit pornographic films to people who had smoked marihuana and to measure the response with sensors attached to the penis.

Marihuana, sex, pornographic films—all in one package, priced at $120,000. The senators smothered the hot potato with a ketchup of colorful oratory and mixed metaphors.

"I am firmly convinced we can do without this combination of red ink, 'blue' movies, and Acapulco 'gold,'" Senator John McClellan of Arkansas opined in a persiflage of purple prose.

(a) The subjects were volunteers who gave properly informed consent. If you were a member of the review board, would you veto this experiment on grounds of "decency" or "good taste"?

(b) Suppose we concede that a free society should permit any legal experiment with volunteer subjects. It is a further step to say that any such experiment is entitled to government funding if the usual review procedure finds it scientifically worthwhile. If you were a member of Congress, would you ever refuse to pay for an experiment on grounds of "decency" or "good taste"?

EXPLORING THE WEB

7.28 Web-based exercise. Is "anonymous" blogging really anonymous? Read the article at **arstechnica.com/tech-policy/news/2009/10/ anonymous-real-estate-critic-on-the-verge-of-being-unmasked.ars** and explain your answer.

7.29 Web-based exercise. The *Journal of Medical Internet Research* contains an article on ethical issues that arise in providing online psychotherapeutic interventions. Visit the Web page **www.jmir.org/2000/1/e5/**, and read the article. Describe at least two of the ethical issues discussed in the article.

7.30 Web-based exercise. The University of Minnesota has a Web page intended to educate readers about informed consent: **www.research.umn .edu/consent/.** Visit the Web page and select either the *Health and Biological Sciences* or the *Social and Behavioral Sciences* link. Next, select the *Overview of Informed Consent* link. Read through the module and take the overview quiz.

7.31 Web-based exercise. For a glimpse at the work of an institutional review board, visit the Web site of the University of Pittsburgh's board, **www.irb.pitt.edu/.** Look at the *Reference Manual for the Use of Human Subjects in Research* (see the links along the left side of the page) to learn how elaborate the review process can be. According to the chapter on the informed consent process, what is the suggested wording of an informed consent document?

7.32 Web-based exercise. The official ethical codes of the American Statistical Association and the American Psychological Association are long documents that address many issues in addition to those discussed in this chapter. From either of these documents, give one issue that is discussed in the document but not in this chapter. The Web sites where these documents can be found are **www.amstat.org/committees/cmtepc/index .cfm?fuseaction=7** and **www.apa.org/ethics/code/index.aspx.**

7.33 Web-based exercise. Some early experiments in psychology were highly unethical and led to the establishment of present-day standards. Read the article at **listverse.com/2008/09/07/top-10-unethical-psychological-experiments/** and determine which of the experiments described you find most repugnant.

NOTES AND DATA SOURCES

Page 137 Case Study: Lycia L. C. Parker, Ian S. Penton-Voak, Angela S. Attwood, and Marcus R. Munafo, "Effects of acute alcohol consumption on ratings of attractiveness of facial stimuli: evidence of long-term encoding," *Alcohol and Alcoholism,* 43 (2008), pp. 636–640.

Page 138 Example 1: John C. Bailar III, "The real threats to the integrity of science," *Chronicle of Higher Education,* April 21, 1995, pp. B1–B2.

Page 139 The quotation is from the University of Pittsburgh's institutional review board *Reference Manual for the Use of Human Subjects in Research*; available online at **www.irb.pitt.edu/manual/.**

Page 139 Example 2: The difficulties of interpreting guidelines for informed consent and for the work of institutional review boards in medical research are a main theme of Beverly Woodward, "Challenges to human subject protections in U.S. medical research," *Journal of the American Medical Association,* 282 (1999), pp. 1947–1952. The references in this paper point to other discussions.

Page 143 Example 4: The quotation is from the *Report of the Tuskegee Syphilis Study Legacy Committee,* May 20, 1996. A detailed history is James H. Jones, *Bad Blood: The Tuskegee Syphilis Experiment,* Free Press, 1993. Another reference is Susan M. Reverby, *More than Fact and Fiction: Cultural Memory and the Tuskegee Syphilis Study,* Hastings Center Report, 0093-0334, September 1, 2001, Vol. 31, Issue 5.

Page 143 The quotation from Dr. Hennekens is from an interview in the Annenberg/ Corporation for Public Broadcasting video series *Against All Odds: Inside Statistics.*

Page 144 The quotations are from Gina Kolata and Kurt Eichenwald, "Business thrives on unproven care leaving science behind," *New York Times,* October 3, 1999. Background and details about the first clinical trials appear in a National Cancer Institute press release: "Questions and answers: high-dose chemotherapy with bone marrow or stem cell transplants for breast cancer," April 15, 1999. That one of the studies reported there involved falsified data is reported by Denise Grady, "Breast cancer researcher admits falsifying data," *New York Times,* February 5, 2000.

Page 145 Example 6: The quotation is from Thomas B. Freeman et al., "Use of placebo surgery in controlled trials of a cellular-based therapy for Parkinson's disease," *New England Journal of Medicine,* 341 (1999), pp. 988–992. Freeman supports the Parkinson's disease trial. The opposition is represented by Ruth Macklin, "The ethical problems with sham surgery in clinical research," *New England Journal of Medicine,* 341 (1999), pp. 992–996.

Page 146 Example 7: R. D. Middlemist, E. S. Knowles, and C. F. Matter, "Personal space invasions in the lavatory: suggestive evidence for arousal," *Journal of Personality and Social Psychology,* 33 (1976), pp. 541–546.

Page 152 Exercise 7.16: Dr. C. Warren Olanow, chief of neurology, Mt. Sinai School of Medicine, quoted in Margaret Talbot, "The placebo prescription," *New York Times Magazine,* January 8, 2000, pp. 34–39, 44, 58–60.

Page 152 Exercise 7.17: For extensive background, see Jon Cohen, "AIDS trials ethics questioned," *Science,* 276 (1997), pp. 520–523. Search the archives at **www.sciencemag.org** for recent episodes in the continuing controversies of Exercises 7.17, 7.18, and 7.19.

Page 155 Exercise 7.27: News item in *Science,* 192 (1976), p. 1086.

Measuring

CASE STUDY Are people with larger brains more intelligent? People have investigated this question throughout history. To answer it, we must **measure** "intelligence." This requires us to reduce the vague idea to a number that can go up or down. The first step is to say what we mean by intelligence. Does a vast knowledge of many subjects constitute intelligence? How about the ability to solve difficult puzzles or do complicated mathematical calculations? Or is it some combination of all of these?

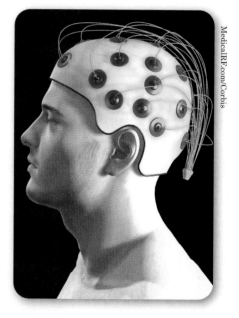

MedicalRF.com/Corbis

Once we decide what intelligence is, we must actually produce the numbers. Should we use the score on a written test? Perhaps a formula that also includes grades in school would be better. Not only is it hard to say exactly what "intelligence" is, but it's hard to attach a number to measure whatever we say it is. And in the end, can we even trust the number we produce?

By the end of this chapter you will have learned principles that will help you understand the process of measurement and determine whether you can trust the resulting numbers. ∎

Measurement basics

Statistics deals with numbers. Planning the production of data through a sample or an experiment does not by itself produce numbers. Once we have our sample respondents or our experimental subjects, we must still *measure* whatever characteristics interest us. First, think broadly: Are we trying to measure the right things? Are we overlooking some outcomes that are important even though they may be hard to measure?

EXAMPLE 1 But what about the patients?

Clinical trials tend to measure things that are easy to measure: blood pressure, tumor size, virus concentration in the blood. They often don't directly measure what matters most to patients—does the treatment really improve their lives? One study found that only 5% of trials published between 1980 and 1997 measured the effect of treatments on patients' emotional well-being or their ability to function in social settings.

Once we have decided what properties we want to measure, we can think about how to do the measurements.

Measurement

We **measure** a property of a person or thing when we assign a number to represent the property.

We often use an **instrument** to make a measurement. We may have a choice of the **units** we use to record the measurements.

The result of measurement is a numerical **variable** that takes different values for people or things that differ in whatever we are measuring.

What are your units?
Not paying attention to units of measurement can get you into trouble. In 1999, the *Mars Climate Orbiter* burned up in the Martian atmosphere. It was supposed to be 93 miles (150 kilometers) above the planet but was in fact only 35 miles (57 kilometers) up. It seems that Lockheed Martin, which built the *Orbiter*, specified important measurements in English units (pounds, miles). The National Aeronautics and Space Administration team who flew the spacecraft thought the numbers were in metric system units (kilograms, kilometers). There went $125 million.

EXAMPLE 2 Length, college readiness, highway safety

To measure the length of a bed, you can use a tape measure as the *instrument*. You can choose either inches or centimeters as the *unit of measurement*. If you choose centimeters, your *variable* is the length of the bed in centimeters.

To measure a student's readiness for college, you might ask the student to take the SAT Reasoning exam. The exam is the *instrument*. The *variable* is the student's score in points, somewhere between 600 and 2400 if you combine the Writing, Critical Reading, and Mathematics sections of the SAT. "Points" are the *units of measurement*, but these are determined by a complicated scoring system described at the SAT Web site (**www.collegeboard.com**).

How can you measure the safety of traveling on the highway? You might decide to use the number of

people who die in motor vehicle accidents in a year as a *variable* to measure highway safety. The government's Fatal Accident Reporting System collects data on all fatal traffic crashes. The *unit of measurement* is the number of people who died, and the Fatal Accident Reporting System serves as our measuring *instrument*.

Here are some questions you should ask about the variables in any statistical study:

1. Exactly how is the variable defined?
2. Is the variable a valid way to describe the property it claims to measure?
3. How accurate are the measurements?

We don't often design our own measuring devices—we use the results of the SAT or the Fatal Accident Reporting System, for example—so we won't go deeply into that aspect of measurement. Any consumer of numbers, however, should know a bit about how they are produced.

Know your variables

Measurement is the process of turning concepts like length or employment status into precisely defined variables. Using a tape measure to turn the idea of "length" into a number is straightforward because we know exactly what we mean by length. Measuring college readiness is controversial because it isn't clear exactly what makes a student ready for college work. Using SAT scores at least says exactly how we will get numbers. Measuring leisure time requires that we first say what time counts as leisure. Even counting highway deaths requires us to say exactly what counts as a highway death: Pedestrians hit by cars? People in cars hit by a train at a crossing? People who die from injuries 6 months after an accident? We can simply accept the government's counts, but someone had to answer those and other questions in order to know what to count. For example, a person must die within 30 days of an accident to count as a traffic death. These details are a nuisance, but they can make a difference.

EXAMPLE 3 Measuring unemployment

Each month the Bureau of Labor Statistics (BLS) announces the *unemployment rate* for the previous month. People who are not available for work (retired people, for example, or students who do not want to work while in school) should not be counted as unemployed just because they don't have a job. To be unemployed, a person must first be in the labor force. That is, she must be available for work and looking for work. The unemployment rate is

$$\text{unemployment rate} = \frac{\text{number of people unemployed}}{\text{number of people in the labor force}}$$

To complete the exact definition of the unemployment rate, the BLS has very detailed descriptions of what it means to be "in the labor force" and what it means to be "employed." For example, if you are on strike but expect to return to the same job, you count as employed. If you are not working and did not look for work in the last two weeks, you are not in the labor force. So people who say they want to work but are too discouraged to keep looking for a job don't count as unemployed. The details matter. The official unemployment rate would be different if the government used a different definition of unemployment.

The BLS estimates the unemployment rate based on interviews with the sample in the monthly Current Population Survey. The interviewer can't simply ask, "Are you in the labor force?" and "Are you employed?" Many questions are needed to classify a person as employed, unemployed, or not in the labor force. Changing the questions can change the unemployment rate. At the beginning of 1994, after several years of planning, the BLS introduced computer-assisted interviewing and improved its questions. Figure 8.1 is a graph of the unemployment rate that appeared on the front page of the BLS monthly news release on the employment situation. There is a gap in the graph before January 1994 because of the change in the

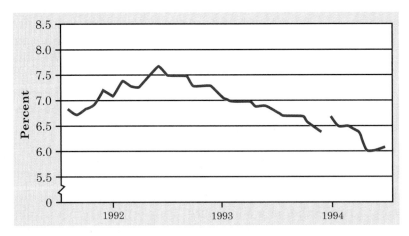

Figure 8.1 The unemployment rate from August 1991 to July 1994. The gap shows the effect of a change in how the government measures unemployment.

interviewing process. The unemployment rate would have been 6.3% under the old system. It was 6.7% under the new system. That's a big enough change to make politicians unhappy.

Measurements, valid and invalid

No one would object to using a tape measure reading in centimeters to measure the length of a bed. Many people object to using SAT scores to measure readiness for college. Let's shortcut that debate: just measure the height in inches of all applicants and accept the tallest. Bad idea, you say. Why? Because height has nothing to do with being prepared for college. In more formal language, height is not a *valid* measure of a student's academic background.

Valid measurement

A variable is a **valid** measure of a property if it is relevant or appropriate as a representation of that property.

It is valid to measure length with a tape measure. It isn't valid to measure a student's readiness for college by recording her height. The BLS unemployment rate is a valid measure, even though changes in the official definitions would give a somewhat different measure. Let's think about measures, both valid and invalid, in some other settings.

EXAMPLE 4   Measuring highway safety

Roads got better. Speed limits increased. Big SUVs replaced cars. Enforcement campaigns reduced drunk driving. How did highway safety change between 1994 and 2007 in this changing environment?

We could just count deaths from motor vehicles. The Fatal Accident Reporting System says there were 40,716 deaths in 1994 and 41,259 deaths 13 years later in 2007. The number of deaths increased. But the number of licensed drivers rose from 175 million in 1994 to 206 million in 2007. The number of miles that people drove rose from 2358 billion to 3032 billion. If more people drive more miles, there may be more deaths even if the roads are safer. The count of deaths alone is not a valid measure of highway safety.

Rather than a *count,* we should use a *rate.* The number of deaths per mile driven takes into account the fact

Mahaux Photography/Getty Images

that more people drive more miles than in the past. In 2007, vehicles drove 3,032,000,000,000 miles in the United States. Because this number is so large, it is usual to measure safety by deaths per 100 million miles driven rather than deaths per mile. For 2007, this death rate is

$$\frac{\text{motor vehicle deaths}}{\text{100s of millions of miles driven}} = \frac{41,259}{30,320}$$
$$= 1.4$$

The death rate fell from 1.7 deaths per 100 million miles in 1994 to 1.4 in 2007. That's a decrease—there were 18% fewer deaths per mile driven in 2007 than in 1994. Driving became safer even though the total number of deaths increased.

Rates and counts

Often a **rate** (a fraction, proportion, or percentage) at which something occurs is a more valid measure than a simple **count** of occurrences.

NOW IT'S YOUR TURN | **8.1 Driver fatigue.** A researcher studied the number of traffic accidents that were attributed to driver fatigue at different times of the day. He noticed that the number of accidents was higher in the late afternoon (between 5 and 6 P.M.) than in the early afternoon (between 1 and 2 P.M.). He concluded that driver fatigue plays a more prominent role in traffic accidents in the late afternoon than in the early afternoon. Do you think this conclusion is justified?

Using height to measure readiness for college and using counts when rates are needed are examples of clearly invalid measures. The tougher questions concern measures that are neither clearly invalid nor obviously valid.

EXAMPLE 5 Achievement tests

When you take a statistics exam, you hope that it will ask you about the main points of the course syllabus. If it does, the exam is a valid measure of how much you know about the course material. The College Board, which administers the SAT, also offers achievement tests in a variety of disciplines (including an Advanced Placement exam in statistics). These achievement tests are not very controversial. Experts can judge validity by comparing the test questions with the syllabus of material the questions are supposed to cover.

EXAMPLE 6 IQ tests

Psychologists would like to measure aspects of the human personality that can't be observed directly, such as "intelligence" or "authoritarian personality." Does an IQ test measure intelligence? Some psychologists say "Yes" rather loudly. There is such a thing as general intelligence, they argue, and the various standard IQ tests do measure it, though not perfectly. Other experts say "No" equally loudly. There is no single intelligence, just a variety of mental abilities (for example, logical, linguistic, spatial, musical, kinesthetic, interpersonal, and intrapersonal) that no one instrument can measure.

The disagreement over the validity of IQ tests is rooted in disagreement over the nature of intelligence. If we can't agree on exactly what intelligence is, we can't agree on how to measure it.

Statistics is little help in these examples. The examples start with an idea like "knowledge of statistics" or "intelligence." If the idea is vague, validity becomes a matter of opinion. However, statistics can help a lot if we refine the idea of validity a bit.

EXAMPLE 7 The SAT again

"SAT bias will illegally cheat thousands of young women out of college admissions and scholarship aid they have earned by superior classroom performance." That's what the organization FairTest said when the 1999 SAT scores were released. The gender gap was larger on the math part of the test, where women averaged 495 and men averaged 531. Twelve years later in 2011 the gap remained. Among high school seniors, women averaged 500 and men 531 on the math part of the test. The federal Office of Civil Rights says that tests on which women and minorities score lower are discriminatory.

The College Board, which administers the SAT, replies that there are many reasons some groups have lower average scores than others. For example, more women than men from families with low incomes and little education sign up for the SAT. Students whose parents have low incomes and little education have, on the average, fewer advantages at home and in school than richer students. They have lower SAT scores because their backgrounds have not prepared them as well for college. The mere fact of lower scores doesn't imply that the test is not valid.

Is the SAT a valid measure of readiness for college? "Readiness for college academic work" is a vague concept that probably combines inborn intelligence (whatever we decide that is), learned knowledge, study and test-taking skills, and motivation to work at academic subjects. Opinions will always differ about whether SAT scores (or any other measure) accurately reflect this vague concept.

Instead, we ask a simpler and more easily answered question: do SAT scores help predict students' success in college? Success in college is a clear concept, measured by whether students graduate and by their college grades. Students with high SAT scores are more likely to graduate and earn (on the average) higher grades than students with low SAT scores. We say that SAT scores have *predictive validity* as measures of readiness for college. This is the only kind of validity that data can assess directly.

Predictive validity

A measurement of a property has **predictive validity** if it can be used to predict success on tasks that are related to the property measured.

Predictive validity is the clearest and most useful form of validity from the statistical viewpoint. "Do SAT scores help predict college grades?" is a much clearer question than "Do IQ test scores measure intelligence?" However, predictive validity is not a yes-or-no idea. We must ask *how accurately* SAT scores predict college grades. Moreover, we must ask *for what groups* the SAT has predictive validity. It is possible, for example, that the SAT predicts college grades well for men but not for women. There are statistical ways to describe "how accurately." The Statistical Controversies feature in this chapter asks you to think about these issues.

Measurements, accurate and inaccurate

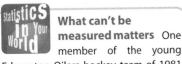

What can't be measured matters One member of the young Edmonton Oilers hockey team of 1981 finished last in almost everything one can measure: strength, speed, reflexes, eyesight. That was Wayne Gretzky, soon to be known as "the Great One." He broke the National Hockey League scoring record that year, then scored yet more points in seven different seasons. Somehow the physical measurements didn't catch what made Gretzky the best hockey player ever. Not everything that matters can be measured.

Using a bathroom scale to measure your weight is valid. If your scale is like many commonly used ones, however, the measurement may not be very accurate. It measures weight, but it may not give the true weight. Let's say that originally your scale always read 3 pounds too high, so

$$\text{measured weight} = \text{true weight} + 3 \text{ pounds}$$

If that is the whole story, the scale will always give the same reading for the same true weight. Most scales vary a bit—they don't always give the same reading when you step off and step right back on. Your scale now is somewhat old and rusty. It still always reads

STATISTICAL CONTROVERSIES

SAT Exams in College Admissions

Susan Stava/The New York Times/Redux

Colleges use a variety of measures to make admissions decisions. The student's record in high school is the most important, but SAT scores do matter, especially at selective colleges. The SAT has the advantage of being a national test. An A in algebra means different things in different high schools, but an SAT Math score of 625 means the same thing everywhere. The SAT can't measure willingness to work hard or creativity, so it won't predict college performance exactly, but most colleges have long found it helpful.

The table at the bottom of this box gives some results about how well SAT scores predict first-year college grades from a sample of 151,316 students in 2006. The numbers in the table say what percentage of the variation among students in college grades can be predicted by SAT scores (Critical Reading, Writing, and Math tests combined), by high school grades, and by SAT and high school grades together. An entry of 0% would mean no predictive validity, and 100% would mean predictions were always exactly correct.

How well do you think SAT scores predict first-year college grades? Should SAT scores be used in deciding college admissions?

	All institutions	Private institutions	Public institutions
SAT	28%	32%	27%
School grades	29%	30%	28%
Both together	38%	42%	37%

3 pounds too high because its aim is off, but now it is also erratic and so readings deviate from 3 pounds. This morning it sticks a bit and reads 1 pound too low for that reason. So the reading is

$$\text{measured weight} = \text{true weight} + 3 \text{ pounds} - 1 \text{ pound}$$

When you step off and step right back on, the scale sticks in a different spot that makes it read 1 pound too high. The reading you get is now

$$\text{measured weight} = \text{true weight} + 3 \text{ pounds} + 1 \text{ pound}$$

You don't like the fact that this second reading is higher than the first, so you again step off and step right back on. The scale again sticks in a different spot and you get the reading

$$\text{measured weight} = \text{true weight} + 3 \text{ pounds} - 1.5 \text{ pounds}$$

If you have nothing better to do than keep stepping on and off the scale, you will keep getting different readings. They center on a reading 3 pounds too high, but they vary about that center.

Your scale has two kinds of errors. If it didn't stick, the scale would always read 3 pounds high. That is true every time anyone steps on the scale. This systematic error that occurs every time we make a measurement is called *bias*. Your scale also sticks—but how much this changes the reading differs every time someone steps on the scale. Sometimes stickiness pushes the scale reading up; sometimes it pulls it down. The result is that the scale weighs 3 pounds too high on the average, but its reading varies when we weigh the same thing repeatedly. We can't predict the error due to stickiness, so we call it *random error*.

Errors in measurement

We can think about errors in measurement this way:

$$\text{measured value} = \text{true value} + \text{bias} + \text{random error}$$

A measurement process has **bias** if it systematically tends to overstate or understate the true value of the property it measures.

A measurement process has **random error** if repeated measurements on the same individual give different results. If the random error is small, we say the measurement is **reliable.**

To determine if the random error is small, we can use a quantity called the **variance.** The variance of n repeated measurements on the same individual is computed as follows:

1. Find the average of these n measurements.
2. Compute the difference between each observation and the mean and square each of these differences.
3. Average the squared differences by dividing their sum by $n - 1$. This average squared difference is the variance.

A reliable measurement process will have a small variance.

For the three measurements on our sticky scale suppose that your true weight is 130 pounds. Then the three measurements are

$$130 + 3 - 1 = 132 \text{ pounds}$$

$$130 + 3 + 1 = 134 \text{ pounds}$$

$$130 + 3 - 1.5 = 131.5 \text{ pounds}$$

The average of these three measurements is

$$(132 + 134 + 131.5)/3 = 397.5/3 = 132.5 \text{ pounds}$$

The differences between each measurement and the average are

$$132 - 132.5 = -0.5$$
$$134 - 132.5 = 1.5$$
$$131.5 - 132.5 = -1$$

The sum of the squares of these differences is

$$(-0.5)^2 + (1.5)^2 + (-1)^2 = 0.25 + 2.25 + 1 = 3.5$$

and so the variance of these random errors is

$$3.5/(3 - 1) = 1.75$$

A scale that always reads the same when it weighs the same item is perfectly reliable even if it is biased. For such a scale, the variance of the measurements will be 0.

Reliability says only that the result is repeatable. Bias means that in repeated measurements the *tendency* is to overstate or understate the true value. It does not necessarily mean that every measurement overstates or understates the true value. Bias and lack of reliability are different kinds of error. And don't confuse reliability with validity just because both sound like good qualities. Using a scale to measure weight is valid even if the scale is not reliable.

Here's an example of a measurement that is reliable but not valid.

EXAMPLE 8 Do big skulls house smart brains?

In the mid-19th century, it was thought that measuring the volume of a human skull would measure the intelligence of the skull's owner. It was difficult to measure a skull's volume reliably, even after it was no longer attached to its owner. Paul Broca, a professor of surgery, showed that filling a skull with small lead shot, then pouring out the shot and weighing it, gave quite reliable measurements of the skull's volume. These accurate measurements do not, however, give a valid measure of intelligence. Skull volume turned out to have no relation to intelligence or achievement.

> **NOW IT'S YOUR TURN**
>
> **8.2 The most popular restaurant.** Each year the Zagat Survey, based on the votes of thousands of diners, lists the most popular restaurants in Los Angeles. In 2010, In-N-Out was ranked 3rd, but in 2009 it was ranked 15th. Cafe Bizou was ranked 7th in 2010 but was 3rd in 2009. Do you think these ratings are biased, unreliable, or both? Explain your answer.

Improving reliability, reducing bias

What time is it? Much modern technology, such as the Global Positioning System, which uses satellite signals to tell you where you are, requires very exact measurements of time. Time starts with the earth's path around the sun, which lasts one year. But the earth is much too erratic. Since 1967, time starts with the standard second, and the second is defined to be the time required for 9,192,631,770 vibrations of a cesium atom. Physical clocks are bothered by changes in temperature, humidity, and air pressure. The cesium atom doesn't care. People who need really accurate time can buy atomic clocks. The National Institute of Standards and Technology (NIST) keeps an even more accurate atomic clock and broadcasts the results (with some loss in transmission) by radio, telephone, and Internet.

EXAMPLE 9 Really accurate time

NIST's atomic clock is very accurate but not perfectly accurate. The world standard is Coordinated Universal Time, compiled by the International Bureau of Weights and Measures (BIPM) in Sèvres, France. BIPM doesn't have a better clock than NIST. It calculates the time by averaging the results of more than 200 atomic clocks around the world. NIST tells us (after the fact) how much it misses the correct time by. Here are the last

Figure 8.2 This atomic clock at the National Institute of Standards and Technology is accurate to 1 second in 6 million years. (Photo courtesy of John Wessels, NIST Time and Frequency Division.)

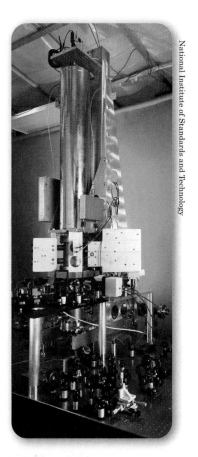

12 errors as we write, in seconds:

0.0000000075	0.0000000012
0.0000000069	−0.0000000020
0.0000000067	−0.0000000045
0.0000000063	−0.0000000046
0.0000000041	−0.0000000042
0.0000000032	−0.0000000036

In the long run, NIST's measurements of time are not biased. The NIST second is sometimes shorter than the BIPM second and sometimes longer, not always off in the same direction. NIST's measurements are very reliable, but the numbers above do show some variation. There is no such thing as a perfectly reliable measurement. The average (mean) of several measurements is more reliable than a single measurement. That's one reason BIPM combines the time measurements of many atomic clocks.

Scientists everywhere repeat their measurements and use the average to get more reliable results. Even students in a chemistry lab often do this. Just as larger samples reduce variation in a sample statistic, averaging over more measurements reduces variation in the final result.

Use averages to improve reliability

No measuring process is perfectly reliable. The **average** of several repeated measurements of the same individual is more reliable (less variable) than a single measurement.

Unfortunately, there is no similarly straightforward way to reduce the bias of measurements. Bias depends on how good the measuring instrument is. To reduce the bias, you need a better instrument. The atomic clock at NIST (Figure 8.2) is accurate to 1 second in 6 million years but is a bit large to put beside your bed.

EXAMPLE 10 Measuring unemployment again

Measuring unemployment is also "measurement." The concepts of bias and reliability apply here just as they do to measuring length or time.

The Bureau of Labor Statistics checks the *reliability* of its measurements of unemployment by having supervisors reinterview about 5% of the sample. This is repeated measurement on the same individual, just as a student in a chemistry lab measures a weight several times.

The BLS attacks *bias* by improving its instrument. That's what happened in 1994, when the Current Population Survey was given its biggest overhaul in more than 50 years. The old system for measuring unemployment, for example, underestimated unemployment among women because the detailed procedures had not kept up with changing patterns of women's work. The new measurement system corrected that bias—and raised the reported rate of unemployment.

Pity the poor psychologist

Statisticians are creatures of habit: they think about measurement much the same way they think about sampling. In both settings, the big idea is to ask, "What would happen if we did this many times?" In sampling we want to estimate a population parameter, and we worry that our estimate may be biased or vary too much from sample to sample. Now we want to measure the true value of some property, and we worry that our measurement may be biased or vary too much when we repeat the measurement on the same individual. Bias is systematic error that happens every time; high variability (low reliability) means that our result can't be trusted because it isn't repeatable.

Thinking of measurement this way is pretty straightforward when you are measuring your weight. To start with, you have a clear idea of what your "true weight" is. You know that there are really good scales around: start at the doctor's office, go to the physics lab, end up at NIST. You can measure your weight as accurately as you wish. This makes it easy to see that your bathroom scale always reads 3 pounds too high. Reliability is also easy to describe—step on and off the scale many times and see how much its readings vary.

Asking "What would happen if we did this many times?" is a lot harder to put into practice when we want to measure "intelligence" or "readiness for college." Consider as an example the poor psychologist who wants to measure "authoritarian personality."

> **EXAMPLE 11** Authoritarian personality?
>
> Do some people have a personality type that disposes them to rigid think-ing and to following strong leaders? Psychologists looking back on the Nazis after World War II thought so. In 1950, a group of psychologists developed the "F-scale" as an instrument to measure "authoritarian per-sonality." The F-scale asks how strongly you agree or disagree with state-ments such as the following:
>
> - *Obedience and respect for authority are the most important virtues children should learn.*
>
> - *Science has its place, but there are many important things that can never be understood by the human mind.*
>
> Strong agreement with such statements marks you as authoritarian. The F-scale and the idea of the authoritarian personality continue to be prominent in psychology, especially in studies of prejudice and right-wing extremist movements.

Here are some questions we might ask about using the F-scale to mea-sure "authoritarian personality." The same questions come to mind when we think about IQ tests or the SAT exam.

1. Just what is an "authoritarian personality"? We understand this much less well than we understand your weight. The answer in practice seems to be "whatever the F-scale measures." Any claim for validity must rest on what kinds of behavior high F-scale scores go along with. That is, we fall back on predictive validity.

2. The name sounds unpleasant, and F stands for Fascist. As the second question in Example 11 suggests, people who hold traditional religious beliefs are likely to get higher F-scale scores than similar people who don't hold those beliefs. Does the instrument reflect the beliefs of those who developed it? That is, would people with different beliefs come up with a quite different instrument?

3. You think you know what your true weight is. What is the true value of your F-scale score? The measuring devices at NIST can help us find a true weight but not a true authoritarianism score. If we suspect that the instrument is biased as a measure of "authoritarian personality" because it penalizes religious beliefs, how can we check that?

4. You can weigh yourself many times to learn the reliability of your bathroom scale. If you take the F-scale test many times, you remember what answers you gave the first time. That is, repeats of the same

psychological measurement are not really repeats. So reliability is hard to check in practice. Psychologists sometimes develop several forms of the same instrument in order to repeat their measurements. But how do we know these forms are really equivalent?

The point is not that psychologists lack answers to these questions. The first two are controversial because not all psychologists think about human personality in the same way. The second two questions have at least partial answers but not simple answers. The point is that "measurement," which seems so straightforward when we measure weight, is complicated indeed when we try to measure human personality.

There is a larger lesson here. Be wary of statistical "facts" about squishy topics like authoritarian personality, intelligence, and even readiness for college. The numbers look solid, as numbers always do. But data are a human product and reflect human desires, prejudices, and weaknesses. If we don't understand and agree on what we are measuring, the numbers may produce more heat than light.

STATISTICS IN SUMMARY

Chapter Specifics

- To **measure** something means to assign a number to some property of an individual.

- When we measure many individuals, we have values of a **variable** that describes them.

- Variables are recorded in **units of measurement.**

- When you work with data or read about a statistical study, ask if the variables are **valid** as numerical measures of the concepts the study discusses.

- Often a **rate** is a more valid measure than a **count.**

- Validity is simple for measurements of physical properties such as length, weight, and time. When we want to measure human personality and other vague properties, **predictive validity** is the most useful way to say whether our measures are valid.

- Also ask if there are **errors in measurement** that reduce the value of the data. You can think about errors in measurement like this:

$$\text{measured value} = \text{true value} + \text{bias} + \text{random error}$$

- Some ways of measuring are **biased,** or systematically wrong in the same direction.

- To reduce bias, you must use a better **instrument** to make the measurements.

- Other measuring processes lack **reliability,** so that measuring the same individuals again would give quite different results due to **random error.**

- A reliable measuring process will have a small **variance** of the measurements. You can improve the reliability of a measurement by repeating it several times and using the **average** result.

 In reasoning from data to a conclusion, we start with the data. In statistics data are ultimately represented by numbers. The planning of the production of data through a sample or experiment does not by itself produce these numbers. The extent to which these numbers represent the characteristics we wish to study affect the quality and relevance of our conclusions. When you work with data or read about a statistical study, ask exactly how the variables are defined and whether they leave out some things you want to know. This chapter presents several ideas one should think about in assessing the variables measured and hence the conclusions based on these measurements.

CASE STUDY The Case Study that opened the chapter is motivated by research con-
EVALUATED ducted in 1991 by Willerman, Schultz, Rutledge, and Bigler. Read about this study in the EESEE story "Brain Size and Intelligence," and use what you have learned in this chapter to answer the following questions.

1. How did the researchers measure brain size? Is this a valid measure of brain size? Is it reliable? Is it biased?

2. How did the researchers measure intelligence? Is this a valid measure of intelligence?

3. The researchers found some evidence that brain size and intelligence are related. However, the study described in Example 8 did not. Discuss the differences in the two studies. ■

CHAPTER 8 EXERCISES

For Exercise 8.1, see page 164; for Exercise 8.2, see page 170.

8.3 Counting the unemployed? We could measure the extent of unemployment by a count (the number of people who are unemployed) or by a rate (the percentage of the labor force that is unemployed). The number of people in the labor force grew from 107 million in 1980, to 126 million in 1990, to 143 million in 2000, to 154 million in 2010. Use these facts to explain why the count of unemployed people is not a valid measure of the extent of unemployment.

8.4 Measuring a healthy lifestyle. You want to measure the "healthiness" of college students' lifestyles. Give an example of a clearly invalid way to measure healthiness. Then briefly describe a measurement process that you think is valid.

8.5 Rates versus counts. Customers returned 36 coats to Sears this holiday season, and only 12 to La Boutique Classique next door. Sears sold 1200 coats this season, while La Boutique sold 200.

(a) Sears had a greater number of coats returned. Why does this not show that Sears's coat customers were less satisfied than those of La Boutique?

(b) What is the rate of returns (percentage of coats returned) at each of the stores?

8.6 Seat belt safety. The National Highway Traffic Safety Administration reports that in 2006 between the hours of 6 A.M. and 6 P.M., 8160 occupants of motor vehicles who were wearing a restraint died in motor vehicle accidents and 7064 who were not wearing a restraint died. These numbers suggest that not using a restraining device is safer than using one. The *counts* aren't fully convincing, however. What *rates* would you like to know to compare the safety of using a restraint with not using one?

8.7 Tough course? A friend tells you, "In the 7:30 A.M. lecture for Statistics 101, 9 students failed, but 20 students failed in the 1:30 P.M. lecture. The 1:30 P.M. prof is a tougher grader than the 7:30 A.M. prof." Explain why the conclusion may not be true. What additional information would you need to compare the classes?

8.8 Obesity. An article in the June 30, 2010, *Columbus Dispatch* reported on the prevalence of obesity among adults in the 50 states. Based on information in the article, California has approximately 6.7 million obese adults, and Texas has approximately 5.2 million. On the other hand, Mississippi has a little over 730,000 obese adults. Do these numbers make a convincing case that California and Texas have a more substantial problem with obesity than Mississippi?

8.9 Capital punishment. Between 1977 and 2010, 1234 convicted criminals were put to death in the United States. Here are data on the number of executions in several states during those years, as well as the April 1, 2010, population of these states:

State	Population (thousands)	Executions	
Alabama	4,780	49	10%
Arkansas	2,916	27	9% 10%
Delaware	898	14	15%
Florida	18,801	69	3%
Indiana	6,484	20	3
Nevada	2,701	12	4%
Oklahoma	3,751	94	25%
Texas	25,146	464	18

Texas and Florida are among the leaders in executions. Because these are large states, we might expect them to have many executions. Find the *rate* of executions for each of the states listed above, in executions per million population. Because population is given in thousands, you can find the rate per million as

$$\text{rate per million} = \frac{\text{executions}}{\text{population in thousands}} \times 1000$$

Arrange the states in order of the number of executions relative to population. Are Florida and Texas still high by this measure?

8.10 Measuring intelligence. "Intelligence" means something like "general problem-solving ability." Explain why it is *not* valid to measure intelligence by a test that asks questions such as

Who wrote "The Star-Spangled Banner"?
Who won the last soccer World Cup?

8.11 Measuring life's quality. Is life in Britain getting better or worse? The usual government data don't say. So the British government announced that it wanted to add measures of such things as housing, traffic, and air pollution. "The quality of life is not simply economic," said a deputy prime minister. Help them out: how would you measure "traffic" and its impact on the quality of life?

8.12 Measuring pain. There are 8 million enrollees in the Department of Veterans Affairs health care system. It wants doctors and nurses to treat pain as a "fifth vital sign," to be recorded along with blood pressure, pulse, temperature, and breathing rate. Help out the VA: how would you measure a patient's pain?

8.13 Fighting cancer. Congress wants the medical establishment to show that progress is being made in fighting cancer. Here are some variables that might be used:

1. Total deaths from cancer. These have risen sharply over time, from 331,000 in 1970, to 505,000 in 1990, and to 572,000 in 2011.

2. The percentage of all Americans who die from cancer. The percentage of deaths due to cancer rose steadily, from 17.2% in 1970 to 23.5% in 1990, then leveled off to 23.2% in 2007.

3. The percentage of cancer patients who survive for 5 years from the time the disease was discovered. These rates are rising slowly. The 5-year survival rate was 50% in the 1975 to 1977 period and 68% from 1999 to 2006.

None of these variables is fully valid as a measure of the effectiveness of cancer treatment. Explain why Variables 1 and 2 could increase even if treatment is getting more effective, and why Variable 3 could increase even if treatment is getting less effective.

8.14 Testing job applicants. The law requires that tests given to job applicants must be shown to be directly job related. The Department of Labor believes that an employment test called the General Aptitude Test Battery (GATB) is valid for a broad range of jobs. As in the case of the SAT, blacks and Hispanics get lower average scores on the GATB than do whites. Describe briefly what must be done to establish that the GATB has predictive validity as a measure of future performance on the job.

8.15 Validity, bias, reliability. This winter I went to a local pharmacy to have my weight and blood pressure measured using a sophisticated electronic machine at the front of the store next to the checkout counter. Will the measurement of my weight be biased? Reliable? Valid? Explain your answer.

8.16 An activity on bias. Let's study bias in an intuitive measurement. Figure 8.3 is a drawing of a tilted glass. Reproduce this drawing on 10 sheets of paper. Choose 10 people: 5 men and 5 women. Explain that the drawing represents a tilted glass of water. Ask each subject to draw the water level when this tilted glass is holding as much water as it can.

Figure 8.3 A tilted glass, for Exercise 8.16. Can you draw the level of water in the glass when it is as full as possible?

The correct level is horizontal (straight back from the lower lip of the glass). Many people make large errors in estimating the level. Use a protractor to measure the angle of each subject's error. Were your subjects systematically wrong in the same direction? How large was the average error? Was there a clear difference between the average errors made by men and by women?

8.17 An activity on bias and reliability. Cut 5 pieces of string having these lengths in inches:

$$2.9 \quad 9.5 \quad 5.7 \quad 4.2 \quad 7.6$$

(a) Show the pieces to another student one at a time, asking the subject to estimate the length to the nearest tenth of an inch by eye. The error your subject makes is measured value minus true value and can be either positive or negative. What is the average of the 5 errors? Explain why this average would be close to 0 if there were no bias and we used many pieces of string rather than just 5.

(b) The following day, ask the subject to again estimate the length of each piece of string. (Present them in a different order on the second day.) Explain why the 5 differences between the first and second guesses would all be 0 if your subject were a perfectly reliable measurer of length. The bigger the differences, the less reliable your subject is. What is the average difference (ignoring whether they are positive or negative) for your subject?

8.18 More on bias and reliability. The previous exercise gives 5 true values for lengths. A subject measures each length twice by eye. Make up a set of results from this activity that matches each of the descriptions below. For simplicity, assume that bias means the same fixed error every time rather than an "on the average" error in many measurements.

(a) The subject has a bias of 0.5 inch too long and is perfectly reliable.

(b) The subject has no bias but is not perfectly reliable, so that the average difference in repeated measurements is 0.5 inch.

8.19 Even more on bias and reliability. Exercise 8.17 gives 5 true values for lengths. A subject measures the first length (true length = 2.9 inches) four times by eye. His measurements are

$$3.0 \quad 2.9 \quad 3.1 \quad 3.0$$

Suppose his measurements have a bias of +0.1 inches.

(a) What are the four random errors for his measurements?

(b) What is the variance of his four measurements?

8.20 Does job training work? To measure the effectiveness of government training programs, it is usual to compare workers' pay before and after training. But many workers sign up for training when their pay drops or they are laid off. So the "before" pay is unusually low and the pay gain looks large.

(a) Is this bias or random error in measuring the effect of training on pay? Why?

(b) How would you measure the success of training programs?

8.21 A recipe for poor reliability. Every month, the government releases data on "personal savings." This number tells us how many dollars individuals saved the previous month. Savings are calculated by subtracting personal spending (an enormously large number) from personal income (another enormous number). The result is one of the government's least reliable statistics.

Give a numerical example to show that small percentage changes in two very large numbers can produce a big percentage change in the difference between those numbers. A variable that is the difference between two big numbers is usually not very reliable.

8.22 Measuring crime. Crime data make headlines. We measure the amount of crime by the number of crimes committed or (better) by crime rates (crimes per 100,000 population). The FBI publishes data on crime in the United States by compiling crimes reported to police departments. The FBI data are recorded in the Uniform Crime Reporting program and are based on reports from approximately 17,000 law enforcement agencies across the United States. The National Crime Victimization Survey publishes data about crimes based on a national probability sample of more than 76,000 households per year. The victim survey shows almost two times as many crimes as the FBI report. Explain why the FBI report has a large downward bias for many types of crime. (Here is a case in which bias in producing data leads to bias in measurement.)

8.23 Measuring crime. Twice each year, the National Crime Victimization Survey asks a random sample of about 40,000 households whether they have been victims of crime and, if so, the details. In all, nearly 135,000 people answer these questions per year. If other people in a household are in the room while one person is answering questions, the measurement of, for example, rape and other sexual assaults could be seriously biased. Why? Would the presence of other people lead to overreporting or underreporting of sexual assaults?

8.24 Measuring pulse rate. You want to measure your resting pulse rate. You might count the number of beats in 5 seconds and multiply by 12 to get beats per minute. Why is this method less reliable than actually measuring the number of beats in a minute?

8.25 Testing job applicants. A company used to give IQ tests to all job applicants. This is now illegal because IQ is not related to the performance of workers in all the company's jobs. Does the reason for the policy change involve the *reliability,* the *bias,* or the *validity* of IQ tests as a measure of future job performance? Explain your answer.

8.26 The best earphones. You are writing an article for a consumer magazine based on a survey of the magazine's readers that asked about satisfaction with mid-priced earphones for the iPod and iPhone. Of 1648 readers who reported owning the Apple in-ear headphone with remote and mic, 347 gave it an outstanding rating. Only 69 outstanding ratings were given by the 134 readers who owned Klipsch Image S4i earphones with microphone. Describe an appropriate variable, which can be computed from these counts, to measure high satisfaction with a make of earphone. Compute the values of this variable for the Apple and Klipsch earphones. Which brand has the better high-satisfaction rating?

8.27 Where to work? Each year, *Forbes* magazine ranks the 2000 largest metropolitan areas in the United States in an article on the best places for businesses and careers. First place in 2010 went to Des Moines, Iowa. Des Moines was ranked 7th in 2009. Second place in 2010 went to Provo, Utah. Provo was ranked 40th in 2009. Durham, North Carolina, was ranked only 23rd in 2010 but was ranked 3rd in 2009. Are these facts evidence that *Forbes*'s ratings are invalid, biased, or unreliable? Explain your choice.

8.28 Validity, bias, reliability. Give your own example of a measurement process that is valid but has large bias. Then give your own example of a measurement process that is invalid but highly reliable.

EXPLORING THE WEB

8.29 Web-based exercise. Each year, *U.S. News & World Report* ranks the nation's colleges and universities. You can find the current rankings on the magazine's Web site, **www.usnews.com/rankings.** Colleges often dispute the validity of the rankings, and some even say that *U.S. News* changes its measurement process a bit each year to produce different "winners" and thus generate news. The magazine describes its methods on the Web site. Give three variables that you would use as indicators of a college's "academic excellence." You may or may not decide to use some of *U.S. News*'s variables.

8.30 Web-based exercise. You can get the time directly from the atomic clock at NIST at **www.time.gov.** There is some error due to Internet delays, but the display even tells you roughly how accurate the time on the screen is. Look up the time for your state. What is the reported accuracy?

8.31 Web-based exercise. *Money* magazine annually publishes a list of its best places to live. You can find the 2011 rankings online at **money.cnn .com/magazines/moneymag/bplive/2011/.** Read how the rankings are determined at **money.cnn.com/magazines/moneymag/bplive/2011/ faq/** and give five variables that are used to determine the rankings. Are there any additional variables that you think should be included? The *Money* magazine Web site retains links to the most recent three years of rankings. If at the time you read this the 2011 rankings are not available, replace 2011 with your current year in the above Web addresses.

NOTES AND DATA SOURCES

Page 159 Case Study: See the EESEE story "Brain Size and Intelligence." This EESEE story is based on L. Willerman, R. Schultz, J. N. Rutledge, and E. Bigler, "In vivo brain size and intelligence," *Intelligence,* 15 (1991), pp. 223–228.

Page 160 Example 1: "Trial and error," *Economist,* October 31, 1998, pp. 87–88.

Page 165 Example 7: Quotation from a FairTest press release dated August 31, 1999, and appearing on the organization's Web page, **www.fairtest.org.** Most of the other information in this example and the Statistical Controversies feature comes from chapter 2 of the National Science Foundation report *Women, Minorities, and Persons with Disabilities in Science and Engineering: 1998,* NSF99-338, 1999. The table reports r^2-values, calculated from correlations given on the College Board Web site, **www.collegeboard.org.** We found data on the most recent SAT scores at **professionals .collegeboard.com/data-reports-research/sat/.**

Page 170 Exercise 8.2: The ratings were found online at **www.docstoc.com/docs/ 25518967/ZAGAT-SURVEY-SUMMARY-2010-Los-Angeles-So–California-Restaurants.**

Page 170 Example 9: The deviations of NIST time from BIPM time are from the Web site of the NIST Time and Frequency Division, **www.nist.gov/pml/div688.**

Page 180 Exercise 8.22: The Web site for the Uniform Crime Reporting program is **www.fbi.gov/ucr/ucr.htm.** The Web site for the National Crime Victimization Survey is **bjs.ojp.usdoj.gov/index.cfm?ty= dcdetail&iid=245.**

Page 180 Exercise 8.27: Recent best places for businesses and careers can be found at **www.forbes.com/lists/.**

Do the Numbers Make Sense?

CASE STUDY Every autumn, *U.S. News & World Report* publishes a story ranking accredited four-year colleges and universities throughout the United States. These ratings by *U.S. News & World Report* are very influential in determining public opinion about the quality of the nation's colleges and universities. However, critics of the ratings question the quality of the data used to rank schools. In the January 2012 article "Gaming the College Rankings," the *New York Times* described several instances of "fudging the numbers" by colleges in order to climb in the rankings.

Business data, advertising claims, debate on public issues—we are assailed daily by numbers intended to prove a point, buttress an argument, or assure us that all is well. Sometimes, as the critics of the *U.S. News & World Report* rankings maintain, we are fed fake data. Sometimes people who use data to argue a cause care more for the cause than for the accuracy of the data. Others simply lack the skills needed to employ numbers carefully. We know that we should always ask

- How were the data produced?
- What exactly was measured?

We also know quite a bit about what good answers to these questions sound like. That's wonderful, but it isn't enough. We also need "number sense," the habit of asking if numbers make sense. Developing number sense is the purpose of this chapter. To help develop number sense, we will look at how bad data, or good data used wrongly, can trick the unwary. ∎

What didn't they tell us?

The most common way to mislead with data is to cite correct numbers that don't quite mean what they appear to say because we aren't told the full story. The numbers are not made up, so the fact that the information is a bit incomplete may be an innocent oversight. Here are some examples. You decide how innocent they are.

EXAMPLE 1   Snow! Snow! Snow!

Crested Butte attracts skiers by advertising that it has the highest average snowfall of any ski town in Colorado. That's true. But skiers want snow on the ski slopes, not in the town—and many other Colorado resorts get more snow on the slopes.

EXAMPLE 2   Yet more snow

News reports of snowstorms say things like "A winter storm spread snow across the area, causing 28 minor traffic accidents." Eric Meyer, a reporter in Milwaukee, Wisconsin, says he often called the sheriff to gather such numbers. One day he decided to ask the sheriff how many minor accidents are typical in good weather: about 48, said the sheriff. Perhaps, says Meyer, the news should say, "Today's winter storm prevented 20 minor traffic accidents."

"Sure your patients have 50% fewer cavities. That's because they have 50% fewer teeth!"

EXAMPLE 3   We attract really good students

Colleges know that many prospective students look at popular guidebooks to decide where to apply for admission. The guidebooks print information supplied by the colleges themselves. Surely no college would simply lie about, say, the average SAT score of its entering students or admission rates. But we do want our scores to look good and admission standards to appear high. How about making SAT scores optional for admissions? Students with low scores will tend not to include them as part of their application so that average scores increase. In addition, the number of applicants increases, admittance rates decrease, and a college appears more selective.

Hobart and William Smith Colleges adopted an SAT-optional policy for fall 2006, and their reported average SAT scores jumped 20 points. At the same time, national average SAT scores declined.

The point of these examples is that numbers have a context. If you don't know the context, the lonely, isolated, naked number doesn't tell you much.

Are the numbers consistent with each other?

Here is an example.

EXAMPLE 4 The case of the missing vans

Auto manufacturers lend their dealers money to help them keep vehicles on their lots. The loans are repaid when the vehicles are sold. A Long Island auto dealer named John McNamara borrowed over $6 billion from General Motors between 1985 and 1991. In December 1990 alone, Mr. McNamara borrowed $425 million to buy 17,000 GM vans customized by an Indiana company, allegedly for sale overseas. GM happily lent McNamara the money because he always repaid the loans.

Let's pause to consider the numbers, as GM should have done but didn't. The entire van-customizing industry produces only about 17,000 customized vans a month. So McNamara was claiming to buy an entire month's production. These large, luxurious, and gas-guzzling vehicles are designed for U.S. interstate highways. The recreational vehicle trade association says that only 1.35% were exported in 1990. It's not plausible to claim that 17,000 vans in a single month are being bought for export. McNamara's claimed purchases were large even when compared with total production of vans. Chevrolet, for example, produced 100,067 full-sized vans in all of 1990.

Having looked at the numbers, you can guess the rest. McNamara admitted in federal court in 1992 that he was defrauding GM on a massive scale. The Indiana company was a shell set up by McNamara, its invoices were phony, and the vans didn't exist. McNamara borrowed vastly from GM, used most of each loan to pay off the previous loan (thus establishing a record as a good credit risk), and skimmed off a bit for himself. The bit he skimmed amounted to over $400 million. GM set aside $275 million to cover its losses. Two executives, who should have looked at the numbers relevant to their business, were fired.

John McNamara fooled General Motors because GM didn't compare his numbers with others. No one asked how a dealer could buy 17,000 vans in a single month for export when the entire custom van industry produces just 17,000 vans a month and only a bit over 1% are exported. Speaking of GM, here's another example in which the numbers don't line up with each other.

EXAMPLE 5 We won!

GM's Cadillac brand was the best-selling luxury car in the United States for 57 years in a row. In 1998, Ford's Lincoln brand seemed to be winning until the last moment. Said the *New York Times,* "After reporting almost unbelievable sales results in December, Cadillac eked out a come-from-behind victory by just 222 cars." The final count was 187,343 for Cadillac, 187,121 for Lincoln. Then GM reported that Cadillac sales dropped 38% in January. How could sales be so different in December and January? Could it be that some January sales were counted in the previous year's total? Just enough, say, to win by 222 cars? Yes, indeed. In May, GM confessed that it sold 4773 fewer Cadillacs in December than it had claimed.

In the General Motors examples, we suspect something is wrong because numbers don't agree as we think they should. Here's an example where we *know* something is wrong because the numbers don't agree. This is part of an article on a cancer researcher at the Sloan-Kettering Institute who was accused of committing the ultimate scientific sin, falsifying data.

EXAMPLE 6 Fake data

"One thing he did manage to finish was a summary paper dealing with the Minnesota mouse experiments. ... That paper, cleared at SKI and accepted by the *Journal of Experimental Medicine,* contains a statistical table that is erroneous in such an elementary way that a bright grammar school pupil could catch the flaw. It lists 6 sets of 20 animals each, with the percentages of successful takes. Although any percentage of 20 has to be a multiple of 5, the percentages that Summerlin recorded were 53, 58, 63, 46, 48, and 67."

Are the numbers plausible?

As the General Motors examples illustrate, you can often detect dubious numbers simply because they don't seem plausible. Sometimes you can check an implausible number against data in reliable sources such as the annual *Statistical Abstract of the United States*. Sometimes, as the next example illustrates, you can do a calculation to show that a number isn't realistic.

EXAMPLE 7   Now that's relief!

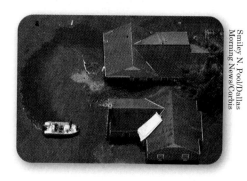

Smiley N. Pool/Dallas Morning News/Corbis

Hurricane Katrina struck the Gulf Coast in August 2005 and caused massive destruction. In September 2005, Senators Mary Landrieu (Democrat) and David Vitter (Republican) of Louisiana introduced the Hurricane Katrina Disaster Relief and Economic Recovery Act in Congress. This bill sought a total of $250 billion in federal funds to provide long-term relief and assistance to the people of New Orleans and the Gulf Coast. Not all of this was to be spent on New Orleans alone and the money was not meant to be distributed directly to residents affected by the hurricane. However, at the time several people noticed that if you were one of the 484,674 residents of New Orleans, $250 billion in federal funds was the equivalent of

$$\text{dollars per resident} = \frac{250,000,000,000}{484,674} = 515,810.6$$

This would mean that a family of four would receive about $2,063,240!

NOW IT'S YOUR TURN   **9.1 The abundant melon field.** The very respectable journal *Science,* in an article on insects that attack plants, mentioned a California field that produces 750,000 melons per acre. Is this plausible? You may want to use the fact that an acre covers 43,560 square feet.

Are the numbers too good to be true?

In Example 6, lack of consistency led to the suspicion that the data were phony. *Too much precision or regularity* can lead to the same suspicion, as when a student's lab report contains data that are exactly as the theory predicts. The laboratory instructor knows that the accuracy of the equipment and the student's laboratory technique are not good enough to give such perfect results. He suspects that the student made them up. Here is an example drawn from an article in *Science* about fraud in medical research.

EXAMPLE 8 More fake data

"Lasker had been asked to write a letter of support. But in reading two of Slutsky's papers side by side, he suspected that the same 'control' animals had been used in both without mention of the fact in either. Identical data points appeared in both articles, but . . . the actual number of animals cited in each case was different. This suggested at best a sloppy approach to the facts. Almost immediately after being asked about the statistical discrepancies, Slutsky resigned and left San Diego."

In this case, suspicious regularity (identical data points) combined with inconsistency (different numbers of animals) led a careful reader to suspect fraud.

Is the arithmetic right?

Conclusions that are wrong or just incomprehensible are often the result of plain old-fashioned blunders. Rates and percentages cause particular trouble.

EXAMPLE 9 Oh, those percents

Here are some examples involving percents. During the December 4, 2009, episode of the TV show *Fox & Friends,* a graphic was displayed with the question heading: "Did scientists falsify research to support their own theories on global warming?" The results, attributed to a Rasmussen Reports Poll on global warming, indicated that 59% of people believed this was "somewhat likely," 35% thought it was "very likely," and 26% considered it "not very likely." That adds up to a whopping 120% of those polled! Turns out that *Fox & Friends* misquoted the actual Rasmussen Reports Poll results but didn't notice the error.

Even smart people have problems with percentages. A newsletter for female university teachers asked, "Does it matter that women are 550% (five and a half times) less likely than men to be appointed to a professional grade?" Now 100% of something is all there is. If you take away 100%, there is nothing left. We have no idea what "550% less likely" might mean. Although we can't be sure, it is possible that the newsletter meant that the likelihood for women is the likelihood for men divided by 5.5. In this case, the percentage decrease would be

$$\text{percentage decrease} = \frac{\text{likelihood for men} - \text{likelihood for women}}{\text{likelihood for men}} \times 100\%$$

$$= \frac{\text{likelihood for men} - (\text{likelihood for men}/5.5)}{\text{likelihood for men}} \times 100\%$$

$$= \frac{1 - (1/5.5)}{1} \times 100\%$$

$$= \frac{4.5}{5.5} \times 100\% = 81.8\%$$

It seems that few people do arithmetic once they leave school. Those who do are less likely to be taken in by meaningless numbers. A little thought and a calculator go a long way.

EXAMPLE 10 Summertime is burglary time

An advertisement for a home security system says, "When you go on vacation, burglars go to work. According to FBI statistics, over 26% of home burglaries take place between Memorial Day and Labor Day."

This is supposed to convince us that burglars are more active in the summer vacation period. Look at your calendar. There are 14 weeks between Memorial Day and Labor Day. As a percentage of the 52 weeks in the year, this is

$$\frac{14}{52} = 0.269 \quad \text{(that is, 26.9\%)}$$

So the ad claims that 26% of burglaries occur in 27% of the year. You should not be impressed.

Just a little arithmetic mistake In 1994, an investment club of grand-motherly women wrote a best-seller, *The Beardstown Ladies' Common-Sense Investment Guide: How We Beat the Stock Market—and How You Can, Too.* On the book cover and in their many TV appearances, the down-home authors claimed a 23.4% annual return, beating the market and most professionals. Four years later, a skeptic discovered that the club treasurer had entered data incorrectly. The Beardstown ladies' true return was only 9.1%, far short of the overall stock market return of 14.9% in the same period. We all make mistakes, but most of them don't earn as much money as this one did.

EXAMPLE 11 The old folks are coming

A writer in *Science* claimed in 1976 that "people over 65, now numbering 10 million, will number 30 million by the year 2000, and will constitute an unprecedented 25 percent of the population." Sound the alarm: the elderly were going to triple in a quarter century to become a fourth of the population.

Let's check the arithmetic. Thirty million is 25% of 120 million, because

$$\frac{30}{120} = 0.25$$

So the writer's numbers make sense only if the population in 2000 is 120 million. The U.S. population in 1975 was already 216 million. Something is wrong.

Thus alerted, we can check the *Statistical Abstract of the United States* to learn the truth. In 1975, there were 22.4 million people over age 65, not 10 million. That's more than 10% of the total population. The estimate of 30 million by the year 2000 was only about 11% of the population of 281 million for that year. Looking back, we now know that people at least 65 years old were 12% of the total U.S. population. As people live longer, the numbers of the elderly are growing. But growth from 10% to 12% over 25 years is far slower than the *Science* writer claimed.

Calculating the percentage increase or decrease in some quantity seems particularly prone to mistakes. The percentage change in a quantity is found by

$$\text{percentage change} = \frac{\text{amount of change}}{\text{starting value}} \times 100$$

EXAMPLE 12 Stocks go up, stocks go down

On September 10, 2001, the NASDAQ composite index of stock prices closed at 1695.38. The next day the September 11 terrorist attacks occurred. A year later, on September 9, 2002, the NASDAQ index closed at 1304.60. What percentage decrease was this?

$$\begin{aligned}
\text{percentage change} &= \frac{\text{amount of change}}{\text{starting value}} \times 100 \\
&= \frac{1304.60 - 1695.38}{1695.38} \times 100 \\
&= \frac{-390.78}{1695.38} \times 100 = -0.230 \times 100 = -23.0\%
\end{aligned}$$

That's a sizable drop. Of course, stock prices go up as well as down. From September 10, 2002, to September 9, 2003, the NASDAQ index rose from 1320.09 to 1873.43. That's a percentage increase of

$$\frac{\text{amount of change}}{\text{starting value}} \times 100 = \frac{553.34}{1320.09} \times 100 = 41.9\%$$

Remember to always use the *starting* value, not the smaller value, in the denominator of your fraction.

> **NOW IT'S YOUR TURN**
>
> **9.2 Percentage increase and decrease.** On the first quiz of the term (worth 20 points total), a student scored a 5. On the second quiz, he scored a 10. Verify that the percentage increase from the first to the second quiz is 100%. On the third quiz, the student again scored a 5. Is it correct to say that the percentage decrease from the second to the third quiz is 100%?

A quantity can increase by any amount—a 100% increase just means it has doubled. But nothing can go down more than 100%—it has then lost 100% of its value, and 100% is all there is.

Is there a hidden agenda?

Lots of people feel strongly about various issues, so strongly that they would like the numbers to support their feelings. Often they can find support in numbers by choosing carefully which numbers to report or by working hard to squeeze the numbers into the shape they prefer. Here are two examples.

EXAMPLE 13   Heart disease in women

A highway billboard says simply, "Half of all heart disease victims are women." What might be the agenda behind this true statement? Perhaps the billboard sponsors just want to make women aware that they do face risks from heart disease. (Surveys show that many women underestimate the risk of heart disease.)

On the other hand, perhaps the sponsors want to fight what some people see as an overemphasis on male heart disease. In that case, we might want to know that although half of heart disease victims are women, they are on the average much older than male victims. Roughly 50,000 women under age 65 and 100,000 men under age 65 die from heart disease each year. The American Heart Association says, "Risk of death due to coronary heart disease in women is roughly similar to that of men 10 years younger."

EXAMPLE 14 Income inequality

During the economic boom of the 1980s and 1990s in the United States, the gap between the highest and lowest earners widened. In 1980, the bottom fifth of households received 4.3% of all income, and the top fifth received 43.7%. By 1998, the share of the bottom fifth had fallen to 3.6% of all income, and the share of the top fifth of households had risen to 49.2%. That is, the top fifth's share was almost 14 times the bottom fifth's share.

Can we massage the numbers to reduce the income gap? An article in *Forbes* (a magazine read mainly by rich folk) tried. First, according to data from the Current Population Survey, household income tends to be larger for larger households, so let's change to income per person. The rich pay more taxes, so look at income after taxes. The poor receive food stamps and other assistance, so let's count that. Finally, high earners work more hours than low earners, so we should adjust for hours worked. After all this, the share of the top fifth is only 3 times that of the bottom fifth. Of course, hours worked are reduced by illness, disability, care of children and aged parents, and so on. If *Forbes*'s hidden agenda is to show that income inequality isn't important, we may not agree.

Yet other adjustments are possible. Income, in these Census Bureau figures, does not include capital gains from, for example, selling stocks that have gone up. Almost all capital gains go to the rich, so including them would widen the income gap. *Forbes* didn't make this adjustment. Making every imaginable adjustment in the meaning of "income," says the Census Bureau, gives the bottom fifth of households 4.7% of total income in 1998 and the top fifth 45.8%.

The gap between the highest and lowest earners continues to widen. In 2010, according to the Census Bureau, the bottom fifth of households received 3.3% of all income and the top fifth received 50.2%.

STATISTICS IN SUMMARY

Chapter Specifics

- Pay attention to voluntary response samples and to confounding. Ask exactly what a number measures and decide if it is a valid measure.

- Look for the context of the numbers and ask if there is important **missing information.**

- Look for **inconsistencies,** numbers that don't agree as they should, and check for **incorrect arithmetic.**

- Compare numbers that are **implausible**—surprisingly large or small—with numbers you know are right.

- Be suspicious when numbers are **too regular or agree too well** with what their author would like to see.

- Look with special care if you suspect the numbers are put forward in support of some **hidden agenda.**

In Chapters 1 to 8 we have seen that we should always ask how the data were produced and exactly what was measured. Both affect the quality of any conclusions drawn. The goal is to gain insight by means of the numbers that make up our data. Numbers are most likely to yield insights to those who examine them closely. We need to develop "number sense," the habit of asking if the numbers make sense. To assist you, in this chapter we have provided you with examples of bad data and of good data used wrongly. If you form the habit of looking at numbers closely, your friends will soon think that you are brilliant. They might even be right.

CASE STUDY How were the abuses reported in the *New York Times* article mentioned **EVALUATED** in the Case Study that opened the chapter discovered? In some cases, unusual changes in rankings prompted a more careful look at the data on which the rankings were based. In other cases, suspicious numbers raised concerns. Examples of such suspicious numbers are discussed more thoroughly in the EESEE story "Quality of College Rankings." This story describes some "suspect" data in the 1995 *U.S. News & World Report*'s annual rankings of accredited four-year colleges and universities. Read the story and then examine the data described in the story. Are any of the numbers suspect? Why? You might look at the questions in the EESEE story to help you identify possible sources of suspicious numbers. ■

CHAPTER 9 EXERCISES

For Exercise 9.1, see page 187; for Exercise 9.2, see page 191.

9.3 Drunk driving. A newspaper article on drunk driving cited data on traffic deaths in Rhode Island: "Forty-two percent of all fatalities occurred on Friday, Saturday, and Sunday, apparently because of increased drinking on the weekends." What percent of the week do Friday, Saturday, and Sunday make up? Are you surprised that 42% of fatalities occur on those days?

9.4 Advertising painkillers. An advertisement for the pain reliever Tylenol was headlined "Why Doctors Recommend Tylenol More Than All Leading Aspirin Brands Combined." The makers of Bayer Aspirin, in a reply headlined "Makers of Tylenol, Shame on You!" accused Tylenol of misleading by giving the truth but not the whole truth. You be the detective. How is Tylenol's claim misleading even if true?

9.5 Advertising painkillers. Anacin was long advertised as containing "more of the ingredient doctors recommend most." Another over-the-counter pain reliever claimed that "doctors specify Bufferin most" over other "leading brands."

Both advertising claims were literally true; the Federal Trade Commission found them both misleading. Explain why. (*Hint:* What is the active pain reliever in both Anacin and Bufferin?)

9.6 Deer in the suburbs. Westchester County is a suburban area covering 433 square miles immediately north of New York City. A garden magazine claimed that the county is home to 800,000 deer. Do a calculation that shows this claim to be implausible.

9.7 Suicides among Vietnam veterans. Did the horrors of fighting in Vietnam drive many veterans of that war to suicide? A figure of 150,000 suicides among Vietnam veterans in the 20 years following the end of the war has been widely quoted. Explain why this number is not plausible. To help you, here are some facts: about 20,000 to 25,000 American men commit suicide each year; about 3 million men served in Southeast Asia during the Vietnam War; there were roughly 93 million adult men in the United States 20 years after the war.

9.8 Trash at sea? A report on the problem of vacation cruise ships polluting the sea by dumping garbage overboard said:

On a seven-day cruise, a medium-size ship (about 1,000 passengers) might accumulate 222,000 coffee cups, 72,000 soda cans, 40,000 beer cans and bottles, and 11,000 wine bottles.

Are these numbers plausible? Do some arithmetic to back up your conclusion. Suppose, for example, that the crew is as large as the passenger list. How many cups of coffee must each person drink every day?

9.9 Funny numbers. Here's a quotation from a book review in a scientific journal:

...a set of 20 studies with 57 percent reporting significant results, of which 42 percent agree on one conclusion while the remaining 15 percent favor another conclusion, often the opposite one.

Do the numbers given in this quotation make sense? Can you decide how many of the 20 studies agreed on "one conclusion," how many favored another conclusion, and how many did not report significant results?

9.10 Airport delays. An article in a midwestern newspaper about flight delays at major airports said:

According to a Gannett News Service study of U.S. airlines' performance during the past five months, Chicago's O'Hare Field scheduled 114,370 flights. Nearly 10 percent, 1,136, were canceled.

Check the newspaper's arithmetic. What percent of scheduled flights from O'Hare were actually canceled?

9.11 How many miles do we drive? Here is an excerpt from Robert Sullivan's "A Slow-Road Movement?" in the Sunday magazine section of the *New York*

Times on June 25, 2006:

According to the Automobile Association of America, in 1956, Americans drove 628 million miles; in 2002, 2.8 billion. The even bigger story is trucks. In 1997, according to the Department of Transportation, the Interstate System handled more than 1 trillion ton-miles of stuff, a feat executed by 21 million truckers driving approximately 412 billion miles.

(a) There were at least 100 million drivers in the United States in 2002. How many miles per driver per year is 2.8 billion miles? Does this seem plausible?

(b) According to the report, on average how many miles per year do truckers drive? Does this seem plausible?

(c) Check the most recent *Statistical Abstract of the United States* at **www.census.gov** and determine how many miles per year Americans actually drive.

9.12 Battered women? A letter to the editor of the *New York Times* complained about a *Times* editorial that said "an American woman is beaten by her husband or boyfriend every 15 seconds." The writer of the letter claimed that "at that rate, 21 million women would be beaten by their husbands or boyfriends every year. That is simply not the case." He cited the National Crime Victimization Survey, which estimated 56,000 cases of violence against women by their husbands and 198,000 by boyfriends or former boyfriends. The survey showed 2.2 million assaults against women in all, most by strangers or someone the woman knew who was not her past or present husband or boyfriend.

(a) First do the arithmetic. Every 15 seconds is 4 per minute. At that rate, how many beatings would take place in an hour? In a day? In a year? Is the letter writer's arithmetic correct?

(b) Is the letter writer correct to claim that the *Times* overstated the number of cases of domestic violence against women?

9.13 We can read, but can we count? The Census Bureau once gave a simple test of literacy in English to a random sample of 3400 people. The *New York Times* printed some of the questions under the headline "113% of Adults in U.S. Failed This Test." Why is the percent in the headline clearly wrong?

9.14 Stocks go down. On September 29, 2008, the Dow Jones Industrial Average dropped 778 points from its opening level of 11,143. This was the biggest one-day decline ever. By what percentage did the Dow drop that day? On October 28, 1929, the Dow Jones Industrial Average dropped 38 points from its opening level of 299. By what percentage did the Dow drop that day? This was the second-biggest one-day percentage drop ever.

9.15 Poverty. The number of Americans living below the official poverty line increased from 24,975,000 to 43,569,000 in the 34 years between 1976 and 2009. What percentage increase was this? You should not conclude from this that poverty grew more common in these years, however. Why not?

9.16 Reducing CO_2 emissions. An online article reported the following.

In order to limit warming to two degrees Celsius and permit development, several developed countries would have to reduce their CO_2 emissions by more than 100 percent.

Explain carefully why it is impossible to reduce anything by more than 100%.

9.17 Are men more promiscuous? On August 12, 2007, the *New York Times* reported the following.

Everyone knows men are promiscuous by nature. It's part of the genetic strategy that evolved to help men spread their genes far and wide. The strategy is different for a woman, who must go through so much just to have a baby and then nurture it. She is genetically programmed to want just one man who will stick with her and help raise their children.

Surveys bear this out. In study after study and in country after country, men report more, often many more, sexual partners than women.

One survey, recently reported by the U.S. government, concluded that men had a median of seven female sex partners. Women had a median of four male sex partners. Another study, by British researchers, stated that men had 12.7 heterosexual partners in their lifetimes and women had 6.5.

But there is just one problem, mathematicians say.

What is this problem?

9.18 Don't dare to drive? A university sends a monthly newsletter on health to its employees. One issue included a column called "What Is the Chance?" that said:

Chance that you'll die in a car accident this year: 1 in 75.

There are about 310 million people in the United States. About 40,000 people die each year from motor vehicle accidents. What is the chance a typical person will die in a motor vehicle accident this year?

9.19 How many miles of highways? *Organic Gardening* magazine once said that "the U.S. Interstate Highway System spans 3.9 million miles and is wearing out 50% faster than it can be fixed. Continuous road deterioration adds $7 billion yearly in fuel costs to motorists." The distance from the east coast to the west coast of the United States is about 3000 miles. How many separate highways across the continent would be needed to account for 3.9 million miles of roads? What do you conclude about the number of miles in the interstate system?

9.20 In the garden. *Organic Gardening* magazine, describing how to improve your garden's soil, said, "Since a 6-inch layer of soil in a 100-square-foot plot weighs about 45,000 pounds, adding 230 pounds of compost will give you an instant 5% organic matter."

(a) What percent of 45,000 is 230?

(b) Water weighs about 62 pounds per cubic foot. There are 50 cubic feet in a garden layer 100 square feet in area and 6 inches deep. What would 50 cubic feet of water weigh? Is it plausible that 50 cubic feet of soil weighs 45,000 pounds?

(c) It appears from (b) that the 45,000 pounds isn't right. In fact, soil weighs about 75 pounds per cubic foot. If we use the correct weight, is the "5% organic matter" conclusion roughly correct?

9.21 No eligible men? A news report quotes a sociologist as saying that for every 233 unmarried women in their 40s in the United States, there are only 100 unmarried men in their 40s. These numbers point to an unpleasant social situation for women of that age. Are the numbers plausible? (*Optional:* The *Statistical Abstract of the United States* has a table titled "Marital status of the population by age and sex" that gives the actual counts.)

9.22 Too good to be true? The late English psychologist Cyril Burt was known for his studies of the IQ scores of identical twins who were raised apart. The high correlation between the IQs of separated twins in Burt's studies pointed to heredity as a major factor in IQ. ("Correlation" measures how closely two variables are connected. We will meet correlation in Chapter 14.) Burt wrote several accounts of his work, adding more pairs of twins over time. Here are his reported correlations as he published them:

Publication date	Twins reared apart	Twins reared together
1955	0.771 (21 pairs)	0.944 (83 pairs)
1966	0.771 (53 pairs)	0.944 (95 pairs)

What is suspicious here?

9.23 Where you start matters. When comparing numbers over time, you can slant the comparison by choosing your starting point. Say the Chicago Cubs lose 5 games, then win 4, then lose 1. You can truthfully say that the Cubs have lost 6 of their last 10 games (sounds bad) or that they have won 4 of their last 5 (sounds good).

The median income of American households (in dollars of 2009 buying power) was $47,637 in 1990, $52,301 in 2000, and $49,777 in 2009. By what percentage did household income increase between 1990 and 2009? Between 2000 and 2009? You see that you can make the income trend sound bad or good by choosing your starting point.

9.24 Being on top also matters. The previous exercise noted that median household income decreased slightly between 2000 and 2009. The top 5% of households earned $180,879 or more in 2000 and $180,001 or more in 2009. (These amounts are in dollars of 2009 buying power.) By what percentage did the income of top earners decrease between 2000 and 2009? How does this compare with the percentage decrease in median household income between 2000 and 2009?

9.25 Boating safety. Data on accidents in recreational boating in the *Statistical Abstract of the United States* show that the number of deaths has dropped from 1360 in 1980 to 736 in 2009. However, the number of injuries reported grew from 2650 in 1980 to 3358 in 2009. Why are there so few deaths in these government data relative to the number of injuries? Which count (deaths or injuries) is probably more accurate? Why might the injury count rise when deaths do not?

9.26 Obesity and income. An article in the November 3, 2009, issue of the *Guardian* reported, "A separate opinion poll yesterday suggested that 50% of obese people earn less than the national average income." Is this evidence that obese people tend to earn less than other workers?

EXPLORING THE WEB

9.27 Web-based exercise. Find an example of one of the following. Explain in detail the statistical shortcomings of your example.

Leaving out essential information

Lack of consistency

Implausible numbers

Faulty arithmetic

One place to look is in the online *Chance News*. The CHANCE Web site contains lots of interesting stuff (at least if you are interested in statistics). In particular, the *Chance News* section, **http://test.causeweb.org/wiki/chance/index.php/Main_Page**, offers a Wiki that keeps track of statistics in the press, including dubious statistics.

9.28 Web-based exercise. In Exercise 9.14 you were asked to compute the percentage drop in the Dow Jones Industrial Average on September 29, 2008. Because the index had gone up so much in previous years, this was not the biggest percentage drop ever. Go to **www.djaverages.com/?view=industrial&page=milestones** and determine if this is even in the top 10 biggest percentage drops. Explain, in plain language, how the drop on September 29, 2008, can be the biggest drop in actual value but not in percentage.

9.29 Web-based exercise. The *Statistical Abstract of the United States* is an essential compilation of data. You can find it online at the Census Bureau Web site. The most recent edition is at **www.census.gov/compendia/statab/**. To find what you want, you can use the search tool at the Web site. Use the *Statistical Abstract* to determine the percentage change in the annual average price of regular gasoline in New York City for the two most recent years reported.

NOTES AND DATA SOURCES

Page 183 Case Study: The *New York Times* article discussed in the Case Study was found online at **www.nytimes.com/2012/ 02/01/education/gaming-the-college-rankings.html.**

Page 184 Example 1: R. J. Newan, "Snow job on the slopes," *US News & World Report,* December 17, 1994, pp. 62–65.

Page 184 Example 2: Reported by Robert Niles of the *Los Angeles Times,* at **www.nilesonline.com.**

Page 184 Example 3: "The other side of 'test optional,' " *New York Times,* July 20, 2009.

Page 185 Example 4: The "missing vans" case is based on news articles in the *New York Times,* April 18, 1992, and September 10, 1992.

Page 186 Example 5: Robyn Meredith, "Oops, Cadillac says, Lincoln won after all," *New York Times,* May 6, 1999.

Page 186 Example 6: B. Yuncker, "The strange case of the painted mice," *Saturday Review/World,* November 30, 1974, p. 53.

Page 188 Example 8: E. Marshall, "San Diego's tough stand on research fraud," *Science,* 234 (1986), pp. 534–535.

Page 188 Example 9: The second item is from *Chance News,* **http://test.causeweb .org/wiki/chance/index.php/Main_Page.**

Page 189 Example 10: Darryl Nester spotted this ad.

Page 189 Example 11: *Science,* 192 (1976), p. 1081.

Page 190 Example 12: Historical prices of the NASDAQ composite index can be found at **finance.yahoo.com/q?s=%5EIXIC.**

Page 191 Example 13: Quotation from the American Heart Association statement "Cardiovascular disease in women," found at **www.americanheart.org**.

Page 192 Example 14: Edwin S. Rubenstein, "Inequality," *Forbes,* November 1, 1999, pp. 158–160; and Bureau of the Census, *Money Income in the United States, 1998.*

Page 193 Exercise 9.3: *Providence (R.I.) Journal,* December 24, 1999. We found the article in *Chance News* 9.02.

Page 194 Exercise 9.6: *Fine Gardening,* September/October 1989, p. 76.

Page 194 Exercise 9.8: *Condé Nast Traveler* magazine, June 1992.

Page 194 Exercise 9.9: *Science,* 189 (1975), p. 373.

Page 194 Exercise 9.10: *Lafayette (Ind.) Journal and Courier,* October 23, 1988.

Page 194 Exercise 9.11: This example is courtesy of Professor Steve Samuels, Purdue University.

Page 195 Exercise 9.12: Letter by L. Jarvik in the *New York Times,* May 4, 1993. The editorial, "Muggings in the kitchen," appeared on April 23, 1993.

Page 195 Exercise 9.13: *New York Times,* April 21, 1986.

Page 196 Exercise 9.16: Found online at **www.dissentmagazine.org/online.php? id=315.**

Page 196 Exercise 9.19: *Organic Gardening,* July 1983.

Page 196 Exercise 9.20: *Organic Gardening,* March 1983.

The first and most important question to ask about any statistical study is "Where do the data come from?" Chapter 1 addressed this question. The distinction between observational and experimental data is a key part of the answer. Good statistics starts with good designs for producing data. Chapters 2, 3, and 4 discussed sampling, the art of choosing part of a population to represent the whole. Figure I.1 summarizes the big idea of a simple random sample. Chapters 5 and 6 dealt with the statistical aspects of designing experiments, studies that impose some treatment in order to learn about the response. The big idea is the randomized comparative experiment. Figure I.2 outlines the simplest design.

Random sampling and randomized comparative experiments are perhaps the most important statistical inventions of the 20th century. Both were slow to gain acceptance, and you will still see many voluntary response samples and uncontrolled experiments. Both random samples and randomized experiments involve the deliberate use of chance to eliminate bias and produce a regular pattern of outcomes. The regular pattern allows us to give margins of error, make confidence statements, and assess the statistical significance of conclusions based on samples or experiments.

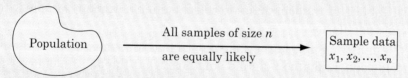

Figure I.1 The idea of a simple random sample.

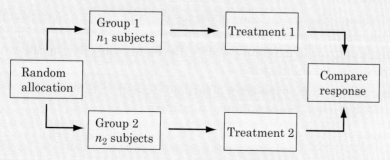

Figure I.2 The idea of a randomized comparative experiment.

When we collect data about people, ethical issues can be important. Chapter 7 discussed these issues and introduced three principles that apply to any study with human subjects. The last step in producing data is to measure the characteristics of interest to produce numbers we can work with. Measurement was the subject of Chapter 8. "Where do the data come from?" is the first question we should ask about a study, and "Do the numbers make sense?" is the second. Chapter 9 encouraged the valuable habit of looking skeptically at numbers before accepting what they seem to say.

PART I SUMMARY

Here are the most important skills you should have acquired after reading Chapters 1 to 9.

A. DATA

1. Recognize the individuals and variables in a statistical study.
2. Distinguish observational from experimental studies.
3. Identify sample surveys, censuses, and experiments.

B. SAMPLING

1. Identify the population in a sampling situation.
2. Recognize bias due to voluntary response samples and other inferior sampling methods.
3. Use Table A of random digits to select a simple random sample (SRS) from a population.
4. Explain how sample surveys deal with bias and variability in their conclusions. Explain in simple language what the margin of error for a sample survey result tells us and what "95% confidence" means.
5. Use the quick method to get an approximate margin of error for 95% confidence.
6. Understand the distinction between sampling errors and nonsampling errors. Recognize the presence of undercoverage and nonresponse as sources of error in a sample survey. Recognize the effect of the wording of questions on the responses.
7. Use random digits to select a stratified random sample from a population when the strata are identified.

C. EXPERIMENTS

1. Identify the explanatory variables, treatments, response variables, and subjects in an experiment.

2. Recognize bias due to confounding of explanatory variables with lurking variables in either an observational study or an experiment.

3. Outline the design of a completely randomized experiment using a diagram like that in Figure I.2. Such a diagram should show the sizes of the groups, the specific treatments, and the response variable.

4. Use Table A of random digits to carry out the random assignment of subjects to groups in a completely randomized experiment.

5. Make use of matched pairs or block designs when appropriate.

6. Recognize the placebo effect. Recognize when the double-blind technique should be used. Be aware of weaknesses in an experiment, especially in the ability to generalize its conclusions.

7. Explain why a randomized comparative experiment can give good evidence for cause-and-effect relationships.

8. Explain the meaning of statistical significance.

D. OTHER TOPICS

1. Explain the three first principles of data ethics. Discuss how they might apply in specific settings.

2. Explain how measuring leads to clearly defined variables in specific settings.

3. Evaluate the validity of a variable as a measure of a given characteristic, including predictive validity.

4. Explain how to reduce bias and improve reliability in measurement.

5. Recognize inconsistent numbers, implausible numbers, numbers so good they are suspicious, and arithmetic mistakes.

6. Calculate percentage increase or decrease correctly.

PART I REVIEW EXERCISES

Review exercises are short and straightforward exercises that help you solidify the basic ideas and skills in each part of this book. We have provided "hints" that indicate where you can find the relevant material for the odd-numbered problems.

I.1 Know these terms. A friend who knows no statistics has encountered some statistical terms in reading for her psychology course. Explain each of the following terms in one or two simple sentences.

(a) Simple random sample. (*Hint:* See page 26.)

(b) 95% confidence. (*Hint:* See pages 45–46.)

(c) Statistically significant. (*Hint:* See page 101.)

(d) Informed consent. (*Hint:* See pages 138–140.)

I.2 Know these terms. A friend who knows no statistics has encountered some statistical terms in her biology course. Explain each of the following terms in one or two simple sentences.

(a) Observational study.

(b) Placebo effect.

(c) Nonsampling error.

(d) Institutional review board.

I.3 A biased sample. You see a woman student standing in front of the student center, now and then stopping other students to ask them questions. She says that she is collecting student opinions for a class assignment. Explain why this sampling method is almost certainly biased. (*Hint:* See pages 23–26.)

I.4 Select an SRS. A student at a large university wants to study the responses that students receive when calling an academic department for information. She selects an SRS of 5 departments from the following list for her study. Use Table A at line 115 to do this.

Accounting	Electrical Engineering	Natural Resources
Architecture	Elementary Education	Nursing
Art	English	Pharmacy
Biology	Foreign Languages	Philosophy
Business Administration	History	Physics
Chemistry	Horticulture	Political Science
Communication	International Studies	Pre-med
Computer Science	Marketing	Psychology
Dance	Mathematics	Sociology
Economics	Music	Veterinary Science

I.5 Select an SRS. The faculty grievance system at a university specifies that a 4-member hearing panel shall be drawn at random from the 25-member grievance committee. Use Table A at line 108 to draw an SRS of size 4 from the following committee members.

01 Allen	06 Frazier	11 Lehman	16 Roy	21 Wei
02 Bose	07 Jacroux	12 Majumdar	17 Svenson	22 Whittinghill
03 Chang	08 Kao	13 Marin	18 Ting	23 Williams
04 Dean	09 Kumar	14 Miller	19 Toman	24 Zhang
05 Donley	10 Lam	15 Quan	20 Tsai	25 Zhu

(*Hint:* See pages 26–31.)

I.6 Errors in surveys. Give an example of a source of nonsampling error in a sample survey. Then give an example of a source of sampling error.

I.7 Errors in surveys. An overnight opinion poll calls randomly selected telephone numbers. This polling method misses all people without a phone. Is this a source of nonsampling error or of sampling error? Does the poll's announced margin of error take this source of error into account? (*Hint:* See pages 62–67.)

I.8 Errors in surveys. A college chooses an SRS of 100 students from the registrar's list of all undergraduates to interview about student life. If it selected two SRSs of 100 students at the same time, the two samples would give somewhat different results. Is this variation a source of sampling error or of nonsampling error? Does the survey's announced margin of error take this source of error into account?

I.9 Errors in surveys. Exercises I.7 and I.8 each mention a source of error in a sample survey. Would each kind of error be reduced by doubling the size of the sample with no other changes in the survey procedures? Explain your answers. (*Hint:* See pages 43–45 and pages 62–67.)

I.10 Errors in surveys. A Gallup Poll found that 30% of adult Americans claim to have postponed medical treatment due to cost. The Gallup press release says:

Results are based on telephone interviews with 1,014 national adults, aged 18 and older, conducted Nov. 11–14, 2007. For results based on the total sample of national adults, one can say with 95% confidence that the maximum margin of error is ±3 percentage points.

The release also points out that this margin of error is due only to sampling error. Give one example of a source of error in the poll result that is *not* included in this margin of error.

I.11 Find the margin of error. A CNN/Opinion Research Corporation survey conducted December 16 to 20, 2010, asked 1160 adult Americans their opinion of Tiger Woods, a few weeks after his car crash and infidelity admission. Only 34% of those surveyed gave Woods a "favorable" rating. In 2005, the star golfer's favorable rating was at 85%.

(a) What is the population for this 2010 survey? (*Hint:* See pages 9–11.)

(b) Assuming the 2010 survey used random sampling, use the quick method to find a margin of error. Then give a complete confidence statement for a conclusion about the population. (*Hint:* See pages 46–49.)

I.12 Find the margin of error. A survey conducted in June 2010 asked 512 American adults, "How would you rate the job British Petroleum, or BP, has done in responding to the Gulf oil spill, as very good, good, poor or very poor?" Four hunded and fifteen said, "Poor" or "Very poor."

(a) What is the population for this sample survey?

(b) Assuming the sample was a random sample, use the quick method to find a margin of error. Make a confidence statement about the opinion of the population.

I.13 What kind of sample? At a party there are 30 students over age 21 and 20 students under age 21. You choose at random 3 of those over 21 and separately choose at random 2 of those under 21 to interview about attitudes toward alcohol. You have given every student at the party the same chance to be interviewed: what is that chance? Why is your sample not an SRS? What is this kind of sample called? (*Hint:* See pages 26–27 and pages 71–72.)

I.14 Design an experiment. A university's Department of Statistics wants to attract more majors. It prepares two advertising brochures. Brochure A stresses the intellectual excitement of statistics. Brochure B stresses how much money statisticians make. Which will be more attractive to first-year students? You have a questionnaire to measure interest in majoring in statistics, and you have 50 first-year students to work with. Outline the design of an experiment to decide which brochure works better.

I.15 Design an experiment. Gary thinks that the way to get a date is to tell a woman about yourself. Greg thinks that getting her to talk about herself works better. You recruit 20 guys who are willing to try either method in conversation and then call the woman a day later to ask for a date. Outline the design of an experiment to decide which method succeeds more often. (*Hint:* See pages 96–98.)

Exercises I.16 to I.19 are based on an article in the *Journal of the American Medical Association* that asks if reducing levels of a certain amino acid, homocysteine, in subjects affects the occurrence of major vascular events (such as major coronary events or strokes) in survivors of myocardial infarction (heart attack). Too much homocysteine in the blood is associated with a higher risk of coronary heart disease, stroke, and peripheral vascular disease, but it is not known if the association is causal. The article reports a study in which folic acid plus vitamin B12 were used to reduce homocysteine levels and subjects were monitored for 10 years. Here is information from the article's summary:

Design Randomized, double-blind, placebo-controlled trial conducted in the United Kingdom between 1998 and 2008.

Participants A total of 12,064 survivors of myocardial infarction in secondary-care hospitals in the United Kingdom.

Intervention Participants were randomized to receive either 2 mg folic acid plus 1 mg vitamin B12 daily ($n = 6033$) or a matching placebo ($n = 6031$).

Results During 6.7 years of follow-up, major vascular events occurred in 1537 of 6033 participants (25.5%) allocated folic acid plus vitamin B12 vs. 1493 of 6031 participants (24.8%) allocated placebo. The numbers of deaths attributed to vascular causes for those receiving the vitamins was 578 (9.6%), vs. 559 (9.3%) for those receiving the placebo.

I.16 Know these terms. Explain in one sentence each what "randomized," "double-blind," and "placebo-controlled" mean in the description of the design of the study.

I.17 Experiment basics. Identify the subjects, the explanatory variable, and several response variables for this study. (*Hint:* See pages 91–93.)

I.18 Design an experiment. Use a diagram to outline the design of the experiment in this medical study.

I.19 Ethics. What are the three first principles of data ethics? Explain briefly what the medical study must do to apply each of these principles. (*Hint:* See pages 137–139.)

I.20 Measuring. Joni wants to measure the degree to which male college students belong to the political left. She decides simply to measure the length of their hair—longer hair will mean more left-wing.

(a) Is this method likely to be reliable? Why?

(b) This measurement appears to be invalid. Why?

(c) Nevertheless, it is possible that measuring politics by hair length might have some predictive validity. Explain how this could happen.

I.21 Reliability. You are laboring through a chemistry laboratory assignment in which you measure the conductivity of a solution. What does it mean for your measurement to be reliable? How can you improve the reliability of your final result? (*Hint:* See pages 166–172.)

I.22 Observation or experiment? The Nurses' Health Study has queried a sample of over 100,000 female registered nurses every two years since 1976. Beginning in 1980, the study asked questions about diet, including alcohol consumption. The researchers concluded that "light-to-moderate drinkers had a significantly lower risk of death" than either nondrinkers or heavy drinkers.

(a) Is the Nurses' Health Study an observational study or an experiment? Why?

(b) What does "significant" mean in a statistical report?

(c) Suggest some lurking variables that might explain why moderate drinkers have lower death rates than nondrinkers. (The study did adjust for these variables.)

I.23 Observation or experiment? In a study of the relationship between physical fitness and personality, middle-aged college faculty who have volunteered for an exercise program are divided into low-fitness and high-fitness groups on the basis of a physical examination. All subjects then take a personality test. The high-fitness group has a higher average score for "self-confidence."

(a) Is this an observational study or an experiment? Why? (*Hint:* See pages 7–8 and 12–14.)

(b) We cannot conclude that higher fitness causes higher self-confidence. Suggest other relationships among these variables and perhaps lurking variables that might explain the higher self-confidence of the high-fitness group. (*Hint:* See pages 7–9 and 93–96.)

I.24 Percents up and down. Between January 12, 2009, and January 16, 2012, the average price of regular gasoline increased from $1.78 per gallon to $3.39 per gallon.

(a) Verify that this is a 90% increase in price.

(b) If the price of gasoline decreases by 90% from its January 16, 2012, level of $3.39 per gallon, what would be the new price? Notice that a 90% increase followed by a 90% decrease does not take us back to the starting point.

I.25 Percentage decrease. On Monday, September 10, 2001 (the day before the September 11 attacks), the NASDAQ stock index closed the day at 1695. By the end of Monday, September 17, 2001 (the first full day of trading after the attacks), the NASDAQ stock index had dropped to 1580. By what percentage did the index drop? (*Hint:* See pages 188–191.)

I.26 An implausible number? *Newsweek* once said in a story that a woman who is not currently married at age 40 has a better chance of being killed by a terrorist than of getting married. Do you think this is plausible? What kind of data would help you check this claim?

PART I PROJECTS

Projects are longer exercises that require gathering information or producing data and that emphasize writing a short essay to describe your work. Many are suitable for teams of students.

Project 1. Design your own sample survey. Choose an issue of current interest to students at your school. Prepare a short (no more than five questions) questionnaire to determine opinions on this issue. Choose a sample of about 25 students, administer your questionnaire, and write a brief description of your findings. Also write a short discussion of your experiences in designing and carrying out the survey.

(Although 25 students are too few for you to be statistically confident of your results, this project centers on the practical work of a survey. You must first identify a population; if it is not possible to reach a wider student population, use students enrolled in this course. Did the subjects find your questions clear? Did you write the questions so that it was easy to tabulate the responses? At the end, did you wish you had asked different questions?)

Project 2. Measuring. Exercise 6.16 (page 132) asks you to outline the design of a matched pairs experiment to see if the right hand is stronger than the left hand in right-handed people. You can measure hand strength by asking a subject to squeeze a bathroom scale that is placed on a shelf so that its end protrudes. Using several subjects, try to determine whether this method of measuring hand strength is reliable. Write an account of your findings. For example, did you find that subjects used different grips, so that careful instructions were needed to get a consistent way of measuring? Prepare written instructions for subjects.

Project 3. Experimenting. After you or other members of your team have refined the measurement of hand strength in the previous project, carry out the matched pairs experiment of Exercise 6.16 (page 132) with at least 10 subjects. Write a report that describes the randomization, gives the data, reports the differences in strength (right hand minus left hand), and says whether your small experiment seems to show that the right hand is stronger, on the average.

Project 4. Describe a medical study. Go to the Web site of the *Journal of the American Medical Association* (**http://jama.ama-assn.org**). Unlike the *New England Journal of Medicine* (**http://content.nejm.org**), *JAMA* makes the full text of some articles freely available online. Select an article from the current issue or from a past issue that describes a study whose topic interests you. Write a newspaper article that summarizes the design and the findings of the study. (Be sure to include statistical aspects, such as observational study versus experiment and any randomization used. News accounts often neglect these facts.)

Project 5. Do you drive a hybrid? Are students at your college more or less likely than staff to drive a hybrid? Design and carry out a study to find out, and write a report describing your design and your findings. You must first be clear about what "hybrid" means so that each car is clearly a hybrid or not. Then you must locate a suitable sample of cars—perhaps from a student parking area and a staff parking area. If the areas are large, you will want to sample the cars rather than look at all of them. Consider using a systematic sample (Exercise 4.27, page 86).

Project 6. Data ethics. Locate a news discussion of an ethical issue that concerns statistical studies. Write your own brief summary of the debate and any conclusions you feel you can reach.

Here is an example of one way to approach this project. Testing new drugs on human subjects continues to be an ongoing concern in medical studies. How does one balance the need for knowledge with protecting subjects from possible harm? Searching the archives at the Web site of the *New York Times* (**www.nytimes.com**) for "experiments and ethics" (to use the *New York Times* search engine you should enter "+experiments +ethics") one finds many articles, including a promising one in the issue of February 21, 2010. To read the article, you may be able to link to it directly online, but if not, you will have to either pay a fee or go to the library. You could also try searching the Web with Google. We entered "drugs and human guinea pigs" and found some possible leads.

Project 7. Measuring income. What is the "income" of a household? Household income may determine eligibility for government programs that assist "low-income" people. Income statistics also have political effects. Political conservatives often argue that government data overstate how many people are poor because the data include only money income, leaving out the value of food stamps and subsidized housing. Political liberals reply that the government should measure money income so it can see how many people need help.

You are on the staff of a member of Congress who is considering new welfare legislation. Write an exact definition of "income" for the purpose of determining which households are eligible for welfare. A short essay will be needed. Will you include nonmoney income such as the value of food stamps or subsidized housing? Will you allow deductions for the cost of child care needed to permit the parent to work? What about assets that are worth a lot but do not produce income, such as a house?

NOTES AND DATA SOURCES

Page 205 Exercise I.10: Gallup polls are described on the Gallup Web site, **www.gallup.com.** The quotation in Exercise I.10 is from Magali Rheault, "Three in 10 have postponed medical treatment due to cost," Gallup press release dated December 14, 2007.

Page 206 Exercises I.16 to I.19: Jane M. Armitage et al., "Effects of homocysteine-lowering with folic acid plus vitamin B12 vs placebo on mortality and major morbidity in myocardial infarction survivors," *Journal of the American Medical Association,* 303 (2010), pp. 2486–2494.

Organizing Data

Words alone don't tell a story. A writer organizes words into sentences and organizes the sentences into a story line. If the words are badly organized, the story isn't clear. Data also need organizing if they are to tell a clear story. Too many words obscure a subject rather than illuminate it. Vast amounts of data are even harder to digest—we often need a brief summary to highlight essential facts. How to organize, summarize, and present data are our topics in the second part of this book.

Organizing and summarizing a large body of facts opens the door to distortions, both unintentional and deliberate. This is no less (but also no more) the case when the facts take the form of numbers rather than words. We will point out some of the traps that data presentations can set for the unwary. Those who picture statistics as primarily a piece of the liar's art concentrate on the part of statistics that deals with summarizing and presenting data. We claim that misleading summaries and selective presentations go back to that after-the-apple conversation among Adam, Eve, and God. Don't blame statistics. Do remember the saying "Figures won't lie, but liars will figure," and beware.

Graphs, Good and Bad

CASE STUDY Americans are reading less and reading skills are declining. "The habit of daily reading, for instance, overwhelmingly correlates with better reading skills and higher academic performance." So says a report by the National Endowment for the Arts. The report included Figure 10.1. The graphic was one among many tables and graphs intended to convince one that reading proficiency is declining. The general trend in the graph is downward. Should we be concerned?

Statistics deals with data and we use tables and graphs to present data. Tables and graphs help us see what the data say. But not all tables and graphs do so accurately or clearly. In this chapter you will learn some basic methods for displaying data and how to assess the quality of the graphics you see in the media. By the end of the chapter you will be able to determine whether Figure 10.1 is a good or a bad graphic. ■

Data tables

Take a look at the *Statistical Abstract of the United States,* an annual volume packed with almost every variety of numerical information. Has the number of private elementary and secondary schools grown over time? What about minority enrollments in these schools? How many college degrees were given in each of the past several years, and how were these degrees divided among fields of study and by the age, race, and sex of the students? You can find all this and more in the education section of the *Statistical Abstract.* The tables *summarize* data. We don't want to see information on every college degree individually, only the counts in categories of interest to us.

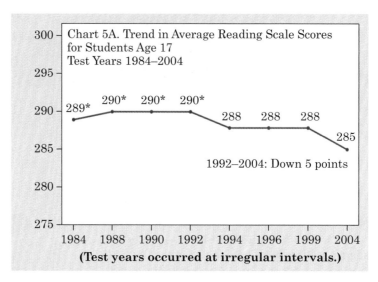

Figure 10.1 Graph that appeared in the National Endowment for the Arts report *To Read or Not to Read,* November 2007.

EXAMPLE 1 What makes a clear table?

How well educated are adults? Table 10.1 presents the data for people aged 25 years and over. This table illustrates some good practices for data tables. It is clearly *labeled* so that we can see the subject of the data at once. The main heading describes the general subject of the data and gives the date because these data will change over time. Labels within the table identify the variables and state the *units* in which they are measured. Notice, for example, that the counts are in thousands. The *source* of the data appears at the foot of the table. This Census Bureau publication in fact presents data from our old friend, the Current Population Survey.

TABLE 10.1 Education of people aged 25 years and over in 2009

Level of education	Number of persons (thousands)	Percent
Less than high school	26,415	13.3
High school graduate	61,626	31.1
Some college, no degree	33,832	17.1
Associate's degree	17,838	9.0
Bachelor's degree	37,635	19.0
Advanced degree	20,938	10.6
Total	198,285	100.0

Source: Census Bureau, *Educational Attainment in the United States: 2009.*

Table 10.1 starts with the *counts* of people aged 25 years and over who have attained each level of education. *Rates* (percentages or proportions) are often clearer than counts—it is more helpful to hear that 13.3% of this age group did not finish high school than to hear that there are 26,415,000 such people. The percentages also appear in Table 10.1. The last two columns of the table present the *distribution* of the variable "level of education" in two alternate forms. Each of these columns gives information about what values the variable takes and how often it takes each value.

Distribution of a variable

The **distribution** of a variable tells us what values it takes and how often it takes these values.

EXAMPLE 2 Roundoff errors

Did you check Table 10.1 for consistency? The total number of people should be

$$26{,}415+61{,}626+33{,}832+17{,}838+37{,}635+20{,}938 = 198{,}284 \text{ (thousands)}$$

But the table gives the total as 198,285. What happened? Each entry is rounded to the nearest thousand. The rounded entries don't quite add to the total, which is rounded separately. Such **roundoff errors** will be with us from now on as we do more arithmetic.

It is not uncommon to see roundoff errors in tables. For example, when table entries are percentages or proportions, the total may sum to a value slightly different from 100% or 1. The percentages in Table 10.1 add to 100.1%, not 100%.

Pie charts and bar graphs

The distribution in Table 10.1 is quite simple because "level of education" has only 6 possible values. To picture this distribution in a graph, we might use a **pie chart.** Figure 10.2 is a pie chart of the level of education of people aged 25 years and over. Pie charts show how a whole is divided into parts. To make a pie chart, first draw a circle. The circle represents the whole, in this case all people aged 25 years and over. Wedges within the circle represent the parts, with the angle spanned by each wedge in proportion to the size of that part. For example, 19.0% of those in this age group and over have a bachelor's degree but not an advanced degree. Because there

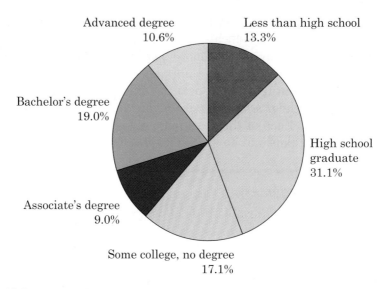

Figure 10.2 Pie chart of the distribution of level of education among persons aged 25 years and over in 2009. (This figure was created using the Minitab software package.)

are 360 degrees in a circle, the "bachelor's degree" wedge spans an angle of

$$0.19 \times 360 = 68.4 \text{ degrees}$$

Pie charts force us to see that the parts do make a whole. But because angles are harder to compare than lengths, a pie chart is not as good a way to compare the sizes of the various parts of the whole.

Figure 10.3 is a **bar graph** of the same data. The height of each bar shows the percentage of people aged 25 years and over who have attained

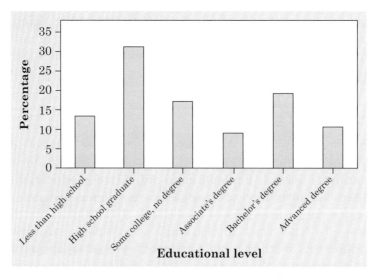

Figure 10.3 Bar graph of the distribution of level of education among persons aged 25 years and over in 2009. (This figure was created using the Minitab software package.)

the level of education marked at the bar's base. The bar graph makes it clear that there are more high school graduates than people who also have some college—the "High school graduate" bar is taller. Because it is hard to see this from the wedges in the pie chart, we added labels to the wedges that include the actual percentages. The bar graph is also easier to draw than the pie chart unless a computer is doing the drawing for you. In addition, if there is a natural ordering of the variable, such as how much education a person has, this order can be displayed along the horizontal axis of the bar graph, but cannot be displayed in an obvious way in a pie chart.

When we think about graphs, it is helpful to distinguish between variables whose values have a meaningful numerical scale (for example, height in centimeters, SAT scores) and variables such as sex, occupation, or level of education that just place individuals into categories. Pie charts and bar graphs are most useful for the second kind of variable.

Categorical and quantitative variables

A **categorical variable** places an individual into one of several groups or categories.

A **quantitative variable** takes numerical values for which arithmetic operations such as adding and averaging make sense.

To display the distribution of a categorical variable, use a pie chart or a bar graph.

Although both pie charts and bar graphs can show the distribution (either counts or percentages) of a categorical variable such as level of education, bar graphs have wider uses as well.

EXAMPLE 3 High taxes?

Figure 10.4 compares the level of taxation in eight democratic nations. The heights of the bars show the percentages of each nation's gross domestic product (GDP, the total value of all goods and services produced) that is taken in taxes. Americans accustomed to complaining about high taxes may be surprised to see that the United States, at 24.0% of GDP, is at the bottom of the group.

We cannot replace Figure 10.4 by a pie chart, because it compares eight separate quantities, not the parts of some whole. A pie chart can compare only parts of a whole. Bar graphs can compare quantities that are not parts of a whole.

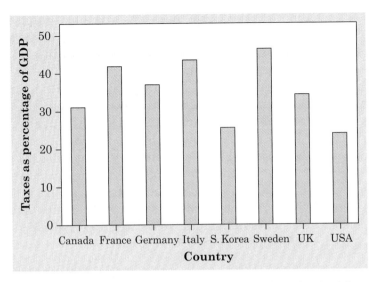

Figure 10.4 Total tax revenue as a percentage of gross domestic product in eight countries in 2009, for Example 3. (Data from the Organisation for Economic Cooperation and Development, **http://stats.oecd.org/index.aspx.** This figure was created using the Minitab software package.)

NOW IT'S YOUR TURN | **10.1 Taxes.** The Retirement Living Information Center compiled data on combined state and local taxes as a percentage of total income for a family of three earning $50,000 in the largest city in each of the 50 states for 2009. The table below gives the values for some selected midwestern states.

State	Tax as percentage of income
Illinois	10.9
Indiana	8.7
Iowa	11.3
Michigan	11.4
Minnesota	10.0
Ohio	9.1
Wisconsin	10.8

Is the variable "state" categorical or quantitative? To show the distribution of these data, would you use a pie chart or a bar graph?

Beware the pictogram

Bar graphs compare several quantities by means of the differing heights of bars that represent the quantities. Our eyes, however, react to the *area* of the bars as well as to their height. When all bars have the same width, the area (width × height) varies in proportion to the height and our eyes receive the right impression. When you draw a bar graph, make the bars equally wide. Artistically speaking, bar graphs are a bit dull. It is tempting to replace the bars with pictures for greater eye appeal.

EXAMPLE 4 A misleading graph

Figure 10.5 is a **pictogram.** It is a bar graph in which pictures replace the bars. The graph is aimed at advertisers deciding where to spend their budgets. It shows that *Time* magazine attracts the lion's share of advertising spending. Or does it? The numbers above the pens show that advertising spending in *Time* is 1.64 times as great as in *Newsweek*. Why does the graph suggest that *Time* is much farther ahead?

To magnify a picture, the artist must increase *both* height and width to avoid distortion. If both the height and width of *Time*'s pen are 1.64 times as large as *Newsweek*'s, the area is 1.64 × 1.64, or 2.7 times as large. Our eyes, responding to the area of the pens, see *Time* as the big winner.

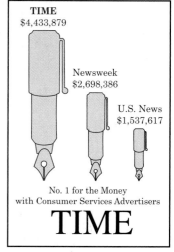

Figure 10.5 A pictogram, for Example 4. This variation of a bar graph is attractive but misleading. (Copyright © 1971 by Time, Inc. Reproduced by permission.)

Change over time: line graphs

Many quantitative variables are measured at intervals over time. We might, for example, measure the height of a growing child or the price of a stock at the end of each month. In these examples, our main interest is change over time. To display change over time, make a *line graph*.

Line graph

A **line graph** of a variable plots each observation against the time at which it was measured. Always put time on the horizontal scale of your plot and the variable you are measuring on the vertical scale. Connect the data points by lines to display the change over time.

EXAMPLE 5 The price of gasoline

How has the price of gasoline at the pump changed over time? Figure 10.6 is a line graph of the average price, across the United States, of regular

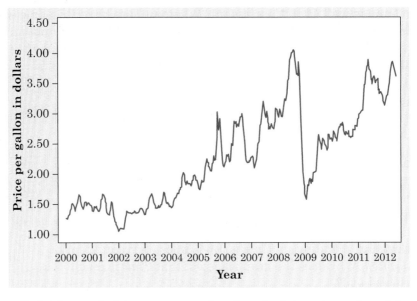

Figure 10.6 A line graph of the average cost of regular unleaded gasoline each week from January 3, 2000, to May 21, 2012, for Example 5. (Data from the Bureau of Labor Statistics. This figure was created using the Minitab software package.)

unleaded gasoline each week from January 3, 2000, to May 21, 2012. For example, the January 3, 2000, price was $1.26 per gallon. The first point in the figure is above the beginning of 2000 (January) and at 1.26 on the vertical scale.

It would be difficult to see patterns in a long table of monthly prices. Figure 10.6 makes the patterns clearer. What should we look for?

- First, look for an **overall pattern.** For example, a **trend** is a long-term upward or downward movement over time. There was no consistent overall up or down trend in gasoline prices from 2000 to the middle of 2001, but from a low in early 2002 to the middle of 2008 there was an overall upward trend with a spike in late 2005 after Hurricane Katrina. This upward trend began as OPEC nations drove the price up by reducing their oil production. A major strike in Venezuela, the U.S. war with Iraq, unrest in Nigeria, and increasing demand in countries such as India and China contributed to the upward trend. In the latter part of 2008 and the beginning of 2009 there was a large drop in price followed again by an upward trend. The drop was due to the financial crisis that began in late 2008 and the subsequent global recession. Demand for oil did not decrease much and prices began to rise after the large drop.

- Next, look for striking **deviations** from the overall pattern. The sharp decreases in prices near the beginning of 2006, the beginning of 2007, and especially at the end of 2008 stand out from the overall upward trend that began in 2002.

- Change over time often has a regular pattern of **seasonal variation** that repeats each year. Gasoline prices are usually highest during the summer driving season and lowest in the winter, when there is less demand for gasoline. You can see this up-in-summer then down-in-autumn pattern throughout Figure 10.6.

The Vietnam effect
Folklore says that during the Vietnam War many men went to college to avoid being drafted. Statistician Howard Wainer looked for traces of this "Vietnam effect" by plotting data against time. He found that scores on the Armed Forces Qualifying Test (an IQ test given to recruits) dipped sharply during the war, then rebounded. Scores on the SAT exams, taken by students applying to college, also dropped at the beginning of the war. It appears that the men who chose college over the army lowered the average test scores in both places.

Because seasonal variation is common, many government statistics are adjusted to remove seasonal effects. For example, the unemployment rate rises every year in January as holiday sales jobs end and outdoor work slows in the north due to winter weather. It would cause confusion (and perhaps political trouble) if the government's official unemployment rate jumped every January. The Bureau of Labor Statistics knows about how much it expects unemployment to rise in January, so it adjusts the published data for this expected change. The published unemployment rate goes up only if actual unemployment rises more than expected. We can then see the underlying changes in the employment situation without being confused by regular seasonal changes.

Seasonal variation, seasonal adjustment

A pattern that repeats itself at known regular intervals of time is called **seasonal variation.** Many series of regular measurements over time are **seasonally adjusted.** That is, the expected seasonal variation is removed before the data are published.

NOW IT'S YOUR TURN   **10.2 The price of milk.** The U.S. Department of Agriculture Web site provides data on the monthly average price of a gallon of whole milk. The table below gives the values for January 2010 to December 2011.

Mark E. Gibson/CORBIS

Date	Price ($)	Date	Price ($)
1/10	3.24	1/11	3.30
2/10	3.20	2/11	3.36
3/10	3.19	3/11	3.50
4/10	3.14	4/11	3.60
5/10	3.18	5/11	3.65
6/10	3.30	6/11	3.62
7/10	3.31	7/11	3.69
8/10	3.30	8/11	3.65
9/10	3.29	9/11	3.66
10/10	3.32	10/11	3.66
11/10	3.33	11/11	3.32
12/10	3.32	12/11	3.32

Make a line graph of these data and comment on any patterns you observe.

Watch those scales!

Because graphs speak so strongly, they can mislead the unwary. The careful reader of a line graph looks closely at the *scales* marked off on the axes.

EXAMPLE 6 Living together

The number of unmarried couples living together has increased in recent years, to the point that some people say that cohabitation is delaying or even replacing marriage. Figure 10.7 presents two line graphs of the number of unmarried-couple households in the United States. The data once again come from the Current Population Survey. The graph on the left suggests a steady but moderate increase. The right-hand graph says that cohabitation is thundering upward.

The secret is in the scales. You can transform the left-hand graph into the right-hand graph by stretching the vertical scale, squeezing the horizontal scale, and cutting off the vertical scale just above and below the

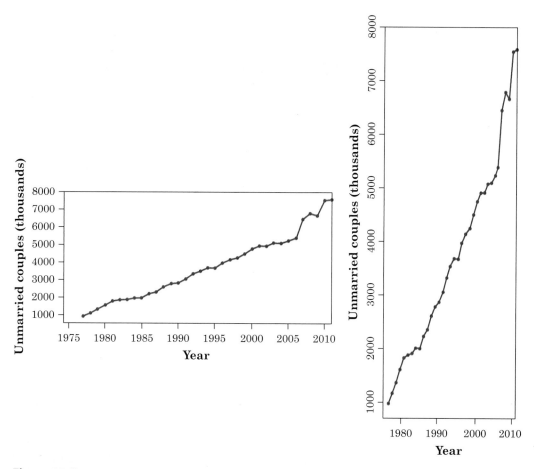

Figure 10.7 The effect of changing the scales in a line graph, for Example 6. Both graphs plot the same data, but the right-hand graph makes the increase appear much more rapid. (These figures were created using the Minitab software package.)

values to be plotted. Now you know how to either exaggerate or play down a trend in a line graph.

Which graph is correct? Both are accurate graphs of the data, but both have scales chosen to create a specific effect. Because there is no one "right" scale for a line graph, correct graphs can give different impressions by their choices of scale. Watch those scales!

Another important issue concerning scales is the following. When examining the change in the price or value of an item over time, plotting the actual increase can be misleading. It is often better to plot the percentage increase from the previous period.

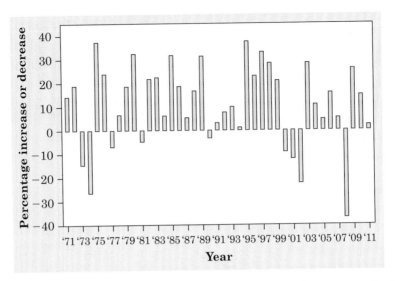

Figure 10.8 Percentage increase or decrease in the S&P 500 index of common stock prices, 1971 to 2011, for Example 7. (This figure was created using the Minitab software package.)

EXAMPLE 7 Getting rich in hindsight

The end of the 20th century saw a great bull market (a period when the value of stocks rises) in U.S. common stocks. How great? Pictures tell the tale more clearly than words.

Look first at Figure 10.8. This shows the percentage increase or decrease in stocks (measured by the Standard & Poor's 500 index) in each year from 1971 to 2011. Until 1982, stock prices bounce up and down. Sometimes they go down a lot—stocks lost 14.7% of their value in 1973 and another 26.5% in 1974. But starting in 1982, stocks go up in 17 of the next 18 years, often by a lot. From 2000 to 2011 stocks again bounce up and down, with a large loss of 37% in the recession that began in 2008.

Figure 10.9 shows, in hindsight, how you could have become rich in the period from 1971 to 1999. If you had invested $1000 in stocks at the end of 1970, the graph shows how much money you would have had at the end of each of the following years. After 1974, your $1000 was down to $853, and at the end of 1981, it had grown to only $2145. That's an increase of only 7.2% a year. Your money would have grown faster in a bank during these years. Then the great bull market begins its work. By the end of 1999, it would have turned your $1000 into $36,108. Unfortunately, over the next 10 years as a whole, stocks gained little value, and by the end of 2011, your $36,108 would have increased slightly to $38,471.

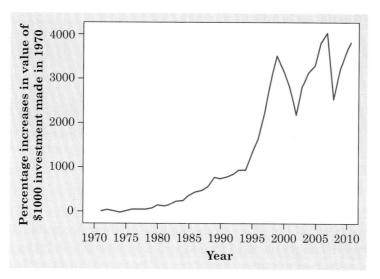

Figure 10.9 Value at the end of each year, 1971 to 2011, of $1000 invested in the S&P 500 index at the end of 1970, for Example 7. (This figure was created using the Minitab software package.)

Figures 10.8 and 10.9 are instructive. Figure 10.9 gives the impression that increases between 1970 and 1980 were negligible but that increases between 1995 and 1999 were dramatic. While it is true that the *actual value* of our investment increased much more between 1995 and 1999 than it did between 1970 and 1980, it would be incorrect to conclude that investments in general increased much more dramatically between 1995 and 1999 than in any of the years between 1970 and 1980.

Figure 10.8 tells a different, and more accurate, story. For example, the percentage increase in 1975 (approximately 37%) rivaled that in any of the years between 1995 and 1999. However, in 1975 the actual value of our investment was relatively small ($1170) and a 37% increase in such a small amount is nearly imperceptible on the scale used in Figure 10.9. By 1995 the actual value of our investment was about $14,000, and a 37% increase appears much more striking.

Making good graphs

Graphs are the most effective way to communicate using data. A good graph frequently reveals facts about the data that would be difficult or impossible to detect from a table. What is more, the immediate visual impression of a graph is much stronger than the impression made by data in numerical form. Here are some principles for making good graphs:

- Make sure **labels and legends** tell what variables are plotted, their units, and the source of the data.

· **Make the data stand out.** Be sure that the actual data, not labels, grids, or background art, catch the viewer's attention. You are drawing a graph, not a piece of creative art.

· **Pay attention to what the eye sees.** Avoid pictograms and be careful choosing scales. Avoid fancy "three-dimensional" effects that confuse the eye without adding information. Ask if a simple change in a graph would make the message clearer.

EXAMPLE 8 The rise in college education

Figure 10.10 shows the rise in the percentage of women aged 25 years and over who have at least a bachelor's degree. There are only 5 data points, so a line graph should be simple. Figure 10.10 isn't simple. The artist couldn't resist a nice background sketch and also cluttered the graph with grid lines. We find it harder than it should be to see the data. Grid lines on a graph serve no purpose—if your audience must know the actual numbers, give them a table along with the graph. A good graph uses no more ink than is needed to present the data clearly. In statistical graphics, decoration is always a distraction from the data, and sometimes an actual distortion of the data.

Figure 10.10 Chart junk: this graph is so cluttered with unnecessary ink that it is hard to see the data.

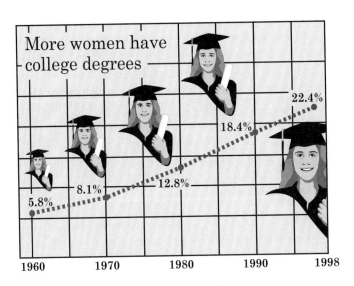

EXAMPLE 9 High taxes, reconsidered

Figure 10.4 (page 218) is a respectable bar graph comparing taxes as a percentage of gross domestic product in eight countries. The countries are arranged alphabetically. Wouldn't it be clearer to arrange them in order

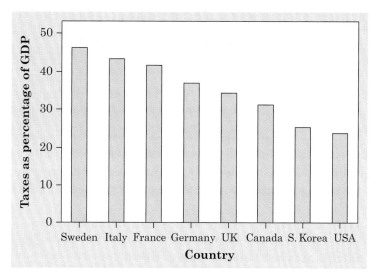

Figure 10.11 Total tax revenue as a percentage of gross domestic product in eight countries in 2009, for Example 9. Changing the order of the bars has improved the graph in Figure 10.4. (This figure was created using the Minitab software package.)

of their tax burdens? Figure 10.11 does this. This simple change improves the graph by making it clearer where each country stands in the group of eight countries.

STATISTICS IN SUMMARY

Chapter Specifics

- To see what data say, start with graphs.

- The choice of graph depends on the type of data. Do you have a **categorical variable,** such as level of education or occupation, which puts individuals into categories? Or do you have a **quantitative variable** measured in meaningful numerical units?

- Check data presented in a table for **roundoff errors.**

- To display the **distribution** of a categorical variable, use a **pie chart** or a **bar graph.** Pie charts always show the parts of some whole, but bar graphs can compare any set of numbers measured in the same units.

- To show how a quantitative variable changes over time, use a **line graph** that plots values of the variable (vertical scale) against time (horizontal scale). Look for **trends** and **seasonal variation** in a line graph, and ask whether the data have been **seasonally adjusted.**

- Graphs can mislead the eye. Avoid **pictograms** that replace the bars of a bar graph by pictures whose height and width both change. Look at the scales of a line graph to see if they have been stretched or squeezed to create a particular impression. Avoid clutter that makes the data hard to see.

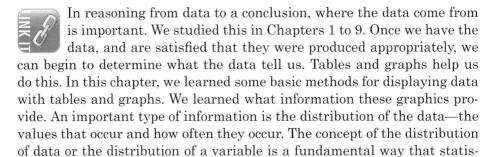 In reasoning from data to a conclusion, where the data come from is important. We studied this in Chapters 1 to 9. Once we have the data, and are satisfied that they were produced appropriately, we can begin to determine what the data tell us. Tables and graphs help us do this. In this chapter, we learned some basic methods for displaying data with tables and graphs. We learned what information these graphics provide. An important type of information is the distribution of the data—the values that occur and how often they occur. The concept of the distribution of data or the distribution of a variable is a fundamental way that statisticians think about data. We will encounter it again and again in future chapters.

Data that are produced badly can mislead us. Likewise, graphs that are produced badly can mislead us. In this chapter we learned how to recognize bad graphics. Developing "graphic sense," the habit of asking if a graphic accurately and clearly displays our data, is as important as developing "number sense," discussed in Chapter 9.

CASE STUDY   Look again at Figure 10.1, described in the Case Study that opened the
EVALUATED   chapter. Based on what you have learned, is Figure 10.1 a fair graphical representation of declines in reading proficiency among seventeen-year-olds? If you believe it is, what features of the graph do you like? If you do not believe it is, prepare a graphic that is more fair.

You can find some comments about this graphic online at **www.futureofthebook .org/blog/archives/2007/11/reading_responsibly_nancy_kaplan.html.** ◾

CHAPTER 10 EXERCISES

For Exercise 10.1, see page 218; for Exercise 10.2, see page 222.

10.3 Lottery sales. States sell lots of lottery tickets. Table 10.2 shows where the money comes from in the state of Illinois. Make a bar graph that shows the distribution of lottery sales by type of game. Is it also proper to make a pie chart of these data? Explain.

TABLE 10.2 Illinois state lottery sales by type of game, fiscal year 2010

Game	Sales (dollars)
Pick Three	301,417,049
Pick Four	191,038,518
Lotto	111,158,528
Little Lotto	119,634,946
Mega Millions	221,809,484
Megaplier5	848,077
Pick N Play	1,549,252
Raffle	19,999,460
Powerball	43,269,461
Power Play	8,469,680
Instants	1,190,109,917
Total	2,209,304,371

Source: Illinois Lottery Fiscal Year 2010 Financial Release.

10.4 Consistency? Table 10.2 shows how Illinois state lottery sales are divided among different types of games. What is the sum of the amounts spent on the eleven types of games? Why is this sum not exactly equal to the total given in the table?

10.5 Marital status. In the U.S. Census Bureau document *America's Families and Living Arrangements: 2011,* we find these data on the marital status of American women aged 15 years and older as of 2011:

Marital status	Count (thousands)
Never married	34,963
Married	65,000
Widowed	11,306
Divorced	13,762

(a) How many women were not married in 2009?

(b) Make a bar graph to show the distribution of marital status.

(c) Would it also be correct to use a pie chart? Explain.

10.6 We pay high interest. Figure 10.12 shows a graph taken from an advertisement for an investment that promises to pay a higher interest rate than bank accounts and other competing investments. Is this graph a correct comparison of the four interest rates? Explain your answer.

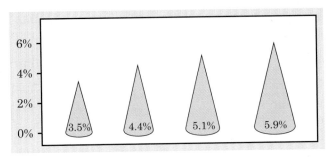

Figure 10.12 Comparing interest rates, for Exercise 10.6.

10.7 We sell CDs. Who sold the most CDs on the Web just before listeners began to download their favorites rather than buy CDs? Figure 10.13 graphs the shares of the market leaders in 1997 and says what percentage of all online sales they had. Does the graph fairly represent the data? Explain your answer.

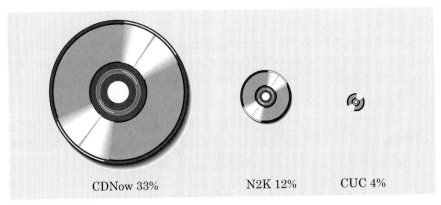

Figure 10.13 Percentage of market share for the three leading online sellers of compact discs in 1997, for Exercise 10.7. (Data from Jupiter Communications.)

10.8 Murder weapons. The 2010 *Statistical Abstract of the United States* reports FBI data on murders for 2007. In that year, 49.6% of all murders were committed with handguns, 18.4% with other firearms, 12.1% with knives or other cutting implements, 5.8% with a part of the body (usually the hands, fists, or feet), and 4.4% with blunt objects. Make a graph to display these data. Do you need an "other methods" category? Why?

10.9 The cost of imported oranges. Figure 10.14 is a line graph of the average cost of imported oranges each month from July 1995 to April 2012. These data are the price in U.S. dollars per metric ton.

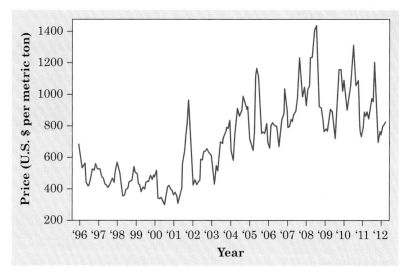

Figure 10.14 The price of oranges, July 1995 to April 2012, for Exercise 10.9. (This figure was created using the Minitab software package.)

(a) The graph shows strong seasonal variation. How is this visible in the graph? Why would you expect the price of oranges to show seasonal variation?

(b) What is the overall trend in orange prices during this period, after we take account of the seasonal variation?

10.10 College freshmen. A survey of college freshmen in 2007 asked what field they planned to study. The results: 12.8%, arts and humanities; 17.7%, business; 9.2%, education; 19.3%, engineering, biological sciences, or physical sciences; 14.5%, professional; and 11.1%, social science.

(a) What percentage of college freshmen plan to study fields other than those listed?

(b) Make a graph that compares the percentages of college freshmen planning to study various fields.

10.11 Exports. Figure 10.15 (next page) compares the value of exports (in billions of dollars) in 2002 from the world's leading exporters: Germany, Japan, and the United States.

(a) Explain why this is not an accurate graph.

(b) Make an accurate graph to compare the exports of these nations.

10.12 Civil disorders. The years around 1970 brought unrest to many U.S. cities. Here are government data on the number of civil disturbances in each 3-month period during the years 1968 to 1972.

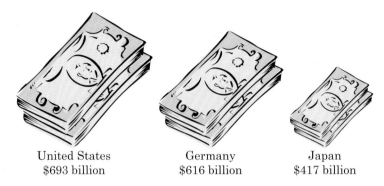

United States
$693 billion

Germany
$616 billion

Japan
$417 billion

Value of exports, 2002

Figure 10.15 Value of goods exported by Germany, Japan, and the United States in 2002, for Exercise 10.11. (Data from the Organization for Economic Cooperation and Development.)

Period	Count	Period	Count
1968, Jan.–Mar.	6	1970, July–Sept.	20
Apr.–June	46	Oct.–Dec.	6
July–Sept.	25	1971, Jan.–Mar.	12
Oct.–Dec.	3	Apr.–June	21
1969, Jan.–Mar.	5	July–Sept.	5
Apr.–June	27	Oct.–Dec.	1
July–Sept.	19	1972, Jan.–Mar.	3
Oct.–Dec.	6	Apr.–June	8
1970, Jan.–Mar.	26	July–Sept.	5
Apr.–June	24	Oct.–Dec.	5

(a) Make a line graph of these data.

(b) The data show both a longer-term trend and seasonal variation within years. Describe the nature of both patterns. Can you suggest an explanation for the seasonal variation in civil disorders?

10.13 Births to unwed mothers. Here are data on births to unwed mothers as a percentage of births in the United States, from the *Statistical Abstract*. These data show a clear increasing trend over time. Make two line graphs of these data: one designed to show only a gradual increase over time, and a second designed to show a shockingly steep increase.

Year:	1960	1965	1970	1975	1980	1985	1990	1995	2000	2005	2008
% unwed mothers:	5.3	7.7	10.7	14.2	18.4	22.0	28.0	32.2	33.2	36.9	40.7

10.14 A bad graph? Figure 10.16 shows a graph that appeared in the *Lexington (Ky.) Herald-Leader* on October 5, 1975. Discuss the correctness of this graph.

10.15 Trends. Which of the following series of data do you expect to show a clear trend? Will the trend be upward or downward? (All data are recorded annually.)

(a) The percentage of students entering a university who bring a typewriter with them.

(b) The percentage of students entering a university who bring a personal computer with them.

(c) The percentage of adult women who do not work outside the home.

10.16 Seasonal variation. You examine the average temperature in Chicago each month for many years. Do you expect a line graph of the data to show seasonal variation? Describe the kind of seasonal variation you expect to see.

10.17 Sales are up. The sales at your new gift shop in December are double the November value. Should you conclude that your shop is growing more popular and will soon make you rich? Explain your answer.

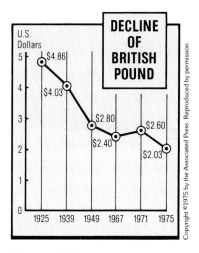

Figure 10.16 A newspaper's graph of the value of the British pound, for Exercise 10.14.

10.18 Counting employed people. A news article says:

More people were working in America in June than in any month since the end of 1990. If that's not what you thought you read in the papers last week, don't worry. It isn't. The report that employment plunged in June, with nonfarm payrolls declining by 117,000, helped to persuade the Federal Reserve to cut interest rates yet again.... In reality, however, there were 457,000 more people employed in June than in May.

What explains the difference between the fact that employment went up by 457,000 and the official report that employment went down by 117,000?

10.19 The sunspot cycle. Some plots against time show **cycles** of up-and-down movements. Figure 10.17 (next page) is a line graph of the average number of sunspots on the sun's visible face for each month from 1900 to 2011. What is the approximate length of the sunspot cycle? That is, how many years are there between the successive valleys in the graph? Is there any overall trend in the number of sunspots?

10.20 Trucks versus cars. Do consumers prefer trucks, SUVs, and minivans to passenger cars? Following are data on sales and leases of new cars and trucks in the United States. (The definition of "truck" includes SUVs and minivans.) Plot two line graphs on the same axes to compare the change in car and truck sales over time. Describe the trend that you see.

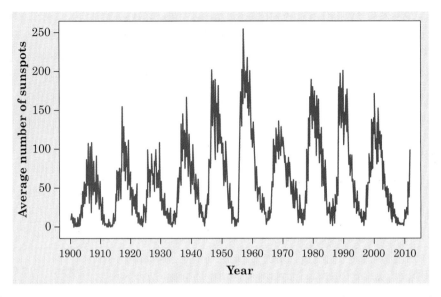

Figure 10.17 The sunspot cycle, for Exercise 10.19. This is a line graph of the average number of sunspots per month for the years 1900 to 2011. (Data from the National Oceanic and Atmospheric Administration. This figure was created using the Minitab software package.)

Year:	1981	1983	1985	1987	1989	1991	1993	1995
Cars (1000s):	8,536	9,182	11,042	10,277	9,772	8,589	8,518	8,636
Trucks (1000s):	2,260	3,129	4,682	4,912	4,941	4,136	5,654	6,469

Year:	1997	1999	2001	2003	2005	2007	2009	
Cars (1000s):	8,273	8,697	8,422	7,615	7,720	7,618	5,456	
Trucks (1000s):	7,217	8,704	9,046	9,356	9,725	8,842	5,145	

10.21 Who sells cars? Figure 10.18 is a pie chart of the percentage of passenger car sales in 1997 by various manufacturers. The artist has tried to make the graph attractive by using the wheel of a car for the "pie." Is the graph still a correct display of the data? Explain your answer.

Figure 10.18 Passenger car sales by several manufacturers in 1997, for Exercise 10.21.

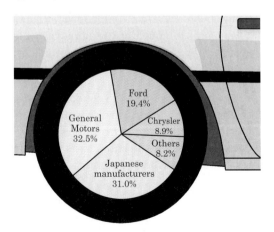

10.22 Who sells cars? Make a bar graph of the data in Exercise 10.21. What advantage does your new graph have over the pie chart in Figure 10.18?

10.23 The Border Patrol. Here are the numbers of deportable aliens caught by the U.S. Border Patrol for 1971 through 2009. Display these data in a graph. What are the most important facts that the data show?

Year :	1971	1973	1975	1977	1979	1981	1983	1985	1987	1989
Count (1000s):	420	656	767	1042	1076	976	1251	1349	1190	954

Year:	1991	1993	1995	1997	1999	2001	2003	2005	2007	2009
Count (1000s):	1198	1327	1395	1536	1714	1266	932	1189	877	556

10.24 Some time series. Sketch line graphs of a series of observations over time having each of the following characteristics. Mark your time axis in years.

(a) A strong downward trend, but no seasonal variation.

(b) Seasonal variation each year, but no clear trend.

(c) A strong downward trend with yearly seasonal variation.

10.25 Bad habits. According to the National Household Survey on Drug Use and Health, 30.6% of adolescents aged 14 or 15 years used alcohol in 2008, 11.6% used marijuana or hashish, 1.0% used cocaine, and 17.7% used cigarettes. Explain why it is *not* correct to display these data in a pie chart.

10.26 Accidental deaths. In 2007 there were 123,706 deaths from accidents in the United States. Among these were 43,945 deaths from motor vehicle accidents, 29,846 from poisoning, 22,631 from falls, 3443 from drowning, and 3286 from fires.

(a) Find the percentage of accidental deaths from each of these causes, rounded to the nearest percent. What percentage of accidental deaths were due to other causes?

(b) Make a well-labeled graph of the distribution of causes of accidental deaths.

10.27 Yields of money market funds. Many people invest in money market funds. These are mutual funds that attempt to maintain a constant price of $1 per share while paying monthly interest. Table 10.3 gives the average annual interest rates (in percent) paid by all taxable money market funds from 1973 (the first full year in which such funds were available) to 2008.

(a) Make a line graph of the interest paid by money market funds for these years.

(b) Interest rates, like many economic variables, show **cycles,** clear but repeating up-and-down movements. In which years did the interest rate cycle reach temporary peaks?

(c) A plot against time may show a consistent **trend** underneath cycles. When did interest rates reach their overall peak during these years? Describe the general trend downward since that year.

TABLE 10.3 Average annual interest rates (in percent) paid by money market funds, 1973–2008

Year	Rate	Year	Rate	Year	Rate	Year	Rate
1973	7.60	1982	12.23	1991	5.71	2000	5.89
1974	10.79	1983	8.58	1992	3.36	2001	3.67
1975	6.39	1984	10.04	1993	2.70	2002	1.29
1976	5.11	1985	7.71	1994	3.75	2003	0.64
1977	4.92	1986	6.26	1995	5.48	2004	0.82
1978	7.25	1987	6.12	1996	4.95	2005	2.66
1979	10.92	1988	7.11	1997	5.10	2006	4.51
1980	12.68	1989	8.87	1998	5.04	2007	4.70
1981	16.82	1990	7.82	1999	4.64	2008	2.05

Source: Albert J. Fredman, "A closer look at money market funds," *American Association of Individual Investors Journal,* February 1997, pp. 22–27; and the 2010 *Statistical Abstract of the United States.*

10.28 The Boston Marathon. Women were first allowed to enter the Boston Marathon in 1972. The time (in minutes, rounded to the nearest minute) for each winning woman from 1972 to 2010 appears in Table 10.4.

(a) Make a graph of the winning times.

(b) Give a brief description of the pattern of Boston Marathon winning times over these years. Have times stopped improving in recent years?

TABLE 10.4 Women's winning times (minutes) in the Boston Marathon, 1972–2010

Year	Time	Year	Time	Year	Time
1972	190	1985	154	1998	143
1973	186	1986	145	1999	143
1974	167	1987	146	2000	146
1975	162	1988	145	2001	144
1976	167	1989	144	2002	141
1977	168	1990	145	2003	145
1978	165	1991	144	2004	144
1979	155	1992	144	2005	145
1980	154	1993	145	2006	144
1981	147	1994	142	2007	149
1982	150	1995	145	2008	145
1983	143	1996	147	2009	152
1984	149	1997	146	2010	146

Source: See the Web site **en.wikipedia.org/wiki/List_of_winners_of_the_Boston_Marathon/.**

EXPLORING THE WEB

10.29 Web-based exercise. Find an example of a poor graphic. One possible source is the CHANCE Web site at Dartmouth College. In particular, see the *Chance News* section, **http://test.causeweb.org/wiki/chance/index.php/Main_Page.** You can also try a Google search for something like "misleading graphical displays in the media."

10.30 Web-based exercise. Find an example of a variable, such as the price of a commodity, that changes over time. One possible source is the Bureau of Labor Statistics Web site. Go to **www.bls.gov/cpi/home.htm** and click on the *Average Price Data* link.

NOTES AND DATA SOURCES

Page 213 Case Study. National Endowment for the Arts, *To Read or Not to Read,* Research Report 47, November 2007; available online at **www.nea.gov/research/ToRead.pdf.**

Page 218 Exercise 10.1: These data are available online at **www.retirementliving.com/RLtaxburdens.html.**

Page 219 Example 5: The prices in Figure 10.6 are from the Department of Energy Web site, where you can find weekly historical prices: **www.eia.doe.gov/oil_gas/petroleum/data_publications/wrgp/mogas_history.html.**

Page 222 Exercise 10.2: These data are available online at **www.ers.usda.gov/Data/.**

Page 230 Exercise 10.9: These data are available online at **www.indexmundi.com/commodities/.**

Page 231 Exercise 10.10: 2009 *Statistical Abstract of the United States,* Table 276.

Page 233 Exercise 10.18: F. Norris, Market Watch column, *New York Times,* July 15, 1992.

Page 233 Exercise 10.20: These data are from the 2010 *Statistical Abstract of the United States;* available online at **www.census.gov/compendia/statab/.**

Page 235 Exercise 10.23: The data for 1971 to 1997 are from the *1997 Statistical Yearbook of the Immigration and Naturalization Service,* U.S. Department of Justice, 1999. The data for 1999 are from the 2004–5 *Statistical Abstract of the United States;* available online at **www.census.gov/compendia/statab/.** The data for 2001 to 2009 are from the 2008, 2010, and 2012 *Statistical Abstract of the United States;* available online at **www.census.gov/compendia/statab/.**

Page 235 Exercise 10.26: 2010 *Statistical Abstract of the United States;* available online at **www.census.gov/compendia/statab/.**

Displaying Distributions with Graphs

CASE STUDY Nutritionists tell us that a healthy diet should include 20 to 35 grams of fiber daily. Cereal manufacturers, hoping to attract health-conscious consumers, advertise their products as "high-fiber" and other healthy options. The food label on the side of a box of cereal (mandated by the Food and Drug Administration) provides information that allows the consumer to choose a healthy breakfast cereal.

If you go to the breakfast cereal aisle of your local grocery store, you will find lots of different cereals displayed. You could examine all the boxes to see how much fiber each contains, but how do you make sense of all the numbers? Is your favorite cereal, Wheaties, with 3 grams of dietary fiber, among those with the highest fiber content? You could make a list of the fiber content of all the cereals, but it can be difficult to see patterns in large lists. As we saw in Chapter 10, graphs are a powerful way to make sense of large collections of numbers.

In this chapter we will study two types of graphics, histograms and stemplots (also called stem-and-leaf plots), that help us make sense of large lists. By the end of this chapter you will know how to make them and what to look for when you study one of these graphs. ■

Histograms

Categorical variables just record group membership, such as the marital status of a man or the race of a college student. We can use a pie chart or bar graph to display the distribution of categorical variables because they have relatively few values. What about quantitative variables such as the SAT scores of students admitted to a college or the income of families? These variables take so many values that a graph of the distribution is clearer if nearby values are grouped together. The most common graph of the distribution of a quantitative variable is a **histogram.**

EXAMPLE 1 How to make a histogram

Table 11.1 presents the percentage of residents aged 65 years and over in each of the 50 states. To make a histogram of this distribution, proceed as follows.

Step 1. Divide the range of the data into classes of equal width. The data in Table 11.1 range from 7.3 to 17.4, so we choose as our classes

$$7.0 \le \text{percentage over 65 } < 8.0$$
$$8.0 \le \text{percentage over 65 } < 9.0$$
$$\vdots$$
$$17.0 \le \text{percentage over 65 } < 18.0$$

Be sure to specify the classes precisely so that each individual falls into exactly one class. In other words, be sure that the classes are *exclusive* (no individual is in more than one class) and *exhaustive* (every individual

TABLE 11.1 Percentage of residents aged 65 and over in the states, July 2008

State	Percent	State	Percent	State	Percent
Alabama	13.8	Louisiana	12.3	Ohio	13.7
Alaska	7.3	Maine	15.1	Oklahoma	13.5
Arizona	13.3	Maryland	12.1	Oregon	13.3
Arkansas	14.3	Massachusetts	13.4	Pennsylvania	15.4
California	11.2	Michigan	13.0	Rhode Island	14.1
Colorado	10.4	Minnesota	12.5	South Carolina	13.3
Connecticut	13.7	Mississippi	12.7	South Dakota	14.4
Delaware	13.9	Missouri	13.6	Tennessee	13.2
Florida	17.4	Montana	14.2	Texas	10.2
Georgia	10.1	Nebraska	13.5	Utah	9.0
Hawaii	14.8	Nevada	11.4	Vermont	14.0
Idaho	12.0	New Hampshire	12.9	Virginia	12.1
Illinois	12.2	New Jersey	13.3	Washington	12.0
Indiana	12.8	New Mexico	13.1	West Virginia	15.7
Iowa	14.8	New York	13.4	Wisconsin	13.3
Kansas	13.1	North Carolina	12.4	Wyoming	12.3
Kentucky	13.3	North Dakota	14.7		

Source: 2010 *Statistical Abstract of the United States*; available online at **www.census.gov/compendia/statab/.**

appears in some class). A state with 7.9% of its residents aged 65 or older would fall into the first class, but 8.0% falls into the second.

Step 2. Count the number of individuals in each class. Here are the counts:

Class	Count	Class	Count	Class	Count
7.0 to 7.9	1	11.0 to 11.9	2	15.0 to 15.9	3
8.0 to 8.9	0	12.0 to 12.9	12	16.0 to 16.9	0
9.0 to 9.9	1	13.0 to 13.9	19	17.0 to 17.9	1
10.0 to 10.9	3	14.0 to 14.9	8		

Step 3. Draw the histogram. Mark on the horizontal axis the scale for the variable whose distribution you are displaying. That's "percentage of residents aged 65 and over" in this example. The scale runs from 5 to 20 because that range spans the classes we chose. The vertical axis contains the scale of counts. Each bar represents a class. The base of the bar covers the class, and the bar height is the class count. There is no horizontal space between the bars unless a class is empty, so that its bar has height zero. Figure 11.1 is our histogram.

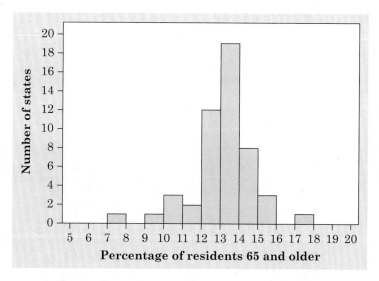

Figure 11.1 Histogram of the percentages of residents aged 65 and older in the 50 states, for Example 1. Note the two outliers. (This figure was created using the Minitab software package.)

Just as with bar graphs, our eyes respond to the area of the bars in a histogram. Be sure that the classes for a histogram have equal widths. There is

What the eye really sees We make the bars in bar graphs and histograms equal in width because the eye responds to their area. That's roughly true. Careful study by statistician William Cleveland shows that our eyes "see" the size of a bar in proportion to the 0.7 power of its area. Suppose, for example, that one figure in a pictogram is both twice as high and twice as wide as another. The area of the bigger figure is 4 times that of the smaller. But we perceive the bigger figure as only 2.6 times the size of the smaller, because 2.6 is 4 to the 0.7 power.

no one right choice for the number of classes. Some people recommend between 10 and 20 classes but suggest using fewer when the size of the data set is small. Too few classes will give a "skyscraper" histogram, with all values in a few classes with tall bars. Too many classes will produce a "pancake" graph, with most classes having one or no observations. Neither choice will give a good picture of the shape of the distribution. You must use your judgment in choosing classes to display the shape. Statistics software will choose the classes for you and may use slightly different rules than those we have discussed. The computer's choice is usually a good one, but you can change it if you want. When using statistical software, it is good practice to check what rules are used to determine the classes.

NOW IT'S YOUR TURN

11.1 Under-18-year-olds. Below are the percentages of residents under the age of 18 in the 50 states in July 2008.

State	Percent	State	Percent	State	Percent
Alabama	24.1	Louisiana	25.1	Ohio	23.8
Alaska	26.2	Maine	20.8	Oklahoma	24.9
Arizona	26.3	Maryland	23.8	Oregon	22.9
Arkansas	24.6	Massachusetts	22.0	Pennsylvania	22.2
California	25.5	Michigan	23.9	Rhode Island	21.8
Colorado	24.4	Minnesota	24.0	South Carolina	23.8
Connecticut	23.2	Mississippi	26.1	South Dakota	24.8
Delaware	23.6	Missouri	24.0	Tennessee	23.8
Florida	21.9	Montana	22.8	Texas	27.6
Georgia	26.3	Nebraska	25.1	Utah	31.1
Hawaii	22.1	Nevada	25.7	Vermont	20.8
Idaho	27.1	New Hampshire	22.3	Virginia	23.5
Illinois	24.6	New Jersey	23.6	Washington	23.5
Indiana	24.9	New Mexico	25.3	West Virginia	21.3
Iowa	23.7	New York	22.6	Wisconsin	23.3
Kansas	25.0	North Carolina	24.3	Wyoming	24.0
Kentucky	23.6	North Dakota	22.3		

Make a histogram of this distribution following the three steps described in Example 1.

Interpreting histograms

Making a statistical graph is not an end in itself. The purpose of the graph is to help us understand the data. After you (or your computer) make a graph, always ask, "What do I see?" Here is a general strategy for looking at graphs.

Pattern and deviations

In any graph of data, look for an **overall pattern** and also for striking **deviations** from that pattern.

We have already applied this strategy to line graphs. Trend and seasonal variation are common overall patterns in a line graph. The decrease in gasoline prices in the latter part of 2008 and beginning of 2009 in Figure 10.6 (page 220), caused by the financial crisis that began in late 2008 and the subsequent global recession, is an example of a deviation from the general pattern that one observes between 2002 and the middle of 2008. In the case of the histogram of Figure 11.1, it is easiest to begin with deviations from the overall pattern of the histogram. Two states stand out as separated from the main body of the histogram. You can find them in the table once the histogram has called attention to them. Alaska has 7.3% and Florida 17.4% of its residents over age 65. These states are clear *outliers*.

Outliers

An **outlier** in any graph of data is an individual observation that falls outside the overall pattern of the graph.

Is Utah, with 9.0% of its population over 65, an outlier? Whether an observation is an outlier is to some extent a matter of judgment, although statisticians have developed some objective criteria for identifying possible outliers. Utah is the smallest of the main body of observations and, unlike Alaska and Florida, is not separated from the general pattern. We would not call Utah an outlier. Once you have spotted outliers, look for an explanation. Many outliers are due to mistakes, such as typing 4.0 as 40. Other outliers point to the special nature of some observations. Explaining outliers usually requires some background information. It is not surprising that Alaska, the northern frontier, has few residents 65 and over and that Florida, a popular state for retirees, has many residents 65 and over.

To see the *overall pattern* of a histogram, ignore any outliers. Here is a simple way to organize your thinking.

> **Overall pattern of a distribution**
>
> To describe the overall pattern of a distribution:
>
> • Give the **center** and the **spread.**
> • See if the distribution has a simple **shape** that you can describe in a few words.

We will learn how to describe center and spread numerically in Chapter 12. For now, we can describe the center of a distribution by its *midpoint,* the value with roughly half the observations taking smaller values and half taking larger values. We can describe the spread of a distribution by giving the *smallest and largest values,* ignoring any outliers.

EXAMPLE 2 Describing distributions

Look again at the histogram in Figure 11.1. **Shape:** The distribution has a *single peak.* It is roughly *symmetric*—that is, the pattern is similar on both sides of the peak. **Center:** The midpoint of the distribution is close to the single peak, at about 13%. **Spread:** The spread is about 9% to 18% if we ignore the outliers.

EXAMPLE 3 Tuition and fees in Illinois

There are 116 colleges and universities in Illinois. Their tuition and fees for the 2009–2010 school year run from $1974 at Moraine Valley Community College to $38,550 at the University of Chicago. Figure 11.2 is a histogram of the tuition and fees charged by all 116 Illinois colleges and universities. We see that many (mostly community colleges) charge less than $4000. The distribution extends out to the right. At the upper extreme, two colleges charge between $36,000 and $40,000.

The distribution of tuition and fees at Illinois colleges, shown in Figure 11.2, has a quite different **shape** from the distribution in Figure 11.1. There is a strong *peak* in the lowest cost class. Most colleges charge less than $8000, but there is a long right tail extending up to almost $40,000.

We call a distribution with a long tail on one side *skewed.* The **center** is roughly $8000 (half the colleges charge less than this). The **spread** is large, from $1974 to more than $38,000. There are no outliers—the colleges with the highest tuition just continue the long right tail that is part of the overall pattern.

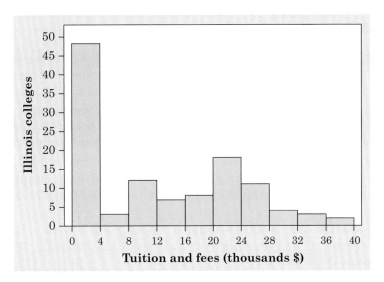

Figure 11.2 Histogram of the tuition and fees charged by 116 Illinois colleges and universities in the 2009–2010 academic year, for Example 3. (Data from the Web site **www.isac.org.** This figure was created using the Minitab software package.)

When you describe a distribution, concentrate on the main features. Look for major peaks, not for minor ups and downs in the bars of the histogram like those in Figure 11.2. Look for clear outliers, not just for the smallest and largest observations. Look for rough *symmetry* or clear *skewness*.

Symmetric and skewed distributions

A distribution is **symmetric** if the right and left sides of the histogram are approximately mirror images of each other.

A distribution is **skewed to the right** if the right side of the histogram (containing the half of the observations with larger values) extends much farther out than the left side. It is **skewed to the left** if the left side of the histogram extends much farther out than the right side.

In mathematics, symmetry means that the two sides of a figure like a histogram are exact mirror images of each other. Data are almost never exactly symmetric, so we are willing to call histograms like that in Figure 11.1 roughly symmetric as an overall description. The tuition distribution in Figure 11.2, on the other hand, is clearly skewed to the right. Here are more examples.

EXAMPLE 4 Sampling again

The values a statistic takes in many random samples from the same population form a distribution with a regular pattern. The histogram in Figure 11.3 displays a distribution that we met in Chapter 3. Take a simple random sample of 2527 adults. Ask each whether they favor a constitutional amendment that would define marriage as being between a man and a woman. The proportion who say "Yes" is the sample proportion $\hat{p}$. Do this 1000 times and collect the 1000 sample proportions $\hat{p}$ from the 1000 samples. Figure 11.3 shows the distribution of 1000 sample proportions when the truth about the population is that 50% would favor such an amendment.

This distribution is quite symmetric about a single peak in the center. The center is at 0.50, reflecting the lack of bias of the statistic. The spread from the smallest to the largest of the 1000 values is from 0.463 to 0.533.

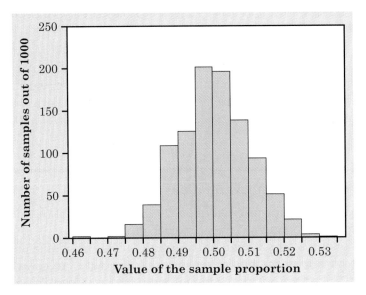

Figure 11.3 Histogram of the sample proportion $\hat{p}$ for 1000 simple random samples from the same population, for Example 4. This is a symmetric distribution. (This figure was created using the SPSS software package.)

EXAMPLE 5 Shakespeare's words

Figure 11.4 shows the distribution of lengths of words used in Shakespeare's plays. This distribution has a single peak and is somewhat

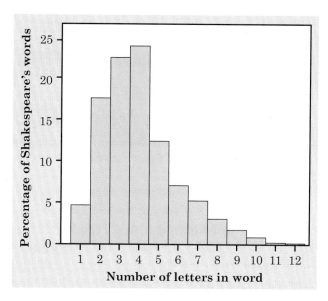

Figure 11.4 The distribution of word lengths used by Shakespeare in his plays, for Example 5. This distribution is skewed to the right. (Data from C. B. Williams, *Style and Vocabulary: Numerical Studies,* Griffin, 1970.)

skewed to the right. There are many short words (3 and 4 letters) and few very long words (10, 11, or 12 letters), so that the right tail of the histogram extends out farther than the left tail. The center of the distribution is about 4. That is, about half of Shakespeare's words have 4 or fewer letters. The spread is from 1 letter to 12 letters.

Notice that the vertical scale in Figure 11.4 is not the *count* of words but the *percentage* of all of Shakespeare's words that have each length. A histogram of percentages rather than counts is convenient when the counts are very large or when we want to compare several distributions. Different kinds of writing have different distributions of word lengths, but all are right-skewed because short words are common and very long words are rare.

The overall shape of a distribution is important information about a variable. Some types of data regularly produce distributions that are symmetric or skewed. For example, the sizes of living things of the same species (like lengths of crickets) tend to be symmetric. Data on incomes (whether of individuals, companies, or nations) are usually strongly skewed to the right. There are many moderate incomes, some large incomes, and a few very large incomes. It is very common for data to be skewed to the right when they have a strict minimum value (often 0). Income and the lengths of Shakespeare's words are examples. Likewise, data that have a strict maximum value (such as 100, as in student test scores) are often skewed to the left. Do remember that many distributions have shapes that are neither symmetric nor skewed. Some data show other patterns. Scores on an exam, for example, may have a cluster near the top of the scale if many students did well. Or they may show two distinct peaks if a tough problem divided the class into those who did and didn't solve it. Use your eyes and describe what you see.

> **NOW IT'S YOUR TURN**
>
> **11.2 85-year-olds and older.** Figure 11.5 is a histogram of the percentages of residents aged 85 and older in the 50 states in July 2008. Describe the shape, center, and spread of this distribution. Are there any outliers?

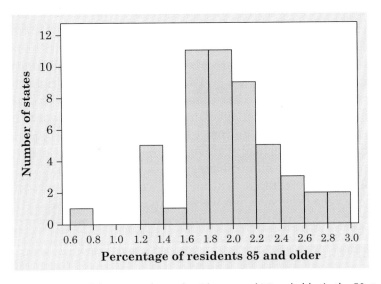

Figure 11.5 Histogram of the percentages of residents aged 85 and older in the 50 states, for Exercise 11.2. (This figure was created using the Minitab software package.)

Stemplots

Histograms are not the only graphical display of distributions. For small data sets, a *stemplot* (sometimes called a *stem-and-leaf plot*) is quicker to make and presents more detailed information.

Stemplot

To make a **stemplot:**

1. Separate each observation into a **stem** consisting of all but the final (rightmost) digit and a **leaf,** the final digit. Stems may have as many digits as needed, but each leaf contains only a single digit.
2. Write the stems in a vertical column with the smallest at the top, and draw a vertical line at the right of this column.
3. Write each leaf in the row to the right of its stem, in increasing order out from the stem.

EXAMPLE 6 Stemplot of the "65 and over" data

For the "65 and over" percentages in Table 11.1, the whole-number part of the observation is the stem, and the final digit (tenths) is the leaf. The Alabama entry, 13.8, has stem 13 and leaf 8. Stems can have as many digits as needed, but each leaf must consist of only a single digit. Figure 11.6 shows the steps in making a stemplot for the data in Table 11.1. First, write the stems. Then go through the table adding a leaf for each observation. Finally, arrange the leaves in increasing order out from each stem.

```
 7 |          7 | 3               7 | 3
 8 |          8 |                 8 |
 9 |          9 | 0               9 | 0
10 |         10 | 412            10 | 124
11 |         11 | 24             11 | 24
12 |         12 | 028315794103   12 | 001123345789
13 |         13 | 83791340653147 13 | 01123333334455567789
              |                53323
14 |         14 | 38827140       14 | 01234788
15 |         15 | 147            15 | 147
16 |         16 |                16 |
17 |         17 | 4              17 | 4
```

Figure 11.6 Making a stemplot of the data in Table 11.1. Whole percents form the stems, and tenths of a percent form the leaves. (This figure was created using the Minitab software package.)

A stemplot looks like a histogram turned on end. The stemplot in Figure 11.6 is just like the histogram in Figure 11.1 because the classes chosen for the histogram are the same as the stems in the stemplot. Figure 11.7 is a stemplot of the Illinois tuition data discussed in Example 3. This stemplot has almost 4 times as many classes as the histogram of the same data in Figure 11.2. We interpret stemplots as we do histograms, looking for the overall pattern and for any outliers.

```
 1 |
 2 | 0011222222222233445555555555556666777888999
 3 | 01123477
 4 | 5
 5 |
 6 |
 7 | 17
 8 | 136
 9 | 0345
10 | 477
11 | 08
12 | 088
13 | 09
14 | 3
15 | 8
16 | 7
17 | 46
18 | 048
19 | 24
20 | 26
21 | 00038
22 | 011347
23 | 0056
24 | 0347
25 | 0
26 | 3459
27 | 134
28 |
29 | 4
30 | 7
31 | 39
32 |
33 |
34 | 002
35 |
36 |
37 |
38 | 16
```

Figure 11.7 Stemplot of the Illinois tuition and fee data. (Data from the Web site **www.isac.org.** This figure was created using the Minitab software package.)

You can choose the classes in a histogram. The classes (the stems) of a stemplot are given to you. You can get more flexibility by **rounding** the data so that the final digit after rounding is suitable as a leaf. Do this when the data have too many digits. For example, the tuition charges of Illinois colleges look like

$9,500 $9,430 $7,092 $10,672   ...

A stemplot would have too many stems if we took all but the final digit as the stem and the final digit as the leaf. To make the stemplot in Figure 11.7, we round these data to the nearest hundred dollars:

$$95 \quad 94 \quad 71 \quad 107 \quad \ldots$$

These values appear in Figure 11.7 on the stems 9, 9, 7, and 10.

The chief advantage of a stemplot is that it displays the actual values of the observations. We can see from the stemplot in Figure 11.7, but not from the histogram in Figure 11.2, that Illinois's most expensive college charges $38,600 (rounded to the nearest hundred dollars). Stemplots are also faster to draw than histograms. A stemplot requires that we use the first digit or digits as stems. This amounts to an automatic choice of classes and can give a poor picture of the distribution. Stemplots do not work well with large data sets, because the stems then have too many leaves.

STATISTICS IN SUMMARY

Chapter Specifics

- The **distribution** of a variable tells us what values the variable takes and how often it takes each value.

- To display the distribution of a quantitative variable, use a **histogram** or a **stemplot.** We usually favor stemplots when we have a small number of observations and histograms for larger data sets.

- When you look at a graph, look for an **overall pattern** and for **deviations** from that pattern, such as **outliers.**

- We can describe the overall pattern of a histogram or stemplot by giving its **shape, center,** and **spread.** Some distributions have simple shapes such as **symmetric** or **skewed,** but others are too irregular to describe by a simple shape.

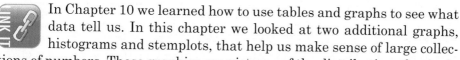 In Chapter 10 we learned how to use tables and graphs to see what data tell us. In this chapter we looked at two additional graphs, histograms and stemplots, that help us make sense of large collections of numbers. These graphics are pictures of the distribution of a single quantitative variable. Although the bar graphs in Chapter 10 look much like histograms, the difference is that bar graphs are used to display the distribution of a categorical variable, while histograms display the distribution of a quantitative variable. The overall pattern (shape, center, and spread) and deviations from this pattern (outliers) are important features of the distribution of a variable. We will look more carefully at these features in future chapters. In addition, these features will figure prominently in some of the conclusions that we will draw about a variable from data.

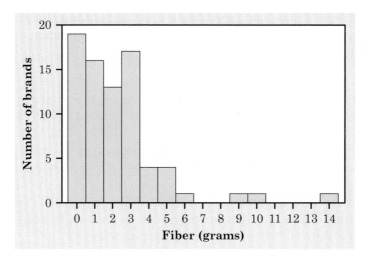

Figure 11.8 Histogram of the amount of dietary fiber (in grams) in 77 different brands of cereal. (This figure was created using the Minitab software package.)

CASE STUDY Figure 11.8 is a histogram of the distribution of the amount of dietary
EVALUATED fiber in 77 brands of cereal found on the shelves of a grocery store. Use
what you have learned in this chapter to answer the following questions.

1. Describe the overall shape, the center, and the spread of the distribution.

2. Are there any outliers?

3. Wheaties has 3 grams of dietary fiber. Is this a low, high, or typical amount? ▪

CHAPTER 11 EXERCISES

For Exercise 11.1, see page 242; for Exercise 11.2, see page 248.

11.3 Lightning storms. Figure 11.9 comes from a study of lightning storms in
Colorado. It shows the distribution of the hour of the day during which the first
lightning flash for that day occurred. Describe the shape, center, and spread of
this distribution. Are there any outliers?

11.4 Where do 18- to 34-year-olds live? Figure 11.10 is a stemplot of the
percentage of residents aged 18 to 34 in each of the 50 states in July 2008. As
in Figure 11.6 (page 249) for older residents, the stems are whole percents and
the leaves are tenths of a percent.

(a) Utah has the largest percentage of young adults. What is the percentage for
this state?

(b) Ignoring Utah, describe the shape, center, and spread of this distribution.

(c) Is the distribution for young adults more or less spread out than the distribution in Figure 11.6 for older adults?

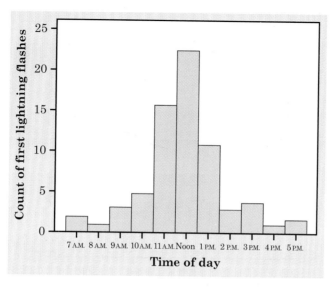

Figure 11.9 Histogram of the time of day at which the day's first lightning flash occurred (from a study in Colorado), for Exercise 11.3. (Based on an episode in the Annenberg/Corporation for Public Broadcasting telecourse *Against All Odds: Inside Statistics.*)

11.5 Minority students in engineering. Figure 11.11 is a histogram of the number of minority students (black, Hispanic, Native American) who earned doctorate degrees in engineering from each of 152 universities in the years 2000 through 2002. Briefly describe the shape, center, and spread of this distribution. The classes for Figure 11.11 are 1–5, 6–10, and so on.

11.6 Returns on common stocks. The total return on a stock is the change in its market price plus any dividend payments made. Total return is usually expressed as a percentage of the beginning price. Figure 11.12 is a histogram of the distribution of total returns for all 1528 common stocks listed on the New York Stock Exchange in one year.

```
19 | 9
20 | 59
21 | 13444
22 | 122255666777899
23 | 00011345555666677888
24 | 01588
25 | 69
26 |
27 |
28 | 4
```

Figure 11.10 Stemplot of the percentage of each state's residents who are 18 to 34 years old, for Exercise 11.4. (This figure was created using the Minitab software package.)

(a) Describe the overall shape of the distribution of total returns.

(b) What is the approximate center of this distribution? Approximately what were the smallest and largest total returns? (This describes the spread of the distribution.)

(c) A return less than zero means that owners of the stock lost money. About what percentage of all stocks lost money?

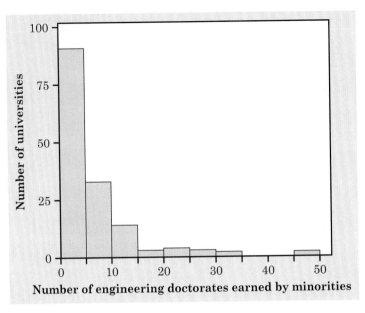

Figure 11.11 The distribution of the number of engineering doctorates earned by minority students at 152 universities, 2000 to 2002, for Exercise 11.5. (Data from the 2003 National Science Foundation Survey of Earned Doctorates, found at the Web site **webcaspar.nsf.gov/.** This figure was created using the SPSS software package.)

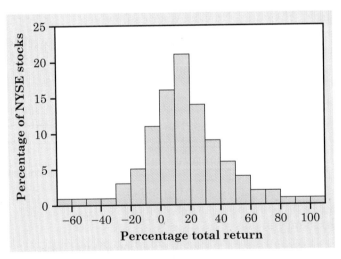

Figure 11.12 The distribution of total returns for all New York Stock Exchange common stocks in one year, for Exercise 11.6. (Based on J. K. Ford, "Diversification: how many stocks will suffice?" *American Association of Individual Investors Journal,* January 1990, pp. 14–16.)

TABLE 11.2 Highway gas mileage for model year 2010 sedans

Model	mpg	Model	mpg
Acura RL	22	Lexus ES350	27
Bentley Azure	15	Lexus GS460	24
BMW 535i	26	Lincoln Town Car	24
Buick Lacrosse	26	Maybach 57	16
Cadillac CTS	27	Mazda 5	27
Cadillac STS	27	Mazda 6	25
Chevrolet Malibu	33	Mercedes-Benz E350	26
Chrysler Sebring	27	Mercedes-Benz E550	23
Dodge Avenger	30	Nissan Maxima	26
Honda Accord	31	Pontiac G6	26
Hyundai Sonata	35	Rolls Royce Phantom	18
Infiniti M37	26	Saturn Aura	26
Infiniti M56	25	Toyota Camry	33
Jaguar XJ	22	Volkswagen CC	27
Kia Forte	31	Volvo S80 AWD	26
Kia Optima	28		

Source: **www.fueleconomy.gov.**

11.7 Histogram or stemplot? Explain why we prefer a histogram to a stemplot for describing the returns on 1528 common stocks.

11.8 Automobile fuel economy. Government regulations require automakers to give the city and highway gas mileages for each model of car. Table 11.2 gives the highway mileages (miles per gallon) for 31 model year 2010 sedans. Make a stemplot of the highway gas mileages of these cars. What can you say about the overall shape of the distribution? Where is the center (the value such that half the cars have better gas mileage and half have worse gas mileage)? Three of these cars are subject to the "gas guzzler tax" because of their low gas mileage. Which three?

11.9 The obesity epidemic. Medical authorities describe the spread of obesity in the United States as an epidemic. Table 11.3 gives the percentage of adults who were obese in each of the 50 states in 2009. Display the distribution in a graph and briefly describe its shape, center, and spread.

11.10 Yankee money. Table 11.4 gives the salaries of the players on the New York Yankees baseball team for the 2011 season. Make a histogram of these data. Is the distribution roughly symmetric, skewed to the right, or skewed to the left? Explain.

11.11 The statistics of writing style. Numerical data can distinguish different types of writing, and sometimes even individual authors. Here are data

TABLE 11.3 Percentage of adult population who are obese, 2009

State	Percent	State	Percent	State	Percent
Alabama	31.0	Louisiana	33.0	Ohio	28.8
Alaska	24.8	Maine	25.8	Oklahoma	31.4
Arizona	25.5	Maryland	26.2	Oregon	23.0
Arkansas	30.5	Massachusetts	21.4	Pennsylvania	27.4
California	24.8	Michigan	29.6	Rhode Island	24.6
Colorado	18.6	Minnesota	24.6	South Carolina	29.4
Connecticut	20.6	Mississippi	34.4	South Dakota	29.6
Delaware	27.0	Missouri	30.0	Tennessee	32.3
Florida	25.2	Montana	23.2	Texas	28.7
Georgia	27.2	Nebraska	27.2	Utah	23.5
Hawaii	22.3	Nevada	25.8	Vermont	22.8
Idaho	24.5	New Hampshire	25.7	Virginia	25.0
Illinois	26.5	New Jersey	23.3	Washington	26.4
Indiana	29.5	New Mexico	25.1	West Virginia	31.1
Iowa	27.9	New York	24.2	Wisconsin	28.7
Kansas	28.1	North Carolina	29.3	Wyoming	24.6
Kentucky	31.5	North Dakota	27.9		

Source: National Centers for Disease Control and Prevention, **www.cdc.gov/obesity/data/trends .html.**

TABLE 11.4 Salaries of the New York Yankees, 2011

Player	Salary ($)	Player	Salary ($)
Alex Rodriguez	32,000,000	Eric Chavez	1,500,000
CC Sabathia	24,285,714	Freddy Garcia	1,500,000
Mark Teixeira	23,125,000	Andruw Jones	1,500,000
A. J. Burnett	16,500,000	Joba Chamberlain	1,400,000
Mariano Rivera	14,911,700	Boone Logan	1,200,000
Derek Jeter	14,729,364	Bartolo Colon	900,000
Jorge Posada	13,100,000	Luis Ayala	650,000
Robinson Cano	10,000,000	Brett Gardner	529,500
Rafael Soriano	10,000,000	David Robertson	460,450
Nick Swisher	9,100,000	Francisco Cervelli	455,700
Curtis Granderson	8,250,000	Gustavo Molina	455,000
Damaso Marte	4,000,000	Ivan Nova	432,900
Russell Martin	4,000,000	Colin Curtis	420,400
Pedro Feliciano	3,750,000	Eduardo Nunez	419,300
Phil Hughes	2,700,000	Reegie Corona	414,000

Source: **content.usatoday.com/sportsdata/baseball/mlb/salaries/team.**

collected by students on the percentages of words of 1 to 15 letters used in articles in *Popular Science* magazine:

Length:	1	2	3	4	5	6	7	8	9	10	11	12	13	14	15
Percent:	3.6	14.8	18.7	16.0	12.5	8.2	8.1	5.9	4.4	3.6	2.1	0.9	0.6	0.4	0.2

(a) Make a histogram of this distribution. Describe its shape, center, and spread.

(b) How does the distribution of lengths of words used in *Popular Science* compare with the similar distribution in Figure 11.4 (page 247) for Shakespeare's plays? Look in particular at short words (2, 3, and 4 letters) and very long words (more than 10 letters).

11.12 Skewed left. Sketch a histogram for a distribution that is skewed to the left. Suppose that you and your friends emptied your pockets of coins and recorded the year marked on each coin. The distribution of dates would be skewed to the left. Explain why.

11.13 What's my shape? Do you expect the distribution of the total player payroll for each of the 30 teams in Major League Baseball to be roughly symmetric, clearly skewed to the right, or clearly skewed to the left? Why?

11.14 Asians in the eastern states. Here are the percentages of the population who are of Asian origin in each state east of the Mississippi River in 2008:

State	Percent	State	Percent	State	Percent
Alabama	1.0	Connecticut	3.6	Delaware	2.9
Florida	2.3	Georgia	2.9	Illinois	4.3
Indiana	1.4	Kentucky	1.0	Maine	0.9
Maryland	5.1	Massachusetts	4.9	Michigan	2.4
Mississippi	0.8	New Hampshire	1.9	New Jersey	7.6
New York	7.0	North Carolina	1.9	Ohio	1.6
Pennsylvania	2.4	Rhode Island	2.8	South Carolina	1.2
Tennessee	1.3	Vermont	1.1	Virginia	4.9
West Virginia	0.7	Wisconsin	2.0		

Make a stemplot of these data. Describe the overall pattern of the distribution. Are there any outliers?

11.15 How many calories does a hot dog have? *Consumer Reports* magazine presented the following data on the number of calories in a hot dog for each of 17 brands of meat hot dogs:

$$173 \quad 191 \quad 182 \quad 190 \quad 172 \quad 147 \quad 146 \quad 139 \quad 175$$
$$136 \quad 179 \quad 153 \quad 107 \quad 195 \quad 135 \quad 140 \quad 138$$

Make a stemplot of the distribution of calories in meat hot dogs and briefly describe the shape of the distribution. Most brands of meat hot dogs contain

TABLE 11.5 Age distribution in the United States, 1950 and 2050 (in millions of persons)

Age group	1950	2050
Under 10 years	29.3	56.2
10 to 19 years	21.8	56.7
20 to 29 years	24.0	56.2
30 to 39 years	22.8	55.9
40 to 49 years	19.3	52.8
50 to 59 years	15.5	49.1
60 to 69 years	11.0	45.0
70 to 79 years	5.5	34.5
80 to 89 years	1.6	23.7
90 to 99 years	0.1	8.1
100 years and over	—	0.6
Total	151.1	438.8

Source: 2010 *Statistical Abstract of the United States,* **www.census.gov/ compendia/statab/;** and Census Bureau, **www.census.gov/population/www/ projections/natproj.html.**

a mixture of beef and pork, with up to 15% poultry allowed by government regulations. The only brand with a different makeup was Eat Slim Veal Hot Dogs. Which point on your stemplot do you think represents this brand?

11.16 The changing age distribution of the United States. The distribution of the ages of a nation's population has a strong influence on economic and social conditions. Table 11.5 shows the age distribution of U.S. residents in 1950 and 2050, in millions of persons. The 1950 data come from that year's census. The 2050 data are projections made by the Census Bureau.

(a) Because the total population in 2050 is much larger than the 1950 population, comparing percentages in each age group is clearer than comparing counts. Make a table of the percentage of the total population in each age group for both 1950 and 2050.

(b) Make a histogram of the 1950 age distribution (in percents). Then describe the main features of the distribution. In particular, look at the percentage of children relative to the rest of the population.

(c) Make a histogram of the projected age distribution for the year 2050. Use the same scales as in (b) for easy comparison. What are the most important changes in the U.S. age distribution projected for the 100-year period between 1950 and 2050?

11.17 Babe Ruth's home runs. Here are the numbers of home runs that Babe Ruth hit in his 15 years with the New York Yankees, 1920 to 1934:

54 59 35 41 46 25 47 60 54 46 49 46 41 34 22

Make a stemplot of these data. Is the distribution roughly symmetric, clearly skewed, or neither? About how many home runs did Ruth hit in a typical year? Is his famous 60 home runs in 1927 an outlier?

11.18 Back-to-back stemplot. The current major league single-season home run record is held by Barry Bonds of the San Francisco Giants. Here are Bonds's home run counts for 1986 to 2007:

16	25	24	19	33	25	34	46	37	33	42
40	37	34	49	73	46	45	45	5	26	28

A **back-to-back stemplot** helps us compare two distributions. Write the stems as usual, but with a vertical line both to their left and to their right. On the right, put leaves for Ruth (Exercise 11.17). On the left, put leaves for Bonds. Arrange the leaves on each stem in increasing order out from the stem. Now write a brief comparison of Ruth and Bonds as home run hitters.

11.19 When it rains, it pours. On July 25 to 26, 1979, 42.00 inches of rain fell on Alvin, Texas. That's the most rain ever recorded in Texas for a 24-hour period. Table 11.6 gives the maximum precipitation ever recorded in 24 hours (through 2010) at any weather station in each state. The record amount varies a great deal from state to state—hurricanes bring extreme rains on the Atlantic coast, and the mountain West is generally dry. Make a graph to display

TABLE 11.6 Record 24-hour precipitation amounts (inches) by state

State	Precip.	State	Precip.	State	Precip.
Alabama	32.52	Louisiana	22.00	Ohio	10.75
Alaska	15.20	Maine	13.32	Oklahoma	15.68
Arizona	11.40	Maryland	14.75	Oregon	11.77
Arkansas	14.06	Massachusetts	18.15	Pennsylvania	13.50
California	25.83	Michigan	9.78	Rhode Island	12.13
Colorado	11.08	Minnesota	15.10	South Carolina	14.80
Connecticut	12.77	Mississippi	15.68	South Dakota	8.74
Delaware	8.50	Missouri	18.18	Tennessee	13.60
Florida	23.28	Montana	11.50	Texas	42.00
Georgia	21.10	Nebraska	13.15	Utah	5.08
Hawaii	38.00	Nevada	7.78	Vermont	9.92
Idaho	7.17	New Hampshire	11.07	Virginia	14.28
Illinois	16.91	New Jersey	14.81	Washington	14.26
Indiana	10.50	New Mexico	11.28	West Virginia	12.02
Iowa	13.18	New York	11.15	Wisconsin	11.72
Kansas	13.53	North Carolina	22.22	Wyoming	6.06
Kentucky	10.40	North Dakota	8.10		

Source: National Oceanic and Atmospheric Administration, **www.noaa.gov.**

the distribution of records for the states. Mark where your state lies in this distribution. Briefly describe the distribution.

EXPLORING THE WEB

11.20 Web-based exercise. The all-time home run leader prior to 2007 was Hank Aaron. You can find his career statistics at the Web site **www .baseball-reference.com.** Make a stemplot of the number of home runs that Hank Aaron hit in his career. Is the distribution roughly symmetric, clearly skewed, or neither? About how many home runs did Aaron hit in a typical year? Are there any outliers?

NOTES AND DATA SOURCES

Page 257 Exercise 11.14: These data are from the 2010 *Statistical Abstract of the United States;* available online at **www .census.gov/compendia/statab/.**

Page 257 Exercise 11.15: *Consumer Reports,* June 1986, pp. 366–367. A more recent study of hot dogs appears in *Consumer Reports,* July 1993, pp. 415–419. The newer data cover few brands of poultry hot dogs and

take calorie counts mainly from the package labels, resulting in suspiciously round numbers.

Page 258 Table 11.5: Source: 2010 *Statistical Abstract of the United States,* **www .census.gov/compendia/statab/;** and Census Bureau, **www.census.gov/population/ www/projections/natproj.html.**

Describing Distributions with Numbers

CASE STUDY Does education pay? We are told that people with more education earn more on the average than people with less education. How much more? How can we answer this question?

Data on income can be found at the Census Bureau Web site. The data are estimates, for the year 2010, of the total incomes of 133,074,000 people aged 25 and over with earnings and are based on the results of the Current Population Survey in 2011. The Web site gives the income distribution for each of several education categories. In particular, it gives the number of people in each of several education categories who earned between $1 and $2499, between $2500 and $4999, up to between $97,500 and $99,999, and $100,000 and over. That is a lot of information. A histogram could be used to display the data, but are there simple ways to summarize the information that allow us to make sensible comparisons?

In this chapter we will learn several ways to summarize large data sets. By the end of this chapter you will be able to provide an answer to whether education really pays. ■

In the summer of 2007, Barry Bonds shattered the career home run record, breaking the previous record set by Hank Aaron. Here are his home run counts for the years 1986 (his rookie year) to 2007:

1986	1987	1988	1989	1990	1991	1992	1993	1994	1995	1996
16	25	24	19	33	25	34	46	37	33	42

1997	1998	1999	2000	2001	2002	2003	2004	2005	2006	2007
40	37	34	49	73	46	45	45	5	26	28

```
0 | 5
1 | 69
2 | 45568
3 | 334477
4 | 0255669
5 |
6 |
7 | 3
```

Figure 12.1 Stemplot of the number of home runs hit by Barry Bonds in his first 22 seasons.

The stemplot in Figure 12.1 displays the data. The shape of the distribution is a bit irregular, but we see that it has one high outlier, and if we ignore this outlier, we might describe it as slightly skewed to the left with a single peak. The outlier is, of course, Bonds's record season in 2001.

A graph and a few words give a good description of Barry Bonds's home run career. But words are less adequate to describe, for example, the incomes of people with a high school education. We need *numbers* that summarize the center and spread of a distribution.

Median and quartiles

A simple and effective way to describe center and spread is to give the **median** and the **quartiles.** The median is the midpoint, the value that separates the smaller half of the observations from the larger half. The first and third quartiles mark off the middle half of the observations. The quartiles get their name because with the median they divide the observations into quarters—one-quarter of the observations lie below the first quartile, half lie below the median, and three-quarters lie below the third quartile. That's the idea. To actually get numbers, we need a rule that makes the idea exact.

EXAMPLE 1 Finding the median

We might compare Bonds's career with that of Hank Aaron, the previous holder of the career record. Here are Aaron's home run counts for his 23 years in baseball.

13	27	26	44	30	39	40	34	45	44	24	32
44	39	29	44	38	47	34	40	20	12	10	

To find the median, first arrange them in order from smallest to largest:

10	12	13	20	24	26	27	29	30	32	34	**34**
38	39	39	40	40	44	44	44	44	45	47	

The bold 34 is the center observation, with 11 observations to its left and 11 to its right. When the number of observations n is odd (here $n = 23$), there is always one observation in the center of the ordered list. This is the median, $M = 34$.

How does this compare with Bonds's record? Here are Bonds's 22 home run counts, arranged in order from smallest to largest:

$$\begin{array}{ccccccccccc}
5 & 16 & 19 & 24 & 25 & 25 & 26 & 28 & 33 & 33 & \mathbf{34} \\
\mathbf{34} & 37 & 37 & 40 & 42 & 45 & 45 & 46 & 46 & 49 & 73
\end{array}$$

When n is even, there is no one middle observation. But there is a middle pair—the bold 34 and 34 have 10 observations on either side. We take the median to be halfway between this middle pair. So Bonds's median is

$$M = \frac{34 + 34}{2} = \frac{68}{2} = 34$$

There is a fast way to locate the median in an ordered list: count up $(n + 1)/2$ places from the beginning of the list. Try it. For Aaron, $n = 23$ and $(23 + 1)/2 = 12$, so the median is the 12th entry in the ordered list. For Bonds, $n = 22$ and $(22 + 1)/2 = 11.5$. This means "halfway between the 11th and 12th" entries, so M is the average of these two entries. This "$(n + 1)/2$ rule" is especially handy when you have many observations. The median of $n = 46,940$ incomes is halfway between the 23,470th and 23,471st in the ordered list. Be sure to note that $(n + 1)/2$ does *not* give the median M, just its position in the ordered list of observations.

The median *M*

The **median *M*** is the midpoint of a distribution, the number such that half the observations are smaller and the other half are larger. To find the median of a distribution:

1. Arrange all observations in order of size, from smallest to largest.

2. If the number of observations n is odd, the median M is the center observation in the ordered list. Find the location of the median by counting $(n + 1)/2$ observations up from the bottom of the list.

3. If the number of observations n is even, the median M is the average of the two center observations in the ordered list. The location of the median is again $(n + 1)/2$ from the bottom of the list.

"Yup, Old Bob drowned due to being ignorant of statistics. He thought it was enough to know the average depth of the river."

The Census Bureau Web site provides data on income inequality. For example, it tells us that in 2008 the median income of Hispanic households was $37,913. That's helpful but incomplete. Do most Hispanic households earn close to this amount, or are the incomes very spread out? The simplest useful description of a distribution consists of both a measure of *center* and a measure of *spread*. If we choose the median (the midpoint) to describe center, the quartiles provide a natural description of spread. Again, the idea is clear: find the points one-quarter and three-quarters up the ordered list of observations. Again, we need a rule to make the idea precise. The rule for calculating the quartiles uses the rule for the median.

The quartiles Q_1 and Q_3

To calculate the **quartiles**:

1. Arrange the observations in increasing order and locate the median M in the ordered list of observations.
2. The **first quartile Q_1** is the median of the observations whose position in the ordered list is to the left of the location of the overall median. The overall median is not included in the observations considered to be to the left of the overall median.
3. The **third quartile Q_3** is the median of the observations whose position in the ordered list is to the right of the location of the overall median. The overall median is not included in the observations considered to be to the right of the overall median.

EXAMPLE 2 Finding the quartiles

Hank Aaron's 23 home run counts are

$$10 \quad 12 \quad 13 \quad 20 \quad 24 \quad 26 \quad 27 \quad 29 \quad 30 \quad 32 \quad 34 \quad \mathbf{34} \quad 38$$

$$\uparrow \qquad\qquad\qquad\qquad\qquad\qquad\qquad\qquad \uparrow$$

$$Q_1 \qquad\qquad\qquad\qquad\qquad\qquad\qquad\qquad M$$

$$39 \quad 39 \quad 40 \quad 40 \quad 44 \quad 44 \quad 44 \quad 44 \quad 45 \quad 47$$

$$\uparrow$$

$$Q_3$$

There is an odd number of observations, so the median is the one in the middle, the bold 34 in the list. To find the quartiles, ignore this central observation. The first quartile is the median of the 11 observations to the left of the bold 34 in the list. That's the 6th, so $Q_1 = 26$. The third quartile is the median of the 11 observations to the right of the bold 34. It is $Q_3 = 44$.

Barry Bonds's 22 home run counts are

$$5 \quad 16 \quad 19 \quad 24 \quad 25 \quad 25 \quad 26 \quad 28 \quad 33 \quad 33 \quad \mathbf{34} \quad \mathbf{34}$$

$$\underset{Q_1}{\uparrow} \qquad\qquad\qquad\qquad\qquad \underset{M}{\uparrow}$$

$$37 \quad 37 \quad 40 \quad 42 \quad 45 \quad 45 \quad 46 \quad 46 \quad 49 \quad 73$$

$$\underset{Q_3}{\uparrow}$$

The median lies halfway between the middle pair. There are 11 observations to the left of this location. The first quartile is the median of these 11 numbers. That's the 6th, so $Q_1 = 25$. The third quartile is the median of the 11 observations to the right of the overall median's location, $Q_3 = 45$.

NOW IT'S YOUR TURN | **12.1 Babe Ruth.** Prior to Hank Aaron, Babe Ruth was the holder of the career record. Here are Ruth's home run counts for his 22 years in Major League Baseball, arranged in order from smallest to largest:

$$0 \quad 2 \quad 3 \quad 4 \quad 6 \quad 11 \quad 22 \quad 25 \quad 29 \quad 34 \quad 35$$
$$41 \quad 41 \quad 46 \quad 46 \quad 46 \quad 47 \quad 49 \quad 54 \quad 54 \quad 59 \quad 60$$

Find the median, first quartile, and third quartile of these counts.

You can use the $(n + 1)/2$ rule to locate the quartiles when there are many observations. The Census Bureau Web site tells us that there were 13,425,000 (rounded off to the nearest 1000) Hispanic households in the United States in 2008. If we ignore the roundoff, the median of these 13,425,000 incomes is halfway between the 6,712,500th and 6,712,501st in the list arranged in order from smallest to largest. So the first quartile is the median of the 6,712,500 incomes below this point in the list. Use the $(n + 1)/2$ rule with $n = 6,712,500$ to locate the quartile:

$$\frac{n + 1}{2} = \frac{6,712,500 + 1}{2} = 3,356,250.5$$

The average of the 3,356,250th and 3,356,251st incomes in the ordered list falls in the range $20,000 to $22,499. We estimate the first quartile to be approximately $20,601 based on the approximate number of people in this range using interpolation.

The third quartile is the median of the 6,712,500 incomes above the median. By the $(n + 1)/2$ rule with 6,712,500, this will be the average of the 3,356,250th and 3,356,251st incomes above the median in the ordered list. We find that this falls in the range $65,000 to $67,999. We estimate the third quartile to be approximately $66,852 based on the approximate number of people in this range using interpolation.

In practice, people use statistical software to compute quartiles. Software can give results that differ from those you will obtain using the method described here. In fact, different software packages use slightly different rules for deciding how to divide the space between two adjacent values between which the quartile is believed to lie. We have chosen to select the point halfway between them, but other rules exist. Two different software packages can give two slightly different answers, depending on the rule employed.

The five-number summary and boxplots

The smallest and largest observations tell us little about the distribution as a whole, but they give information about the tails of the distribution that is missing if we know only the median and the quartiles. To get a quick summary of both center and spread, combine all five numbers.

The five-number summary

The **five-number summary** of a distribution consists of the smallest observation, the first quartile, the median, the third quartile, and the largest observation, written in order from smallest to largest. In symbols, the five-number summary is

$$\text{Minimum} \quad Q_1 \quad M \quad Q_3 \quad \text{Maximum}$$

These five numbers offer a reasonably complete description of center and spread. The five-number summaries of home run counts are

$$10 \ \ 26 \ \ 34 \ \ 44 \ \ 47$$

for Aaron and

$$5 \ \ 25 \ \ 34 \ \ 45 \ \ 73$$

for Bonds. The five-number summary of a distribution leads to a new graph, the *boxplot*. Figure 12.2 shows boxplots for the home run comparison.

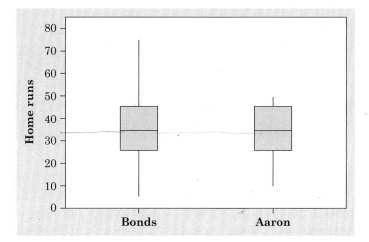

Figure 12.2 Boxplots comparing the yearly home run production of Barry Bonds and Hank Aaron. (This figure was created using the Minitab software package.)

Boxplot

A **boxplot** is a graph of the five-number summary.

- A central box spans the quartiles.
- A line in the box marks the median.
- Lines extend from the box out to the smallest and largest observations.

You can draw boxplots either horizontally or vertically. Be sure to include a numerical scale in the graph. When you look at a boxplot, first locate the median, which marks the center of the distribution. Then look at the spread. The quartiles show the spread of the middle half of the data, and the extremes (the smallest and largest observations) show the spread of the entire data set. We see from Figure 12.2 that Bonds's usual performance, as indicated by the median and the box that marks the middle half of the distribution, is similar to that of Aaron. We also see that the distribution for Aaron is less spread out than the distribution for Bonds.

NOW IT'S YOUR TURN | **12.2 Babe Ruth.** Here are Babe Ruth's home run counts for his 22 years in Major League Baseball, arranged in order from smallest to largest:

0	2	3	4	6	11	22	25	29	34	35
41	41	46	46	46	47	49	54	54	59	60

Draw a boxplot of this distribution. How does it compare with those of Barry Bonds and Hank Aaron in Figure 12.2?

Because boxplots show less detail than histograms or stemplots, they are best used for side-by-side comparison of more than one distribution, as in Figure 12.2. For such small numbers of observations, a back-to-back stemplot is better yet (see Exercise 11.18, page 259). It would make clear, as the boxplot cannot, that Bonds's record 73 home runs in 2001 is an outlier in his career. Let us look at an example where boxplots are more genuinely useful.

EXAMPLE 3 Income inequality

To investigate income inequality, we compare household incomes of Hispanics, blacks, and whites. The Census Bureau Web site provides information on income distribution by race. Figure 12.3 compares the income distributions for Hispanics, blacks, and whites in 2008. This figure is a variation on the boxplot idea. The largest income among several million people will surely be very large. Figure 12.3 uses the 95% points (the values representing where the top 5% of incomes start) in the distributions instead of the single largest incomes. So, for example, the line above the box for the Hispanic group extends only to $137,568 rather than to the highest income. Many statistical software packages allow you to produce boxplots that suppress extreme values, but the rules for what constitutes an extreme value usually do not use the 95% point in the distribution instead of the single largest value.

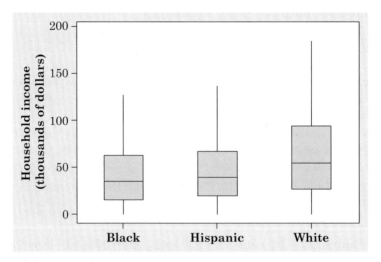

Figure 12.3 Boxplots comparing the distributions of income among Hispanics, blacks, and whites. The ends of each plot are at 0 and at the 95% point in the distribution. (This figure was created using the Minitab software package.)

Figure 12.3 gives us a clear and simple visual comparison. We see that the median and middle half are slightly greater for Hispanics than for blacks, and that for whites the median and middle half are greater than for both blacks and Hispanics. The income of the bottom 5% stays small because there are some people in each group with no income or even negative income, perhaps due to illness or disability. The 95% point, marking off the top 5% of incomes, is greater for whites than for either blacks and Hispanics, and the 95% point of incomes for Hispanics is greater than for blacks. Overall, incomes for whites tend to be larger than those for Hispanics and blacks, highlighting racial inequities in income.

Figure 12.3 also illustrates how boxplots often indicate the symmetry or skewness of a distribution. In a symmetric distribution, the first and third quartiles are equally distant from the median. In most distributions that are skewed to the right, on the other hand, the third quartile will be farther above the median than the first quartile is below it. The extremes behave the same way. Even with the top 5% not present, we can see the right-skewness of incomes for all three races.

STATISTICAL CONTROVERSIES

Income Inequality

During the prosperous 1980s and 1990s, the incomes of all American households went up, but the gap between rich and poor grew. Figures 12.4 and 12.5 give two views of increasing inequality. Figure 12.4 is a line graph of household income, in dollars adjusted to have the same buying power every year. The lines show the 20th and 80th percentiles of income, which mark off the bottom fifth and the top fifth of households. The 80th percentile (up 48% between 1967 and 2010) is pulling away from the 20th percentile (up about 17%).

Figure 12.5 looks at the *share* of all income that goes to the top fifth and the bottom fifth. The bottom fifth's share has drifted down, to 3.3% of all income in 2010.

The share of the top fifth grew to 50.2% (up 15.1% between 1967 and 2010). Although not displayed in the figures, the share of the top 5% grew even faster, from 17.2% in 1967 to 21.3% of the income of all households in the country in 2010. This is a 23.8% increase between 1967 and 2010. Income inequality in the United States is greater than in other developed nations and has been increasing.

Are these numbers cause for concern? And do they accurately reflect the disparity between the wealthy and the poor? For example, as people get older, their income increases. Perhaps these numbers reflect only the disparity between younger and older wage earners. What do you think?

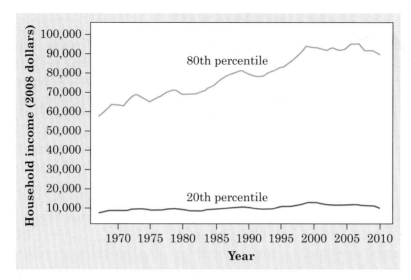

Figure 12.4 The change over time of two points in the distribution of incomes for American households. Eighty percent of households have incomes below the 80th percentile, and 20% have incomes below the 20th percentile. In 2010, the 20th percentile was $20,000, and the 80th percentile was $100,065. (This figure was created using the Minitab software package.)

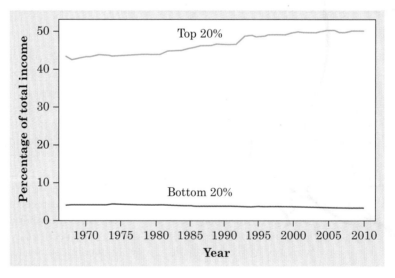

Figure 12.5 The change over time of the shares of total household income that go to the highest-income 20% and to the lowest-income 20% of households. In 2010, the top 20% of households received half of all income. (This figure was created using the Minitab software package.)

Mean and standard deviation

The five-number summary is not the most common numerical description of a distribution. That distinction belongs to the combination of the *mean* to measure center and the *standard deviation* to measure spread. The mean is familiar—it is the ordinary average of the observations. The idea of the standard deviation is to give the average distance of observations from the mean. The "average distance" in the standard deviation is found in a rather obscure way. We will give the details, but you may want to just think of the standard deviation as "average distance from the mean" and leave the details to your calculator or other technology.

Mean and standard deviation

The **mean** $\overline{x}$ (pronounced "x-bar") of a set of observations is their average. To find the mean of n observations, add the values and divide by n:

$$\overline{x} = \frac{\text{sum of the observations}}{n}$$

The **standard deviation** s measures the average distance of the observations from their mean. It is calculated by finding an average of the squared distances and then taking the square root. To find the standard deviation of n observations:

1. Find the distance of each observation from the mean and square each of these distances.

2. Average the squared distances by dividing their sum by $n - 1$. This average squared distance is called the **variance.**

3. The standard deviation s is the square root of this average squared distance.

EXAMPLE 4 Finding the mean and standard deviation

The numbers of home runs Barry Bonds hit in his first 22 major league seasons are

$$
\begin{array}{ccccccccccc}
16 & 25 & 24 & 19 & 33 & 25 & 34 & 46 & 37 & 33 & 42 \\
40 & 37 & 34 & 49 & 73 & 46 & 45 & 45 & 5 & 26 & 28
\end{array}
$$

To find the mean of these observations,

$$\overline{x} = \frac{\text{sum of observations}}{n}$$

$$= \frac{16 + 25 + \cdots + 28}{22}$$

$$= \frac{762}{22} = 34.6$$

Figure 12.6 displays the data as points above the number line, with their mean marked by a line. The arrow shows one of the distances from the mean. The idea behind the standard deviation s is to average the 22 distances. To find the standard deviation by hand, you can use a table layout:

Observation	Squared distance from mean
16	$(16 - 34.6)^2 = (-18.6)^2 = 345.96$
25	$(25 - 34.6)^2 = (-9.6)^2 = 92.16$
$\vdots$	
28	$(28 - 34.6)^2 = (-6.6)^2 = 43.56$
	sum $= 4139.12$

The average is

$$\frac{4139.12}{21} = 197.1$$

Notice that we "average" by dividing by *one less* than the number of observations. Finally, the standard deviation is the square root of this number:

$$s = \sqrt{197.1} = 14.04$$

In practice, you can key the data into your calculator and hit the mean key and the standard deviation key. Or you can enter the data into a

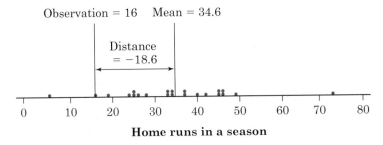

Figure 12.6 Barry Bonds's home run counts, for Example 4, with their mean and the distance of one observation from the mean indicated. Think of the standard deviation as an average of these distances.

spreadsheet or other software to find $\bar{x}$ and s. It is usual, for good but somewhat technical reasons, to average the squared distances by dividing their total by $n - 1$ rather than by n. Many calculators have two standard deviation buttons, giving you a choice between dividing by n and dividing by $n - 1$. Be sure to choose $n - 1$.

NOW IT'S YOUR TURN | **12.3 Hank Aaron.** Here are Aaron's home run counts for his 23 years in baseball.

| 13 | 27 | 26 | 44 | 30 | 39 | 40 | 34 | 45 | 44 | 24 | 32 |
| 44 | 39 | 29 | 44 | 38 | 47 | 34 | 40 | 20 | 12 | 10 |

Find the mean and standard deviation of the number of home runs Aaron hit in each season of his career. How do the mean and median compare?

More important than the details of the calculation are the properties that show how the standard deviation measures spread.

Properties of the standard deviation s

- s measures spread about the mean $\bar{x}$. Use s to describe the spread of a distribution only when you use $\bar{x}$ to describe the center.

- $s = 0$ only when there is *no spread*. This happens only when all observations have the same value. So standard deviation zero means no spread at all. Otherwise, $s > 0$. As the observations become more spread out about their mean, s gets larger.

EXAMPLE 5 Investing 101

Enough examples about income. Here is an example about what to do with it once you've earned it. One of the first principles of investing is that taking more risk brings higher returns, at least on the average in the long run. People who work in finance define risk as the variability of returns from an investment (greater variability means higher risk) and measure risk by how unpredictable the return on an investment is. A bank account that is insured by the government and has a fixed rate of interest has no risk—its return is known exactly. Stock in a new company may soar one week and plunge the next. It has high risk because you can't predict what it will be worth when you want to sell.

Investors should think statistically. You can assess an investment by thinking about the distribution of (say) yearly returns. That means asking about both the center and the spread of the pattern of returns. Only naive investors look for a high average return without asking about risk, that is, about how spread out or variable the returns are. Financial experts use the mean and standard deviation to describe returns on investments. The standard deviation was long considered too complicated to mention to the public, but now you will find standard deviations appearing regularly in mutual funds reports.

Here by way of illustration are the means and standard deviations of the yearly returns on three investments over the second half of the 20th century (the 50 years from 1950 to 1999), a time of rapid economic growth:

Investment	Mean return	Standard deviation
Treasury bills	5.34%	2.96%
Treasury bonds	6.12%	10.73%
Common stocks	14.62%	16.32%

You can see that risk (variability) goes up as the mean return goes up, just as financial theory claims. Treasury bills and bonds are ways of loaning money to the U.S. government. Treasury bills are paid back in one year, so their return changes from year to year depending on interest rates. Bonds are 30-year loans. They are riskier because the value of a bond you own will drop if interest rates go up. Stocks are even riskier. They give higher returns (on the average in the long run) but at the cost of lots of sharp ups and downs along the way. As the stemplot in Figure 12.7 shows, stocks went up by as much as 50% and down by as much as 26% in one year during the 50 years covered by our data.

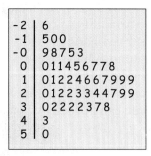

```
-2 | 6
-1 | 500
-0 | 98753
 0 | 011456778
 1 | 01224667999
 2 | 01223344799
 3 | 02222378
 4 | 3
 5 | 0
```

Figure 12.7 Stemplot of the yearly returns on common stocks for the 50 years 1950 to 1999, for Example 5. The returns are rounded to the nearest whole percent. The stems are 10s of percents and the leaves are single percents.

Choosing numerical descriptions

The five-number summary is easy to understand and is the best short description for most distributions. The mean and standard deviation are harder to understand but are more common. How can we decide which of these two descriptions of center and spread to use? Let's start by comparing the mean and the median. "Midpoint" and "arithmetic average" are both reasonable ideas for describing the center of a set of data, but they are different ideas with different uses. The most important distinction is that the mean (the average) is strongly influenced by a few extreme observations and the median (the midpoint) is not.

TABLE 12.1 Salaries of the Los Angeles Lakers, 2011–2012 season

Player	Salary ($)	Player	Salary ($)
Kobe Bryant	25.2 million	Derek Fisher	3.4 million
Pau Gasol	18.7 million	Matt Barnes	1.9 million
Andrew Bynum	15.2 million	Troy Murphy	1.4 million
Lamar Odom	8.9 million	Jason Kapono	1.2 million
Metta World Peace	6.8 million	Derrick Caracter	0.8 million
Luke Walton	5.7 million	Devin Ebanks	0.8 million
Steve Blake	4.0 million		

Source: The salaries are estimates are from **www.sportscity.com/NBA/ Los-Angeles-Lakers-Salaries.**

EXAMPLE 6 Mean versus median

Table 12.1 gives the approximate salaries (in millions of dollars) of the 13 members of the Los Angeles Lakers basketball team for the 2011–2012 season. You can calculate that the mean is $\bar{x} = \$7.2$ million and that the median is $M = \$4.0$ million. No wonder professional basketball players have big houses.

Why is the mean so much higher than the median? Figure 12.8 is a stemplot of the salaries, with millions as stems. The distribution is skewed to the right and there are three high outliers. The very high salaries of Kobe Bryant, Pau Gasol, and Andrew Bynum pull up the sum of the salaries and so pull up the mean. If we drop the outliers, the mean for the other 10 players is only $3.5 million. The median doesn't change nearly as much: it drops from $4.0 million to $2.7 million.

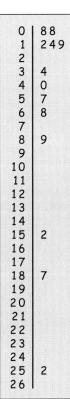

Figure 12.8 Stemplot of the salaries of Los Angeles Lakers players, from Table 12.1. (This figure was created using the Minitab software package.)

We can make the mean as large as we like by just increasing Kobe Bryant's salary. The mean will follow one outlier up and up. But to the median, Kobe's salary just counts as one observation at the upper end of the distribution. Moving it from $25.2 million to $252 million would not change the median at all.

The mean and median of a symmetric distribution are close to each other. In fact, $\bar{x}$ and M are exactly equal if the distribution is exactly symmetric. In skewed distributions, however, the mean runs away from the median toward the long tail. Many distributions of monetary values—incomes, house prices, wealth—are strongly skewed to the right. The mean may be much larger than the median. For example, we saw in Example 3 that the distribution of incomes for blacks, Hispanics, and whites is skewed to the right. The Census Bureau Web site gives the mean incomes for 2008 as $45,127 for blacks, $50,575 for Hispanics, and $69,107 for whites. Compare these with the corresponding medians of $31,969, $37,781, and $50,673. Because monetary data often have a few extremely high observations, descriptions of these distributions usually employ the median.

Poor New York? Is New York a rich state? New York's mean income per person ranks seventh among the states, right up there with its rich neighbors Connecticut and New Jersey, which rank first and second. But while Connecticut and New Jersey rank third and second in median household income, New York stands 17th. What's going on? Just another example of mean versus median. New York has many very highly paid people, who pull up its mean income per person. But it also has a higher proportion of poor households than do Connecticut and New Jersey, and this brings the median down. New York is not a rich state—it's a state with extremes of wealth and poverty.

You should think about more than symmetry versus skewness when choosing between the mean and the median. The distribution of selling prices for homes in Middletown is no doubt skewed to the right—but if the Middletown City Council wants to estimate the total market value of all houses in order to set tax rates, the mean and not the median helps them out because the mean will be larger. (The total market value is just the number of houses times the mean market value and has no connection with the median.)

The standard deviation is pulled up by outliers or the long tail of a skewed distribution even more strongly than the mean. The standard deviation of the Lakers' salaries is $s = \$7.8$ million for all 13 players and only $s = \$2.8$ million when the three outliers are removed. The quartiles are much less sensitive to a few extreme observations. There is another reason to avoid the standard deviation in describing skewed distributions. Because the two sides of a strongly skewed distribution have different spreads, no single number such as s describes the spread well. The five-number summary, with its two quartiles and two extremes, does a better job. In most situations, it is wise to use $\bar{x}$ and s only for distributions that are roughly symmetric.

Choosing a summary

The mean and standard deviation are strongly affected by outliers or by the long tail of a skewed distribution. The median and quartiles are less affected.

The five-number summary is usually better than the mean and standard deviation for describing a skewed distribution or a distribution with outliers. Use $\bar{x}$ and s only for reasonably symmetric distributions that are free of outliers.

Why do we bother with the standard deviation at all? One answer appears in the next chapter: the mean and standard deviation are the natural measures of center and spread for an important kind of symmetric distribution, called the Normal distribution.

Do remember that a graph gives the best overall picture of a distribution. Numerical measures of center and spread report specific facts about a distribution, but they do not describe its entire shape. Numerical summaries do not disclose the presence of multiple peaks or gaps, for example. *Always start with a graph of your data.*

STATISTICS IN SUMMARY

Chapter Specifics

- If we have data on a single quantitative variable, we start with a histogram or stemplot to display the distribution. Then we add numbers to describe the **center and spread** of the distribution.

- There are two common descriptions of center and spread: the **five-number summary** and the **mean and standard deviation.**

- The five-number summary consists of the **median** M, the midpoint of the observations, to measure center and the two **quartiles** Q_1 and Q_3 and the smallest and largest observations to describe spread.

- A **boxplot** is a graph of the five-number summary.

- The **mean** $\bar{x}$ is the average of the observations.

- The **standard deviation** *s* measures spread as a kind of average distance from the mean, so use it only with the mean. The **variance** is the square of the standard deviation.

- The mean and standard deviation can be changed a lot by a few outliers. The mean and median are the same for symmetric distributions, but the mean moves farther toward the long tail of a skewed distribution.

- In general, use the five-number summary to describe most distributions and the mean and standard deviation only for roughly symmetric distributions.

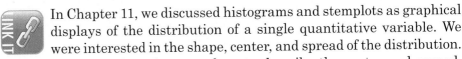 In Chapter 11, we discussed histograms and stemplots as graphical displays of the distribution of a single quantitative variable. We were interested in the shape, center, and spread of the distribution. In this chapter we introduce numbers to describe the center and spread. For symmetric distributions, the mean and standard deviation are used to describe the center and spread. For distributions that are not roughly symmetric, we use the five-number summary to describe the center and spread.

In most of the examples, we used graphical displays and numbers to describe the distribution of data on a single quantitative variable. These data are typically a sample from some population. Thus, the numbers that describe features of the distribution are statistics as discussed in Chapter 3. In the next chapter we begin to think about distributions of populations. Thus, the numbers that describe features of these distributions are parameters. In later chapters we will use statistics to draw conclusions, or make inferences, about parameters. Drawing conclusions about parameters that describe the center of a distribution of a single quantitative variable will be an important type of inference.

CASE STUDY Find the data on income by education at the Census Bureau Web site
EVALUATED listed in the Notes and Data Sources section at the end of this chapter.
Use what you have learned in this chapter to answer the following questions.

1. What are the median incomes for people 25 years old and over who are high school graduates only, have some college but no degree, have a bachelor's degree, have a master's degree, and have a doctorate degree? At the bottom of the table you will find median earnings in dollars.

2. From the distribution given in the tables, can you find the (approximately) first and third quartiles?

3. Do people with more education earn more than people with less education? Discuss. ■

CHAPTER 12 EXERCISES

For Exercise 12.1, see page 265; for Exercise 12.2, see page 267; for Exercise 12.3, see page 273.

12.4 Median income. You read that the median income of U.S. households in 2010 was $49,455. Explain in plain language what "the median income" is.

12.5 What's the average? The Census Bureau Web site gives several choices for "average income" in its historical income data. In 2010, the median income of American households was $49,455. The mean household income was $67,530. The median income of families was $60,395, and the mean family income was $78,361. The Census Bureau says, "Households consist of all people who occupy a housing unit. The term 'family' refers to a group of two or more people related by birth, marriage, or adoption who reside together." Explain carefully why mean incomes are higher than median incomes and why family incomes are higher than household incomes.

12.6 Rich magazine readers. Echo Media reports (January 20, 2012) that the average income for readers of the business magazine *Forbes* is $217,000. Is the median wealth of these readers greater or less than $217,000? Why?

12.7 College tuition. Figure 11.7 (page 250) is a stemplot of the tuition charged by 116 colleges in Illinois. The stems are thousands of dollars and the leaves are hundreds of dollars. For example, the highest tuition is $38,600 and appears as leaf 6 on stem 38.

(a) Find the five-number summary of Illinois college tuitions. You see that the stemplot already arranges the data in order.

(b) Would the mean tuition be clearly smaller than the median, about the same as the median, or clearly larger than the median? Why?

12.8 Where are the young more likely to live? Figure 11.10 (page 253) is a stemplot of the percentage of residents aged 18 to 34 in each of the 50 states. The stems are whole percents and the leaves are tenths of a percent.

(a) The shape of the distribution suggests that the mean will be larger than the median. Why?

(b) Find the mean and median of these data and verify that the mean is larger than the median.

12.9 Gas mileage. Table 11.2 (page 255) gives the highway gas mileages for some model year 2010 sedans.

(a) Make a stemplot of these data if you did not do so in Exercise 11.8.

(b) Find the five-number summary of gas mileages. Which cars are in the bottom quarter of gas mileages?

(c) The stemplot shows a fact about the overall shape of the distribution that the five-number summary cannot describe. What is it?

12.10 Yankee money. Table 11.4 (page 256) gives the salaries of the New York Yankees baseball team. What shape do you expect the distribution to have? Do you expect the mean salary to be close to the median, clearly higher, or clearly lower? Verify your choices by making a graph and calculating the mean and median.

12.11 The richest 1%. The distribution of individual incomes in the United States is strongly skewed to the right. In 2008, the mean and median incomes of the top 1% of Americans were \$558,726 and \$1,137,680. Which of these numbers is the mean and which is the median? Explain your reasoning.

12.12 How many calories does a hot dog have? *Consumer Reports* magazine presented the following data on the number of calories in a hot dog for each of 17 brands of meat hot dogs:

173	191	182	190	172	147	146	139	175
136	179	153	107	195	135	140	138	

Make a stemplot (if you did not already do so in Exercise 11.15 (page 257)), and find the five-number summary. The stemplot shows important facts about the distribution that the numerical summary does not tell us. What are these facts?

12.13 Returns on common stocks. Example 5 informs us that financial theory uses the mean and standard deviation to describe the returns on investments. Figure 11.12 (page 254) is a histogram of the returns of all New York Stock Exchange common stocks in one year. Are the mean and standard deviation suitable as a brief description of this distribution? Why?

12.14 Minority students in engineering. Figure 11.11 (page 254) is a histogram of the number of minority students (black, Hispanic, Native American) who earned doctorate degrees in engineering from each of 152 universities in the years 2000 through 2002. The classes for Figure 11.11 are 1–5, 6–10, and so on.

(a) What is the position of each number in the five-number summary in a list of 152 observations arranged from smallest to largest?

(b) Even without the actual data, you can use your answer to (a) and the histogram to give the five-number summary approximately. Do this. About how many minority engineering PhD's must a university graduate to be in the top quarter?

12.15 The statistics of writing style. Here are data on the percentages of words of 1 to 15 letters used in articles in *Popular Science* magazine. Exercise 11.11 (page 255) asked you to make a histogram of these data.

Length:	1	2	3	4	5	6	7	8	9	10	11	12	13	14	15
Percent:	3.6	14.8	18.7	16.0	12.5	8.2	8.1	5.9	4.4	3.6	2.1	0.9	0.6	0.4	0.2

Find the five-number summary of the distribution of word lengths from this table. (*Hint:* Half, or 50%, of the values are less than the median; 25% are less than the first quartile; and 75% are less than the third quartile.)

12.16 Immigrants in the eastern states. Here are the number of legal immigrants (in thousands) who settled in each state east of the Mississippi River from 2000 to 2007:

Alabama	60.4	Connecticut	158.6	Delaware	23.4
Florida	997.9	Georgia	316.8	Illinois	459.9
Indiana	95.3	Kentucky	49.6	Maine	11.7
Maryland	227.3	Massachusetts	274.2	Michigan	186.4
Mississippi	21.6	New Hampshire	19.7	New Jersey	444.9
New York	992.6	North Carolina	253.3	Ohio	134.9
Pennsylvania	204.2	Rhode Island	32.9	South Carolina	73.0
Tennessee	102.0	Vermont	4.5	Virginia	268.4
West Virginia	7.0	Wisconsin	86.2		

Make a graph of the distribution. Describe its overall shape and any outliers. Then choose and calculate a suitable numerical summary.

12.17 Immigrants in the eastern states. New York and Florida are high outliers in the distribution of the previous exercise. Find the mean and the median for these data with and without New York and Florida. Which measure changes more when we omit the outliers?

12.18 State SAT scores. Figure 12.9 is a histogram of the average scores on the SAT Mathematics exam for college-bound senior students in the 50 states

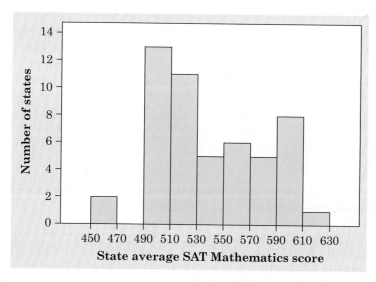

Figure 12.9 Histogram of the average scores on the SAT Mathematics exam for college-bound senior students in the 50 states and the District of Columbia in 2009, for Exercise 12.18. (This figure was created using the Minitab software package.)

and the District of Columbia in 2009. The distinctive overall shape of this distribution implies that a single measure of center such as the mean or the median is of little value in describing the distribution. Explain why this is true.

12.19 Highly paid athletes. A news article reported that of the 411 players on National Basketball Association rosters in February 1998, only 139 "made more than the league average salary" of $2.36 million. Was $2.36 million the mean or median salary for NBA players? How do you know?

12.20 Mean or median? Which measure of center, the mean or the median, should you use in each of the following situations? Why?

(a) Middletown is considering imposing an income tax on citizens. The city government wants to know the average income of citizens so that it can estimate the total tax base.

(b) In a study of the standard of living of typical families in Middletown, a sociologist estimates the average family income in that city.

12.21 Mean or median? You are planning a party and want to know how many cans of soda to buy. A genie offers to tell you either the mean number of cans guests will drink or the median number of cans. Which measure of center should you ask for? Why? To make your answer concrete, suppose there will be 30 guests and the genie will tell you either $\bar{x} = 5$ cans or $M = 3$ cans. Which of these two measures would best help you determine how many cans you should have on hand?

12.22 State SAT scores. We want to compare the distributions of average SAT Math and Writing scores for the states and the District of Columbia. We enter these data into a computer with the names SATM for Math scores and SATW for Writing scores. Here is output from the statistical software package Minitab. (Other software produces similar output. Some software uses rules for finding the quartiles that differ slightly from ours. So software may not give exactly the answer you would get by hand.)

Variable	N	Mean	Median	StDev	Minimum	Maximum	Q1	Q3
SATW	51	521.37	510.00	39.44	455.00	588.00	484.00	559.00
SATM	51	538.82	526.00	41.24	451.00	615.00	505.00	572.00

Use this output to make boxplots of SAT Math and Writing scores for the states. Briefly compare the two distributions in words.

12.23 Do SUVs waste gas? Table 11.2 (page 255) gives the highway fuel consumption (in miles per gallon) for 31 model year 2010 sedans. You found the five-number summary for these data in Exercise 12.9. Here are the highway gas mileages for 26 four-wheel-drive model year 2010 sport utility vehicles:

Model	mpg	Model	mpg
BMW X5	21	Kia Borrego	20
Chevrolet Tahoe	21	Land Rover LR4	17
Chevrolet Traverse	23	Lexus LX570	18
Dodge Journey	23	Lincoln MKX	23
Ford Escape	26	Mazda Tribute	26
Ford Explorer	19	Mercedes-Benz ML350	20
Honda Pilot	22	Mitsubishi Outlander	25
Hummer H3	18	Nissan Pathfinder	20
Infiniti QX56	17	Nissan Xterra	20
Jeep Commander	19	Subaru Forester	27
Jeep Grand Cherokee	20	Suzuki Grand Vitara	23
Jeep Liberty	21	Toyota RAV4	26
Jeep Wrangler	19	Toyota 4Runner	22

(a) Give a graphical and numerical description of highway fuel consumption for SUVs. What are the main features of the distribution?

(b) Make boxplots to compare the highway fuel consumption of the sedans in Table 11.2 and SUVs. What are the most important differences between the two distributions?

12.24 How many calories in a hot dog? Some people worry about how many calories they consume. *Consumer Reports* magazine, in a story on hot dogs, measured the calories in 20 brands of beef hot dogs, 17 brands of meat hot dogs, and 17 brands of poultry hot dogs. Here is computer output describing the beef hot dogs,

```
Mean = 156.8    Standard deviation = 22.64    Min = 111    Max = 190
N = 20    Median = 152.5    Quartiles = 140, 178.5
```

the meat hot dogs,

```
Mean = 158.7    Standard deviation = 25.24    Min = 107    Max = 195
N = 17    Median = 153    Quartiles = 139, 179
```

and the poultry hot dogs,

```
Mean = 122.5    Standard deviation = 25.48    Min = 87    Max = 170
N = 17    Median = 129    Quartiles = 102, 143
```

(Some software uses rules for finding the quartiles that differ slightly from ours. So software may not give exactly the answer you would get by hand.) Use this information to make boxplots of the calorie counts for the three types of hot dogs. Write a brief comparison of the distributions. Will eating poultry hot dogs usually lower your calorie consumption compared with eating beef or meat hot dogs? Explain.

12.25 Finding the standard deviation. The level of various substances in the blood influences our health. Here are measurements of the level of phosphate in the blood of a patient, in milligrams of phosphate per deciliter of blood, made on 6 consecutive visits to a clinic:

$$5.6 \quad 5.2 \quad 4.6 \quad 4.9 \quad 5.7 \quad 6.4$$

A graph of only 6 observations gives little information, so we proceed to compute the mean and standard deviation.

(a) Find the mean from its definition. That is, find the sum of the 6 observations and divide by 6.

(b) Find the standard deviation from its definition. That is, find the distance of each observation from the mean, square the distances, then calculate the standard deviation. Example 4 shows the method.

(c) Now enter the data into your calculator and use the mean and standard deviation keys to obtain $\bar{x}$ and s. Do the results agree with your hand calculations?

12.26 What s measures. Use a calculator to find the mean and standard deviation of these two sets of numbers:

(a) 4 0 1 4 3 6

(b) 5 3 1 3 4 2

Which data set is more spread out?

12.27 What s measures. Add 2 to each of the numbers in data set (a) in the previous exercise. The data are now 6 2 3 6 5 8.

(a) Use a calculator to find the mean and standard deviation and compare your answers with those for data set (a) in the previous exercise. How does adding 2 to each number change the mean? How does it change the standard deviation?

(b) Without doing the calculation, what would happen to $\bar{x}$ and s if we added 10 to each value in data set (a) of the previous exercise? (This exercise demonstrates that the standard deviation measures only spread about the mean and ignores changes in where the data are centered.)

12.28 Cars and SUVs. Use the mean and standard deviation to compare the gas mileages of sedans (Table 11.2, page 255) and SUVs (Exercise 12.23). Do these numbers catch the main points of your more detailed comparison in Exercise 12.23?

12.29 A contest. This is a standard deviation contest. You must choose four numbers from the whole numbers 0 to 9, with repeats allowed.

(a) Choose four numbers that have the smallest possible standard deviation.

(b) Choose four numbers that have the largest possible standard deviation.

(c) Is more than one choice correct in either (a) or (b)? Explain.

12.30 $\bar{x}$ and s are not enough. The mean $\bar{x}$ and standard deviation s measure center and spread but are not a complete description of a distribution. Data sets with different shapes can have the same mean and standard deviation. To demonstrate this fact, use your calculator to find $\bar{x}$ and s for these two small data sets. Then make a stemplot of each and comment on the shape of each distribution.

Data A:	9.14	8.14	8.74	8.77	9.26	8.10	6.13	3.10	9.13	7.26	4.74
Data B:	6.58	5.76	7.71	8.84	8.47	7.04	5.25	5.56	7.91	6.89	12.50

12.31 Raising pay. A school system employs teachers at salaries between $30,000 and $60,000. The teachers' union and the school board are negotiating the form of next year's increase in the salary schedule. Suppose that every teacher is given a flat $1000 raise.

(a) How much will the mean salary increase? The median salary?

(b) Will a flat $1000 raise increase the spread as measured by the distance between the quartiles? Explain.

(c) Will a flat $1000 raise increase the spread as measured by the standard deviation of the salaries? Explain. (See Exercise 12.27 if you need help.)

12.32 Raising pay. Suppose that the teachers in the previous exercise each receive a 5% raise. The amount of the raise will vary from $1500 to $3000, depending on present salary. Will a 5% across-the-board raise increase the spread of the distribution as measured by the distance between the quartiles? Do you think it will increase the standard deviation? Explain your reasoning.

12.33 Making colleges look good. Colleges announce an "average" SAT score for their entering freshmen. Usually the college would like this "average" to be as high as possible. A *New York Times* article noted, "Private colleges that buy lots of top students with merit scholarships prefer the mean, while open-enrollment public institutions like medians." Use what you know about the behavior of means and medians to explain these preferences.

12.34 What graph to draw? We now understand three kinds of graphs to display distributions of quantitative variables: histograms, stemplots, and boxplots. Give an example (just words, no data) of a situation in which you would prefer each kind of graph.

EXPLORING THE WEB

12.35 Web-based exercise. Willie Mays is fourth on the career home run list, behind Barry Bonds, Hank Aaron, and Babe Ruth. You can find Willie Mays's home run statistics at the Web site **www.baseball-reference .com.** Construct four side-by-side boxplots comparing the yearly home run production of Barry Bonds, Hank Aaron, Babe Ruth, and Willie Mays. Describe any differences that you observe. It is worth noting that in his first four seasons, Babe Ruth was primarily a pitcher. If these four seasons are ignored, how does Babe Ruth compare with Barry Bonds, Hank Aaron, and Willie Mays?

12.36 Web-based exercise. You can compare the behavior of the mean and median by using the *Mean and Median* applet at the *Statistics: Concepts and Controversies* Web site, **www.whfreeman.com/scc8e.** Click to enter data, then use the mouse to drag an outlier up and watch the mean chase after it.

NOTES AND DATA SOURCES

Pages 261 and 278 Data on income and education are from the Census Bureau Web site. Tables for 2010 can be found at **www.census .gov/hhes/www/cpstables/032011/perinc/ new03_000.htm.**

Page 270 The data for Figures 12.4 and 12.5 come from historical income tables that can be found at the Census Bureau Web site **www.census.gov/hhes/www/income/ data/historical/household/index.html.**

Page 279 Exercise 12.6: Found online at **www.echo-media.com/mediaDetail.php? ID=4519.**

Pages 280 and 283 Exercises 12.12 and 12.24: Data from *Consumer Reports,* June 1986, pp. 366–367.

Page 281 Exercise 12.16: Data from the 2010 *Statistical Abstract of the United States.*

Pages 281 and 282 The raw data behind Exercises 12.18 and 12.22 come from College Board Online, **professionals. collegeboard.com/data-reports- research/sat/cb-seniors-2009.**

Page 282 Exercise 12.23: Data from **www.fueleconomy.gov.**

Page 285 Exercise 12.30: Part of the larger set of data in Table 15.2. See the source note for that table.

Page 285 Exercise 12.33: Quotation from the *New York Times,* May 31, 1989.

Normal Distributions

CASE STUDY Bar graphs and histograms are definitely old technology. Using bars to display data goes back to William Playfair (1759–1823), an English economist who was an early pioneer of data graphics. Histograms require that we choose classes, and their appearance can change with different choices. Surely modern software offers a better way to picture distributions?

Software can replace the separate bars of a histogram with a smooth curve that represents the overall shape of a distribution. Look at Figure 13.1. The data are the numbers of minority-group members who earned doctorates in engineering from 152 universities between 2000 and 2002. We met these data in Chapter 11, and the histogram in Figure 13.1 repeats Figure 11.11. The curve is the new-technology replacement for the histogram. The software doesn't start from the histogram—it starts with the actual observations and cleverly draws a curve to describe their distribution.

In Figure 13.1, the software has caught the overall shape and shows the ripples in the long right tail more effectively than does the histogram. It struggles a bit with the peak: it has extended the curve below zero in an attempt to smooth out the sharp peak. In Figure 13.2, we apply the same software to a larger set of data with a more regularly shaped distribution. These are the values of the sample proportion $\hat{p}$ for 1000 SRSs of size 2527 from a population in which the population proportion is $p = 0.5$. We also met these data in Chapter 11, and the histogram here repeats Figure 11.3. The software draws a curve that shows a distinctive symmetric, single-peaked, bell shape.

For the irregular distribution in Figure 13.1, we can't do better. In the case of the very symmetric sampling data in Figure 13.2, however, there is another way to get a smooth curve. It's a mathematical fact that this distribution can be described by a specific kind of smooth curve called a *Normal curve*. Figure 13.3 shows the Normal curve for these data. The curve looks a lot like the one in Figure 13.2, but a close look shows that it is smoother. The Normal curve is much easier to work with and does not require clever software.

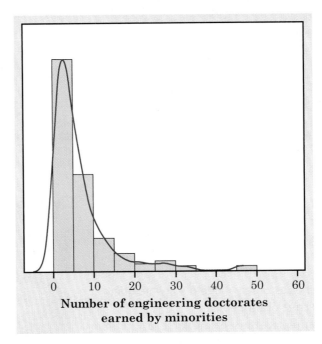

Figure 13.1 A histogram and a computer-drawn curve. Both picture the distribution of the number of engineering doctorates earned by members of minority groups at 152 universities. This distribution is skewed to the right. (This figure was created using the Stata software package.)

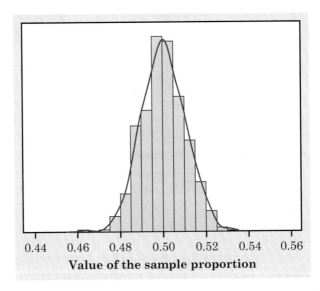

Figure 13.2 A histogram and a computer-drawn curve. Both picture the distribution of the sample proportion in 1000 simple random samples from the same population. This distribution is quite symmetric. (This figure was created using the Stata software package.)

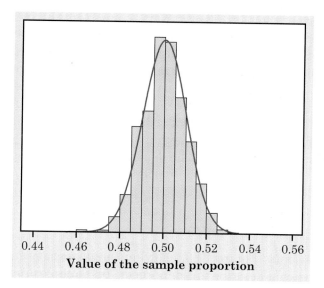

0.44   0.46   0.48   0.50   0.52   0.54   0.56

Value of the sample proportion

Figure 13.3 A perfectly symmetric Normal curve used to describe the distribution of sample proportions. (This figure was created using the Stata software package.)

In this chapter we will learn that Normal curves have special properties that help us use them and think about them. By the end of this chapter you will be able to use these properties to answer questions about the underlying distributions represented in Figures 13.2 and 13.3 that cannot easily be determined from the histograms. ■

We now have a kit of graphical and numerical tools for describing distributions. What is more, we have a clear strategy for exploring data on a single quantitative variable:

1. Always plot your data: make a graph, usually a histogram or a stemplot.
2. Look for the overall pattern (shape, center, spread) and for striking deviations such as outliers.
3. Choose either the five-number summary or the mean and standard deviation to briefly describe center and spread in numbers.

Here is one more step to add to this strategy:

4. Sometimes the overall pattern of a large number of observations is so regular that we can describe it by a smooth curve.

Density curves

Figures 13.1 and 13.2 show curves used in place of histograms to picture the overall shape of a distribution of data. You can think of drawing a curve through the tops of the bars in a histogram, smoothing out the irregular ups and downs of the bars. There are two important distinctions between histograms and these curves. First, most histograms show the *counts* of observations in each class by the heights of their bars and therefore by the areas of the bars. We set up curves to show the *proportion* of observations in any region by areas under the curve. To do that, we choose the scale so that the total area under the curve is exactly 1. We then have a **density curve.** Second, a histogram is a plot of data obtained from a sample. We use this histogram to understand the actual distribution of the population from which the sample was selected. The density curve is intended to reflect the idealized shape of the population distribution.

EXAMPLE 1 Using a density curve

Figure 13.4 copies Figure 13.3, showing the histogram and the Normal density curve that describe this data set of 1000 sample proportions. What proportion of the observations are greater than 0.51? From the actual 1000 observations, we can count that exactly 171 are greater than 0.51. So the proportion is 171/1000, or 0.171. Because 0.51 is one of the break points between the classes in the histogram, the area of the shaded bars in Figure 13.4(a) makes up 0.171 of the total area of all the bars.

Now concentrate on the density curve drawn through the histogram. The total area under this curve is 1, and the shaded area in Figure 13.4(b)

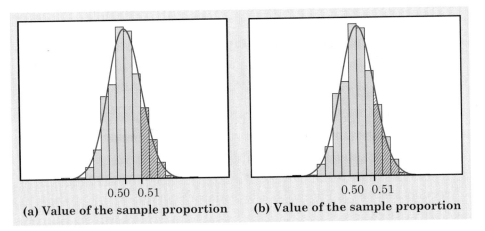

(a) Value of the sample proportion   (b) Value of the sample proportion

Figure 13.4 A histogram and a Normal density curve, for Example 1. (a) The area of the shaded bars in the histogram represents observations greater than 0.51. These make up 171 of the 1000 observations. (b) The shaded area under the Normal curve represents the proportion of observations greater than 0.51. This area is 0.1667. (This figure was created using the Stata software package.)

represents the proportion of observations that are greater than 0.51. This area is 0.1667. You can see that the density curve is a quite good approximation—0.1667 is quite close to 0.171.

The area under the density curve in Example 1 is not exactly equal to the true proportion, because the curve is an idealized picture of the distribution. For example, the curve is exactly symmetric but the actual data are only approximately symmetric. Because density curves are smoothed-out idealized pictures of the overall shapes of distributions, they are most useful for describing large numbers of observations.

The center and spread of a density curve

Density curves help us better understand our measures of center and spread. The median and quartiles are easy. Areas under a density curve represent proportions of the total number of observations. The median is the point with half the observations on either side. So *the median of a density curve is the equal-areas point,* the point with half the area under the curve to its left and the remaining half of the area to its right. The quartiles divide the area under the curve into quarters. One-fourth of the area under the curve is to the left of the first quartile, and three-fourths of the area is to the left of the third quartile. You can roughly locate the median and quartiles of any density curve by eye by dividing the area under the curve into four equal parts.

Because density curves are idealized patterns, a symmetric density curve is exactly symmetric. The median of a symmetric density curve is therefore at its center. Figure 13.5(a) shows the median of a symmetric curve. We can roughly locate the equal-areas point on a skewed curve like that in Figure 13.5(b) by eye.

What about the mean? The mean of a set of observations is their arithmetic average. If we think of the observations as weights stacked on a seesaw, the mean is the point at which the seesaw would balance. This fact is also true of density curves. *The mean is the point at which the curve would balance if made of solid material.* Figure 13.6 illustrates this fact about the mean. A symmetric curve balances at its center because the two sides are identical. *The mean and median of a symmetric density curve are equal,* as in Figure 13.5(a). We know that the mean of a skewed distribution is pulled toward the long tail. Figure 13.5(b) shows how the mean of a skewed density curve is pulled toward the long tail more than is the median.

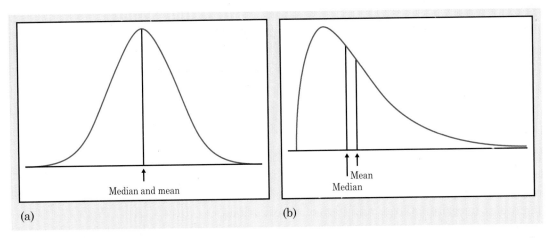

Figure 13.5 The median and mean for two density curves: (a) a symmetric Normal curve and (b) a curve that is skewed to the right.

Figure 13.6 The mean of a density curve is the point at which it would balance.

Median and mean of a density curve

The **median** of a density curve is the equal-areas point, the point that divides the area under the curve in half.

The **mean** of a density curve is the balance point, or center of gravity, at which the curve would balance if made of solid material.

The median and mean are the same for a symmetric density curve. They both lie at the center of the curve. The mean of a skewed curve is pulled away from the median in the direction of the long tail.

Normal distributions

The density curves in Figures 13.3 and 13.4 belong to a particularly important family: the Normal curves. (We capitalize "Normal" to remind you that these curves are special.) Figure 13.7 presents two more Normal density curves. Normal curves are symmetric, single-peaked, and bell-shaped. Their tails fall off quickly, so that we do not expect outliers. Because Normal distributions are symmetric, the mean and median lie together at the peak in the center of the curve.

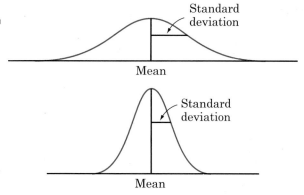

Figure 13.7 Two Normal curves. The standard deviation fixes the spread of a Normal curve.

Normal curves also have the special property that we can locate the standard deviation of the distribution by eye on the curve. This isn't true for most other density curves. Here's how to do it. Imagine that you are skiing down a mountain that has the shape of a Normal curve. At first, you descend at an ever-steeper angle as you go out from the peak:

Fortunately, before you find yourself going straight down, the slope begins to grow flatter rather than steeper as you go out and down:

The points at which this change of curvature takes place are located one standard deviation on either side of the mean. The standard deviations are marked on the two curves in Figure 13.7. You can feel the change as you run a pencil along a Normal curve, and so find the standard deviation.

Normal curves have the special property that giving the mean and the standard deviation completely specifies the curve. The mean fixes the center of the curve, and the standard deviation determines its shape. Changing the mean of a Normal distribution does *not* change its shape, only its location on the axis. Changing the standard deviation *does* change the shape of a Normal curve, as Figure 13.7 illustrates. The distribution with the smaller standard deviation is less spread out and more sharply peaked. Here is a summary of basic facts about Normal curves.

> ### Normal density curves
>
> The **Normal curves** are symmetric, bell-shaped curves that have these properties:
>
> - A specific Normal curve is completely described by giving its mean and its standard deviation.
> - The mean determines the center of the distribution. It is located at the center of symmetry of the curve.
> - The standard deviation determines the shape of the curve. It is the distance from the mean to the change-of-curvature points on either side.

Why are the Normal distributions important in statistics? First, Normal distributions are good descriptions for some distributions of *real data*. Normal curves were first applied to data by the great mathematician Carl Friedrich Gauss (1777–1855), who used them to describe the small errors made by astronomers and surveyors in repeated careful measurements of the same quantity. You will sometimes see Normal distributions labeled "Gaussian" in honor of Gauss. For much of the 19th century Normal curves were called "error curves" because they were first used to describe the distribution of measurement errors. As it became clear that the distributions of some biological and psychological variables were at least roughly Normal, the "error curve" terminology was dropped. The curves were first called "Normal" by Francis Galton in 1889. Galton, a cousin of Charles Darwin, pioneered the statistical study of biological inheritance.

The bell curve? Does the distribution of human intelligence follow the "bell curve" of a Normal distribution? Scores on IQ tests do roughly follow a Normal distribution. That is because a test score is calculated from a person's answers in a way that is designed to produce a Normal distribution. To conclude that intelligence follows a bell curve, we must agree that the test scores directly measure intelligence. Many psychologists don't think there is one human characteristic that we can call "intelligence" and can measure by a single test score.

Normal curves also describe the distribution of *statistics such as sample proportions (when the sample size is large and the value of the proportion is moderate) and sample means* when we take many samples from the same population. We used Normal curves this way in Figures 13.3 and 13.4. The margins of error for the results of sample surveys are usually calculated from Normal curves. However, even though many sets of data follow a Normal distribution, many do not. Most income distributions, for example, are skewed to the right and so are not Normal. Non-Normal data, like nonnormal people, not only are common but are sometimes more interesting than their normal counterparts.

The 68–95–99.7 rule

There are many Normal curves, each described by its mean and standard deviation. All Normal curves share many properties. In particular, the standard deviation is the natural unit of measurement for Normal distributions. This fact is reflected in the following rule.

The 68–95–99.7 rule

In any Normal distribution, approximately

- **68%** of the observations fall within one standard deviation of the mean.
- **95%** of the observations fall within two standard deviations of the mean.
- **99.7%** of the observations fall within three standard deviations of the mean.

Figure 13.8 illustrates the 68–95–99.7 rule. By remembering these three numbers, you can think about Normal distributions without constantly making detailed calculations. Remember also, though, that no set of data is exactly described by a Normal curve. The 68–95–99.7 rule will be only approximately true for SAT scores or the lengths of crickets.

Figure 13.8 The 68–95–99.7 rule for Normal distributions.

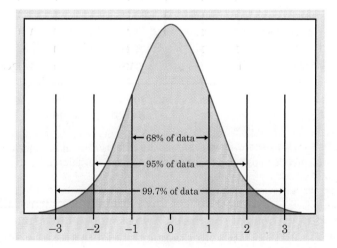

EXAMPLE 2 Heights of young women

The distribution of heights of women aged 18 to 24 is approximately Normal with mean 65 inches and standard deviation 2.5 inches. To use the

68–95–99.7 rule, always start by drawing a picture of the Normal curve. Figure 13.9 shows what the rule says about women's heights.

Half of the observations in any Normal distribution lie above the mean, so half of all young women are taller than 65 inches.

The central 68% of any Normal distribution lies within one standard deviation of the mean. Half of this central 68%, or 34%, lies above the mean. So 34% of young women are between 65 inches and 67.5 inches tall. Adding the 50% who are shorter than 65 inches, we see that 84% of young women have heights less than 67.5 inches. That leaves 16% who are taller than 67.5 inches.

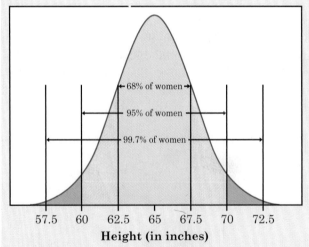

Figure 13.9 The 68–95–99.7 rule for heights of young women, for Example 2. This Normal distribution has mean 65 inches and standard deviation 2.5 inches.

The central 95% of any Normal distribution lies within two standard deviations of the mean. Two standard deviations is 5 inches here, so the middle 95% of young women's heights are between 60 inches (that's 65−5) and 70 inches (that's 65 + 5).

The other 5% of young women have heights outside the range from 60 to 70 inches. Because the Normal distributions are symmetric, half of these women are on the short side. The shortest 2.5% of young women are less than 60 inches (5 feet) tall.

Almost all (99.7%) of the observations in any Normal distribution lie within three standard deviations of the mean. Almost all young women are between 57.5 and 72.5 inches tall.

NOW IT'S YOUR TURN | **13.1 Heights of young men.** The distribution of heights of young men is approximately Normal with mean 70 inches and standard deviation 2.5 inches. Between which heights do the middle 95% of men fall?

Standard scores

Jennie scored 600 on the SAT Mathematics college entrance exam. How good a score is this? That depends on where a score of 600 lies in the distribution of all scores. The SAT exams are scaled so that scores should roughly follow the Normal distribution with mean 500 and standard deviation 100. Jennie's 600 is one standard deviation above the mean. The 68–95–99.7 rule now tells us just where she stands (Figure 13.10). Half of all scores are below 500, and another 34% are between 500 and 600. So Jennie did better

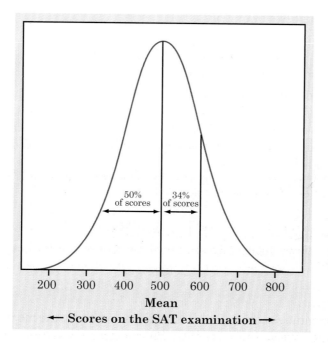

68–1
95–2
99–3

Figure 13.10 The 68–95–99.7 rule shows that 84% of any Normal distribution lies to the left of the point one standard deviation above the mean. Here, this fact is applied to SAT scores.

than 84% of the students who took the SAT. Her score report not only will say she scored 600 but will add that this is at the "84th percentile." That's statistics speak for "You did better than 84% of those who took the test."

Because the standard deviation is the natural unit of measurement for Normal distributions, we restated Jennie's score of 600 as "one standard deviation above the mean." Observations expressed in standard deviations above or below the mean of a distribution are called *standard scores*.

Standard scores

The **standard score** for any observation is

$$\text{standard score} = \frac{\text{observation} - \text{mean}}{\text{standard deviation}}$$

A standard score of 1 says that the observation in question lies one standard deviation above the mean. An observation with standard score -2 is two standard deviations below the mean. Standard scores can be used to compare values in different distributions. Of course, you should not use standard scores unless you are willing to use the standard deviation to describe the spread of the distributions. That requires that the distributions be at least roughly symmetric.

EXAMPLE 3 ACT versus SAT scores

Jennie scored 600 on the SAT Mathematics exam. Her friend Gerald took the American College Testing (ACT) test and scored 21 on the math part. ACT scores are Normally distributed with mean 18 and standard deviation 6. Assuming that both tests measure the same kind of ability, who has the higher score?

Jennie's standard score is

$$\frac{600 - 500}{100} = \frac{100}{100} = 1.0$$

Compare this with Gerald's standard score, which is

$$\frac{21 - 18}{6} = \frac{3}{6} = 0.5$$

Because Jennie's score is 1 standard deviation above the mean and Gerald's is only 0.5 standard deviation above the mean, Jennie's performance is better.

> **NOW IT'S YOUR TURN** | **13.2 Heights of young men.** The distribution of heights of young men is approximately Normal with mean 70 inches and standard deviation 2.5 inches. What is the standard score of a height of 72 inches (6 feet)?

Percentiles of Normal distributions*

For Normal distributions, but not for other distributions, standard scores translate directly into *percentiles*.

Percentiles

The **cth percentile** of a distribution is a value such that c percent of the observations lie below it and the rest lie above.

The median of any distribution is the 50th percentile, and the quartiles are the 25th and 75th percentiles. In any Normal distribution, the point one standard deviation above the mean (standard score 1) is the 84th percentile. Figure 13.10 shows why. Every standard score for a Normal distribution translates into a specific percentile, which is the same no matter what the mean and standard deviation of the original Normal distribution are. Table B at the back of this book gives the percentiles corresponding to various standard scores. This table enables us to do calculations in greater detail than does the 68–95–99.7 rule.

EXAMPLE 4   Percentiles for college entrance exams

Jennie's score of 600 on the SAT translates into a standard score of 1.0. We saw that the 68–95–99.7 rule says that this is the 84th percentile. Table B is a bit more precise: it says that standard score 1 is the 84.13 percentile of a Normal distribution. Gerald's 21 on the ACT is a standard score of 0.5. Table B says that this is the 69.15 percentile. Gerald did well, but not as well as Jennie. The percentile is easier to understand than either the raw score or the standard score. That's why reports of exams such as the SAT usually give both the score and the percentile.

---

*This material is not needed to read the rest of the book.

Jose Luis Pelaez, Inc./CORBIS

EXAMPLE 5 Finding the observation that matches a percentile

How high must a student score on the SAT to fall in the top 10% of all scores? That requires a score at or above the 90th percentile. Look in the body of Table B for the percentiles closest to 90. You see that standard score 1.2 is the 88.49 percentile and standard score 1.3 is the 90.32 percentile. The percentile in the table closest to 90 is 90.32, so we conclude that a standard score of 1.3 is approximately the 90th percentile *of any Normal distribution.*

To go from the standard score back to the scale of SAT scores, "undo" the standard score calculation as follows:

$$\text{observation} = \text{mean} + \text{standard score} \times \text{standard deviation}$$
$$= 500 + (1.3)(100) = 630$$

A score of 630 or higher will be in the top 10%. (More exactly, these scores are in the top 9.68% because 630 is exactly the 90.32 percentile.)

NOW IT'S YOUR TURN | **13.3 SAT scores.** How high must a student score on the SAT Mathematics test to fall in the top 25% of all scores?

STATISTICS IN SUMMARY

Chapter Specifics

- Stemplots, histograms, and boxplots all describe the distributions of quantitative variables.

- **Density curves** also describe distributions. A density curve is a curve with area exactly 1 underneath it whose shape describes the overall pattern of a distribution.

- An area under the curve gives the proportion of the observations that fall in an interval of values.

- You can roughly locate the median (equal-areas point) and the mean (balance point) by eye on a density curve.

- Stemplots, histograms, and boxplots are created from samples. Density curves are intended to display the idealized shape of the distribution of the population from which the samples are taken.

- **Normal curves** are a special kind of density curve that describes the overall pattern of some sets of data. Normal curves are symmetric and bell-shaped. A specific Normal curve is completely described by its mean and standard deviation. You can locate the mean (center point) and the standard deviation (distance from the mean to the change-of-curvature points) on a Normal curve. All Normal distributions obey the **68–95–99.7 rule.**

· **Standard scores** express observations in standard deviation units about the mean, which has standard score 0. A given standard score corresponds to the same **percentile** in any Normal distribution. Table B gives percentiles of Normal distributions.

 Chapters 10, 11, and 12 provide us with a strategy for exploring data on a single quantitative variable.

· Make a graph, usually a histogram or stemplot.

· Look for the overall pattern (shape, center, spread) and striking deviations from the pattern.

· Choose the five-number summary or the mean and standard deviation to briefly describe the center and spread in numbers.

In this chapter we added another step: sometimes the overall pattern of a large number of observations is so regular that we can describe it by a smooth density curve, such as the Normal curve. This step also allows us to identify "a large number of observations" as a population and use density curves to describe the distribution of a population. We did precisely this when we used the Normal distribution to describe the distribution of the heights of all young women or the scores of all students on the SAT exam.

Using a density curve to describe the distribution of a population is a convenient summary, allowing us to determine percentiles of the distribution without having to see a list of all the values in the population. It also suggests the nature of the conclusions we might draw about a single quantitative variable. Use statistics that describe the distribution of the sample to draw conclusions about parameters that describe the distribution of a population. We will explore this in future chapters.

CASE STUDY   The Normal curve that best approximates the distribution of the
EVALUATED   sample proportion in Figures 13.2 and 13.3 has mean 0.50 and standard deviation 0.01. Use what you have learned in this chapter to answer the following questions.

1. What is the area to the right of 0.51 under a Normal curve with mean 0 and standard deviation 0.01? Compare this area with the values reported in Example 1.

2. In the Chapter 3 Case Study we discussed a Gallup Poll of 2527 adults, conducted from July 2003 to February 2004, concerning same-sex marriages. In that poll, 51%

said that they would support a constitutional amendment defining marriage as being between a man and a woman. In Example 2 of Chapter 3 we investigated what would happen if, in fact, only 50% of all Americans supported such an amendment and we repeatedly took simple random samples of size 2527 and recorded the results. Figures 13.2 and 13.3 represent the results of 1000 such simple random samples. If, in fact, only 50% of all Americans supported such an amendment, what is the probability that we would obtain a sample in which at least 51% said they would support such an amendment? ■

CHAPTER 13 EXERCISES

For Exercise 13.1, see page 297; for Exercise 13.2, see page 299; for Exercise 13.3, see page 300.

13.4 Density curves

(a) Sketch a density curve that is symmetric but has a shape different from that of the Normal curves.

(b) Sketch a density curve that is strongly skewed to the right.

13.5 Mean and median. Figure 13.11 shows density curves of several shapes. Briefly describe the overall shape of each distribution. Two or more points are marked on each curve. The mean and the median are among these points. For each curve, which point is the median and which is the mean?

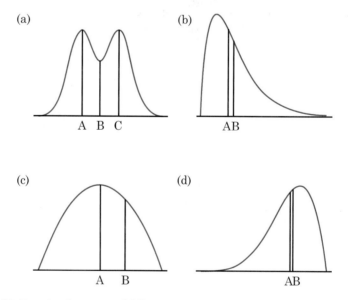

Figure 13.11 Four density curves of different shapes, for Exercise 13.5. In each case, the mean and the median are among the marked points.

Figure 13.12 The density curve of a uniform distribution, for Exercise 13.6. Observations from this distribution are spread "at random" between 0 and 1.

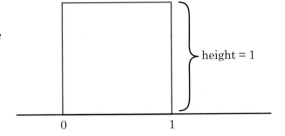

height = 1

0   1

13.6 Random numbers. If you ask a computer to generate "random numbers" between 0 and 1, you will get observations from a *uniform distribution*. Figure 13.12 shows the density curve for a uniform distribution. This curve takes the constant value 1 between 0 and 1 and is zero outside that range. Use this density curve to answer these questions.

(a) Why is the total area under the curve equal to 1?

(b) The curve is symmetric. What is the value of the mean and median?

(c) What percentage of the observations lie between 0 and 0.1?

(d) What percentage of the observations lie between 0.6 and 0.9?

IQ test scores. Figure 13.13 is a stemplot of the IQ test scores of 74 seventh-grade students. This distribution is very close to Normal with mean 111 and standard deviation 11. It includes all the seventh-graders in a rural Midwest school except for 4 low outliers who were dropped because they may have been ill or otherwise not paying attention when taking the test. Take the Normal distribution with mean 111 and standard deviation 11 as a description of the IQ test scores of all rural Midwest seventh-grade students. Use this distribution and the 68–95–99.7 rule to answer Exercises 13.7 to 13.9.

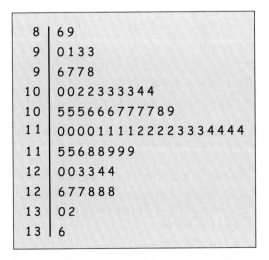

```
 8 | 6 9
 9 | 0 1 3 3
 9 | 6 7 7 8
10 | 0 0 2 2 3 3 3 3 4 4
10 | 5 5 5 6 6 6 7 7 7 7 8 9
11 | 0 0 0 0 1 1 1 1 2 2 2 2 3 3 3 4 4 4 4
11 | 5 5 6 8 8 9 9 9
12 | 0 0 3 3 4 4
12 | 6 7 7 8 8 8
13 | 0 2
13 | 6
```

Figure 13.13 Stemplot of the IQ scores of 74 seventh-grade students, for Exercises 13.7 to 13.9.

13.7 Between what values do the IQ scores of 68% of all rural Midwest seventh-graders lie?

13.8 What percentage of IQ scores for rural Midwest seventh-graders are more than 100?

13.9 What percentage of all students have IQ scores below 78? None of the 74 students in our sample school had scores this low. Are you surprised at this? Why?

13.10 Length of pregnancies. The length of human pregnancies from conception to birth varies according to a distribution that is approximately Normal with mean 266 days and standard deviation 16 days. Use the 68–95–99.7 rule to answer the following questions.

(a) Almost all (99.7%) pregnancies fall in what range of lengths?

(b) How long are the longest 2.5% of all pregnancies?

13.11 A Normal curve. What are the mean and standard deviation of the Normal curve in Figure 13.14?

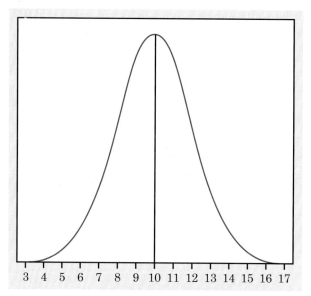

Figure 13.14 What are the mean and standard deviation of this Normal density curve? For Exercise 13.11.

13.12 Horse pregnancies. Bigger animals tend to carry their young longer before birth. The length of horse pregnancies from conception to birth varies according to a roughly Normal distribution with mean 336 days and standard deviation 3 days. Use the 68–95–99.7 rule to answer the following questions.

(a) Between what values do the lengths of the middle 95% of all horse pregnancies fall?

(b) What percentage of horse pregnancies are less than 333 days?

13.13 Three great hitters. Three landmarks of baseball achievement are Ty Cobb's batting average of .420 in 1911, Ted Williams's .406 in 1941, and George Brett's .390 in 1980. These batting averages cannot be compared directly because the distribution of major league batting averages has changed over the years. The distributions are quite symmetric and (except for outliers such as Cobb, Williams, and Brett) reasonably Normal. While the mean batting average has been held roughly constant by rule changes and the balance between hitting and pitching, the standard deviation has dropped over time. Here are the facts:

Decade	Mean	Std. dev.
1910s	.266	.0371
1940s	.267	.0326
1970s	.261	.0317

Compute the standard scores for the batting averages of Cobb, Williams, and Brett to compare how far each stood above his peers.

13.14 Comparing IQ scores. The Wechsler Adult Intelligence Scale (WAIS) is an IQ test. Scores on the WAIS for the 20 to 34 age group are approximately Normally distributed with mean 110 and standard deviation 15. Scores for the 60 to 64 age group are approximately Normally distributed with mean 90 and standard deviation 15. Sarah, who is 30, scores 130 on the WAIS. Her mother, who is 60, takes the test and scores 110.

(a) Express both scores as standard scores that show where each woman stands within her own age group.

(b) Who scored higher relative to her age group, Sarah or her mother? Who has the higher absolute level of the variable measured by the test?

13.15 Men's heights. The distribution of heights of young men is approximately Normal with mean 70 inches and standard deviation 2.5 inches. Sketch a Normal curve on which this mean and standard deviation are correctly located. (*Hint:* Draw the curve first, locate the points where the curvature changes, then mark the horizontal axis.)

13.16 More on men's heights. The distribution of heights of young men is approximately Normal with mean 70 inches and standard deviation 2.5 inches. Use the 68–95–99.7 rule to answer the following questions.

(a) What percentage of men are taller than 77.5 inches?

(b) Between what heights do the middle 68% of men fall?

(c) What percentage of men are shorter than 65 inches?

13.17 Heights of men and women. The heights of young women are approximately Normal with mean 65 inches and standard deviation 2.5 inches. The heights of men in the same age group have mean 70 inches and standard deviation 2.5 inches. What percentage of women are taller than a man of average (mean) height?

13.18 Heights of young adults. The mean height of men aged 18 to 24 is about 70 inches. Women that age have a mean height of about 65 inches. Do you think that the distribution of heights for all Americans aged 18 to 24 is approximately Normal? Explain your answer.

13.19 Sampling. Suppose that the proportion of all adult Americans who rate economic conditions in the United States as poor is $p = 0.47$. This number is consistent with a Gallup Poll conducted during the week ending September 12, 2010. If we took many SRSs of size 3000 (the sample size used by the Gallup Poll), the sample proportion $\hat{p}$ would vary from sample to sample following a Normal distribution with mean 0.47 and standard deviation 0.01. Use this fact and the 68–95–99.7 rule to answer these questions.

(a) In many samples, what percentage of the values of $\hat{p}$ fall above 0.47? Above 0.50?

(b) In a large number of samples, what range contains the central 95% of values of $\hat{p}$?

13.20 Am I winning? Congressman Floyd commissions a sample survey of voters to learn what percentage favor him in his race for reelection. To avoid spending too much, he samples only 50 voters. Suppose that in fact only 43% of the voters support Floyd. The percentage favoring Floyd in a random sample of size 50 will vary from sample to sample according to a Normal distribution with mean 43% and standard deviation 7.0%. What percentage of all such samples will (wrongly) show that half or more of the voters favor Floyd?

The following optional exercises require use of Table B of Normal distribution percentiles.

13.21 NCAA rules for athletes. The National Collegiate Athletic Association (NCAA) requires Division II athletes to get a combined score of at least 820 on the Mathematics and Critical Reading sections of the SAT exam in order to compete in their first college year. In 2011, the combined scores of the millions of college-bound seniors taking the SATs were approximately Normal with mean 1012 and standard deviation approximately 213. What percentage of all college-bound seniors had scores less than 820?

13.22 More NCAA rules. For Division I athletes the NCAA uses a sliding scale, based on both core GPA and the combined Mathematics and Critical Reading SAT score, to determine eligibility to compete in the first year of college. For athletes with a core GPA of 3.0, a score of at least 620 on the combined Mathematics and Critical Reading sections of the SAT exam is required. Use the information in the previous exercise to find the percentage of all SAT scores of college-bound seniors that are less than 620.

13.23 800 on the SAT. It is possible to score higher than 800 on the SAT, but scores above 800 are reported as 800. (That is, a student can get a reported score of 800 without a perfect paper.) In 2011, the scores of college-bound senior men on the SAT Math test followed a Normal distribution with mean 531 and standard deviation 119. What percentage of scores were above 800 (and so reported as 800)?

13.24 Women's SAT scores. The average performance of women on the SAT, especially the math part, is lower than that of men. The reasons for this gender gap are controversial. In 2011, college-bound senior women's scores on the SAT Math test followed a Normal distribution with mean 500 and standard deviation 113. The mean for men was 531. What percentage of women scored higher than the male mean?

13.25 Are we getting smarter? When the Stanford-Binet IQ test came into use in 1932, it was adjusted so that scores for each age group of children followed roughly the Normal distribution with mean 100 and standard deviation 15. The test is readjusted from time to time to keep the mean at 100. If present-day American children took the 1932 Stanford-Binet test, their mean score would be about 120. The reasons for the increase in IQ over time are not known but probably include better childhood nutrition and more experience in taking tests.

(a) IQ scores above 130 are often called "very superior." What percentage of children had very superior scores in 1932?

(b) If present-day children took the 1932 test, what percentage would have very superior scores? (Assume that the standard deviation 15 does not change.)

13.26 Japanese IQ scores. The Wechsler Intelligence Scale for Children is used (in several languages) in the United States and Europe. Scores in each case are approximately Normally distributed with mean 100 and standard deviation 15. When the test was standardized in Japan, the mean was 111. To what percentile of the American-European distribution of scores does the Japanese mean correspond?

13.27 The stock market. The annual rate of return on stock indexes (which combine many individual stocks) is very roughly Normal. Since 1945, the Standard & Poor's 500 index has had a mean yearly return of 12.5%, with a standard deviation of 17.8%. Take this Normal distribution to be the distribution of yearly returns over a long period.

(a) In what range do the middle 95% of all yearly returns lie?

(b) The market is down for the year if the return on the index is less than zero. In what proportion of years is the market down?

(c) In what proportion of years does the index gain 25% or more?

13.28 Locating the quartiles. The quartiles of any distribution are the 25th and 75th percentiles. About how many standard deviations from the mean are the quartiles of any Normal distribution?

13.29 Young women's heights. The heights of women aged 18 to 24 are approximately Normal with mean 65 inches and standard deviation 2.5 inches.

How tall are the tallest 10% of women? (Use the closest percentile that appears in Table B.)

13.30 High IQ scores. Scores on the Wechsler Adult Intelligence Scale for the 20 to 34 age group are approximately Normally distributed with mean 110 and standard deviation 15. How high must a person score to be in the top 10% of all scores?

EXPLORING THE WEB

13.31 Web-based exercise. Tables of areas under a Normal curve, like Table B at the back of this book, are still common but are also giving way to "applets" that let you find areas visually. Go to the *Statistics: Concepts and Controversies* Web site, **www.whfreeman.com/scc8e,** and look at the *Normal Curve* applet. This applet can be used for Normal curve calculations. Use the applet to answer the following question. Scores on the Wechsler Adult Intelligence Scale for the 20 to 34 age group are approximately Normally distributed with mean 110 and standard deviation 15. How high must a person score to be in the top 10% of all scores? The top 1% of all scores?

13.32 Web-based exercise. If you did any of Exercises 13.21 to 13.30, check your calculations using the *Normal Curve* applet described in the previous exercise.

NOTES AND DATA SOURCES

Page 303 The IQ scores in Figure 13.13 were collected by Darlene Gordon, Purdue University School of Education.

Page 305 Exercise 13.13: Stephen Jay Gould, "Entropic homogeneity isn't why no one hits .400 anymore," *Discover,* August 1986, pp. 60–66.

Page 306 Exercise 13.19: The Gallup Poll is available online at **www.gallup.com/poll/142988/Economic-Confidence-Negative-Year-Ago.aspx.**

Page 306 Exercise 13.21: Information on SAT scores of college-bound seniors can be found at the Web site **media.collegeboard.com/digitalServices/pdf/SAT-Percentile-Ranks-Composite-CR-M_2011.pdf.** This is also the source for the information in Exercise 13.22.

Page 306 Exercise 13.23: Information on SAT scores of college-bound seniors can be found at the Web site **professionals.collegeboard.com/profdownload/cbs2011_total_group_report.pdf.** This is also the source for the information in Exercise 13.24.

Page 307 Exercise 13.25: Ulric Neisser, "Rising scores on intelligence tests," *American Scientist,* September–October 1997, online edition, **www.americanscientist.org.**

Describing Relationships: Scatterplots and Correlation

CASE STUDY The news media have a weakness for lists. Best places to live, best colleges, healthiest foods, worst-dressed women . . . a list of best or worst is sure to find a place in the news. When the state-by-state SAT scores come out each year, it's therefore no surprise that we find news articles ranking the states from best (Illinois in 2011) to worst (District of Columbia in 2011) according to the average SAT score achieved by their high school seniors. Unfortunately, such reports leave readers believing that schools in the District of Columbia must be much worse than those in Illinois. Where does your home state rank? And do you believe the ranking reflects the quality of education you received?

The College Board, which sponsors the SAT exams, doesn't like this practice at all. "Comparing or ranking states on the basis of SAT scores alone is invalid and strongly discouraged by the College Board," says the heading on their table of state average SAT scores. To see why, let's look at the data.

Figure 14.1 shows the distribution of average scores on the SAT Mathematics exam for the 50 states and the District of Columbia. Illinois leads at 617, and the District of Columbia trails at 457 on the SAT scale of 200 to 800. The distribution has an unusual shape: it has one clear peak and perhaps a second, small one. This may be a clue that the data mix two distinct groups. But we need to explore the data further to be sure that this is the case.

In this chapter we will learn that to understand one variable, such as SAT scores, we must look at how it is related to other variables. By the end of this chapter you will be able to use what you have learned to understand why Figure 14.1 has such an unusual shape and to appreciate why the College Board discourages ranking states on SAT scores alone. ∎

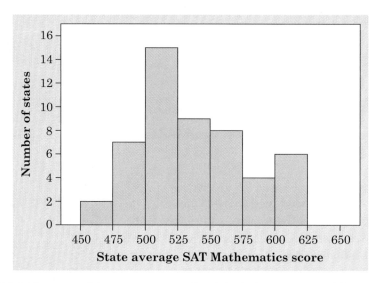

Figure 14.1 Histogram of the average scores of students in the 50 states and the District of Columbia on the SAT Mathematics exam. (This figure was created using the Minitab software package.)

A medical study finds that short women are more likely to have heart attacks than women of average height, while tall women have the fewest heart attacks. An insurance group reports that heavier cars are involved in fewer fatal accidents per 10,000 vehicles registered than are lighter cars. These and many other statistical studies look at the relationship between two variables. To understand such a relationship, we must often examine other variables as well. To conclude that shorter women have higher risk from heart attacks, for example, the researchers had to eliminate the effect of other variables such as weight and exercise habits. Our topic in this and the following chapters is relationships between variables. One of our main themes is that the relationship between two variables can be strongly influenced by other variables that are lurking in the background.

Most statistical studies examine data on more than one variable. Fortunately, statistical analysis of several-variable data builds on the tools we used to examine individual variables. The principles that guide our work also remain the same:

- First plot the data, then add numerical summaries.

- Look for overall patterns and deviations from those patterns.

- When the overall pattern is quite regular, there is sometimes a way to describe it very briefly.

Scatterplots

The most common way to display the relation between two quantitative variables is a *scatterplot*.

EXAMPLE 1   The Big Bang

NASA/JPL-Caltech and The Hubble Heritage Team (STScI/AURA)

How did the universe begin? One popular theory is known as the "Big Bang." The universe began with a big bang and matter expanded outward, like a balloon inflating. If the Big Bang theory is correct, galaxies farthest away from the origin of the bang must be moving faster than those closest to the origin. This also means that galaxies close to the earth must be moving at a similar speed to that of earth, and galaxies far from earth must be moving at very different speeds from earth. Hence, relative to earth, the farther away a galaxy is, the faster it appears to be moving away from earth. Are data consistent with this theory? The answer is "Yes."

In 1929 Edwin Hubble investigated the relationship between the distance from the earth and the recession velocity (the speed at which an object is moving away from an observer) of galaxies. Using data he had collected, Hubble estimated the distance, in megaparsecs, from the earth to 24 galaxies. One parsec equals 3.26 light-years (the distance light travels in one year), and a megaparsec is one million parsecs. The recession velocities, in kilometers per second, of the galaxies were also measured. Figure 14.2 is a scatterplot that shows how recession velocity is related to distance from the earth. We think that "distance from the earth" will help explain "recession velocity." That is, "distance from the earth" is the *explanatory variable,* and "recession velocity" is the *response variable.* We want to see how recession velocity changes when distance from the earth changes, so we put distance from the earth (the explanatory variable) on the horizontal axis. We can then see that, as distance from the earth goes up, recession velocity goes up. Each point on the plot represents one galaxy. For example, the point with a different plotting symbol corresponds to a galaxy that is 1.7 megaparsecs from the earth and that has a recession velocity of 960 kilometers per second.

Hubble's discovery turned out to be one of the most important discoveries in all of astronomy. The data helped establish Hubble's law, which is recession velocity = $H_0 \times$ Distance where H_0 is the value known as the Hubble constant. Hubble's law says that the apparent recession velocities

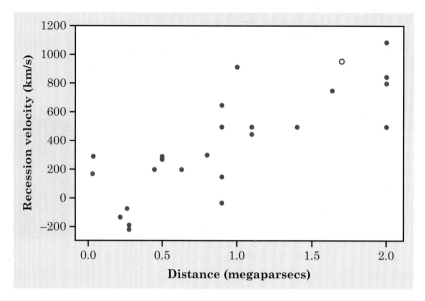

Figure 14.2 Scatterplot of recession velocity against distance from the earth, for Example 1. (This figure was created using the Minitab software package.)

of galaxies are directly proportional to their distances. This relationship is the key evidence for the idea of the expanding universe, as suggested by the Big Bang.

Scatterplot

A **scatterplot** shows the relationship between two quantitative variables measured on the same individuals. The values of one variable appear on the horizontal axis, and the values of the other variable appear on the vertical axis. Each individual in the data appears as the point in the plot fixed by the values of both variables for that individual.

Always plot the explanatory variable, if there is one, on the horizontal axis (the x axis) of a scatterplot. As a reminder, we usually call the explanatory variable x and the response variable y. If there is no explanatory-response distinction, either variable can go on the horizontal axis.

EXAMPLE 2 Health and wealth

Figure 14.3 is a scatterplot of data from the World Bank. The individuals are all the world's nations for which data are available. The explanatory

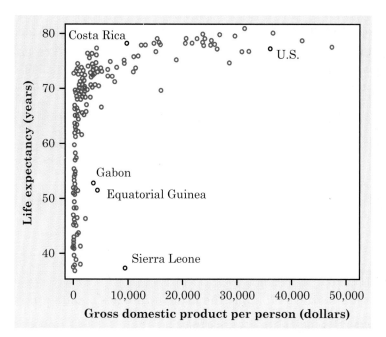

Figure 14.3 Scatterplot of the life expectancy of people in many nations against each nation's gross domestic product per person, for Example 2. (This figure was created using the Minitab software package.)

variable is a measure of how rich a country is: the gross domestic product (GDP) per person. GDP is the total value of the goods and services produced in a country, converted into dollars. The response variable is life expectancy at birth.

We expect people in richer countries to live longer. The overall pattern of the scatterplot does show this, but the relationship has an interesting shape. Life expectancy tends to rise very quickly as GDP increases, then levels off. People in very rich countries such as the United States typically live no longer than people in poorer but not extremely poor nations. Some of these countries, such as Costa Rica, even do better than the United States.

Three African nations are outliers. Their life expectancies are similar to those of their neighbors but their GDPs are higher. Equatorial Guinea and Gabon produce oil, and Sierra Leone produces diamonds. It may be that income from mineral exports goes mainly to a few people and so pulls up GDP per person without much effect on either the income or the life expectancy of ordinary citizens. That is, GDP per person is a mean, and we know that mean income can be much higher than median income.

NOW IT'S YOUR TURN | **14.1 Brain size and intelligence.** For centuries people have associated intelligence with brain size. A recent study used magnetic resonance imaging to measure the brain size of several individuals. The IQ and brain size (in units of 10,000 pixels) of six individuals are as follows:

Brain size:	100	90	95	92	88	106
IQ:	140	90	100	135	80	103

Is there an explanatory variable? If so, what is it and what is the response variable? Make a scatterplot of these data.

Interpreting scatterplots

To interpret a scatterplot, apply the usual strategies of data analysis.

Examining a scatterplot

In any graph of data, look for the **overall pattern** and for striking **deviations** from that pattern.

You can describe the overall pattern of a scatterplot by the **direction, form,** and **strength** of the relationship.

An important kind of deviation is an **outlier,** an individual value that falls outside the overall pattern of the relationship.

After you plot your data, think! Abraham Wald (1902–1950), like many statisticians, worked on war problems during World War II. Wald invented some statistical methods that were military secrets until the war ended. Here is one of his simpler ideas. Asked where extra armor should be added to airplanes, Wald studied the location of enemy bullet holes in planes returning from combat. He plotted the locations on an outline of the plane. As data accumulated, most of the outline filled up. Put the armor in the few spots with no bullet holes, said Wald. That's where bullets hit the planes that didn't make it back.

Both Figures 14.2 and 14.3 have a clear *direction:* recession velocity goes up as distance from the earth increases, and life expectancy generally goes up as GDP increases. We say that Figures 14.2 and 14.3 show a *positive association.* Figure 14.4 is a scatterplot of the gas mileages (in miles per gallon) and the weights (in thousands of pounds) of 38 cars. The response variable is gas mileage and the explanatory variable is weight. We see that gas mileage decreases as weight goes up. We say that Figure 14.4 shows a *negative association.*

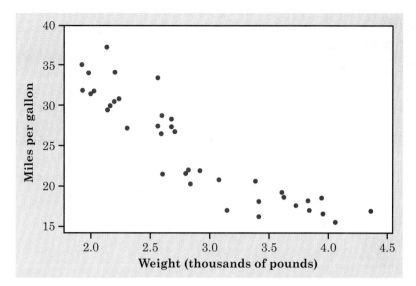

Figure 14.4 Scatterplot of miles per gallon against weight for 38 cars. (This figure was created using the Minitab software package.)

Positive association, negative association

Two variables are **positively associated** when above-average values of one tend to accompany above-average values of the other and below-average values also tend to occur together. The scatterplot slopes upward as we move from left to right.

Two variables are **negatively associated** when above-average values of one tend to accompany below-average values of the other, and vice versa. The scatterplot slopes downward from left to right.

Each of our scatterplots has a distinctive *form*. Figure 14.2 shows a roughly straight-line trend, and Figure 14.3 shows a *curved relationship*. Figure 14.4 shows a slightly curved relationship. The *strength* of a relationship in a scatterplot is determined by how closely the points follow a clear form. The relationships in Figures 14.2 and 14.3 are not strong. Galaxies with similar distances from the earth show quite a bit of scatter in their recession velocities, and nations with similar GDPs can have quite different life expectancies. The relationship in Figure 14.4 is moderately strong. Here is an example of a stronger relationship with a simple form.

EXAMPLE 3 Classifying fossils

Archaeopteryx is an extinct beast having feathers like a bird but teeth and a long bony tail like a reptile. Only six fossil specimens are known. Because these fossils differ greatly in size, some scientists think they are different species rather than individuals from the same species. We will examine data on the lengths in centimeters of the femur (a leg bone) and the humerus (a bone in the upper arm) for the five fossils that preserve both bones. Here are the data:

Femur:	38	56	59	64	74
Humerus:	41	63	70	72	84

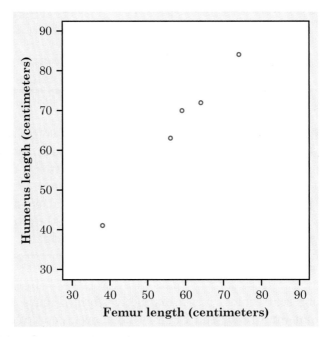

Figure 14.5 Scatterplot of the lengths of two bones in 5 fossil specimens of the extinct beast archaeopteryx, for Example 3.

Because there is no explanatory-response distinction, we can put either measurement on the *x* axis of a scatterplot. The plot appears in Figure 14.5.

The plot shows a *strong, positive, straight-line association.* The straight-line form is important because it is common and simple. The association is strong because the points lie close to a line. It is positive because as the length of one bone increases, so does the length of the other bone. These data suggest that all five fossils belong to the same species and differ in size because some are younger than others. We expect that a different species would have a different relationship between the lengths of the two bones, so that it would appear as an outlier.

NOW IT'S YOUR TURN   **14.2 Brain size and intelligence.** For centuries people have associated intelligence with brain size. A recent study used magnetic resonance imaging to measure the brain size of several individuals. The IQ and brain size (in units of 10,000 pixels) of six individuals are as follows:

| Brain size: | 100 | 90 | 95 | 92 | 88 | 106 |
| IQ: | 140 | 90 | 100 | 135 | 80 | 103 |

Make a scatterplot of these data if you have not already done so. What is the form, direction, and strength of the association? Are there any outliers?

Correlation

A scatterplot displays the direction, form, and strength of the relationship between two variables. Straight-line relations are particularly important because a straight line is a simple pattern that is quite common. A straight-line relation is strong if the points lie close to a straight line, and weak if they are widely scattered about a line. Our eyes are not good judges of how strong a relationship is. The two scatterplots in Figure 14.6 depict the same data, but the right-hand plot is drawn smaller in a large field. The right-hand plot seems to show a stronger straight-line relationship. Our eyes can be fooled by changing the plotting scales or the amount of blank space around the cloud of points in a scatterplot. We need to follow our strategy for data analysis by using a numerical measure to supplement the graph. *Correlation* is the measure we use.

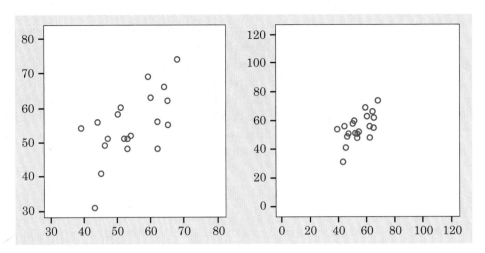

Figure 14.6 Two scatterplots of the same data. The right-hand plot suggests a stronger relationship between the variables because of the surrounding space.

Correlation

The **correlation** describes the direction and strength of a straight-line relationship between two quantitative variables. Correlation is usually written as r.

Calculating a correlation takes a bit of work. You can usually think of r as the result of pushing a calculator button or giving a command in software and concentrate on understanding its properties and use. Knowing how we obtain r from data, however, does help us understand how correlation works, so here we go.

EXAMPLE 4 Calculating correlation

We have data on two variables, x and y, for n individuals. For the fossil data in Example 3, x is femur length, y is humerus length, and we have data for $n = 5$ fossils.

Step 1. Find the mean and standard deviation for both x and y. For the fossil data, a calculator tells us that

Femur:	$\bar{x} = 58.2$ cm	$s_x = 13.20$ cm
Humerus:	$\bar{y} = 66.0$ cm	$s_y = 15.89$ cm

We use s_x and s_y to remind ourselves that there are two standard deviations, one for the values of x and the other for the values of y.

Step 2. Using the means and standard deviations from Step 1, find the standard scores for each x-value and for each y-value:

Value of x	Standard score $(x - \bar{x})/s_x$	Value of y	Standard score $(y - \bar{y})/s_y$
38	$(38 - 58.2)/13.20 = -1.530$	41	$(41 - 66.0)/15.89 = -1.573$
56	$(56 - 58.2)/13.20 = -0.167$	63	$(63 - 66.0)/15.89 = -0.189$
59	$(59 - 58.2)/13.20 = 0.061$	70	$(70 - 66.0)/15.89 = 0.252$
64	$(64 - 58.2)/13.20 = 0.439$	72	$(72 - 66.0)/15.89 = 0.378$
74	$(74 - 58.2)/13.20 = 1.197$	84	$(84 - 66.0)/15.89 = 1.133$

Step 3. The correlation is the average of the products of these standard scores. As with the standard deviation, we "average" by dividing by $n - 1$, one fewer than the number of individuals:

$$r = \frac{1}{4}\Big[(-1.530)(-1.573) + (-0.167)(-0.189) + (0.061)(0.252)$$

$$+ (0.439)(0.378) + (1.197)(1.133)\Big]$$

$$= \frac{1}{4}(2.4067 + 0.0316 + 0.0154 + 0.1659 + 1.3562)$$

$$= \frac{3.9758}{4} = 0.994$$

The algebraic shorthand for the set of calculations in Example 4 is

$$r = \frac{1}{n-1}\sum\left(\frac{x - \bar{x}}{s_x}\right)\left(\frac{y - \bar{y}}{s_y}\right)$$

The symbol $\sum$, called "sigma," means "add them all up."

Understanding correlation

More important than calculating r (a task for technology) is understanding how correlation measures association. Here are the facts:

- **Positive r indicates positive association between the variables, and negative r indicates negative association.** The scatterplot in Figure 14.5 shows strong positive association between femur length and humerus length. In three fossils, both bones are longer than their average values, so their standard scores are positive for both x and y. In the other two fossils, the bones are shorter than their averages, so both standard scores are negative. The products are all positive, giving a positive r.

· **The correlation r always falls between -1 and 1.** Values of r near 0 indicate a very weak straight-line relationship. The strength of the relationship increases as r moves away from 0 toward either -1 or 1. Values of r close to -1 or 1 indicate that the points lie close to a straight line. The extreme values $r = -1$ and $r = 1$ occur only when the points in a scatterplot lie exactly along a straight line.

The result $r = 0.994$ in Example 4 reflects the strong positive straight-line pattern in Figure 14.5. The scatterplots in Figure 14.7 illustrate how r measures both the direction and the strength of a straight-line relationship. Study them carefully. Note that the sign of r matches the direction of the

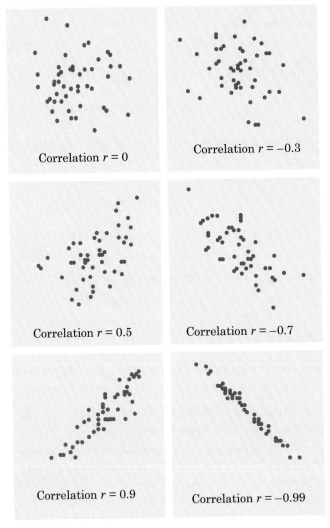

Correlation $r = 0$

Correlation $r = -0.3$

Correlation $r = 0.5$

Correlation $r = -0.7$

Correlation $r = 0.9$

Correlation $r = -0.99$

Figure 14.7 How correlation measures the strength of a straight-line relationship. Patterns closer to a straight line have correlations closer to 1 or -1.

slope in each plot, and that r approaches -1 or 1 as the pattern of the plot comes closer to a straight line.

- Because r uses the standard scores for the observations, **the correlation does not change when we change the units of measurement** of x, y, or both. Measuring length in inches rather than centimeters in Example 4 would not change the correlation $r = 0.994$.

Our descriptive measures for one variable all share the same units as the original observations. If we measure length in centimeters, the median, quartiles, mean, and standard deviation are all in centimeters. The correlation between two variables, however, has no unit of measurement; it is just a number between -1 and 1.

- **Correlation ignores the distinction between explanatory and response variables.** If we reverse our choice of which variable to call x and which to call y, the correlation does not change.

- **Correlation measures the strength of only straight-line association between two variables.** Correlation does not describe curved relationships between variables, no matter how strong they are.

- Like the mean and standard deviation, **the correlation is strongly affected by a few outlying observations.** Use r with caution when outliers appear in the scatterplot. Look, for example, at Figure 14.8. We changed the femur length of the first fossil from 38 to 60 centimeters.

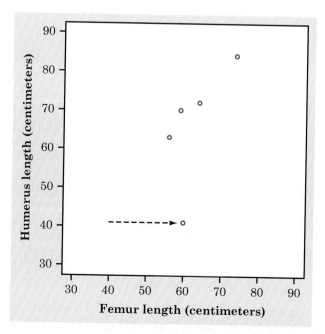

Figure 14.8 Moving one point reduces the correlation from $r = 0.994$ to $r = 0.640$.

Rather than falling in line with the other fossils, the first is now an outlier. The correlation drops from $r = 0.994$ for the original data to $r = 0.640$.

NOW IT'S YOUR TURN

14.3 Brain size and intelligence. For centuries people have associated intelligence with brain size. A recent study used magnetic resonance imaging to measure the brain size of several individuals. The IQ and brain size (in units of 10,000 pixels) of six individuals are as follows:

Brain size:	100	90	95	92	88	106
IQ:	140	90	100	135	80	103

Make a scatterplot of these data if you have not already done so. Compare your plot with those in Figure 14.7 (page 320). What would you estimate the correlation r to be?

There are many kinds of relationships between variables and many ways to measure them. Although correlation is very common, remember its limitations. Correlation makes sense only for quantitative variables—we can speak of the relationship between the sex of voters and the political party they prefer, but not of the correlation between these variables. Even for quantitative variables such as the length of bones, correlation measures only straight-line association.

Remember also that correlation is not a complete description of two-variable data, even when there is a straight-line relationship between the variables. You should give the means and standard deviations of both x and y along with the correlation. Because the formula for correlation uses the means and standard deviations, these measures are the proper choice to accompany a correlation.

STATISTICS IN SUMMARY

Chapter Specifics

- A **scatterplot** is a graph of the relationship between two quantitative variables. If you have an explanatory and a response variable, put the explanatory variable on the x (horizontal) axis of the scatterplot.

- When you examine a scatterplot, look for the **direction, form,** and **strength** of the relationship and also for possible **outliers.**

- If there is a clear direction, is it positive (the scatterplot slopes upward from left to right) or negative (the plot slopes downward)?

- Is the form straight or curved? Are there clusters of observations? Is the relationship strong (a tight pattern in the plot) or weak (the points scatter widely)?

- The **correlation** r measures the direction and strength of a straight-line relationship between two quantitative variables.

- Correlation is a number between -1 and 1. The sign of r shows whether the association is positive or negative. The value of r gets closer to -1 or 1 as the points cluster more tightly about a straight line. The extreme values -1 and 1 occur only when the scatterplot shows a perfectly straight line.

 Chapters 11 to 13 discussed graphical and numerical summaries suitable for a single quantitative variable. In practice, most statistical studies examine relationships between two or more variables. In this chapter we learn about scatterplots, a type of graph that displays the relationship between two quantitative variables, and correlation, a number that measures the direction and strength of a straight-line relationship between two quantitative variables.

   As with other graphics and numbers that summarize data, scatterplots and correlations help us see what the data are telling us, in this case about the possible relationship between two quantitative variables. The goal is to draw conclusions about whether the relationships observed in our data are true in general. In the next chapter, we discuss this in detail.

CASE STUDY Figure 14.9 is a scatterplot of each state's average SAT Mathematics **EVALUATED** score and the proportion of that state's high school seniors who took the SAT exam. SAT score is the response variable, and the proportion of a state's high school seniors who took the SAT exam is the explanatory variable.

1. Describe the overall pattern in words. Is the association positive or negative? Is the relationship strong?

2. The plot shows two groups of states. In one group, no more than about one-third took the SAT. In the other, at least 45% took the exam and the average scores tend to be lower. There are two common college entrance exams, the SAT and the ACT. In ACT states, only students applying to selective colleges take the SAT. Which group of states in the plot corresponds to states in which most students take the ACT exam?

3. Write a paragraph, in language that someone who knows no statistics would understand, explaining why comparing states on the basis of average SAT scores alone would be misleading as a way of comparing the quality of education in the states. ∎

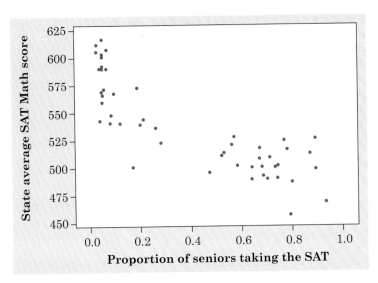

Figure 14.9 Scatterplot of average SAT Mathematics score for each state against the proportion of the state's high school seniors who took the SAT. (This figure was created using the Minitab software package.)

CHAPTER 14 EXERCISES

For Exercise 14.1, see page 314; for Exercise 14.2, see page 317; for Exercise 14.3, see page 322.

14.4 What number can I be?

(a) What are all the values that a correlation r can possibly take?

(b) What are all the values that a standard deviation s can possibly take?

(c) What are all the values that a mean $\bar{x}$ can possibly take?

14.5 Measuring mice. For a biology project, you measure the tail length (millimeters) and weight (grams) of 10 mice.

(a) Explain why you expect the correlation between tail length and weight to be positive.

(b) If you measured tail length in centimeters, how would the correlation change?

14.6 Living on campus. A February 2, 2008, article in the *Columbus Dispatch* reported a study on the distances students lived from campus and average GPA. Here is a summary of the results:

Residence	Avg. GPA
Residence hall	3.33
Walking distance	3.16
Near campus, long walk or short drive	3.12
Within the county, not near campus	2.97
Outside the county	2.94

Based on these data, is the association between the distance a student lives from campus and average GPA positive, negative, or near 0?

14.7 IQ and GPA. Figure 14.10 is a scatterplot of school GPA versus IQ score for all 78 seventh-grade students in a rural Midwest school. Points A, B, and C might be called outliers.

(a) Describe the overall pattern of the relationship in words.

(b) About what are the IQ and GPA for Student A?

(c) For each point A, B, and C, say how it is unusual (for example, "low GPA but a moderately high IQ score").

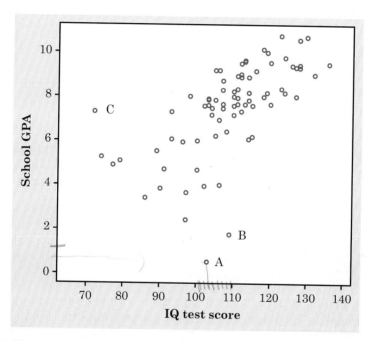

Figure 14.10 School grade point averages and IQ test scores for 78 seventh-grade students, for Exercise 14.7.

14.8 Calories and salt in hot dogs. Figure 14.11 shows the calories and sodium content in 17 brands of meat hot dogs. Describe the overall pattern of these data. In what way is the point marked A unusual?

14.9 IQ and GPA. Is the correlation r for the data in Figure 14.10 near -1, clearly negative but not near -1, near 0, clearly positive but not near 1, or near 1? Explain your answer.

14.10 Calories and salt in hot dogs. Is the correlation r for the data in Figure 14.11 near -1, clearly negative but not near -1, near 0, clearly positive but not near 1, or near 1? Explain your answer.

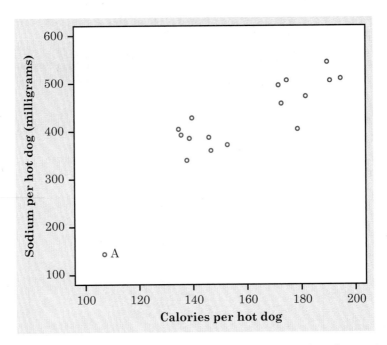

Figure 14.11 Calories and sodium content for 17 brands of meat hot dogs, for Exercise 14.8.

14.11 Comparing correlations. Which of Figures 14.10 and 14.11 has a correlation closer to 1? Explain your answer.

14.12 Outliers and correlation. Figure 14.10 contains outliers marked A, B, and C. In Figure 14.11 the point marked A is an outlier. Removing the outliers will *increase* the correlation *r* in one figure and *decrease* *r* in the other figure. What happens in each figure, and why?

14.13 The professor swims. Professor Moore swims 2000 yards regularly in a vain attempt to undo middle age. Here are his times (in minutes) and his pulse rate (in beats per minute) after swimming for 23 sessions in the pool:

Time:	34.12	35.72	34.72	34.05	34.13	35.72	36.17	35.57
Pulse:	152	124	140	152	146	128	136	144

Time:	35.37	35.57	35.43	36.05	34.85	34.70	34.75	33.93
Pulse:	148	144	136	124	148	144	140	156

Time:	34.60	34.00	34.35	35.62	35.68	35.28	35.97
Pulse:	136	148	148	132	124	132	139

(a) Make a scatterplot. (Which is the explanatory variable?)

(b) Is the association between these variables positive or negative? Explain why you expect the relationship to have this direction.

(c) Describe the form and strength of the relationship.

TABLE 14.1 Lean body mass and metabolic rate

Subject	Sex	Mass (lb)	Rate (cal)	Subject	Sex	Mass (lb)	Rate (cal)
1	M	136.4	1792	11	F	88.7	1189
2	M	138.4	1666	12	F	72.8	913
3	F	79.4	995	13	M	114.2	1460
4	F	120.1	1425	14	F	93.3	1124
5	F	106.7	1396	15	F	75.9	1052
6	F	92.4	1418	16	F	112.4	1347
7	M	104.3	1362	17	F	90.6	1204
8	F	111.3	1502	18	M	114.1	1867
9	F	92.4	1256	19	M	103.2	1439
10	M	107.1	1614				

14.14 Who burns more energy? Metabolic rate, the rate at which the body consumes energy, is important in studies of weight gain, dieting, and exercise. Table 14.1 gives data on the lean body mass and resting metabolic rate for 12 women and 7 men who are subjects in a study of dieting. Lean body mass, given in pounds, is a person's weight leaving out all fat. Metabolic rate is measured in calories burned per 24 hours, the same calories used to describe the energy content of foods. The researchers believe that lean body mass is an important influence on metabolic rate.

(a) Make a scatterplot of the data for the female subjects. Which is the explanatory variable?

(b) Is the association between these variables positive or negative? What is the form of the relationship? How strong is the relationship?

(c) Now add the data for the male subjects to your graph, using a different color or a different plotting symbol. Does the pattern of the relationship that you observed in (b) hold for men also? How do the male subjects as a group differ from the female subjects as a group?

14.15 Marriage. Suppose that men always married women 3 years younger than themselves. Draw a scatterplot of the ages of 6 married couples, with the husband's age as the explanatory variable. What is the correlation r for your data? Why?

14.16 Stretching a scatterplot. Changing the units of measurement can greatly alter the appearance of a scatterplot. Return to the fossil data from Example 3:

Femur:	38	56	59	64	74
Humerus:	41	63	70	72	84

These measurements are in centimeters. Suppose a deranged scientist measured the femur in meters and the humerus in millimeters. The data would then be

Femur:	0.38	0.56	0.59	0.64	0.74
Humerus:	410	630	700	720	840

(a) Draw an x axis extending from 0 to 75 and a y axis extending from 0 to 850. Plot the original data on these axes. Then plot the new data on the same axes in a different color. The two plots look very different.

(b) Nonetheless, the correlation is exactly the same for the two sets of measurements. Why do you know that this is true without doing any calculations?

14.17 The professor swims. Exercise 14.13 gives data on the time to swim 2000 yards and the pulse rate after swimming for a middle-aged professor.

(a) Use a calculator to find the correlation r. Explain from looking at the scatterplot why this value of r is reasonable.

(b) Suppose that the times had been recorded in seconds. For example, the time 34.12 minutes would be 2047 seconds. How would the value of r change?

14.18 Who burns more energy? Table 14.1 gives data on the lean body mass and metabolic rate for 12 women and 7 men. You made a scatterplot of these data in Exercise 14.14.

(a) Do you think the correlation will be about the same for men and women or quite different for the two groups? Why?

(b) Calculate r for women alone and also for men alone. (Use your calculator.)

14.19 Strong association but no correlation. The gas mileage of an automobile first increases and then decreases as the speed increases. Suppose that this relationship is very regular, as shown by the following data on speed (miles per hour) and mileage (miles per gallon):

Speed:	25	35	45	55	65
Mileage:	20	24	26	24	20

Make a scatterplot of mileage versus speed. Use a calculator to show that the correlation between speed and mileage is $r = 0$. Explain why the correlation is 0 even though there is a strong relationship between speed and mileage.

14.20 Body mass and metabolic rate. The body mass data in Table 14.1 are given in pounds. There are 2.2 pounds in a kilogram. If we changed the data from pounds to kilograms, how would the mean body mass change? How would the correlation between body mass and metabolic rate change?

14.21 What are the units? How sensitive to changes in water temperature are coral reefs? To find out, measure the growth of corals in aquariums (where growth is the change in weight, in pounds, of the coral before and after the experiment) when the water temperature (in degrees Fahrenheit) is controlled at different levels. In what units are each of the following descriptive statistics measured?

(a) The mean growth of the coral.

(b) The standard deviation of the growth of the coral.

(c) The correlation between weight gain and temperature.

(d) The median growth of the coral.

14.22 Teaching and research. A college newspaper interviews a psychologist about student ratings of the teaching of faculty members. The psychologist says, "The evidence indicates that the correlation between the research productivity and teaching rating of faculty members is close to zero." The paper reports this as "Professor McDaniel said that good researchers tend to be poor teachers, and vice versa." Explain why the paper's report is wrong. Write a statement in plain language (don't use the word "correlation") to explain the psychologist's meaning.

14.23 Sloppy writing about correlation. Each of the following statements contains a blunder. Explain in each case what is wrong.

(a) "There is a high correlation between the manufacturer of a car and the gas mileage of the car."

(b) "We found a high correlation ($r = 1.09$) between the horsepower of a car and the gas mileage of the car."

(c) "The correlation between the weight of a car and the gas mileage of the car was found to be $r = 0.53$ miles per gallon."

14.24 Guess the correlation. Measurements in large samples show that the correlation

(a) between this semester's GPA and the previous semester's GPA of an upper-class student is about _____.

(b) between IQ and the scores on a test of the reading ability of seventh-grade students is about _____.

(c) between the number of hours a student spends studying per week and the average number of hours spent studying by his or her roommates is about _____.

The answers (in scrambled order) are

$$r = 0.2 \quad r = 0.5 \quad r = 0.8$$

Match the answers to the statements and explain your choice.

14.25 Guess the correlation. For each of the following pairs of variables, would you expect a substantial negative correlation, a substantial positive correlation, or a small correlation?

(a) The cost of a cable TV service and the number of channels provided by the service.

(b) The weight of a road-racing bicycle and the cost of the bicycle.

(c) The number of hours a student spends on Facebook and the student's GPA.

(d) The heights and salaries of faculty members at your university.

14.26 Investment diversification. A mutual funds company's newsletter says, "A well-diversified portfolio includes assets with low correlations." The newsletter includes a table of correlations between the returns on various classes of investments. For example, the correlation between municipal bonds and large-cap stocks is 0.50, and the correlation between municipal bonds and small-cap stocks is 0.21.

(a) Rachel invests heavily in municipal bonds. She wants to diversify by adding an investment whose returns do not closely follow the returns on her bonds. Should she choose large-cap stocks or small-cap stocks for this purpose? Explain your answer.

(b) If Rachel wants an investment that tends to increase when the return on her bonds drops, what kind of correlation should she look for?

14.27 Take me out to the ball game. What is the relationship between the price charged for a hot dog and the price charged for a 16-ounce soda in Major League Baseball stadiums? Table 14.2 gives some data. Make a scatterplot appropriate for showing how soda price helps explain hot dog price. Describe the relationship that you see. Are there any outliers?

14.28 When it rains, it pours. Figure 14.12 plots the highest *yearly* precipitation ever recorded in each state against the highest *daily* precipitation ever

TABLE 14.2 Hot dog and soda prices (in dollars) at some Major League Baseball stadiums

Team	Hot dog	Soda	Team	Hot dog	Soda	Team	Hot dog	Soda
Angels	3.00	3.43	Giants	4.50	4.00	Rays	5.00	3.00
Astros	4.75	3.24	Indians	4.25	3.33	Reds	1.00	1.33
Blue Jays	5.00	2.86	Marlins	5.00	2.00	Red Sox	4.50	3.20
Braves	4.25	3.09	Mets	5.00	2.67	Rockies	3.25	2.89
Brewers	3.25	3.33	Padres	4.00	2.91	Royals	4.00	2.91
Cardinals	4.00	3.81	Phillies	3.75	3.00	Tigers	3.00	2.40
Diamondbacks	2.75	2.50	Pirates	2.50	2.75	Twins	3.75	3.20
Dodgers	5.00	3.50	Rangers	4.75	3.55	White Sox	3.25	3.43

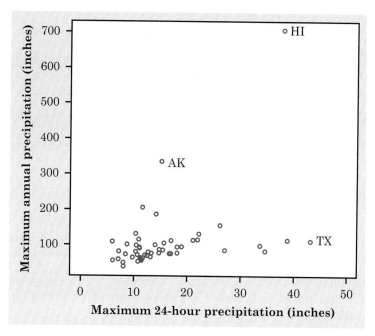

Figure 14.12 Record-high yearly precipitation recorded at any weather station in each state plotted against record-high daily precipitation for the state, for Exercise 14.28.

recorded in that state. The points for Alaska (AK), Hawaii (HI), and Texas (TX) are marked on the scatterplot.

(a) About what are the highest daily and yearly precipitation values for Alaska?

(b) Alaska and Hawaii have very high yearly maximums relative to their daily maximums. Omit these two states as outliers. Describe the nature of the relationship for the other states. Would knowing a state's highest daily precipitation be a great help in predicting that state's highest yearly precipitation?

14.29 How many corn plants are too many? How much corn per acre should a farmer plant to obtain the highest yield? To find the best planting rate, do an experiment: plant at different rates on several plots of ground and measure the harvest. Here are data from such an experiment:

Plants per acre	Yield (bushels per acre)			
12,000	150.1	113.0	118.4	142.6
16,000	166.9	120.7	135.2	149.8
20,000	165.3	130.1	139.6	149.9
24,000	134.7	138.4	156.1	
28,000	119.0	150.5		

(a) Is yield or planting rate the explanatory variable? Why?

(b) Make a scatterplot of yield and planting rate.

(c) Describe the overall pattern of the relationship. Is it a straight line? Is there a positive or negative association, or neither? Explain why increasing the number of plants per acre of ground has the effect that your graph shows.

14.30 Why so small? Make a scatterplot of the following data:

x	1	2	3	4	9	10
y	12	2	3	5	9	11

Use your calculator to show that the correlation is about 0.4. What feature of the data is responsible for reducing the correlation to this value despite a strong straight-line association between x and y in most of the observations?

14.31 Ecological correlation. Many studies reveal a positive correlation between income and number of years of education. To investigate this, a researcher makes two plots.

Plot 1: Plot the number of years of education (the explanatory variable) versus the average annual income of all adults having that many years of education (the response variable).

Plot 2: Plot the number of years of education (the explanatory variable) versus the individual annual incomes of all adults (the response variable).

Which plot will display a stronger correlation? (*Hint:* Which plot will display a greater amount of scatter? In particular, will the variation from individual to individual having the same number of years of education create more or less scatter in Plot 2 compared with plotting the average incomes in Plot 1? What effect will increased scatter have on the strength of the association we observe?)

   Note: A correlation based on averages rather than on individuals is called an **ecological correlation**. Correlations based on averages can be misleading if they are interpreted to be about individuals.

14.32 Ecological correlation again. In Exercise 14.6 (page 324), would the association be stronger, weaker, or the same if the data given listed the GPAs of individual students (rather than averages) and the distance they lived from campus?

EXPLORING THE WEB

14.33 Web-based exercise. A popular saying in golf is "You drive for show but you putt for dough." You can find this season's Professional Golfers Association (PGA) Tour statistics at the PGA tour Web site: **www.pgatour.com/r/stats/.** You can also find these statistics at the ESPN Web site: **espn.go.com/golf/leaders.** Look at the most recent putting, driving, and money earnings data for the current season on the PGA Tour.

(a) Make a scatterplot of earnings and putting average. Use earnings as the response variable. Describe the direction, form, and strength of the relationship in the plot. Are there any outliers?

(b) Make a scatterplot of earnings and driving distance. Use earnings as the response variable. Describe the direction, form, and strength of the relationship in the plot. Are there any outliers?

(c) Do your plots support the maxim "You drive for show but you putt for dough"?

14.34 Web-based exercise. The best way to grasp how the correlation reflects the pattern of the points on a scatterplot is to use an "applet" that allows you to plot data points and watch the correlation change. Go to the *Statistics: Concepts and Controversies* Web site, **www.whfreeman.com/scc8e,** and look at the *Correlation and Regression* applet. The correlation is constantly recalculated as you click to add points. Create a plot with a correlation near 1 and then add a point to make the correlation negative. Where did you add the point?

NOTES AND DATA SOURCES

Pages 310 and 324 The data for Figures 14.1 and 14.9 come from the College Board Web site, **www.collegeboard.org.**

Page 312 Example 2: Data extracted from the World Bank's Web site, **databank.worldbank.org/data/home.aspx.** Life expectancy and GDP per capita (purchasing-power parity basis) are estimated for 2002.

Page 316 Example 3: M. A. Houck et al., "Allometric scaling in the earliest fossil bird, *Archaeopteryx lithographica,*" *Science,* 247 (1990), pp. 195–198. The authors conclude

from a variety of evidence that all specimens represent the same species.

Page 330 Exercise 14.26: Quotation from the *T. Rowe Price Report,* Winter 1997, p. 4.

Page 330 Exercise 14.27: These data are for 2010 and were found at the Web site **teammarketing.com.ismmedia.com/ISM3/std-content/repos/Top/News/2010_mlb_fci.pdf.** Because the sodas served vary in size, we have converted soda prices to the price of a 16-ounce soda at each price per ounce.

Page 330 Exercise 14.28: These data can be found at the Web site of the National Climatic Data Center of the National Oceanic and Atmospheric Administration, **lwf.ncdc .noaa.gov/oa/climate/severeweather/ extremes.html.**

Page 331 Exercise 14.29: W. L. Colville and D. P. McGill, "Effect of rate and method of planting on several plant characters and yield of irrigated corn," *Agronomy Journal,* 54 (1962), pp. 235–238.

Describing Relationships: Regression, Prediction, and Causation

CASE STUDY Predicting the future course of the stock market could make you rich. No wonder lots of people and lots of computers pore over market data looking for patterns.

There are some surprising methods. The "Super Bowl Indicator" says that the football Super Bowl, played in January or early February, predicts how stocks will behave each year. The current National Football League (NFL) was formed by merging the original NFL with the American Football League (AFL). The current NFL consists of two conferences, the National Football Conference (NFC) and the American Football Conference (AFC). The indicator claims that stocks go up in years when a team from the NFC (or from the old NFL) wins and down when an AFC team wins. The indicator was right in 35 of 45 years between the first Super Bowl in 1967 and 2011. (For purposes of the legend, we will regard the Baltimore Ravens as an old NFL team because they were the Cleveland Browns before the franchise moved to Baltimore in 1996. We will also regard the Tampa Bay Buccaneers as an NFC team, although they were neither a pre-merger team nor an old NFL team and started out as an AFC team.) The indicator is right over 75% of the time, which seems impressive.

Yesterday (February 5, 2012) the Giants, an NFC team, won the Super Bowl. According to the Super Bowl Indicator, stocks will rise this year. Should I invest?

In this chapter we will study statistical methods to predict one variable from others that go well beyond just counting ups and downs. We will also distinguish between the ability to predict one variable from others and the issue of whether changes in one variable are caused by changes in others. By the end of this chapter, you will be able to critically evaluate the Super Bowl Indicator. ■

David Shluka/AP Photo

Regression lines

If a scatterplot shows a straight-line relationship between two quantitative variables, we would like to summarize this overall pattern by drawing a line on the graph. A *regression line* summarizes the relationship between two variables, but only in a specific setting: one of the variables helps explain or predict the other. That is, regression describes a relationship between an explanatory variable and a response variable.

Regression line

A **regression line** is a straight line that describes how a response variable y changes as an explanatory variable x changes. We often use a regression line to predict the value of y for a given value of x.

EXAMPLE 1 Fossil bones

We saw that the lengths of two bones in fossils of the extinct beast archaeopteryx closely follow a straight-line pattern. Figure 15.1 plots the lengths for the 5 available fossils. The regression line on the plot gives a quick summary of the overall pattern.

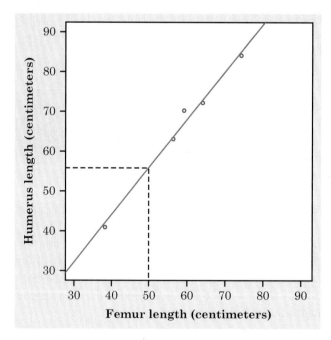

Figure 15.1 Using a straight-line pattern for prediction, for Example 1. The data are the lengths of two bones in 5 fossils of the extinct beast archaeopteryx.

Another archaeopteryx fossil is incomplete. Its femur is 50 centimeters long, but the humerus is missing. Can we predict how long the humerus is? The straight-line pattern connecting humerus length to femur length is so strong that we feel quite safe in using femur length to predict humerus length. Figure 15.1 shows how: starting at the femur length (50 cm), go up to the line, then over to the humerus length axis. We predict a length of about 56 cm. This is the length the humerus would have if this fossil's point lay exactly on the line. All the other points are close to the line, so we think the missing point would also be close to the line. That is, we think this prediction will be quite accurate.

EXAMPLE 2 Presidential elections

Republican Ronald Reagan was elected president twice, in 1980 and in 1984. His economic policy of tax cuts to stimulate the economy, eventually leading to increases in tax revenue, was still advocated by Republican presidential candidates in 2008. Figure 15.2 plots the percentage of voters in each state who voted for Reagan's Democratic opponents: Jimmy Carter in 1980 and Walter Mondale in 1984. The plot shows a positive straight-line relationship. We expect this because some states tend to vote

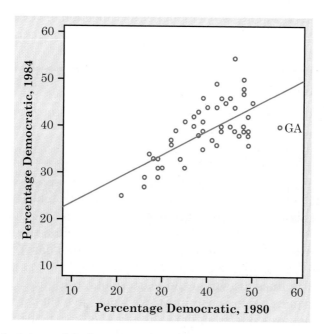

Figure 15.2 A weaker straight-line pattern, for Example 2. The data are the percentage in each state who voted Democratic in the two Reagan presidential elections.

Democratic and others tend to vote Republican. There is one outlier: Georgia, President Carter's home state, voted 56% for the Democrat Carter in 1980 but only 40% Democratic in 1984.

We could use the regression line drawn in Figure 15.2 to predict a state's 1984 vote from its 1980 vote. The points in this figure are more widely scattered about the line than are the points in the fossil bone plot in Figure 15.1. The correlations, which measure the strength of the straight-line relationships, are $r = 0.994$ for Figure 15.1 and $r = 0.704$ for Figure 15.2. The scatter of the points makes it clear that predictions of voting will be generally less accurate than predictions of bone length.

Regression equations

When a plot shows a straight-line relationship as strong as that in Figure 15.1, it is easy to draw a line close to the points by eye. In Figure 15.2, however, different people might draw quite different lines by eye. Because we want to predict y from x, we want a line that is close to the points in the *vertical* (y) direction. It is hard to concentrate on just the vertical distances when drawing a line by eye. What is more, drawing by eye gives us a line on the graph but not an equation for the line. We need a way to find from the data the equation of the line that comes closest to the points in the vertical direction. There are many ways to make the collection of vertical distances "as small as possible." The most common is the *least-squares* method.

Least-squares regression line

The **least-squares regression line** of y on x is the line that makes the sum of the squares of the vertical distances of the data points from the line as small as possible.

Figure 15.3 illustrates the least-squares idea. This figure magnifies the center part of Figure 15.1 to focus on 3 of the points. We see the vertical distances of these 3 points from the regression line. To find the least-squares line, look at these vertical distances (all 5 for the fossil data), square them, and move the line until the sum of the squares is the smallest it can be for any line. The lines drawn on the scatterplots in Figures 15.1 and 15.2 are the least-squares regression lines. We won't give the formula for finding the least-squares line from data—that's a job for a calculator or computer. You should, however, be able to use the equation that the machine produces.

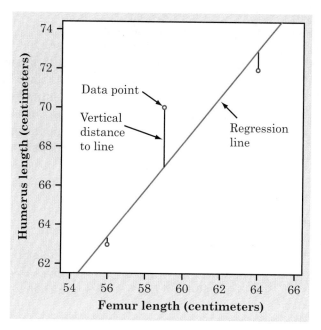

Figure 15.3 A regression line aims to predict y from x. So a good regression line makes the vertical distances from the data points to the line small.

In writing the equation of a line, x stands as usual for the explanatory variable and y for the response variable. The equation of a line has the form

$$y = a + bx$$

The number b is the **slope** of the line, the amount by which y changes when x increases by one unit. The number a is the **intercept,** the value of y when $x = 0$. To use the equation for prediction, just substitute your x-value into the equation and calculate the resulting y-value.

EXAMPLE 3 Using a regression equation

In Example 1, we used the "up-and-over" method in Figure 15.1 to predict the humerus length for a fossil whose femur length is 50 cm. The equation of the least-squares line is

$$\text{humerus length} = -3.66 + (1.197 \times \text{femur length})$$

The *slope* of this line is $b = 1.197$. This means that for these fossils, humerus length goes up by 1.197 cm when femur length goes up 1 cm. The slope of a regression line is usually important for understanding the data. The slope is the rate of change, the amount of change in the predicted y when x increases by 1.

Regression toward the mean   To "regress" means to go backward. Why are statistical methods for predicting a response from an explanatory variable called "regression"? Sir Francis Galton (1822–1911), who was the first to apply regression to biological and psychological data, looked at examples such as the heights of children versus the heights of their parents. He found that the taller-than-average parents tended to have children who were also taller than average, but not as tall as their parents. Galton called this fact "regression toward the mean" and the name came to be applied to the statistical method.

The *intercept* of the least-squares line is $a = -3.66$. This is the value of the predicted y when $x = 0$. Although we need the intercept to draw the line, it is statistically meaningful only when x can actually take values close to zero. Here femur length 0 is impossible (recall that the femur is a bone in the leg), so the intercept has no statistical meaning.

To use the equation for *prediction*, substitute the value of x and calculate y. The predicted humerus length for a fossil with a femur 50 cm long is

$$\text{humerus length} = -3.66 + (1.197)(50)$$
$$= 56.2 \text{ cm}$$

To *draw the line* on the scatterplot, predict y for two different values of x. This gives two points. Plot them and draw the line through them.

NOW IT'S YOUR TURN

15.1 Fossil bones. Use the equation of the least-squares line

$$\text{humerus length} = -3.66 + (1.197 \times \text{femur length})$$

to predict the humerus length for a fossil with a femur 70 cm long.

Understanding prediction

Computers make prediction easy and automatic, even from very large sets of data. Anything that can be done automatically is often done thoughtlessly. Regression software will happily fit a straight line to a curved relationship, for example. Also, the computer cannot decide which is the explanatory variable and which is the response variable. This is important, because the same data give two different lines depending on which is the explanatory variable.

In practice, we often use several explanatory variables to predict a response. As part of its admissions process, a college might use SAT Math and Verbal scores and high school grades in English, math, and science (5 explanatory variables) to predict first-year college grades. Although the details are messy, all statistical methods of predicting a response share some basic properties of least-squares regression lines.

- **Prediction is based on fitting some "model" to a set of data.** In Figures 15.1 and 15.2, our model is a straight line that we draw through the points in a scatterplot. Other prediction methods use more elaborate models.

- **Prediction works best when the model fits the data closely.** Compare again Figure 15.1, where the data closely follow a line, with Figure 15.2, where they do not. Prediction is more trustworthy in Figure 15.1. Also, it is not so easy to see patterns when there are many variables, but if the data do not have strong patterns, prediction may be very inaccurate.

- **Prediction outside the range of the available data is risky.** Suppose that you have data on a child's growth between 3 and 8 years of age. You find a strong straight-line relationship between age x and height y. If you fit a regression line to these data and use it to predict height at age 25 years, you will predict that the child will be 8 feet tall. Growth slows down and stops at maturity, so extending the straight line to adult ages is foolish. No one would make this mistake in predicting height. But almost all economic predictions try to tell us what will happen next quarter or next year. No wonder economic predictions are often wrong. Prediction outside the range of available data is referred to as extrapolation. Beware of **extrapolation**!

EXAMPLE 4 Predicting the national surplus

The Congressional Budget Office is required to submit annual reports that predict the federal budget and its deficit or surplus for the next 5 years. These forecasts depend on future economic trends (unknown) and on what Congress will decide about taxes and spending (also unknown). Even the prediction of the state of the budget if current policies are not changed has been wildly inaccurate. The forecast made in 2004 for 2008, for example, missed by nearly $177 billion. The 2005 forecast for 2009 was $1193 billion off! As Senator Everett Dirksen once said, "A billion here and a billion there and pretty soon you are talking real money." In 1999, the Budget Office was predicting a surplus (ignoring Social Security) of $996 billion over the following 10 years. Politicians debated what to do with the money, but no one else believed the prediction (correctly, as it turned out). In 2008 there was a $455 billion deficit and in 2009 a $1400 billion deficit.

Correlation and regression

Correlation measures the direction and strength of a straight-line relationship. Regression draws a line to describe the relationship. Correlation and regression are closely connected, even though regression requires choosing an explanatory variable and correlation does not.

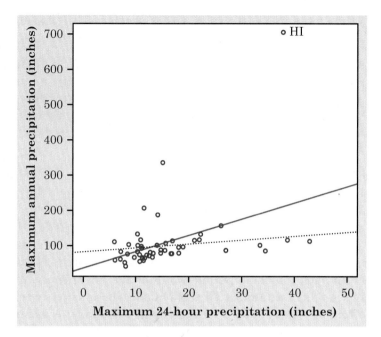

Figure 15.4 Least-squares regression lines are strongly influenced by outliers. The solid line is based on all 50 data points. The dotted line leaves out Hawaii.

Both correlation and regression are strongly affected by outliers. Be wary if your scatterplot shows strong outliers. Figure 15.4 plots the record-high yearly precipitation in each state against that state's record-high 24-hour precipitation. Hawaii is a high outlier, with a yearly record of 704.83 inches of rain recorded at Kukui in 1982. The correlation for all 50 states in Figure 15.4 is 0.510. If we leave out Hawaii, the correlation drops to $r = 0.248$. The solid line in the figure is the least-squares line for predicting the annual record from the 24-hour record. If we leave out Hawaii, the least-squares line drops down to the dotted line. This line is nearly flat—there is little relation between yearly and 24-hour record precipitation once we decide to ignore Hawaii.

The usefulness of the regression line for prediction depends on the strength of the association. That is, the usefulness of a regression line depends on the correlation between the variables. It turns out that the *square* of the correlation is the right measure.

---

r^2 in regression

The **square of the correlation, r^2,** is the proportion of the variation in the values of y that is explained by the least-squares regression of y on x.

The idea is that when there is a straight-line relationship, some of the variation in y is accounted for by the fact that as x changes it pulls y along with it.

EXAMPLE 5 Using r^2

Look again at Figure 15.1. There is a lot of variation in the humerus lengths of these 5 fossils, from a low of 41 cm to a high of 84 cm. The scatterplot shows that we can explain almost all of this variation by looking at femur length and at the regression line. As femur length increases, it pulls humerus length up with it along the line. There is very little leftover variation in humerus length, which appears in the scatter of points about the line. Because $r = 0.994$ for these data, $r^2 = (0.994)^2 = 0.988$. So the variation "along the line" as femur length pulls humerus length with it accounts for 98.8% of all the variation in humerus length. The scatter of the points about the line accounts for only the remaining 1.2%. Little leftover scatter says that prediction will be accurate.

Contrast the voting data in Figure 15.2. There is still a straight-line relationship between the 1980 and 1984 Democratic votes, but there is also much more scatter of points about the regression line. Here, $r = 0.704$ and so $r^2 = 0.496$. Only about half the observed variation in the 1984 Democratic vote is explained by the straight-line pattern. You would still guess a higher 1984 Democratic vote for a state that was 45% Democratic in 1980 than for a state that was only 30% Democratic in 1980. But lots of variation remains in the 1984 votes of states with the same 1980 vote. That is the other half of the total variation among the states in 1984. It is due to other factors, such as differences in the main issues in the two elections and the fact that President Reagan's two Democratic opponents came from different parts of the country.

In reporting a regression, it is usual to give r^2 as a measure of how successful the regression was in explaining the response. When you see a correlation, square it to get a better feel for the strength of the association. Perfect correlation ($r = -1$ or $r = 1$) means the points lie exactly on a line. Then $r^2 = 1$ and all of the variation in one variable is accounted for by the straight-line relationship with the other variable. If $r = -0.7$ or $r = 0.7$, $r^2 = 0.49$ and about half the variation is accounted for by the straight-line relationship. In the r^2 scale, correlation ± 0.7 is about halfway between 0 and ± 1.

Did the vote counters cheat? Republican Bruce Marks was ahead of Democrat William Stinson when the voting-machines results were tallied in their 1993 Pennsylvania election. But Stinson was ahead after absentee ballots were counted by the Democrats, who controlled the election board. A court fight followed. The court called in a statistician, who used regression with data from past elections to predict the counts of absentee ballots from the voting-machine results. Marks's lead of 564 votes from the machines predicted that he would get 133 more absentee votes than Stinson. In fact, Stinson got 1025 more absentee votes than Marks. Did the vote counters cheat?

> **NOW IT'S YOUR TURN**
>
> **15.2 At the ballpark.** Table 14.2 (page 330) gives data on the prices charged for a 16-ounce soda and for a hot dog at Major League Baseball stadiums. The correlation between the prices is $r = 0.45$. What proportion of the variation in hot dog prices is explained by the least-squares regression of hot dog prices on 16-ounce soda prices?

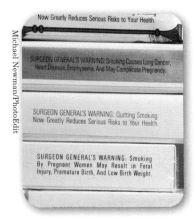

Michael Newman/PhotoEdit

The question of causation

There is a strong relationship between cigarette smoking and death rate from lung cancer. Does smoking cigarettes *cause* lung cancer? There is a strong association between the availability of handguns in a nation and that nation's homicide rate from guns. Does easy access to handguns *cause* more murders? It says right on the pack that cigarettes cause cancer. Whether more guns cause more murders is hotly debated. Why is the evidence for cigarettes and cancer better than the evidence for guns and homicide?

We already know three big facts about statistical evidence for cause and effect.

Statistics and causation

1. A strong relationship between two variables does not always mean that changes in one variable cause changes in the other.
2. The relationship between two variables is often influenced by other variables lurking in the background.
3. The best evidence for causation comes from randomized comparative experiments.

EXAMPLE 6 Does television extend life?

Measure the number of television sets per person x and the life expectancy y for the world's nations. There is a high positive correlation: nations with many TV sets have higher life expectancies.

The basic meaning of causation is that by changing x we can bring about a change in y. Could we lengthen the lives of people in Botswana by shipping them TV sets? No. Rich nations have more TV sets than poor nations. Rich nations also have longer life expectancies because they offer better nutrition, clean water, and better health care. There is no cause-and-effect tie between TV sets and length of life.

"In a new attack on third-world poverty, aid organizations today began delivery of 100,000 television sets."

Example 6 illustrates our first two big facts. Correlations such as this are sometimes called "nonsense correlations." The correlation is real. What is nonsense is the conclusion that changing one of the variables causes changes in the other. A lurking variable—such as national wealth in Example 6—that influences both x and y can create a high correlation even though there is no direct connection between x and y. We might call this *common response:* both the explanatory and the response variable are responding to some lurking variable.

EXAMPLE 7 Obesity in mothers and daughters

What causes obesity in children? Inheritance from parents, overeating, lack of physical activity, and too much television have all been named as explanatory variables.

The results of a study of Mexican American girls aged 9 to 12 years are typical. Researchers measured body mass index (BMI), a measure of weight relative to height, for both the girls and their mothers. People with high BMI are overweight or obese. They also measured hours of television watched, minutes of physical activity, and intake of several kinds of food. The result: the girls' BMIs were weakly correlated with physical activity ($r = -0.18$), diet, and television. The strongest correlation ($r = 0.506$) was between the BMI of daughters and the BMI of their mothers.

Body type is in part determined by heredity. Daughters inherit half their genes from their mothers. There is therefore a direct causal link between the BMI of mothers and daughters. Of course, the causal link is far from perfect. The mothers' BMIs explain only 25.6% (that's r^2 again) of the variation among the daughters' BMIs. Other factors, some measured in the

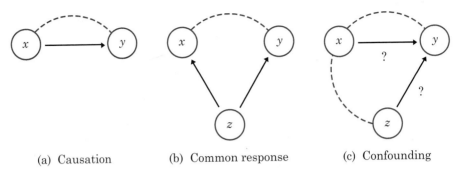

(a) Causation   (b) Common response   (c) Confounding

Figure 15.5 Some explanations for an observed association. A dashed line shows an association. An arrow shows a cause-and-effect link. Variable *x* is explanatory, *y* is a response variable, and *z* is a lurking variable.

study and others not measured, also influence BMI. *Even when direct causation is present, it is rarely a complete explanation of an association between two variables.*

Can we use r or r^2 from Example 7 to say how much inheritance contributes to the daughters' BMIs? No. Remember *confounding*. It may well be that mothers who are overweight also set an example of little exercise, poor eating habits, and lots of television. Their daughters pick up these habits to some extent, so the influence of heredity is mixed up with influences from the girls' environment. We can't say how much of the correlation between mother and daughter BMIs is due to inheritance.

Figure 15.5 shows in outline form how a variety of underlying links between variables can explain association. The dashed line represents an observed association between the variables x and y. Some associations are explained by a *direct cause-and-effect* link between the variables. The first diagram in Figure 15.5 shows "x causes y" by an arrow running from x to y. The second diagram illustrates *common response*. The observed association between the variables x and y is explained by a lurking variable z. Both x and y change in response to changes in z. This common response creates an association even though there may be no direct causal link between x and y. The third diagram in Figure 15.5 illustrates *confounding*. Both the explanatory variable x and the lurking variable z may influence the response variable y. Variables x and z are themselves associated, so we cannot distinguish the influence of x from the influence of z. We cannot say how strong the direct effect of x on y is. In fact, it can be hard to say if x influences y at all.

In Example 7, there is a causal link between the BMI of mothers and daughters. However, other factors, some measured in the study and some not measured, also influence the BMI of daughters. This is an example of confounding, illustrated in Figure 15.5(c). The x in the figure corresponds to the BMI of the mother, the z to one of the other factors, and the y to the BMI of the daughter.

Both common response and confounding involve the influence of a lurking variable or variables z on the response variable y. We won't belabor the distinction between the two kinds of relationships. Just remember that "beware the lurking variable" is good advice in thinking about relationships between variables. Here is another example of common response, in a setting where we want to do prediction.

EXAMPLE 8 SAT scores and college grades

High scores on the SAT examinations in high school certainly do not *cause* high grades in college. The moderate association (r^2 is about 27%) is no doubt explained by common response variables such as academic ability, study habits, and staying sober. Figure 15.5(b) illustrates this. In the figure z might correspond to academic ability, x to SAT scores, and y to grades in college.

The ability of SAT scores to partly predict college performance doesn't depend on causation. We need only believe that the relationship between SAT scores and college grades that we see in past years will continue to hold for this year's high school graduates. Think once more of our fossils, where femur length predicts humerus length very well. The strong relationship is explained by common response to the overall age and size of the beasts whose fossils we now examine. *Prediction doesn't require causation.*

Discussion of these examples has brought to light two more big facts about causation:

More about statistics and causation

4. The observed relationship between two variables may be due to **direct causation, common response,** or **confounding.** Two or more of these factors may be present together.

5. An observed relationship can, however, be used for prediction without worrying about causation as long as the patterns found in past data continue to hold true.

STATISTICAL CONTROVERSIES

Gun Control and Crime

Do strict controls on guns, especially handguns, reduce crime? To many people, the answer must be "Yes." More than half of all murders in the United States are committed with handguns. The U.S. murder rate (per 100,000 population) is 1.7 times that of Canada, and the rate of murders with handguns is 15 times higher. Surely guns help bad things happen. Then John Lott, a University of Chicago economist, did an elaborate statistical study using data from all 3054 counties in the United States over the 18-year period from 1977 to 1994. Lott found that as states relaxed gun laws to allow adults to carry guns, the crime rate dropped. He argued that guns reduce crime by allowing citizens to defend themselves and by making criminals hesitate.

Lott used regression methods to determine the relationship between crime and many explanatory variables and to isolate the effect of permits to carry concealed guns after adjusting for other explanatory variables. You can find a link to a copy of Lott's study at **www2.lib .uchicago.edu/~llou/guns.html.**

The resulting debate, still going on, has been loud. People feel strongly about gun control. Most reacted to Lott's work based on whether or not they liked his conclusion. Gun supporters painted Lott as Moses revealing truth at last; opponents knew he must be both wrong and evil.

Is Lott right? What do you see as the weaknesses of his study based on what you have learned about statistics?

NOW IT'S YOUR TURN

15.3 At the ballpark. Table 14.2 (page 330) gives data on the prices charged for a 16-ounce soda and for a hot dog at Major League Baseball stadiums. The correlation between the prices is $r = 0.45$. Do you think the observed relationship is due to direct causation, common response, confounding, or some combination of these? Explain your answer.

Evidence for causation

Despite the difficulties, it is sometimes possible to build a strong case for causation in the absence of experiments. The evidence that smoking causes lung cancer is about as strong as nonexperimental evidence can be.

Doctors had long observed that most lung cancer patients were smokers. Observational studies comparing smokers and "similar" (in the sense of characteristics such as age, gender, and overall health) nonsmokers showed a strong association between smoking and death from lung cancer. Could the association be explained by lurking variables that the studies could not measure? Might there be, for example, a genetic factor that predisposes people both to nicotine addiction and to lung cancer? Smoking and lung cancer would then be positively associated even if smoking had no direct effect on the lungs. How were these objections overcome?

Let's answer this question in general terms. What are the criteria for establishing causation when we cannot do an experiment?

- **The association is strong.** The association between smoking and lung cancer is very strong.

- **The association is consistent.** Many studies of different kinds of people in many countries link smoking to lung cancer. That reduces the chance that a lurking variable specific to one group or one study explains the association.

- **Higher doses are associated with stronger responses.** People who smoke more cigarettes per day or who smoke over a longer period get lung cancer more often. People who stop smoking reduce their risk.

- **The alleged cause precedes the effect in time.** Lung cancer develops after years of smoking. The number of men dying of lung cancer rose as smoking became more common, with a lag of about 30 years. Lung cancer kills more men than any other form of cancer. Lung cancer was rare among women until women began to smoke. Lung cancer in women rose along with smoking, again with a lag of about 30 years, and has now passed breast cancer as the leading cause of cancer death among women.

- **The alleged cause is plausible.** Experiments with animals show that tars from cigarette smoke do cause cancer.

Medical authorities do not hesitate to say that smoking causes lung cancer. The U.S. Surgeon General has long stated that cigarette smoking is "the largest avoidable cause of death and disability in the United States." The evidence for causation is overwhelming—but it is not as strong as the evidence provided by well-designed experiments.

STATISTICS IN SUMMARY

Chapter Specifics

- **Regression** is the name for statistical methods that fit some model to data in order to predict a response variable from one or more explanatory variables.

- The simplest kind of regression fits a straight line on a scatterplot for use in predicting y from x. The most common way to fit a line is the **least-squares** method, which finds the line that makes the sum of the squared vertical distances of the data points from the line as small as possible.

- The **squared correlation** r^2 tells us what fraction of the variation in the responses is explained by the straight-line tie between y and x.

- **Extrapolation,** or prediction outside the range of the data, is risky because the pattern may be different there. Beware of extrapolation!

- A strong relationship between two variables is not always evidence that changes in one variable **cause** changes in the other. Lurking variables can create relationships through **common response** or **confounding.**

- If we cannot do experiments, it is often difficult to get convincing evidence for causation.

In Chapter 14 we used scatterplots and the correlation to explore and describe the relationship between two quantitative variables. In this chapter we looked carefully at fitting a straight line to data in a scatterplot when there appears to be a straight-line trend, and then we used this line to predict the response from the explanatory variable. In doing this, we have used data to draw conclusions. We assume that the straight line that we fit to our data describes the actual relationship between the response and the explanatory variable, and thus that conclusions (predictions) about additional values of the response based on other values of the explanatory variable are valid.

Are these conclusions (predictions) justified? The squared correlation provides information about the likelihood of a successful prediction. Small values of the squared correlation suggest that our predictions are not likely to be accurate. Extrapolation is another setting in which our predictions are not likely to be accurate.

Finally, when there is a strong relationship between two variables, it is tempting to draw an additional conclusion: namely, that changes in one variable cause changes in another. However, the case for causation requires more than a strong relationship. Unless our data are produced by a proper experiment, the case for causation is difficult to prove.

CASE STUDY   What should we conclude about the Super Bowl Indicator described in
EVALUATED   the Case Study at the beginning of this chapter? To evaluate the Super
Bowl Indicator, answer the following questions.

1. We wrote this Case Study on February 6, 2012, one day after the Giants won the Super Bowl. The Super Bowl Indicator predicts stocks should go up in 2012. Did they go up?

2. Stocks went down only 11 times in the 45 years between 1967 and 2011. If you simply predicted "up" every year, how would you have performed?

3. There are 19 original NFL and NFC teams and only 13 AFC teams. How often would you expect "NFL wins" to occur if one assumes that the chance of winning is proportional to the number of teams? How does this compare with simply predicting "up" every year?

4. Write a paragraph, in language that someone who knows no statistics would understand, explaining why the association between the Super Bowl Indicator and stock prices is not surprising and why it would be incorrect to conclude that the Super Bowl outcome causes changes in stock prices. ■

CHAPTER 15 EXERCISES

For Exercise 15.1, see page 340; for Exercise 15.2, see page 344; for Exercise 15.3, see page 348.

15.4 Obesity in mothers and daughters. The study in Example 7 found that the correlation between the body mass index of young girls and their hours of physical activity in a day was $r = -0.18$. Why might we expect this correlation to be negative? What percentage of the variation in BMI among the girls in the study can be explained by the straight-line relationship with hours of activity?

15.5 State SAT scores. Figure 14.9 (page 324) plots the average SAT Mathematics score of each state's high school seniors against the proportion of each state's seniors who took the exam. In addition to two clusters, the plot shows an overall roughly straight-line pattern. The least-squares regression line for predicting average SAT Math score from proportion taking is

$$\text{average Math SAT score} = 580.0 - (109.7 \times \text{proportion taking})$$

(a) What does the slope $b = -109.7$ tell us about the relationship between these variables?

(b) In New York State, the proportion of high school seniors who took the SAT was 0.89. Predict their average score. (The actual average score in New York was 499.)

(c) On page 341 we mention that using least-squares regression to do prediction outside the range of available data is risky. For what range of data is it reasonable to use the least-squares regression line for predicting average SAT Math score from proportion taking?

15.6 IQ and school GPA. Figure 14.10 (page 325) plots school grade point average (GPA) against IQ test score for 78 seventh-grade students. There is a roughly straight-line pattern with quite a bit of scatter. The correlation between these variables is $r = 0.634$. What percentage of the observed variation among the GPAs of these 78 students is explained by the straight-line relationship between GPA and IQ score? What percentage of the variation is explained by differences in GPA among students with similar IQ scores?

15.7 State SAT scores. The correlation between the average SAT Mathematics score in the states and the proportion of high school seniors who take the SAT is $r = -0.843$.

(a) The correlation is negative. What does that tell us?

(b) How well does proportion taking predict average score? (Use r^2 in your answer.)

15.8 IQ and school GPA. The least-squares line for predicting school GPA from IQ score, based on the 78 students plotted in Figure 14.10 (page 325), is

$$\text{GPA} = -3.56 + (0.101 \times \text{IQ})$$

Explain in words the meaning of the slope $b = 0.101$. Then predict the GPA of a student whose IQ score is 115.

15.9 The professor swims. Here are Professor Moore's times (in minutes) to swim 2000 yards and his pulse rate (in beats per minute) after swimming for 23 sessions in the pool:

Time:	34.12	35.72	34.72	34.05	34.13	35.72	36.17	35.57
Pulse:	152	124	140	152	146	128	136	144
Time:	35.37	35.57	35.43	36.05	34.85	34.70	34.75	33.93
Pulse:	148	144	136	124	148	144	140	156
Time:	34.60	34.00	34.35	35.62	35.68	35.28	35.97	
Pulse:	136	148	148	132	124	132	139	

You made a scatterplot of these data in Exercise 14.13 (page 326). The least-squares regression line is

$$\text{pulse rate} = 479.9 - (9.695 \times \text{time})$$

The next day's time is 34.30 minutes. Predict the professor's pulse rate. In fact, his pulse rate was 152. How accurate is your prediction?

15.10 Wine and heart disease. Drinking moderate amounts of wine may help prevent heart attacks. Let's look at data for entire nations. Table 15.1 gives data on yearly wine consumption (liters of alcohol from drinking wine, per person) and yearly deaths from heart disease (deaths per 100,000 people) in 19 developed countries in 2001.

(a) Make a scatterplot that shows how national wine consumption helps explain heart disease death rates.

(b) Describe in words the direction, form, and strength of the relationship.

(c) The correlation for these variables is $r = -0.645$. Why does this value agree with your description in part (b)?

TABLE 15.1 Wine consumption and heart disease

Country	Alcohol from wine[a]	Heart disease death rate[b]	Country	Alcohol from wine	Heart disease death rate
Australia	3.25	80	Italy	7.50	60
Austria	4.75	100	Netherlands	2.75	70
Belgium	2.75	60	New Zealand	2.50	100
Canada	1.50	80	Norway	1.75	80
Denmark	4.50	90	Spain	5.00	50
Finland	3.00	120	Sweden	2.50	90
France	8.50	40	Switzerland	6.00	70
Germany	3.75	90	United Kingdom	2.75	120
Iceland	1.25	110	United States	1.25	120
Ireland	2.00	130			

[a]Liters of alcohol from drinking wine, per person.
[b]Deaths per 100,000 people, ischemic heart disease.

15.11 The 2000 and 2004 presidential elections. Republican George W. Bush was elected president in 2000 and 2004. Figure 15.6 plots the percentage who voted for Bush in 2000 and 2004 for each of the 50 states and the District of Columbia.

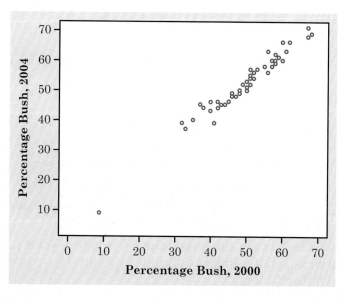

Figure 15.6 Scatterplot of the percentage who voted for Bush in 2000 and 2004 for each of the 50 states and the District of Columbia, for Exercise 15.11. (This figure was created using the Minitab software package.)

(a) Describe in words the direction, form, and strength of the relationship between the percentage of votes for Bush in 2000 and the percentage in 2004. Are there any unusual features in the plot?

(b) The least-squares regression line is

$$\text{percentage in 2004} = 3.82 + (0.98 \times \text{percentage in 2000})$$

Draw this line on a separate sheet of paper. (To draw the line, use the equation to predict y for $x = 10$ and for $x = 50$. Plot the two (x, y) points and draw the line through them.)

(c) The correlation between these variables is $r = 0.973$. What percentage of the observed variation in 2004 percentages can be explained by straight-line dependence on 2000 percentages?

15.12 Beavers and beetles. Ecologists sometimes find rather strange relationships in our environment. One study seems to show that beavers benefit beetles. The researchers laid out 23 circular plots, each 4 meters in diameter, in an area where beavers were cutting down cottonwood trees. In each plot, they counted the number of stumps from trees cut by beavers and the number of clusters of beetle larvae. Here are the data:

Stumps:	2	2	1	3	3	4	3	1	2	5	1	3
Larvae clusters:	10	30	12	24	36	40	43	11	27	56	18	40

Stumps:	2	1	2	2	1	1	4	1	2	1	4	
Larvae clusters:	25	8	21	14	16	6	54	9	13	14	50	

(a) Make a scatterplot that shows how the number of beaver-caused stumps influences the number of beetle larvae clusters. What does your plot show? (Ecologists think that the new sprouts from stumps are more tender than other cottonwood growth, so that beetles prefer them.)

(b) The least-squares regression line is

$$\text{larvae clusters} = -1.286 + (11.894 \times \text{stumps})$$

Draw this line on your plot. (To draw the line, use the equation to predict y for $x = 1$ and for $x = 5$. Plot the two (x, y) points and draw the line through them.)

(c) The correlation between these variables is $r = 0.916$. What percentage of the observed variation in beetle larvae counts can be explained by straight-line dependence on stump counts?

(d) Based on your work in (a), (b), and (c), do you think that counting stumps offers a quick and reliable way to predict beetle larvae clusters?

15.13 Wine and heart disease. Table 15.1 gives data on wine consumption and heart disease death rates in 19 countries in 2001. A scatterplot (Exercise 15.10) shows a moderately strong relationship. The least-squares regression line for predicting heart disease death rate from wine consumption, calculated from the data in Table 15.1, is

$$y = 115.86 - 8.05x$$

Use this equation to predict the heart disease death rate in a country where adults average 1 liter of alcohol from wine each year and in a country that averages 8 liters per year. Use these two results to draw the least-squares line on your scatterplot.

15.14 Strong association but no correlation. Exercise 14.19 gives these data on the speed (miles per hour) and mileage (miles per gallon) of a car:

Speed:	25	35	45	55	65
Mileage:	20	24	26	24	20

The least-squares line for predicting mileage from speed is

$$\text{mileage} = 22.8 + (0 \times \text{speed})$$

(a) Make a scatterplot of the data and draw this line on the plot.

(b) The correlation between mileage and speed is $r = 0$. What does this say about the usefulness of the regression line in predicting mileage? *not useful at all*

15.15 Wine and heart disease. In Exercises 15.10 and 15.13, you examined data on wine consumption and heart disease deaths from Table 15.1. Suggest some differences among nations that may be confounded with wine-drinking habits. (*Note:* What is more, data about nations may tell us little about individual people. So these data alone are not evidence that you can lower your risk of heart disease by drinking more wine.)

15.16 Correlation and regression. If the correlation between two variables x and y is $r = 0$, there is no straight-line relationship between the variables. It turns out that the correlation is 0 exactly when the slope of the least-squares regression line is 0. Explain why slope 0 means that there is no straight-line relationship between x and y. Start by drawing a line with slope 0 and explaining why in this situation x has no value for predicting y.

15.17 Acid rain. Researchers studying acid rain measured the acidity of precipitation in a Colorado wilderness area for 150 consecutive weeks. Acidity is measured by pH. Lower pH values show higher acidity. The acid rain researchers observed a straight-line pattern over time. They reported that the least-squares regression line

$$pH = 5.43 - (0.0053 \times weeks)$$

fit the data well.

(a) Draw a graph of this line. Is the association positive or negative? Explain in plain language what this association means.

(b) According to the regression line, what was the pH at the beginning of the study (weeks = 1)? At the end (weeks = 150)?

(c) What is the slope of the regression line? Explain clearly what this slope says about the change in the pH of the precipitation in this wilderness area.

(d) Is it reasonable to use this least-squares regression line to predict the pH of precipitation after 200 weeks? Explain your answer.

15.18 Review of straight lines. Fred keeps his savings in his mattress. He began with $1000 from his mother and adds $250 each year. His total savings y after x years are given by the equation

$$y = 1000 + 250x$$

(a) Draw a graph of this equation. (Choose two values of x, such as 0 and 10. Compute the corresponding values of y from the equation. Plot these two points on graph paper and draw the straight line joining them.)

(b) After 20 years, how much will Fred have in his mattress?

(c) If Fred had added $300 instead of $250 each year to his initial $1000, what is the equation that describes his savings after x years?

15.19 Review of straight lines. During the period after birth, a male white rat gains exactly 39 grams (g) per week. (This rat is unusually regular in his growth, but 39 g per week is a realistic rate.)

(a) If the rat weighed 110 g at birth, give an equation for his weight after x weeks. What is the slope of this line?

(b) Draw a graph of this line between birth and 10 weeks of age.

(c) Would you be willing to use this line to predict the rat's weight at age 2 years? Do the prediction and think about the reasonableness of the result. (There are 454 grams in a pound. An average cat weighs about 10 pounds.)

15.20 More on correlation and regression. In Exercises 15.6 and 15.8, the correlation and the slope of the least-squares line for IQ and school GPA are both positive. In Exercises 15.10 and 15.13, both the correlation and the slope for wine consumption and heart disease deaths are negative. Is it possible for these two quantities (the correlation and the slope) to have opposite signs? Explain your answer.

TABLE 15.2 Four data sets for exploring correlation and regression

Data Set A

x	10	8	13	9	11	14	6	4	12	7	5
y	8.04	6.95	7.58	8.81	8.33	9.96	7.24	4.26	10.84	4.82	5.68

Data Set B

x	10	8	13	9	11	14	6	4	12	7	5
y	9.14	8.14	8.74	8.77	9.26	8.10	6.13	3.10	9.13	7.26	4.74

Data Set C

x	10	8	13	9	11	14	6	4	12	7	5
y	7.46	6.77	12.74	7.11	7.81	8.84	6.08	5.39	8.15	6.42	5.73

Data Set D

x	8	8	8	8	8	8	8	8	8	8	19
y	6.58	5.76	7.71	8.84	8.47	7.04	5.25	5.56	7.91	6.89	12.50

Source: Frank J. Anscombe, "Graphs in statistical analysis," *The American Statistician, 27* (1973), pp. 17–21.

15.21 Always plot your data! Table 15.2 presents four sets of data prepared by the statistician Frank Anscombe to illustrate the dangers of calculating without first plotting the data. *All four sets have the same correlation and the same least-squares regression line* to several decimal places. The regression equation is

$$y = 3 + 0.5x$$

(a) Make a scatterplot for each of the four data sets and draw the regression line on each of the plots. (To draw the regression line, substitute $x = 5$ and $x = 10$ into the equation. Find the predicted y for each x. Plot these two points and draw the line through them on all four plots.)

(b) In which of the four cases would you be willing to use the regression line to predict y given that $x = 10$? Explain your answer in each case.

15.22 Going to class helps. A study of class attendance and grades among first-year students at a state university showed that in general students who attended a higher percentage of their classes earned higher grades. Class attendance explained 25% of the variation in grade index among the students. What is the numerical value of the correlation between percentage of classes attended and grade index?

15.23 The declining farm population. The number of people living on American farms declined steadily during the past century. Here are data on the farm population (millions of persons) from 1935 to 2000:

Year:	1935	1940	1945	1950	1955	1960	1965	1970	1975	1980	1990	2000
Population:	32.1	30.5	24.4	23.0	19.1	15.6	12.4	9.7	8.9	7.2	3.9	3.0

(a) Make a scatterplot of these data. Draw by eye a regression line for predicting a year's farm population.

(b) Extend your line to predict the number of people living on farms in 2020. Is this result reasonable? Why?

15.24 Lots of wine. Exercise 15.13 gives us the least-squares line for predicting heart disease deaths per 100,000 people from liters of alcohol from wine consumed, per person. The line is based on data from 19 rich countries. The equation is $y = 115.86 - 8.05x$. What is the predicted heart disease death rate for a country where wine consumption is 150 liters of alcohol per person? Explain why this result can't be true. Explain why using the regression line for this prediction is not intelligent.

15.25 Do emergency personnel make injuries worse? Someone says, "There is a strong positive correlation between the number of emergency personnel at the scene of an accident and the extent of injuries of those in the accident. So sending lots of emergency personnel just causes more severe injuries." Explain why this reasoning is wrong.

15.26 Facebook and grades. A September 2010 article on **msnbc.com** reported on a study that found that college students who are on Facebook while studying or doing homework wind up getting lower grades. Perhaps limiting time on Facebook will improve grades. Can you think of explanations for the association between time on Facebook and grades other than "time on Facebook causes a drop in grades"?

15.27 Freeway exhaust and atherosclerosis. A February 2010 news story on **cnet.com** reported that the artery walls of people living close to a freeway thicken faster than the walls of those who don't. Researchers correlated changes in artery wall thickness of subjects with estimates of outdoor particulate levels at each subject's home. Does this mean that you can reduce atherosclerosis (the thickening and calcification of arteries) by avoiding living near a freeway? Why?

15.28 Health and wealth. An article entitled "The Health and Wealth of Nations" says:

The positive correlation between health and income per capita is one of the best-known relations in international development. This correlation is commonly thought to reflect a causal link running from income to health. ... Recently, however, another intriguing possibility has emerged: that the health-income correlation is partly explained by a causal link running the other way—from health to income.

Explain how higher income in a nation can cause better health. Then explain how better health can cause higher income. There is no simple way to determine the direction of the link.

15.29 Is math the key to success in college? Here is the opening of a newspaper account of a College Board study of 15,941 high school graduates:

Minority students who take high school algebra and geometry succeed\in college at almost the same rate as whites, a new study says.

The link between high school math and college graduation is "almost magical," says College Board President Donald Stewart, suggesting "math is the gatekeeper for success in college."

"These findings," he says, "justify serious consideration of a national policy to ensure that all students take algebra and geometry."

What lurking variables might explain the association between taking several math courses in high school and success in college? Explain why requiring algebra and geometry may have little effect on who succeeds in college.

15.30 Does low-calorie salad dressing cause weight gain? People who use low-calorie salad dressing in place of regular dressing tend to be heavier than people who use regular dressing. Does this mean that low-calorie salad dressings cause weight gain? Give a more plausible explanation for this association.

15.31 Internet use and school grades. Children who spend many hours on the Internet get lower grades in school, on average, than those who spend less time on the Internet. Suggest some lurking variables that may explain this relationship because they contribute to both heavy Internet use and poor grades.

15.32 Correlation again. The correlation between IQ score and school GPA (Exercise 15.6) is $r = 0.634$. The correlation between wine consumption and heart disease deaths (Exercise 15.10) is $r = -0.645$. Which of these two correlations indicates a stronger straight-line relationship? Explain your answer.

15.33 Magic Mozart. In 1998, the Kalamazoo (Michigan) Symphony advertised a "Mozart for Minors" program with this statement: "Question: Which students scored 51 points higher in verbal skills and 39 points higher in math? Answer: Students who had experience in music." What do you think of the claim that "experience in music" causes higher test scores?

15.34 Living on campus. A February 2, 2008, article in the *Columbus Dispatch* reported a study on the distances students lived from campus and average GPA. Here is a summary of the results:

Residence	Avg. GPA
Residence hall	3.33
Walking distance	3.16
Near campus, long walk or short drive	3.12
Within the county, not near campus	2.97
Outside the county	2.94

Based on these data, the association between the distance a student lives from campus and GPA is negative. Many universities require freshmen to live on campus, but these data have prompted some to suggest that sophomores should also be required to live on campus in order to improve grades. Do these data imply that living closer to campus improves grades? Why?

15.35 Calculating the least-squares line. Like to know the details when you study something? Here is the formula for the least-squares regression line for predicting y from x. Start with the means $\bar{x}$ and $\bar{y}$ and the standard deviations s_x and s_y of the two variables and the correlation r between them. The least-squares line has equation $y = a + bx$ with

$$\text{slope:} \quad b = r\frac{s_y}{s_x} \qquad\qquad \text{intercept:} \quad a = \bar{y} - b\bar{x}$$

Example 4 in Chapter 14 (page 318) gives the means, standard deviations, and correlation for the fossil bone length data. Use these values in the formulas just given to verify the equation of the least-squares line given on page 339:

$$\text{humerus length} = -3.66 + (1.197 \times \text{femur length})$$

The remaining exercises require a two-variable statistics calculator or software that will calculate the least-squares regression line from data.

15.36 The professor swims. Return to the swimming data in Exercise 15.9.

(a) Verify the equation given for the least-squares line in that exercise.

(b) Suppose you were told only that the pulse rate was 152. You now want to predict swimming time. Find the equation of the least-squares regression line that is appropriate for this purpose. What is your prediction?

(c) The two lines in (a) and (b) are different. Explain clearly why there are two different regression lines.

15.37 Is wine good for your heart? Table 15.1 gives data on wine consumption and heart disease death rates in 19 countries. Verify the equation of the least-squares line given in Exercise 15.13.

15.38 Always plot your data! A skeptic might wonder if the four very different data sets in Table 15.2 really do have the same correlation and least-squares line. Verify that (to a close approximation) the least-squares line is $y = 3 + 0.5x$, as given in Exercise 15.21.

EXPLORING THE WEB

15.39 Web-based exercise. Find an example of a study in which the issue of association and causation is present. This can be either an example in which association is confused with causation or an example in which the association is not confused with causation. Summarize the study and its conclusions in your own words. The CHANCE Web site at **www .causeweb.org/wiki/chance/index.php/Main_Page** is a good place to look for examples.

15.40 Web-based exercise. The best way to develop some feeling for how a regression line fits the points on a scatterplot is to use an applet that allows you to plot data points and watch the least-squares line move as the points change. Go to the *Statistics: Concepts and Controversies* Web site, **www.whfreeman.com/scc8e,** and look at the *Correlation and Regression* applet. Click in the "Show least-squares line" box to see the regression line. Where should one place points in order to have the largest effect on the least-squares line?

15.41 Web-based exercise. Go to the Congressional Budget Office Web site, **www.cbo.gov.** What is the current prediction for the federal budget in five years' time? Is a surplus or a deficit predicted?

NOTES AND DATA SOURCES

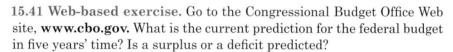

Page 341 Example 4: The Congressional Budget Office annual reports can be found at the Congressional Budget Office Web site, **www.cbo.gov.**

Page 345 Example 7: Laura L. Calderon et al., "Risk factors for obesity in Mexican-American girls: dietary factors, anthropometric factors, physical activity, and hours of television viewing," *Journal of the American Dietetic Association,* 96 (1996), pp. 1177–1179.

Page 348 Studies mentioned in the Statistical Controversies feature are John R. Lott, Jr., *More Guns, Less Crime: Understanding Crime and Gun Control Laws,* University of Chicago Press, 1998; Andrés Villaveces et al., "Effect of a ban on carrying firearms on homicide rates in 2 Colombian cities," *Jour-*

nal of the American Medical Association, 283 (2000), pp. 1205–1209; see also the editorial by Lawrence W. Sherman in the same issue. See also Lawrence W. Sherman, James W. Shaw, and Dennis P. Rogan, "The Kansas City gun experiment," National Institute of Justice, 1995.

Page 348 Information about cancer statistics can be obtained from the U.S. Cancer Statistics Working Group, **apps.nccd .cdc.gov/uscs/index.aspx.**

Page 352 Exercise 15.10: Data estimated from graphs found at the World Health Organization Web site, **www.who.int/ substance_abuse/publications/ globalstatusreportalcoholprofiles/en/ index.html.**

Page 354 Exercise 15.12: Data estimated from a graph in G. D. Martinsen, E. M. Driebe, and T. G. Whitham, "Indirect interactions mediated by changing plant chemistry: beaver browsing benefits beetles," *Ecology,* 79 (1998), pp. 192–200.

Page 355 Exercise 15.17: W. M. Lewis and M. C. Grant, "Acid precipitation in the western United States," *Science,* 207 (1980), pp. 176–177.

Page 358 Exercise 15.28: David E. Bloom and David Canning, "The health and wealth of nations," *Science,* 287 (2000), pp. 1207–1208.

Page 359 Exercise 15.29: Quotation from a Gannett News Service article appearing in the *Lafayette (Ind.) Journal and Courier,* April 23, 1994.

Page 359 Marigene Arnold of Kalamazoo College contributed the orchestra ad in Exercise 15.33.

The Consumer Price Index and Government Statistics

CASE STUDY The three career home run leaders are Barry Bonds (762 career home runs at the end of the 2007 season), Hank Aaron (755 career home runs), and Babe Ruth (714 career home runs). In Chapter 12 we compared the careers of these sluggers using the five-number summary and boxplots. We pay for what we value, so another way to compare these players is by their salaries. Bonds's highest salary was $22,000,000 in 2005. Aaron's highest salary was $250,000 in 1976. And Ruth's highest salary was $80,000 in 1931.

John Medina/WireImage/Getty Images

Bonds's highest salary is by far the largest. Does this mean he is clearly the best of the three? We know that a dollar in 1931 bought a lot more than a dollar in 1976, and both bought more than a dollar in 2005. Maybe in terms of buying power, Aaron's or Ruth's salary is highest.

In this chapter we discuss methods for comparing the buying power of the dollar across different years. By the end of this chapter you will be able to determine whether Bonds, Aaron, or Ruth had the highest salary in terms of buying power. ∎

We all notice the high salaries paid to professional athletes. In Major League Baseball, for example, the mean salary rose from $329,408 in 1984 to $3,095,183 in 2011. That's a big jump. Not as big as it first appears, however. *A dollar in 2011 did not buy as much as a dollar in 1984, so 1984 salaries cannot be directly compared with 2011 salaries.* The hard fact that the dollar has steadily lost buying power over time means that we must make an adjustment whenever we compare dollar values from different years. The adjustment is easy. What is not easy is measuring the changing buying

"Now this here's a genuine 1980 dollar. They don't make 'em like that anymore."

power of the dollar. The government's Consumer Price Index (CPI) is the tool we need.

Index numbers

The CPI is a new kind of numerical description, an *index number*. We can attach an index number to any quantitative variable that we measure repeatedly over time. The idea of the index number is to give a picture of changes in a variable much like that drawn by saying, "The average cost of hospital charges rose 90% between 1997 and 2005." That is, an index number describes the percentage change from a *base period*.

Index number

An **index number** measures the value of a variable relative to its value at a **base period.** To find the index number for any value of the variable:

$$\text{index number} = \frac{\text{value}}{\text{base value}} \times 100$$

EXAMPLE 1 Calculating an index number

A gallon of unleaded regular gasoline cost $0.992 during the first week of January 1994 and $3.213 during the last week of December 2011. (These are national average prices calculated by the U.S. Department of Energy.) The gasoline price index number for the last week in December 2011, with the first week in January 1994 as the base period, is

$$\text{index number} = \frac{\text{value}}{\text{base value}} \times 100$$

$$= \frac{3.213}{0.992} \times 100 = 323.9$$

The gasoline price index number for the base period, January 1994, is

$$\text{index number} = \frac{0.992}{0.992} \times 100 = 100$$

Knowing the base period is essential to making sense of an index number. Because the index number for the base period is always 100, it is usual to identify the base period as 1994 by writing "1994 = 100." In news reports concerning the CPI, you will notice the mysterious equation "1982–84 = 100." That's shorthand for the fact that the years 1982 to 1984 are the base period for the CPI. An index number just gives the current value as a percentage of the base value. Index number 323.9 means 323.9% of the base value, or a 223.9% increase from the base value. Index number 57 means that the current value is 57% of the base, a 43% decrease.

Fixed market basket price indexes

It may seem that index numbers are little more than a plot to disguise simple statements in complex language. Why say, "The Consumer Price Index (1982–84 = 100) stood at 225.7 in December 2011," instead of "Consumer prices rose 125.7% between the 1982–84 average and December 2011"? In fact, the term "index number" usually means more than a measure of change relative to a base. It also tells us the kind of variable whose change we measure. That variable is a weighted average of several quantities, with fixed weights. Let's illustrate the idea by a simple price index.

EXAMPLE 2 The Mountain Man Price Index

Bill Smith lives in a cabin in the mountains and strives for self-sufficiency. He buys only salt, kerosene, and the services of a professional welder. Here are Bill's purchases in 1990, the base period. His cost, in the last column, is the price per unit multiplied by the number of units he purchased.

Good or service	1990 quantity	1990 price	1990 cost
Salt	100 pounds	$0.50/pound	$50.00
Kerosene	50 gallons	0.80/gallon	40.00
Welding	10 hours	13.00/hour	130.00
		Total cost =	$220.00

The total cost of Bill's collection of goods and services in 1990 was $220. To find the "Mountain Man Price Index" for 2011, we use 2011 prices to calculate the 2011 cost of this *same* collection of goods and services. Here is the calculation:

Good or service	1990 quantity	2011 price	2011 cost
Salt	100 pounds	$0.60/pound	$60.00
Kerosene	50 gallons	1.40/gallon	70.00
Welding	10 hours	22.00/hour	220.00
		Total cost =	$350.00

The same goods and services that cost $220 in 1990 cost $350 in 2011. So the Mountain Man Price Index (1990 = 100) for 2011 is

$$\text{index number} = \frac{350}{220} \times 100 = 159.1$$

The point of Example 2 is that we follow the cost of the *same* collection of goods and services over time. It may be that Bill refused to hire the welder in 2011 because he could not afford him. No matter—the index number uses the 1990 quantities, ignoring any changes in Bill's purchases between 1990 and 2011. We call the collection of goods and services whose total cost we follow a *market basket*. The index number is then a *fixed market basket price index*.

Fixed market basket price index

A **fixed market basket price index** is an index number for the total cost of a fixed collection of goods and services.

The basic idea of a fixed market basket price index is that the weight given to each component (salt, kerosene, welding) remains fixed over time. The CPI is in essence a fixed market basket price index, with several hundred items that represent all consumer purchases. Holding the market basket fixed allows a legitimate comparison of prices because we compare the prices of exactly the same items at each time. As we will see, it also poses severe problems for the CPI.

Using the CPI

For now, think of the CPI as an index number for the cost of everything that American consumers buy. That the CPI for December 2011 was 225.7 means that we must spend $225.7 in December 2011 to buy goods and services that cost $100 in the 1982 to 1984 base period. An index number for "the cost of everything" lets us compare dollar amounts from different years by converting all the amounts into dollars of the same year. You will find tables in the *Statistical Abstract,* for example, with headings such as "Median Household Income, in Constant (2000) Dollars." That table has restated all incomes in dollars that will buy as much as the dollar would buy in 2000. Watch for the term *constant dollars* and for phrases like *real income* or *real terms.* They mean that all dollar amounts represent the same buying power even though they may describe different years.

Table 16.1 gives annual average CPIs from 1915 to 2011. Figure 16.1 is a line graph of the *annual percentage increase* in CPI values. It shows that the periods from 1915 to 1920, the 1940s, and from 1975 to the present experienced high inflation. Although there is considerable variation, the annual percentage increase is positive in most years. In general, the 20th century was a time of inflation. Faced with this depressing fact, it would be foolish to think about dollars without adjusting for their decline in buying power. Following is the recipe for converting dollars of one year into dollars of another year.

TABLE 16.1 Annual average Consumer Price Index, 1982–84 = 100

Year	CPI	Year	CPI	Year	CPI
1915	10.1	1980	82.4	1997	160.5
1920	20.0	1981	90.9	1998	163.0
1925	17.5	1982	96.5	1999	166.6
1930	16.7	1983	99.6	2000	172.2
1935	13.7	1984	103.9	2001	177.1
1940	14.0	1985	107.6	2002	179.9
1945	18.0	1986	109.6	2003	184.0
1950	24.1	1987	113.6	2004	188.9
1955	26.8	1988	118.3	2005	195.3
1960	29.6	1989	124.0	2006	201.6
1965	31.5	1990	130.7	2007	207.3
1970	38.8	1991	136.2	2008	215.3
1975	53.8	1992	140.3	2009	214.5
1976	56.9	1993	144.5	2010	218.1
1977	60.6	1994	148.2	2011	224.9
1978	65.2	1995	152.4		
1979	72.6	1996	156.9		

Source: Bureau of Labor Statistics.

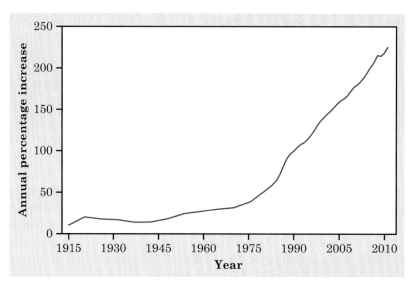

Figure 16.1 The annual percentage increase in the Consumer Price Index (1982–84 = 100) from 1915 to 2011. Percentage increases are generally positive in the 20th century. (This figure was created using the Minitab software package.)

Adjusting for changes in buying power

To convert an amount in dollars at time A to the amount with the same buying power at time B:

$$\text{dollars at time B} = \text{dollars at time A} \times \frac{\text{CPI at time B}}{\text{CPI at time A}}$$

Notice that the CPI for the time you are *going to* appears on the top in the ratio of CPIs in this recipe. Here are some examples.

EXAMPLE 3 Salaries of professional athletes

The mean salary of Major League Baseball players rose from $329,408 in 1984 to $3,095,183 in 2011. How big was the increase in real terms? Let's convert the 1984 average into 2011 dollars. Table 16.1 gives the annual average CPIs that we need.

$$
\begin{aligned}
\text{2011 dollars} &= \text{1984 dollars} \times \frac{\text{2011 CPI}}{\text{1984 CPI}} \\
&= \$329,408 \times \frac{224.9}{103.9} \\
&= \$713,030
\end{aligned}
$$

That is, it took $713,030 in 2011 to buy what $329,408 would buy in 1984. We can now compare the 1984 mean salary of $329,408 *in 2011 dollars* with the actual 2011 mean salary, $3,095,183. Today's athletes earn much more than 1984 athletes even after adjusting for the fact that the dollar buys less now. (Of course, the mean salary is pulled up by the very high salaries paid to a few star players. The 1984 median salary was $229,750, and the 2011 median salary was $1,100,000.)

> **NOW IT'S YOUR TURN** | **16.1 Baseball salaries.** Refer to Example 3. Convert the 1984 median baseball salary into 2011 dollars.

EXAMPLE 4 Rising incomes?

For a more serious example, let's leave the pampered world of professional athletes and look at the incomes of ordinary people. The median annual income of all American households was $22,415 in 1984. By 2011 (the most recent year for which data were available at the time of this writing) the median income had risen to $49,445. Dollar income more than doubled, but we know that much of that rise is an illusion because of the dollar's declining buying power. To compare these incomes, we must express them in dollars of the same year. Let's express the 1984 median household income in 2011 dollars:

$$\text{2011 dollars} = \$22,415 \times \frac{224.9}{103.9} = \$48,519$$

Real household incomes rose only from $48,519 to $49,445 in the 27 years between 1984 and 2011. That's a 1.9% increase.

The picture is different at the top. The 5% of households with the highest incomes earned $68,500 or more in 1984. In 2011 dollars, this is

$$\text{2011 dollars} = \$68,500 \times \frac{224.9}{103.9} = \$148,274$$

In fact, the top 5% of households earned $180,810 or more in 2011. That is, the real income of the highest earners increased by 21.9%.

Finally, let's look at the bottom. The 20% of households with the lowest incomes earned $9,500 or less in 1984. In 2011 dollars, this is

$$\text{2011 dollars} = \$9,500 \times \frac{224.9}{103.9} = \$20,564$$

In fact, the bottom 20% of households earned $20,000 or less in 2011. That is, the real income of the lowest earners decreased by 2.7%.

NOW IT'S YOUR TURN

16.2 Production workers. Let's look at production workers, the traditional "working men" (and women). Their median annual earnings were $17,281 in 1984 and $28,701 in 2010. Compute the 1984 earnings in 2010 dollars. By what percentage did the real earnings of production workers change between 1984 and 2010?

So you think that's inflation? Americans were unhappy when oil price increases in 1973 set off a round of inflation that saw the CPI almost double in the following decade. That's nothing. In Argentina, prices rose 127% in a single month, July 1989. The Turkish lira went from 14 to the dollar in 1970 to 579,000 to the dollar in 2000. There were 65 German marks to a dollar in January 1920, and 4,200,000,000,000 marks to a dollar in November 1923. Now *that's* inflation.

Example 4 illustrates how using the CPI to compare dollar amounts from different years brings out truths that are otherwise hidden. In this case, the truth is that the fruits of prosperity since the 1980s went mostly to those at the top of the income distribution and that very little real progress was made by those at the bottom. Put another way, people with the highest pay (usually those with skills and education) did much better than people with the lowest pay (usually those who lack special skills and college educations). Economists suggest several reasons: the "new economy" that rewards knowledge, high immigration leading to competition for less-skilled jobs, more competition from abroad, and so on. Exactly why incomes at the top have increased so much, and what we should do about the stagnant incomes of those at the bottom, are controversial questions.

Understanding the CPI

The idea of the CPI is that it is an index number for the cost of everything American consumers buy. That idea needs lots of adjusting to be practical. Much of the fiddling uses the results of large sample surveys.

Who is covered? The official name for the common version of the CPI (there are others, but we will ignore them) is the Consumer Price Index for All Urban Consumers. The CPI market basket represents the purchases of people living in urban areas. The official definition of "urban" is broad: about 80% of the U.S. population is covered. But if you live on a farm, the CPI doesn't apply to you.

How is the market basket chosen? Different households buy different things, so how can we get a single market basket? From a sample survey. The Consumer Expenditure Survey gathers detailed data on the spending of more than 30,000 households. The Bureau of Labor Statistics (BLS) breaks spending into categories such as "fresh fruits and vegetables," "new and used motor vehicles," and "hospital and related services." Then it chooses specific items, such as "fresh oranges," to represent each group in the market basket. The items in the market basket get weights that represent their category's proportion of all spending. The weights, and even the specific market basket items, are updated regularly to keep up with changing buying habits. So the market basket isn't actually fixed.

How are the prices determined? From more sample surveys. The BLS must discover the price of "fresh oranges" every month. That price differs from city to city and from store to store in the same city. Each month, the BLS records 80,000 prices in 87 cities at a sample of stores. The Point of Purchase Survey of 16,800 households keeps the BLS up-to-date on where consumers shop for each category of goods and services (supermarkets, convenience stores, discount stores, and so on).

Does the CPI measure changes in the cost of living? A fixed market basket price index measures the cost of *living the same* over time, as Example 2 illustrated. In fact, we don't keep buying the same market basket of goods and services over time. We switch from LP records to tapes and CDs and then to music downloads. We don't buy new 1991 cars in 2001 or 2011. As prices change, we change what we buy—if beef becomes expensive, we buy less beef and more chicken or more tofu. A fixed market basket price index can't accurately measure changes in the cost of living when the economy itself is changing.

The BLS tries hard to keep its market basket up-to-date and to compensate for changes in quality. Every year, for example, the BLS must decide how much of the increase in new-car prices is paying for better quality. Only what's left counts as a genuine price increase in calculating the CPI. Between December 1967 and December 1994, actual car prices went up 313.4%, but the new-car price in the CPI went up only 172.1%. In 1995, adjustments for better quality reduced the overall rise in the prices of goods and services from 4.7% to only 2.2%. Prices of goods and services make up about 70% of the CPI. Most of the rest is the cost of shelter—renting an apartment or buying a house. House prices are another problem for the BLS. People buy houses partly to live in and partly because they think owning a house is a good investment. If we pay more for a house because we think it's a good investment, the full price should not go into the CPI.

By now it is clear that the CPI is *not* a fixed market basket price index, though that is the best way to start thinking about it. The BLS must constantly change the market basket as new products appear and our buying habits change. It must adjust the prices its sample surveys record to take account of better quality and the investment component of house prices. Yet the CPI still does not measure all changes in our cost of living. It leaves out taxes, for example, which are certainly part of our cost of living.

Even if we agree that the CPI should look only at the goods and services we buy, it doesn't perfectly measure changes in our cost of living. In principle, a true "cost-of-living index" would measure the cost of the *same*

STATISTICAL CONTROVERSIES

Does the CPI Overstate Inflation?

Alex Wong/Getty Images

In 1995, Federal Reserve Chairman Alan Greenspan estimated that the CPI overstates inflation by somewhere between 0.5% and 1.5% per year. Mr. Greenspan was unhappy about this, because increases in the CPI automatically drive up federal spending. At the end of 1996, a group of outside experts appointed by the Senate Finance Committee estimated that the CPI had in the past overstated the rate of inflation by about 1.1% per year. The Bureau of Labor Statistics (BLS) agreed that the CPI overstates inflation but thought that the experts' guess of 1.1% per year was too high.

The reasons the CPI shows the value of a dollar falling faster than is true are due partly to the nature of the CPI and partly to limits on how quickly the BLS can adjust the details of the enormous machine that lies behind the CPI. Think first about the details. The prices of new products, such as digital cameras and flat-screen televisions, often start high and drop rapidly. The CPI market basket changes too slowly to capture the drop in price. Discount stores with lower prices also enter the CPI sample slowly. Although the BLS tries hard to adjust for better product quality, the outside experts thought these adjustments were often too little and too late. The BLS has made many improvements in these details. The improved CPI would have grown about 0.5% per year more slowly than the actual CPI between 1978 and 1998.

The wider issue is the nature of the CPI as essentially a fixed market basket index. What sort of bias does such an index create? Does it produce an upward bias; that is, does it overstate the cost of living? Or does it create a downward bias; that is, does it understate the cost of living? Why?

standard of living over time. That's why we start with a fixed market basket price index, which also measures the cost of living the same over time but takes the simple view that "the same" means buying exactly the same things. If we are just as satisfied after switching from beef to tofu to avoid paying more for beef, our standard of living hasn't changed and a cost-of-living index should ignore the higher price of beef. If we are willing to pay more for products that keep our environment clean, we are paying for a higher standard of living and the index should treat this just like an improvement in the quality of a new car. The BLS says that it would like the CPI to track changes in the cost of living, but that a true cost-of-living index isn't possible in the real world.

The place of government statistics

Modern nations run on statistics. Economic data, in particular, guide government policy and inform the decisions of private business and individuals. Price indexes and unemployment rates, along with many other, less publicized series of data, are produced by government statistical offices.

Some countries have a single statistical office, such as Statistics Canada (**www.stacan.gc.ca**). Others attach smaller offices to various branches of government. The United States is an extreme case: there are 72 federal statistical offices, with relatively weak coordination among them. The Census Bureau and the Bureau of Labor Statistics are the largest, but you may at times use the products of the Bureau of Economic Analysis, the National Center for Health Statistics, the Bureau of Justice Statistics, or others in the federal government's collection of statistical agencies.

A 1993 ranking of government statistical agencies by the heads of these agencies in several nations put Canada at the top, with the United States tied with Britain and Germany for sixth place. The top spots generally went to countries with a single, independent statistical office. In 1996, Britain combined its main statistical agencies to form a new Office for National Statistics (**www.statistics.gov.uk**). U.S. government statistical agencies remain fragmented.

What do citizens need from their government statistical agencies? First of all, they need data that are *accurate* and *timely* and that *keep up with changes in society and the economy*. Producing accurate data quickly demands considerable resources. Think of the large-scale sample surveys that produce the unemployment rate and the CPI. The major U.S. statistical offices have a good reputation for accuracy and lead the world in getting data to the public quickly. Their record for keeping up with changes is less impressive. The struggle to adjust the CPI for changing buying habits and changing quality is one issue. Another is the failure of U.S. economic statistics to keep up with trends such as the shift from manufacturing to services as the major type of economic activity. Business organizations have expressed strong dissatisfaction with the overall state of our economic data.

Much of the difficulty stems from lack of money. In the years after 1980, reducing federal spending was a political priority. Government statistical agencies lost staff and cut programs. Lower salaries made it hard to attract the best economists and statisticians to government. The level of government spending on data also depends on our view of what data the government should produce. In particular, should the government produce data that are used mainly by private business rather than by the government's own policymakers? Perhaps such data should be either compiled by private concerns or produced only for those who are willing to pay. This is a question of political philosophy rather than statistics, but it helps determine what level of government statistics we want to pay for.

Freedom from political influence is as important to government statistics as accuracy and timeliness. When a statistical office is part of a government ministry, it can be influenced by the needs and desires of that ministry. The Census Bureau is in the Department of Commerce, which serves business interests. The BLS is in the Department of Labor. Thus, business and labor each have "their own" statistical office. The professionals in the statistical offices successfully resist direct political interference—a poor unemployment report is never delayed until after an election, for example. But indirect influence is clearly present. The BLS must compete with other Department of Labor activities for its budget, for example. Political interference with statistical work seems to be increasing, as when Congress refused to allow the Census Bureau to use sample surveys to correct for undercounting in the 2000 census. Such corrections are convincing to statisticians, but not necessarily to the general public. Congress, not without justification, considers the public legitimacy of the census to be as important as its technical perfection. Thus, from Congress's point of view, political interference was justified even though the decision was disappointing to statisticians.

"Yes sir, I know that we have to know where the economy is going. But do we have to publish the statistics so that everyone else does too?"

The 1996 reorganization of Britain's statistical offices was prompted in part by a widespread feeling that political influence was too strong. The details of how unemployment is measured in Britain were changed many times in the 1980s, for example, and almost all the changes had the effect of reducing the reported unemployment rate—just what the government wanted to see.

We favor a single "Statistics USA" office not attached to any other government ministry, as in Canada. Such unification might also help the money problem by eliminating duplication. It would at least allow a central decision about which programs deserve a larger share of limited resources. Unification is unlikely, but stronger coordination of the many federal statistical offices could achieve many of the same ends.

The question of social statistics

National economic statistics are well established with the government, the media, and the public. The government also produces many data on social issues such as education, health, housing, and crime. Social statistics are less complete than economic statistics. We have good data about how much money is spent on food but less information about how many people are poorly nourished. Social data are also less carefully produced than economic data. Economic statistics are generally based on larger samples, are compiled more often, and are published with a shorter time lag. The reason is clear: economic data are used by the government to guide economic policy month by month. Social data help us understand our society and address its problems but are not needed for short-term management.

There are other reasons the government is reluctant to produce social data. Many people don't want the government to ask about their sexual behavior or religion. Many people feel that the government should avoid asking about our opinions—apparently it's OK to ask, "When did you last visit a doctor?" but not "How satisfied are you with the quality of your health care?" These hesitations reflect the American suspicion of government intrusion. Yet issues such as sexual behaviors that contribute to the spread of HIV and satisfaction with health care are clearly important to citizens. Both facts and opinions on these issues can sway elections and influence policy. How can we get accurate information about social issues, collected consistently over time, and yet not entangle the government with sex, religion, and other touchy subjects?

The solution in the United States has been government funding of university sample surveys. After first deciding to undertake a sample survey asking people about their sexual behavior, in part to guide AIDS policy, the government backed away. Instead, it funded a much smaller survey of 3452 adults by the University of Chicago's National Opinion Research Center (NORC). NORC's General Social Survey (GSS), funded by the government's National Science Foundation, belongs with the Current Population Survey and the samples that undergird the CPI on any list of the most important sample surveys in the United States. The GSS includes both "fact" and "opinion" items. Respondents answer questions about their job security, their job satisfaction, and their satisfaction with their city, their friends, and their families. They talk about race, religion, and sex. Many Americans would object if the government asked whether they had seen an X-rated movie in the past year, but they reply when the GSS asks this question. The Web site for the GSS is **www.norc.org/GSS+Website/**.

This indirect system of government funding of a university-based sample survey fits the American feeling that the government itself should not be unduly invasive. It also insulates the survey from most political pressure. Alas, the government's budget cutting extends to the GSS, which now describes itself as an "almost annual" survey because lack of funds has prevented taking samples in some years. The GSS is, we think, a bargain.

STATISTICS IN SUMMARY

Chapter Specifics

- An **index number** describes the value of a variable relative to its value at some **base period.**

- A **fixed market basket price index** is an index number that describes the total cost of a collection of goods and services.

- Think of the government's **Consumer Price Index** (CPI) as a fixed market basket price index for the collection of all the goods and services that consumers buy.

- Because the CPI shows how consumer prices change over time, we can use it to change a dollar amount at one time into the amount at another time that has the same buying power. This is needed to compare dollar values from different times in **real terms.**

- The details of the CPI are complex. It uses data from several large sample surveys. It is not a true fixed market basket price index because of adjustments for changing buying habits, new products, and improved quality.

- **Government statistical offices** produce data needed for government policy and decisions by businesses and individuals. The data should be accurate, timely, and free from political interference. Citizens therefore have a stake in the competence and independence of government statistical offices.

 In Chapters 10 to 15 we studied methods for summarizing large amounts of data to help us see what the data are telling us. In this chapter we discussed numbers that summarize large amounts of data on consumers to help us see what these data are telling us about the costs of goods and services. Because these index numbers, in particular the CPI, are used by the government and media to describe the cost of living, understanding how they are computed and what they represent will help us be better-informed citizens.

CASE STUDY The average CPI for 1931 was 15.2. The average CPI for 1976 and 2005 **EVALUATED** can be found in Table 16.1. Use what you have learned in this chapter to convert Babe Ruth's 1931 salary and Hank Aaron's 1976 salary, given in the Case Study at the beginning of this chapter, to 2005 dollars. Who had the largest salary in terms of 2005 dollars: Bonds, Aaron, or Ruth? ■

CHAPTER 16 EXERCISES

For Exercise 16.1, see page 369. For Exercise 16.2, see page 370.

When you need the CPI for a year that does not appear in Table 16.1 (page 367), use the table entry for the year that most closely follows the year you want.

16.3 The price of gasoline. The yearly average price of unleaded regular gasoline has fluctuated as follows:

1991:	$1.10 per gallon
2001:	$1.42 per gallon
2011:	$3.52 per gallon

Give the gasoline price index numbers (2001 = 100) for 1991, 2001, and 2011.

16.4 The cost of college. The part of the CPI that measures the cost of college tuition (1982–84 = 100) was 691.8 in December 2011. The overall CPI was 225.7 that month.

(a) Explain exactly what the index number 691.8 tells us about the rise in college tuition between the base period and December 2011.

(b) College tuition has risen much faster than consumer prices in general. How do you know this?

16.5 The price of gasoline. Use your results from Exercise 16.3 to answer these questions.

(a) By how many points did the gasoline price index number change between 1991 and 2011? What percentage change was this?

(b) By how many points did the gasoline price index number change between 2001 and 2011? What percentage change was this?

You see that the point change and the percentage change in an index number are the same if we start in the base period, but not otherwise.

16.6 Toxic releases. The Environmental Protection Agency requires industry to report releases of any of a list of toxic chemicals. The total amounts released (in thousands of pounds) were 3,006,577 in 1988, 1,736,461 in 1998, and 1,265,879 in 2005. Give an index number for toxic chemical releases in each of these years, with 1988 as the base period. By what percentage did releases increase or decrease between 1988 and 2005?

16.7 How much can a dollar buy? The buying power of a dollar changes over time. The Bureau of Labor Statistics measures the cost of a "market basket" of goods and services to compile its Consumer Price Index (CPI). If the CPI is 120, goods and services that cost $100 in the base period now cost $120. Here are the average values of the CPI for the years between 1970 and 2010. The base period is the years 1982 to 1984.

Year	CPI	Year	CPI	Year	CPI
1970	38.8	1984	103.9	1998	163.0
1972	41.8	1986	109.6	2000	172.2
1974	49.3	1988	118.3	2002	179.9
1976	56.9	1990	130.7	2004	188.9
1978	65.2	1992	140.3	2006	201.6
1980	82.4	1994	148.2	2008	215.3
1982	96.5	1996	156.9	2010	218.1

(a) Make a graph that shows how the CPI has changed over time.

(b) What was the overall trend in prices during this period? Were there any years in which this trend was reversed?

(c) In which years were prices rising fastest, in terms of percentage increase? In what period were they rising slowest?

16.8 Los Angeles and New York. The Bureau of Labor Statistics publishes separate consumer price indexes for major metropolitan areas in addition to the national CPI. The CPI (1982–84 = 100) in December 2011 was 231.6 in Los Angeles and 249.3 in New York.

(a) These numbers tell us that prices rose faster in New York than in Los Angeles between the base period and the end of 2011. Explain how we know this.

(b) These numbers do *not* tell us that prices in December 2011 were higher in New York than in Los Angeles. Explain why.

16.9 The Food Faddist Price Index. A food faddist eats only steak, rice, and ice cream. In 1995, he bought:

Item	1995 quantity	1995 price
Steak	200 pounds	$5.45/pound
Rice	300 pounds	0.49/pound
Ice cream	50 gallons	5.08/gallon

After a visit from his mother, he adds oranges to his diet. Oranges cost $0.56/pound in 1995. Here are the food faddist's food purchases in 2011:

Item	2011 quantity	2011 price
Steak	100 pounds	$7.29/pound
Rice	350 pounds	0.60/pound
Ice cream	75 gallons	6.99/gallon
Oranges	100 pounds	0.67/pound

Find the fixed market basket Food Faddist Price Index (1995 = 100) for the year 2011.

16.10 The Guru Price Index. A guru purchases only olive oil, loincloths, and copies of the *Atharva Veda,* from which he selects mantras for his disciples. Here are the quantities and prices of his purchases in 1985 and 2011:

Item	1985 quantity	1985 price	2011 quantity	2011 price
Olive oil	20 pints	$2.50/pint	18 pints	$4.60/pint
Loincloth	2	2.75 each	3	3.00 each
Atharva Veda	1	10.95	1	17.95

From these data, find the fixed market basket Guru Price Index (1985 = 100) for 2011.

16.11 The curse of the Bambino. In 1920 the Boston Red Sox sold Babe Ruth to the New York Yankees for $125,000. Between 1920 and 2004, the Yankees won 26 World Series and the Red Sox won 1 (and that occurred in 2004). The Red Sox victory in 2004 supposedly broke the curse. How much is $125,000 in 2004 dollars?

16.12 Dream on. When Julie started college in 2007, she set a goal of making $50,000 when she graduated. Julie graduated in 2011. What must Julie earn in 2011 in order to have the same buying power that $50,000 had in 2007?

16.13 Living too long? If both husband and wife are alive at age 65, in half the cases at least one will still be alive at age 93, 28 years later. Myrna and Bill retired in 1983 with an income of $21,000 per year. They were quite comfortable—that was about the median family income in 1983. How much income did they need 28 years later, in 2011, to have the same buying power?

16.14 Microwaves on sale. The prices of new gadgets often start high and then fall rapidly. The first home microwave oven cost $1300 in 1955. You can now buy a better microwave oven for $100. Find the latest value of the CPI (it's on the BLS Web site, **www.bls.gov/cpi/home.htm**) and use it to restate $100 in present-day dollars in 1955 dollars. Compare this with $1300 to see how much microwave oven real prices have come down.

16.15 Good golfers. In 2009, Tiger Woods won $10,508,163 on the Professional Golfers Association tour. The leading money winner in 1940 was Ben Hogan, at $10,655. Jack Nicklaus, the leader in 1976, won $266,438 that year. How do these amounts compare in real terms?

16.16 Joe DiMaggio. Yankee center fielder Joe DiMaggio was paid $32,000 in 1940 and $100,000 in 1950. Express his 1940 salary in 1950 dollars. By what percentage did DiMaggio's real income change in the decade?

16.17 Calling London. A 10-minute telephone call to London via AT&T cost $12 in 1976 and $14.10 in December 2011 using an occasional calling plan. Compare the real costs of these calls. By what percentage did the real cost go down between 1976 and December 2011? The CPI in December 2011 was 225.7.

16.18 Paying for Harvard. Harvard charged $5900 for tuition, room, and board in 1976. The 2011 charge was $52,650. Express Harvard's 1976 charges in 2011 dollars. Did the cost of going to Harvard go up faster or slower than consumer prices in general? How do you know?

16.19 The minimum wage. The federal government sets the minimum hourly wage that employers can pay a worker. Labor wants a high minimum wage, but many economists argue that too high a minimum makes employers reluctant to hire workers with low skills and so increases unemployment. Here is information on changes in the federal minimum wage, in dollars per hour:

Year:	1960	1965	1970	1975	1980	1985	1990	1995	2005	2007	2008	2009
Min. wage ($):	1.00	1.25	1.60	2.10	3.10	3.35	3.80	4.25	5.15	5.85	6.55	7.25

Use annual average CPIs from Table 16.1 to restate the minimum wage in constant 1960 dollars. Make two line graphs on the same axes, one showing the actual minimum wage during these years and the other showing the minimum wage in constant dollars. Explain carefully to someone who knows no statistics what your graphs show about the history of the minimum wage.

16.20 College tuition. Tuition for Colorado residents at Colorado State University has increased as follows (use tuition at your own college if you have those data):

Year:	1983	1985	1987	1989	1991	1993	1995	1997	1999	2001	2003	2005	2007	2009	2011
Tuition ($):	1008	1275	1474	1636	1855	2022	2174	2258	2340	2502	2908	3381	4040	4822	5419

Use annual average CPIs from Table 16.1 to restate each year's tuition in constant 1983 dollars. Make two line graphs on the same axes, one showing actual dollar tuition for these years and the other showing constant dollar tuition. Then explain to someone who knows no statistics what your graphs show.

16.21 Rising incomes? In Example 4, we saw that the median real income (in 2011 dollars) of all households rose from $48,519 in 1984 to $49,445 in 2011. The real income that marks off the top 5% of households rose from $148,274 to $180,810 in the same period. Verify the claim in Example 4 that median income rose 1.9% and that the real income of top earners rose 21.9%.

16.22 Cable TV. Suppose that cable television systems across the country add channels to their lineup and raise the monthly fee they charge subscribers. The part of the CPI that tracks cable TV prices might not go up at all, even though consumers must pay more. Explain why.

16.23 Item weights in the CPI. The cost of buying a house (with the investment component removed) makes up about 22% of the CPI. The cost of renting a place to live makes up about 6%. Where do the weights 22% and 6% come from? Why does the cost of buying get more weight in the index?

16.24 The CPI doesn't fit me. The CPI may not measure your personal experience with changing prices. Explain why the CPI will not fit each of these people:

(a) Marcia lives on a cattle ranch in Montana.

(b) Jim heats his home with a wood stove and does not have air-conditioning.

(c) Luis and Maria were in a serious auto accident and spent much of last year in a rehabilitation center.

16.25 Seasonal adjustment. Like many government data series, the CPI is published in both unadjusted and seasonally adjusted forms. The BLS says that it "strongly recommends using indexes unadjusted for seasonal variation (i.e., not seasonally adjusted indexes) for escalation." "Escalation" here means adjusting wage or other payments to keep up with changes in the CPI. Why is the unadjusted CPI preferred for this purpose?

16.26 More CPIs. In addition to the national CPI, the BLS publishes CPIs for 4 regions and for 37 local areas. Each regional and local CPI is based on just the part of the national sample of prices that applies to the region or local area. The BLS says that the local CPIs should be used with caution because they are much more variable than national or regional CPIs. Why is this?

16.27 The poverty line. The federal government announces "poverty lines" each year for households of different sizes. Households with income below the announced levels are considered to be living in poverty. An economist looked at the poverty lines over time and said that they "show a pattern of getting higher in real terms as the real income of the general population rises." What does "getting higher in real terms" say about the official poverty level?

16.28 Real wages (optional). In one of the many reports on stagnant incomes in the United States, we read this: "Practically every income group faced a decline in real wages during the 1980s. However, workers at the 33rd percentile experienced a 14 percent drop in the real wage, workers at the 66th percentile experienced only a 6 percent drop, and workers in the upper tail of the distribution experienced a 1 percent wage increase."

(a) What is meant by "the 33rd percentile" of the income distribution?

(b) What does "real wages" mean?

16.29 Saving money? One way to cut the cost of government statistics is to reduce the sizes of the samples. We might, for example, cut the Current Population Survey from 50,000 households to 20,000. Explain clearly, to someone who knows no statistics, why such cuts reduce the accuracy of the resulting data.

16.30 The General Social Survey. The General Social Survey places much emphasis on asking many of the same questions year after year. Why do you think it does this?

16.31 Measuring the effects of crime. We wish to include, as part of a set of social statistics, measures of the amount of crime and of the impact of crime on people's attitudes and activities. Suggest some possible measures in each of the following categories:

(a) Statistics to be compiled from official sources such as police records.

(b) Factual information to be collected using a sample survey of citizens.

(c) Information on opinions and attitudes to be collected using a sample survey.

16.32 Statistical agencies. Write a short description of the work of one of these government statistical agencies. You can find information by starting at the FedStats Web site (**www.fedstats.gov**) and going to *Agencies*.

(a) Bureau of Economic Analysis (Department of Commerce).

(b) National Center for Education Statistics.

(c) National Center for Health Statistics.

EXPLORING THE WEB

16.33 Web-based exercise. Go to the Bureau of Labor Statistics Web site and find the most recent index number for the cost of college tuition. You can find this by conducting a public-data query. Go to the Web address **www.bls.gov/cpi.** Under *CPI Databases* click on the link *ONE-SCREEN-DATA-SEARCH* in the row labeled "All Urban Consumers (Current Series)." Select the area "U.S. City Average" and the item "College tuition and fees" from the two menus. Then click on the button "Get Data." How does this index number compare with the CPI for the corresponding period?

16.34 Web-based exercise. If you like data, you can go to the BLS data page, **www.bls.gov/cpi,** and under *CPI Databases* choose *Average Price Data* and see for yourself how the prices of such things as white bread and gasoline have changed over time. How does the current price of white bread compare with the price 10 years ago? 20 years ago?

16.35 Web-based exercise. Search the Internet to find a recent news report about the CPI. For example, a Google search of "latest CPI" turned up a couple of articles. Summarize what the article you choose says about any changes in the CPI.

NOTES AND DATA SOURCES

Page 369 Example 4: The data come from the Web site of the U.S. Census Bureau. Go to **www.census.gov/hhes/www/income/ data/historical/household/index.html.**

Page 371 Information about the effects of changes in the Consumer Price Index are from Bureau of Labor Statistics, "Updated response to the recommendations of the advisory commission to study the Consumer Price Index," June 1998. The entire December 1996 issue of the BLS *Monthly Labor Review* is devoted to the major revision of the CPI that became effective in January 1998. The effect of changes in the CPI is discussed by Kenneth J. Stewart and Stephen B. Reed, "CPI research series using current methods, 1978–98," *Monthly Labor Review,* 122 (1999), pp. 29–38. All of these are available on the BLS Web site.

Page 373 The survey of government statistics offices is reported in "The good statistics guide," *Economist,* September 13, 1993, p. 65.

Page 377 Exercise 16.3: The data are from the U.S. Department of Energy Web site, **www.eia.gov/petroleum/gasdiesel/.**

Page 377 Exercise 16.6: The data are from the U.S. Environmental Protection Agency's Toxic Release Inventory (TRI) Program Web site, **www.epa.gov/tri/tridata/ tri05/index.htm.**

Page 379 Exercise 16.15: The data are from the Web site **en.wikipedia.org/wiki/PGA_ Tour.**

Page 381 Exercise 16.27: Gordon M. Fisher, "Is there such a thing as an absolute poverty line over time? Evidence from the United States, Britain, Canada, and Australia on the income elasticity of the poverty line," U.S. Census Bureau Poverty Measurement Working Papers, 1995.

Page 381 Exercise 16.28: G. J. Borjas, "The internationalization of the U.S. labor market and the wage structure," *Federal Reserve Bank of New York Economic Policy Review,* 1, No. 1 (1995), pp. 3–8. The quotation appears on p. 3. This entire issue is devoted to articles seeking to explain stagnant earnings and the income gap. The consensus: we don't know.

Data analysis is the art of describing data using graphs and numerical summaries. The purpose of data analysis is to help us see and understand the most important features of a set of data. Chapter 10 commented on basic graphs, especially pie charts, bar graphs, and line graphs. Chapters 11 to 13 showed how data analysis works by presenting statistical ideas and tools for describing the distribution of one variable. Figure II.1 organizes the big ideas. We plot our data, then describe their center and spread using either the mean and standard deviation or the five-number summary. The last step, which makes sense only for some data, is to summarize the data in compact form by using a Normal curve as a model for the overall pattern. The question marks at the last two stages remind us that the usefulness of numerical summaries and Normal distributions depends on what we find when we examine graphs of our data. No short summary does justice to irregular shapes or to data with several distinct clusters.

Chapters 14 and 15 applied the same ideas to relationships between two quantitative variables. Figure II.2 retraces the big ideas from Figure II.1, with details that fit the new setting. We always begin by making graphs of our data. In the case of a scatterplot, we have learned a numerical summary only for data that show a roughly straight-line pattern on the scatterplot. The summary is then the means and standard deviations of the two variables and their correlation. A regression line drawn on the plot gives

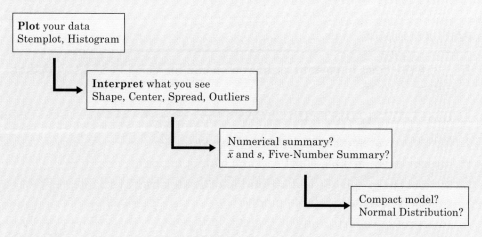

Plot your data
Stemplot, Histogram

Interpret what you see
Shape, Center, Spread, Outliers

Numerical summary?
$\bar{x}$ and s, Five-Number Summary?

Compact model?
Normal Distribution?

Figure II.1

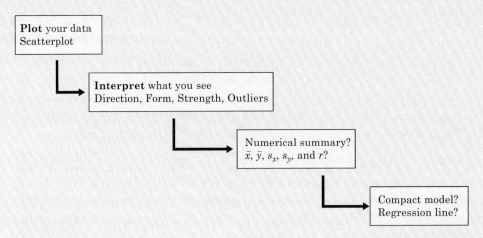

Figure II.2

us a compact model of the overall pattern that we can use for prediction. Once again there are question marks at the last two stages to remind us that correlation and regression describe only straight-line relationships.

Relationships often raise the question of causation. We know that evidence from randomized comparative experiments is the "gold standard" for deciding that one variable causes changes in another variable. Chapter 15 reminded us in more detail that strong associations can appear in data even when there is no direct causation. We must always think about the possible effects of variables lurking in the background. In Chapter 16 we met a new kind of description, index numbers, with the Consumer Price Index as the leading example. Chapter 16 also discussed government statistical offices, a quiet but important part of the statistical world.

PART II SUMMARY

Here are the most important skills you should have acquired after reading Chapters 10 to 16.

A. DISPLAYING DISTRIBUTIONS

1. Recognize categorical and quantitative variables.
2. Recognize when a pie chart can and cannot be used.
3. Make a bar graph of the distribution of a categorical variable, or in general to compare related quantities.
4. Interpret pie charts and bar graphs.
5. Make a line graph of a quantitative variable over time.

6. Recognize patterns such as trends and seasonal variation in line graphs.

7. Be aware of graphical abuses, especially pictograms and distorted scales in line graphs.

8. Make a histogram of the distribution of a quantitative variable.

9. Make a stemplot of the distribution of a small set of observations. Round data as needed to make an effective stemplot.

B. DESCRIBING DISTRIBUTIONS (QUANTITATIVE VARIABLE)

1. Look for the overall pattern of a histogram or stemplot and for major deviations from the pattern.

2. Assess from a histogram or stemplot whether the shape of a distribution is roughly symmetric, distinctly skewed, or neither. Assess whether the distribution has one or more major peaks.

3. Describe the overall pattern by giving numerical measures of center and spread in addition to a verbal description of shape.

4. Decide which measures of center and spread are more appropriate: the mean and standard deviation (especially for symmetric distributions) or the five-number summary (especially for skewed distributions).

5. Recognize outliers and give plausible explanations for them.

C. NUMERICAL SUMMARIES OF DISTRIBUTIONS

1. Find the median M and the quartiles Q_1 and Q_3 for a set of observations.

2. Give the five-number summary and draw a boxplot; assess center, spread, symmetry, and skewness from a boxplot.

3. Find the mean $\bar{x}$ and (using a calculator) the standard deviation s for a small set of observations.

4. Understand that the median is less affected by extreme observations than the mean. Recognize that skewness in a distribution moves the mean away from the median toward the long tail.

5. Know the basic properties of the standard deviation: $s \geq 0$ always; $s = 0$ only when all observations are identical and increases as the spread increases; s has the same units as the original measurements; s is greatly increased by outliers or skewness.

D. NORMAL DISTRIBUTIONS

1. Interpret a density curve as a description of the distribution of a quantitative variable.

2. Recognize the shape of Normal curves, and estimate by eye both the mean and the standard deviation from such a curve.

3. Use the 68–95–99.7 rule and symmetry to state what percentage of the observations from a Normal distribution fall between two points when the points lie at the mean or one, two, or three standard deviations on either side of the mean.

4. Find and interpret the standard score of an observation.

5. (Optional) Use Table B to find the percentile of a value from any Normal distribution and the value that corresponds to a given percentile.

E. SCATTERPLOTS AND CORRELATION

1. Make a scatterplot to display the relationship between two quantitative variables measured on the same subjects. Place the explanatory variable (if any) on the horizontal scale of the plot.

2. Describe the direction, form, and strength of the overall pattern of a scatterplot. In particular, recognize positive or negative association and straight-line patterns. Recognize outliers in a scatterplot.

3. Judge whether it is appropriate to use correlation to describe the relationship between two quantitative variables. Use a calculator to find the correlation r.

4. Know the basic properties of correlation: r measures the strength and direction of only straight-line relationships; r is always a number between -1 and 1; $r = \pm 1$ only for perfect straight-line relations; r moves away from 0 toward ± 1 as the straight-line relation gets stronger.

F. REGRESSION LINES

1. Explain what the slope b and the intercept a mean in the equation $y = a + bx$ of a straight line.

2. Draw a graph of the straight line when you are given its equation.

3. Use a regression line, given on a graph or as an equation, to predict y for a given x. Recognize the danger of prediction outside the range of the available data.

4. Use r^2, the square of the correlation, to describe how much of the variation in one variable can be accounted for by a straight-line relationship with another variable.

G. STATISTICS AND CAUSATION

1. Understand that an observed association can be due to direct causation, common response, or confounding.

2. Give plausible explanations for an observed association between two variables: direct cause and effect, the influence of lurking variables, or both.

3. Assess the strength of statistical evidence for a claim of causation, especially when experiments are not possible.

H. THE CONSUMER PRICE INDEX AND RELATED TOPICS

1. Calculate and interpret index numbers.

2. Calculate a fixed market basket price index for a small market basket.

3. Use the CPI to compare the buying power of dollar amounts from different years. Explain phrases such as "real income."

PART II REVIEW EXERCISES

Review exercises are short and straightforward exercises that help you solidify the basic ideas and skills in each part of this book. We have provided "hints" that indicate where you can find the relevant material for the odd-numbered problems.

II.1 Poverty in the states. Table II.1 gives the percentages of people living below the poverty line in the 26 states east of the Mississippi River. Make a stemplot of these data. Is the distribution roughly symmetric, skewed to the right, or skewed to the left? Which states (if any) are outliers? (*Hint:* See page 245.)

TABLE II.1 Percentages of state residents living in poverty, 2007

State	Percent	State	Percent	State	Percent
Alabama	16.9	Connecticut	7.9	Delaware	10.5
Florida	12.1	Georgia	14.3	Illinois	11.9
Indiana	12.3	Kentucky	17.3	Maine	12.0
Maryland	8.3	Massachusetts	9.9	Michigan	14.0
Mississippi	20.6	New Hampshire	7.1	New Jersey	8.6
New York	13.7	North Carolina	14.3	Ohio	13.1
Pennsylvania	11.6	Rhode Island	12.0	South Carolina	15.0
Tennessee	15.9	Vermont	10.1	Virginia	9.9
West Virginia	16.9	Wisconsin	10.8		

Source: 2010 *Statistical Abstract of the United States.*

II.2 Quarterbacks. Table II.2 gives the total passing yards for National Football League starting quarterbacks during the 2011 season. (These are the quarterbacks with the most passing attempts on each team.) Make a histogram of these data. Does the distribution have a clear shape: roughly symmetric, clearly

TABLE II.2 Passing yards for NFL quarterbacks in 2011

Quarterback	Yards	Quarterback	Yards
Sam Bradford	2164	Cam Newton	4051
Tom Brady	5235	Curtis Painter	1541
Drew Brees	5476	Carson Palmer	2753
Matt Cassel	1713	Christian Ponder	1853
Jay Cutler	2319	Philip Rivers	4624
Andy Dalton	3398	Aaron Rodgers	4643
Ryan Fitzpatrick	3832	Ben Roethlisberger	4077
Joe Flacco	3610	Tony Romo	4184
Josh Freeman	3592	Matt Ryan	4177
Blaine Gabbert	2214	Mark Sanchez	3474
Rex Grossman	3151	Matt Schaub	2479
Matt Hasselbeck	3571	John Skelton	1913
Tavaris Jackson	3091	Alex Smith	3144
Eli Manning	4933	Matthew Stafford	5038
Colt McCoy	2733	Tim Tebow	1729
Matt Moore	2497	Michael Vick	3303

Source: **espn.go.com.**

skewed to the left, clearly skewed to the right, or none of these? Which quarterbacks (if any) are outliers?

II.3 Poverty in the states. Give the five-number summary for the data on poverty from Table II.1. (*Hint:* See page 266.)

II.4 Quarterbacks. Give the five-number summary for the data on passing yards for NFL quarterbacks from Table II.2.

II.5 Poverty in the states. Find the mean percentage of state residents living in poverty from the data in Table II.1. If we removed Mississippi from the data, would the mean increase or decrease? Why? Find the mean for the 25 remaining states to verify your answer. (*Hint:* See page 271.)

II.6 Big heads? The army reports that the distribution of head circumference among male soldiers is approximately Normal with mean 22.8 inches and standard deviation 1.1 inches. Use the 68–95–99.7 rule to answer these questions.

(a) Between what values do the middle 95% of head circumferences fall?

(b) What percentage of soldiers have head circumferences greater than 23.9 inches?

II.7 SAT scores. The scale for SAT exam scores is set so that the distribution of scores is approximately Normal with mean 500 and standard deviation 100. Answer these questions without using a table.

(a) What is the median SAT score? (*Hint:* See page 263.)

(b) You run a tutoring service for students who score between 400 and 600 and hope to attract many students. What percentage of SAT scores are between 400 and 600? (*Hint:* See page 295.)

II.8 Explaining correlation. You have data on the yearly wine consumption (liters of alcohol from drinking wine per person) and yearly deaths from cirrhosis of the liver for several developed countries. Say as specifically as you can what the correlation r between yearly wine consumption and yearly deaths from cirrhosis of the liver measures.

II.9 Data on snakes. For a biology project, you measure the length (inches) and weight (ounces) of 12 snakes of the same variety. What units of measurement do each of the following have?

(a) The mean length of the snakes. (*Hint:* See page 271.)

(b) The first quartile of the snake lengths. (*Hint:* See page 264.)

(c) The standard deviation of the snake lengths. (*Hint:* See page 271.)

(d) The correlation between length and snake weight. (*Hint:* See page 318.)

II.10 More data on snakes. For a biology project, you measure the length (inches) and weight (ounces) of 12 snakes of the same variety.

(a) Explain why you expect the correlation between length and weight to be positive.

(b) The mean length turns out to be 20.8 inches. What is the mean length in centimeters? (There are 2.54 centimeters in an inch.)

(c) The correlation between length and weight turns out to be $r = 0.6$. If you measured length in centimeters instead of inches, what would be the new value of r?

Figure II.3 plots the average brain weight in grams versus average body weight in kilograms for many species of mammals. There are many small mammals whose points at the lower left overlap. Exercises II.11 to II.16 are based on this scatterplot.

II.11 Dolphins and hippos. The points for the dolphin and hippopotamus are labeled in Figure II.3. Read from the graph the approximate body weight and brain weight for these two species. (*Hint:* See page 312.)

II.12 Dolphins and hippos. One reaction to this scatterplot is "Dolphins are smart, hippos are dumb." What feature of the plot lies behind this reaction?

II.13 Outliers. African elephants are much larger than any other mammal in the data set but lie roughly in the overall straight-line pattern. Dolphins, humans, and hippos lie outside the overall pattern. The correlation between body weight and brain weight for the entire data set is $r = 0.86$.

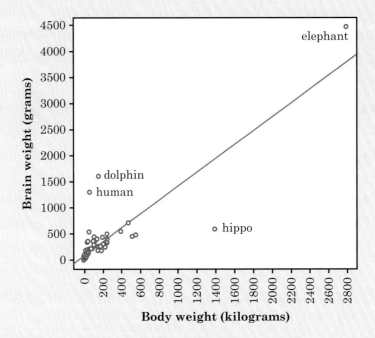

Figure II.3 Scatterplot of the average brain weight (grams) against the average body weight (kilograms) for 96 species of mammals, for Exercises II.11 to II.16.

(a) If we removed elephants, would this correlation increase or decrease or not change much? Explain your answer. (*Hint:* See page 321.)

(b) If we removed dolphins, hippos, and humans, would this correlation increase or decrease or not change much? Explain your answer. (*Hint:* See page 321.)

II.14 Brain and body. The correlation between body weight and brain weight is $r = 0.86$. How well does body weight explain brain weight for mammals? Compute r^2 to answer this question, and briefly explain what r^2 tells us.

II.15 Prediction. The line on the scatterplot in Figure II.3 is the least-squares regression line for predicting brain weight from body weight. Suppose that a new mammal species with body weight 600 kilograms is discovered hidden in the rain forest. Predict the brain weight for this species. (*Hint:* See page 336.)

II.16 Slope. The line on the scatterplot in Figure II.3 is the least-squares regression line for predicting brain weight from body weight. The slope of this line is one of the numbers below. Which number is the slope? Why?

(a) $b = 0.5$

(b) $b = 1.3$

(c) $b = 3.2$

From Rex Boggs in Australia comes an unusual data set: before showering in the morning, he weighed the bar of soap in his shower stall. The weight goes down as the soap is used. The data appear in Table II.3 (weights in grams). Notice that Mr. Boggs forgot to weigh the soap on some days. Exercises II.17 to II.19 are based on the soap data set.

TABLE II.3 Weight (grams) of a bar of soap used to shower

Day	Weight	Day	Weight	Day	Weight
1	124	8	84	16	27
2	121	9	78	18	16
5	103	10	71	19	12
6	96	12	58	20	8
7	90	13	50	21	6

Source: Rex Boggs.

II.17 Scatterplot. Plot the weight of the bar of soap against day. Is the overall pattern roughly straight-line? Based on your scatterplot, is the correlation between day and weight close to 1, positive but not close to 1, close to 0, negative but not close to -1, or close to -1? Explain your answer. (*Hint:* See page 320.)

II.18 Regression. The equation for the least-squares regression line for the data in Table II.3 is

$$\text{weight} = 133.2 - 6.31 \times \text{day}$$

(a) Explain carefully what the slope $b = -6.31$ tells us about how fast the soap lost weight.

(b) Mr. Boggs did not measure the weight of the soap on Day 4. Use the regression equation to predict that weight.

(c) Draw the regression line on your scatterplot from the previous exercise.

II.19 Prediction? Use the regression equation in the previous exercise to predict the weight of the soap after 30 days. Why is it clear that your answer makes no sense? What's wrong with using the regression line to predict weight after 30 days? (*Hint:* See page 341.)

II.20 Keeping up with the Joneses. The Jones family had a household income of $30,000 in 1980, when the average CPI (1982–84 = 100) was 82.4. The average CPI for 2011 was 224.9. How much must the Joneses earn in 2011 to have the same buying power they had in 1980?

II.21 Affording a Mercedes. A Mercedes-Benz 190 cost $24,000 in 1981, when the average CPI (1982–84 = 100) was 90.9. The average CPI for 2011 was 224.9. How many 2011 dollars must you earn to have the same buying power as $24,000 had in 1981? (*Hint:* See page 368.)

II.22 Affording a Steinway. A Steinway concert grand piano cost $13,500 in 1976. A similar Steinway cost $147,995 in August 2010. Has the cost of the piano gone up or down in real terms? Using Table 16.1 and the fact that the August 2010 CPI was 218.3, give a calculation to justify your answer.

II.23 The price of gold. Some people recommend that investors buy gold "to protect against inflation." Here are the prices of an ounce of gold at the end of the year for the years between 1985 and 2011. Using Table 16.1, make a graph that shows how the price of gold changed in real terms over this period. Would an investment in gold have protected against inflation by holding its value in real terms?

Year:	1985	1987	1989	1991	1993	1995	1997	1999	2001	2003	2005	2007	2009	2011
Gold price:	$327	$484	$399	$353	$392	$387	$290	$290	$277	$416	$513	$834	$1083	$1567

(*Hint:* See page 368.)

II.24 Never on Sunday? The Canadian province of Ontario carries out statistical studies to monitor how Canada's national health care system is working in the province. The bar graphs in Figure II.4 come from a study of admissions and discharges from community hospitals in Ontario. They show the number of heart attack patients admitted and discharged on each day of the week during a two-year period.

(a) Explain why you expect the number of patients admitted with heart attacks to be roughly the same for all days of the week. Do the data show that this is true?

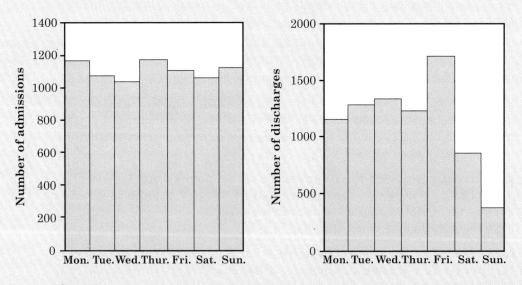

Figure II.4 Bar graphs of the number of heart attack victims admitted and discharged from hospitals in Ontario, Canada, on each day of the week, for Exercise II.24.

(b) Describe how the distribution of the day on which patients are discharged from the hospital differs from that of the day on which they are admitted. What do you think explains the difference?

II.25 Drive time. Professor Moore, who lives a few miles outside a college town, records the time he takes to drive to the college each morning. Here are the times (in minutes) for 42 consecutive weekdays, with the dates in order along the rows:

8.25	7.83	8.30	8.42	8.50	8.67	8.17	9.00	9.00	8.17	7.92
9.00	8.50	9.00	7.75	7.92	8.00	8.08	8.42	8.75	8.08	9.75
8.33	7.83	7.92	8.58	7.83	8.42	7.75	7.42	6.75	7.42	8.50
8.67	10.17	8.75	8.58	8.67	9.17	9.08	8.83	8.67		

(a) Make a histogram of these drive times. Is the distribution roughly symmetric, clearly skewed, or neither? Are there any clear outliers? (*Hint:* See pages 243 and 245.)

(b) Make a line graph of the drive times. (Label the horizontal axis in days, 1 to 42.) The plot shows no clear trend, but it does show one unusually low drive time and two unusually high drive times. Circle these observations on your plot. (*Hint:* See page 219.)

II.26 Drive time outliers. In the previous exercise, there are three outliers in Professor Moore's drive times to work. All three can be explained. The low time is the day after Thanksgiving (no traffic on campus). The two high times reflect delays due to an accident and icy roads. Remove these three observations. To summarize normal drive times, use a calculator to find the mean $\bar{x}$ and standard deviation s of the remaining 39 times.

II.27 House prices. A January 18, 2012, article in the *Los Angeles Times* reported that the median housing price in Southern California was about $270,000. Would the mean housing price be higher, about the same, or lower? Why? (*Hint:* See page 275.)

II.28 The 2008 election. Barack Obama was elected president in 2008 with 52.9% of the popular vote. His Republican opponent, John McCain, received 45.6% of the vote, with minor candidates taking the remaining votes. Table II.4 gives the percentage of the popular vote won by President Obama in each state. Describe these data with a graph, a numerical summary, and a brief verbal description.

II.29 Statistics for investing. Joe's retirement plan invests in stocks through an "index fund" that follows the behavior of the stock market as a whole, as measured by the Standard & Poor's 500 index. Joe wants to buy a mutual fund that does not track the index closely. He reads that monthly returns from Fidelity Technology Fund have correlation $r = 0.77$ with the S&P 500 index and that Fidelity Real Estate Fund has correlation $r = 0.37$ with the index.

f votes for President Obama in 2008

State	Percent	State	Percent
		Louisiana	39.9	Ohio	51.2
		Maine	57.6	Oklahoma	34.4
		Maryland	62.0	Oregon	57.1
		Massachusetts	62.0	Pennsylvania	54.7
Cam..	)	Michigan	57.4	Rhode Island	63.1
Colorado	...5	Minnesota	54.2	South Carolina	44.9
Connecticut	60.6	Mississippi	42.8	South Dakota	44.7
Delaware	61.9	Missouri	49.3	Tennessee	41.8
Florida	50.9	Montana	47.2	Texas	43.8
Georgia	47.0	Nebraska	41.7	Utah	34.7
Hawaii	71.8	Nevada	55.1	Vermont	67.8
Idaho	36.1	New Hampshire	54.3	Virginia	52.7
Illinois	61.8	New Jersey	56.8	Washington	57.5
Indiana	49.9	New Mexico	56.7	West Virginia	42.6
Iowa	54.0	New York	62.2	Wisconsin	56.3
Kansas	41.7	North Carolina	49.9	Wyoming	32.7
Kentucky	41.1	North Dakota	44.7		

Source: **uselectionatlas.org/.**

(a) Which of these funds has the closer relationship to returns from the stock market as a whole? How do you know? (*Hint:* See page 320.)

(b) Does the information given tell Joe anything about which fund has had higher returns? (*Hint:* See page 322.)

PART II PROJECTS

Projects are longer exercises that require gathering information or producing data and that emphasize writing a short essay to describe your work. Many are suitable for teams of students.

Project 1. Statistical graphics in the press. Graphs good and bad fill the news media. Some publications, such as *USA Today,* make particularly heavy use of graphs to present data. Collect several graphs (at least five) from newspapers and magazines (not from advertisements). Include some graphs that, in your opinion, represent good style and some that represent poor style or are misleading. Use your collection as examples in a brief essay about the clarity, accuracy, and attractiveness of graphs in the press.

Project 2. Roll your own regression. Choose two quantitative variables that you think have a roughly straight-line relationship. Gather data on these variables and do a statistical analysis: make a scatterplot, find the correlation, find the regression line (use a statistical calculator or software), and draw the line on your plot. Then write a report on your work. Some examples of suitable pairs of variables are the following:

(a) The height and arm span of a group of people.

(b) The height and walking stride length of a group of people.

(c) The price per ounce and bottle size in ounces for several brands of shampoo and several bottle sizes for each brand.

Project 3. High school dropouts. Write a factual report on high school dropouts in the United States. The following are examples of questions you might address: Which states have the highest percentages of adults who did not finish high school? How do the earnings and employment rates of dropouts compare with those of other adults? Is the percentage who fail to finish high school higher among blacks and Hispanics than among whites?

The *Statistical Abstract* will supply you with data. Look in the index under "education" for an entry on "high school dropouts." You may want to look at other parts of the *Statistical Abstract* for data on earnings and other variables broken down by education.

Project 4. Association is not causation. Write a snappy, attention-getting article on the theme that "association is not causation." Use pointed but not-too-serious examples like those in Example 6 (page 344) and Exercise 15.25 (page 358) of Chapter 15, or this one: there is an association between long hair and height (because women tend to have longer hair than men but also tend to be shorter), but cutting a person's hair will not make him or her taller. Be clear, but don't be technical. Imagine that you are writing for high school students.

Project 5. Military spending. Here are data on U.S. spending for national defense for the fiscal years between 1940 and 2010 from the *Statistical Abstract*. You may want to look in the latest volume for data from the most recent year. You can also find the amounts for every year between 1940 and the present at **www.whitehouse.gov/omb/budget/.** See the pdf file available by clicking on *Historical Tables*. Look in Section 3 of this pdf file. The units are billions of dollars (this is serious money).

Year:	1940	1945	1950	1955	1960	1965	1970	
Military spending:	1.7	83.0	13.7	42.7	48.1	50.6	81.7	

Year:	1975	1980	1985	1990	1995	2000	2005	2010
Military spending:	86.5	134.0	252.7	299.3	272.1	294.5	495.3	693.6

Write an essay that describes the changes in military spending in real terms during this period from just before World War II until a decade after the end of the cold war. Do the necessary calculations and write a brief description that ties military spending to the major military events of this period: World War II (1941–1945), the Korean War (1950–1953), the Vietnam War (roughly 1964–1975), the end of the cold war after the fall of the Berlin Wall in 1989, and the

U.S. war with Iraq (beginning in March 2003). You may want to look at years not included in the above table to help you as you write your essay.

Project 6. Your pulse rate. What is your "resting pulse rate"? Of course, even if you measure your pulse rate while resting, it may vary from day to day and with the time of day. Measure your resting pulse rate at least six times each day (spaced throughout the day) for at least four days. Write a discussion that includes a description of how you made your measurements and an analysis of your data. Based on the data, what would you say when someone asks you what your resting pulse rate is? (If several students do this project, you can discuss variation in pulse rate among a group of individuals as well.)

Project 7. The dates of coins. Coins are stamped with the year in which they were minted. Collect data from at least 50 coins of each denomination: pennies, nickels, dimes, and quarters. Write a description of the distribution of dates on coins now in circulation, including graphs and numerical descriptions. Are there differences among the denominations? Did you find any outliers?

NOTES AND DATA SOURCES

Page 392 The data plotted in Figure II.3 come from G. A. Sacher and E. F. Staffelt, "Relation of gestation time to brain weight for placental mammals: implications for the theory of vertebrate growth," *American Naturalist,* 108 (1974), pp. 593–613. We found them in Fred L. Ramsey and Daniel W. Schafer, *The Statistical Sleuth: A Course in Methods of Data Analysis,* Duxbury Press, 1997, p. 228.

Page 394 Exercise II.23: For the gold prices, see **www.kitco.com/charts/historicalgold.html.**

Page 394 Exercise II.24: Antoni Basinski, "Almost never on Sunday: implications of the patterns of admission and discharge for common conditions," Institute for Clinical Evaluative Sciences in Ontario, October 18, 1993.

Page 395 Exercise II.28: Official final election results are from the Web site of the Federal Election Commission, **www.fec.gov.**

Page 395 Exercise II.29: Correlations for 36 monthly returns ending in December 1999, reported in the *Fidelity Insight* newsletter for January 2000.

Chance

"If chance will have me king, why, chance will crown me." So said Macbeth in Shakespeare's great play. Chance does indeed play with us all, and we can do little to understand or manage it. Sometimes, however, chance is tamed. A roll of dice, a simple random sample, even the inheritance of eye color or blood type, represent chance tied down so that we can understand and manage it. Unlike Macbeth's life or ours, we can roll the dice again. And again, and again. The outcomes are governed by chance, but in many repetitions a pattern emerges. Chance is no longer mysterious, because we can describe its pattern.

We humans use mathematics to describe regular patterns, whether the circles and triangles of geometry or the movements of the planets. We use mathematics to understand the regular patterns of chance behavior when chance is tamed in a setting where we can repeat the same chance phenomenon again and again. The mathematics of chance is called probability. Probability is the topic of this part of the book, though we will go light on the math in favor of experimenting and thinking.

Thinking about Chance

CASE STUDY On February 29, 2012, a woman in Provo, Utah, gave birth on a third consecutive Leap Day, tying a record set in the 1960s. The Associated Press picked up the story, and it was run in newspapers around the country as an amazing feat. If birth dates are random and independent, a statistician can show that the chance that three children, selected at random, are all born on Leap Day is about 1 in 3 billion. The rarity of the event is what made the story newsworthy.

Just how amazing is this event? In this chapter you will learn how to interpret probabilities like 1 in 3 billion. By the end of this chapter you will be able to assess coincidences such as having three children born on Leap Day. Are these events as surprising as they seem? ■

ROB & SAS/Corbis

The idea of probability

Chance is a slippery subject. We start by thinking about "what would happen if we did this many times." We will also start with examples like the 1-in-2 chance of a head in tossing a coin before we try to think about more complicated situations.

Even the rules of football agree that tossing a coin avoids favoritism. Favoritism in choosing subjects for a sample survey or allotting patients to treatment and placebo groups in a medical experiment is as undesirable as it is in awarding first possession of the ball in football. That's why statisticians recommend random samples and randomized experiments, which are fancy versions of tossing a coin. A big fact emerges when we watch coin tosses or the results of random samples closely: **chance behavior is unpredictable in the short run but has a regular and predictable pattern in the long run.**

Toss a coin, or choose a simple random sample. The result can't be predicted in advance, because the result will vary when you toss the coin or choose the sample repeatedly. But there is still a regular pattern in the results, a pattern that emerges clearly only after many repetitions. This remarkable fact is the basis for the idea of probability.

EXAMPLE 1 Coin tossing

When you toss a coin, there are only two possible outcomes, heads or tails. Figure 17.1 shows the results of tossing a coin 1000 times. For each number of tosses from 1 to 1000, we have plotted the proportion of those tosses that gave a head. The first toss was a head, so the proportion of heads starts at 1. The second toss was a tail, reducing the proportion of heads to 0.5 after two tosses. The next four tosses were tails followed by a head, so the proportion of heads after seven tosses is 2/7, or 0.286.

The proportion of tosses that produce heads is quite variable at first, but it settles down as we make more and more tosses. Eventually this proportion gets close to 0.5 and stays there. We say that 0.5 is the *probability* of a head. The probability 0.5 appears as a horizontal line on the graph.

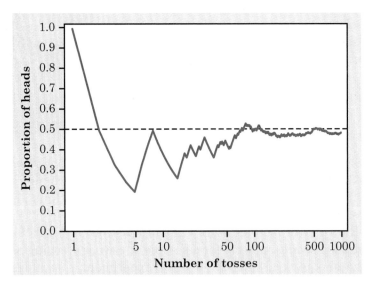

Figure 17.1 Toss a coin many times. The proportion of heads changes as we make more tosses but eventually gets very close to 0.5. This is what we mean when we say, "The probability of a head is one-half."

"Random" in statistics is a description of events that are unpredictable in the short run, but that exhibit a kind of order that emerges only in the long run. It is not a synonym for "haphazard," which is defined as lacking any principle of organization. We encounter the unpredictable side of randomness in our everyday experience, but we rarely see enough repetitions of the same random phenomenon to observe the long-term regularity that probability describes. You can see that regularity emerging in Figure 17.1. In the very long run, the proportion of tosses that give a head is 0.5. This is the intuitive idea of probability. Probability 0.5 means "occurs half the time in a very large number of trials."

We might suspect that a coin has probability 0.5 of coming up heads just because the coin has two sides. We might be tempted to theorize that for events with two seemingly equally likely outcomes, each outcome should have probability 0.5 of occurring. But babies must have one of the two sexes, and the probabilities aren't equal—the probability of a boy is about 0.51, not 0.50. The idea of probability is empirical. That is, it is based on data rather than theorizing alone. Probability describes what happens in very many trials, and we must actually observe many coin tosses or many babies to pin down a probability. In the case of tossing a coin, some diligent people have in fact made thousands of tosses.

EXAMPLE 2 Some coin tossers

The French naturalist Count Buffon (1707–1788) tossed a coin 4040 times. Result: 2048 heads, or proportion $2048/4040 = 0.5069$ for heads.

Around 1900, the English statistician Karl Pearson heroically tossed a coin 24,000 times. Result: 12,012 heads, a proportion of 0.5005.

While imprisoned by the Germans during World War II, the South African mathematician John Kerrich tossed a coin 10,000 times. Result: 5067 heads, a proportion of 0.5067.

Randomness and probability

We call a phenomenon **random** if individual outcomes are uncertain but there is nonetheless a regular distribution of outcomes in a large number of repetitions.

The **probability** of any outcome of a random phenomenon is a number between 0 and 1 that describes the proportion of times the outcome would occur in a very long series of repetitions.

An outcome with probability 0 never occurs. An outcome with probability 1 happens on every repetition. An outcome with probability 1/2 happens

Does God play dice? Few things in the world are truly random in the sense that no amount of information will allow us to predict the outcome. We could in principle apply the laws of physics to a tossed coin, for example, and calculate whether it will land heads or tails. But randomness does rule events inside individual atoms. Albert Einstein didn't like this feature of the new quantum theory. "God does not play dice with the universe," said the great scientist. Many years later, it appears that Einstein was wrong.

half the time in a very long series of trials. Of course, we can never observe a probability exactly. We could always continue tossing the coin, for example. Mathematical probability is an idealization based on imagining what would happen in an infinitely long series of trials.

We aren't thinking deeply here. That some things are random is simply an observed fact about the world. Probability just gives us a language to describe the long-term regularity of random behavior. The outcome of a coin toss, the time between emissions of particles by a radioactive source, and the sexes of the next litter of lab rats are all random. So is the outcome of a random sample or a randomized experiment. The behavior of large groups of individuals is often as random as the behavior of many coin tosses or many random samples. Life insurance, for example, is based on the fact that deaths occur at random among many individuals.

EXAMPLE 3 The probability of dying

We can't predict whether a particular person will die in the next year. But if we observe millions of people, deaths are random. The National Center for Health Statistics says that the proportion of men aged 20 to 24 years who die in any one year is 0.0014. This is the *probability* that a young man will die next year. For women that age, the probability of death is about 0.0005.

If an insurance company sells many policies to people aged 20 to 24, it knows that it will have to pay off next year on about 0.14% of the policies sold on men's lives and on about 0.05% of the policies sold on women's lives. It will charge more to insure a man because the probability of having to pay is higher.

The ancient history of chance

Randomness is most easily noticed in many repetitions of games of chance: rolling dice, dealing shuffled cards, spinning a roulette wheel. Chance devices similar to these have been used from remote antiquity to discover the will of the gods. The most common method of randomization in ancient times was "rolling the bones"—that is, tossing several astragali. The astragalus (Figure 17.2) is a six-sided animal heel bone that, when thrown, will come to rest on one of four sides (the other two sides are rounded). Cubical

Figure 17.2 Animal heel bones (astragali), actual size. (From F. N. David, *Games, Gods and Gambling*, Charles Griffin & Company, 1962. Reproduced by permission of the publishers.)

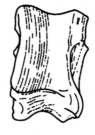

Sheep   Dog

dice, made of pottery or bone, came later, but even dice existed before 2000 B.C. Gambling on the throw of astragali or dice is, compared with divination, almost a modern development. There is no clear record of this vice before about 300 B.C. Gambling reached flood tide in Roman times, then temporarily receded (along with divination) in the face of Christian displeasure.

Chance devices such as astragali have been used from the beginning of recorded history. Yet none of the great mathematicians of antiquity studied the regular pattern of many throws of bones or dice. Perhaps this is because astragali and most ancient dice were so irregular that each had a different pattern of outcomes. Or perhaps the reasons lie deeper, in the classical reluctance to engage in systematic experimentation.

Professional gamblers, who are not as inhibited as philosophers and mathematicians, did notice the regular pattern of outcomes of dice or cards and tried to adjust their bets to the odds of success. "How should I bet?" is the question that launched mathematical probability. The systematic study of randomness began (we oversimplify, but not too much) when 17th-century French gamblers asked French mathematicians for help in figuring out the "fair value" of bets on games of chance. *Probability theory,* the mathematical study of randomness, originated with Pierre de Fermat and Blaise Pascal in the 17th century and was well developed by the time statisticians took it over in the 20th century.

Myths about chance behavior

The idea of probability seems straightforward. It answers the question "What would happen if we did this many times?" In fact, both the behavior of random phenomena and the idea of probability are a bit subtle. We meet chance behavior constantly, and psychologists tell us that we deal with it poorly.

The myth of short-run regularity

The idea of probability is that randomness is regular *in the long run.* Unfortunately, our intuition about randomness tries to tell us that random phenomena should also be regular in the short run. When they aren't, we look for some explanation other than chance variation.

EXAMPLE 4 What looks random?

Toss a coin six times and record heads (H) or tails (T) on each toss. Which of these outcomes is more probable?

HTHTTH   TTTHHH

Almost everyone says that HTHTTH is more probable, because TTTHHH does not "look random." In fact, both are equally probable. That heads and tails are equally probable says only that about half of a very long sequence of tosses will be heads. It doesn't say that heads and tails must come close to alternating in the short run. The coin has no memory. It doesn't know what past outcomes were, and it can't try to create a balanced sequence.

The outcome TTTHHH in tossing six coins looks unusual because of the runs of 3 straight heads and 3 straight tails. Runs seem "not random" to our intuition but are quite common. Here's an example more striking than tossing coins.

Zoonar/Thinkstock/Getty Images

EXAMPLE 5 The hot hand in basketball

Belief that runs must result from something other than "just chance" influences behavior. If a basketball player makes several consecutive shots, both the fans and his teammates believe that he has a "hot hand" and is more likely to make the next shot. This is wrong. Careful study has shown that runs of baskets made or missed are no more frequent in basketball than would be expected if each shot is independent of the player's previous shots. Players perform consistently, not in streaks. If a player makes half her shots in the long run, her hits and misses behave just like tosses of a coin—and that means that runs of hits and misses are more common than our intuition expects.

NOW IT'S YOUR TURN

17.1 Coin tossing and randomness. Toss a coin 10 times and record heads (H) or tails (T) on each toss. Which of these outcomes is most probable? Least probable?

HTHTTHHTHT   TTTTTHHHHH   HHHHHHHHHH

The myth of the surprising coincidence

On November 18, 2006, Ohio State beat Michigan in football by a score of 42 to 39. Later that day, the winning numbers in the Pick 4 Ohio lottery were 4239. What an amazing coincidence!

Well, maybe not. It is certainly unlikely that the Pick 4 lottery would match the Ohio State versus Michigan score that day, but it is not so unlikely that sometime during the 2006 season the winning number of some state lottery would match the recent score of some professional, college, or high school football game involving a team in the state. There are 32 NFL teams, 235 NCAA Division I teams, 150 NCAA Division II teams, and 231 NCAA Division III teams. There are also over 25,000 high school football teams. All play a number of games during the season. There are 38 states with a Pick 3 or Pick 4 lottery game, with winning numbers often drawn multiple times per week. That's a lot of opportunities to match a Pick 3 or Pick 4 lottery number that has digits that could conceivably be a football score like 217 or 4239.

When something unusual happens, we look back and say, "Wasn't that unlikely?" We would have said the same if any of thousands of other unlikely things had happened. Here's an example where it was possible to actually calculate the probabilities.

Emile Wamsteker/Bloomberg News/Landov

EXAMPLE 6   Winning the lottery twice

In 1986, Evelyn Marie Adams won the New Jersey State lottery for the second time, adding $1.5 million to her previous $3.9 million jackpot. The *New York Times* (February 14, 1986) claimed that the odds of one person winning the big prize twice were about 1 in 17 trillion. Nonsense, said two statistics professors in a letter that appeared in the *Times* two weeks later. The chance that Evelyn Marie Adams would win twice in her lifetime is indeed tiny, but it is almost certain that *someone* among the millions of regular lottery players in the United States would win two jackpot prizes. The statisticians estimated even odds (a probability of 1/2) of another double winner within seven years. Sure enough, Robert Humphries won his second Pennsylvania lottery jackpot ($6.8 million total) in May 1988.

Unusual events—especially distressing events—bring out the human desire to pinpoint a reason, a *cause*. Here's a sequel to our earlier discussion of causation: sometimes it's just the play of chance.

EXAMPLE 7 Cancer clusters

Between 1997 and 2004, 16 children were diagnosed with cancer and 3 died in Fallon, Nevada, a farming community of 8300 some 60 miles southeast of Reno. This is an unusual number of cases for such a small town. Residents were concerned that perhaps high levels of naturally occurring arsenic in Fallon's water supply, a pipeline carrying jet fuel to the local Navy base, local pesticide spraying, high tungsten levels, or an underground nuclear test conducted 30 miles away about 40 years ago might be responsible. However, scientists were unable to link any of these to the cancers. Residents were disappointed by the scientists' findings.

In 1984, residents of a neighborhood in Randolph, Massachusetts, counted 67 cancer cases in their 250 residences. This cluster of cancer cases seemed unusual, and the residents expressed concern that runoff from a nearby chemical plant was contaminating their water supply and causing cancer.

In 1979, two of the eight town wells serving Woburn, Massachusetts, were found to be contaminated with organic chemicals. Alarmed citizens began counting cancer cases. Between 1964 and 1983, 20 cases of childhood leukemia were reported in Woburn. This is an unusual number of cases of this rather rare disease. The residents believed that the well water had caused the leukemia and proceeded to sue two companies held responsible for the contamination.

statistics in Your World **The probability of rain is ...** You work all week. Then it rains on the weekend. Can there really be a statistical truth behind our perception that the weather is against us? At least on the east coast of the United States, the answer is "Yes." Going back to 1946, it seems that Sundays receive 22% more precipitation than Mondays. The likely explanation is that the pollution from all those workday cars and trucks forms the seeds for raindrops—with just enough delay to cause rain on the weekend.

Cancer is a common disease, accounting for more than 23% of all deaths in the United States. That cancer cases sometimes occur in clusters in the same neighborhood is not surprising; there are bound to be clusters *somewhere* simply by chance. But when a cancer cluster occurs in *our* neighborhood, we tend to suspect the worst and look for someone to blame. State authorities get several thousand calls a year from people worried about "too much cancer" in their area. But the National Cancer Institute finds that the majority of reported cancer clusters are simply the result of chance.

Both of the Massachusetts cancer clusters were investigated by statisticians from the Harvard School of Public Health. The investigators tried to obtain complete data on everyone who had lived in the neighborhoods in the periods in question and to estimate their exposure to the suspect drinking water. They also tried to obtain data on other factors that might explain cancer, such as smoking and occupational exposure to toxic substances. The verdict: chance is the likely explanation of the Randolph cluster, but there is evidence of an association between drinking water from the two Woburn wells and developing childhood leukemia.

The myth of the law of averages

Roaming the gambling floors in Las Vegas, watching money disappear into the drop boxes under the tables, is revealing. You can see some interesting human behavior in a casino. When the shooter in the dice game craps rolls several winners in a row, some gamblers think she has a "hot hand" and bet that she will keep on winning. Others say that "the law of averages" means that she must now lose so that wins and losses will balance out. Believers in the law of averages think that if you toss a coin six times and get TTTTTT, the next toss must be more likely to give a head. It's true that in the long run heads must appear half the time. What is myth is that future outcomes must make up for an imbalance like six straight tails.

Coins and dice have no memories. A coin doesn't know that the first six outcomes were tails, and it can't try to get a head on the next toss to even things out. Of course, things do even out *in the long run*. After 10,000 tosses, the results of the first six tosses don't matter. They are overwhelmed by the results of the next 9994 tosses, not compensated for.

EXAMPLE 8 We want a boy

Belief in this phony "law of averages" can lead to consequences close to disastrous. A few years ago, "Dear Abby" published in her advice column a letter from a distraught mother of eight girls. It seems that she and her husband had planned to limit their family to four children. When all four were girls, they tried again—and again, and again. After seven straight girls, even her doctor had assured her that "the law of averages was in our favor 100 to 1." Unfortunately for this couple, having children is like tossing coins. Eight girls in a row is highly unlikely, but once seven girls have been born, it is not at all unlikely that the next child will be a girl— and it was.

What is the law of averages?

Is there a "law of averages"? There is, although it is sometimes referred to as the "law of large numbers." It states that in a large number of "independent" repetitions of a random phenomenon (such as coin tossing), averages or proportions are likely to become more stable as the number of trials increases, whereas sums or counts are likely to become more variable. This does not happen by compensation for a run of bad luck, because by "independent"

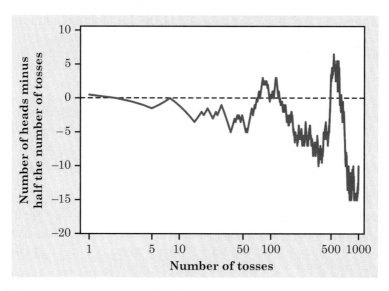

Figure 17.3 Toss a coin many times. The difference between the observed number of heads and exactly one-half the number of tosses becomes more variable as the number of tosses increases.

we mean that knowing the outcome of one trial does not change the probabilities for the outcomes of any other trials. The trials have no memory.

Figures 17.1 and 17.3 show what happens when we toss a coin repeatedly many times. In Figure 17.1 we see that the *proportion* of heads gradually becomes closer and closer to 0.5 as the number of tosses increases. This illustrates the law of large numbers. However, in Figure 17.3 we see that the *total* number of heads for these same tosses varies more and more as the number of tosses increases. The law of large numbers does not apply to sums or counts.

It is not uncommon to see the law of averages misstated in terms of the sums or counts rather than means or proportions. For example, assuming that the birth rates for boys and girls in the United States are equal, you may hear someone state that the *total* number of males and females in the United States should be nearly equal rather than stating that the *proportion* of males and females in the United States should be nearly equal.

> **NOW IT'S YOUR TURN** | **17.2 Coin tossing and the law of averages.** The author C. S. Lewis once wrote the following, referring to the law of averages: "If you tossed a coin a billion times, you could predict a nearly equal number of heads and tails." Is this a correct statement of the law of averages? If not, how would you rewrite the statement so that it is correct?

Personal probabilities

Joe sits staring into his beer as his favorite baseball team, the Chicago Cubs, loses another game. The Cubbies have some good young players, so let's ask Joe, "What's the chance that the Cubs will go to the World Series next year?" Joe brightens up. "Oh, about 10%," he says.

Does Joe assign probability 0.10 to the Cubs' appearing in the World Series? The outcome of next year's pennant race is certainly unpredictable, but we can't reasonably ask what would happen in many repetitions. Next year's baseball season will happen only once and will differ from all other seasons in players, weather, and many other ways. The answer to our question seems clear: if probability measures "what would happen if we did this many times," Joe's 0.10 is not a probability. Probability is based on data about many repetitions of the same random phenomenon. Joe is giving us something else, his personal judgment.

Yet we often use the term "probability" in a way that includes personal judgments of how likely it is that some event will happen. We make decisions based on these judgments—we take the bus downtown because we think the probability of finding a parking spot is low. More serious decisions also take judgments about likelihood into account. A company deciding whether to build a new plant must judge how likely it is that there will be high demand for its products three years from now when the plant is ready. Many companies express "How likely is it?" judgments as numbers—probabilities—and use these numbers in their calculations. High demand in three years, like the Cubs' winning next year's pennant, is a one-time event that doesn't fit the "do it many times" way of thinking. What is more, several company officers may give several different probabilities, reflecting differences in their individual judgment. We need another kind of probability, *personal probability*.

Personal probability

A **personal probability** of an outcome is a number between 0 and 1 that expresses an individual's judgment of how likely the outcome is.

Personal probabilities have the great advantage that they aren't limited to repeatable settings. They are useful because we base decisions on them: "I think the probability that the Patriots will win the Super Bowl is 0.75, so I'm going to bet on the game." Just remember that personal probabilities are different in kind from probabilities as "proportions in many repetitions." Because they express individual opinion, they can't be said to be right or wrong.

This is true even in a "many repetitions" setting. If Craig has a gut feeling that the probability of a head on the next toss of this coin is 0.7, that's what Craig thinks and that's all there is to it. Tossing the coin many times may show that the proportion of heads is very close to 0.5, but that's another matter. **There is no reason why a person's degree of confidence in the outcome of one try must agree with the results of many tries.** We stress this because it is common to say that "personal probability" and "what happens in many trials" are somehow two interpretations of the same idea. In fact, they are quite different ideas.

Why do we even use the word "probability" for personal opinions? There are two good reasons. First, we usually do base our personal opinions on data from many trials when we have such data. Data from Buffon, Pearson, and Kerrich (Example 2) and perhaps from our own experience convince us that coins come up heads very close to half the time in many tosses. When we say that a coin has probability 1/2 of coming up heads *on this toss,* we are applying to a single toss a measure of the chance of a head based on what would happen in a long series of tosses. Second, personal probability and probability as long-term proportion both obey the same mathematical rules. Both kinds of probabilities are numbers between 0 and 1. We will look at more of the rules of probability in the next chapter. These rules apply to both kinds of probability.

Although "personal probability" and "what happens in many trials" are different ideas, what happens in many trials often causes us to revise our personal probability of an event. If Craig has a gut feeling that the probability of a head when he tosses a particular coin is 0.7, that's what Craig thinks. If he tosses it 20 times and gets 9 heads, he may continue to believe that the probability of heads is 0.7—because personal probabilities need not agree with the results of many trials. But he may also decide to revise his personal probability downward based on what he has observed. Is there a sensible way to do this, or is this also just a matter of personal opinion?

In statistics there are formal methods for using data to adjust personal probabilities. These are called *Bayes's procedures.* The basic rule, called *Bayes's theorem,* is attributed to the Reverend Thomas Bayes, who discussed the rule in "An Essay towards Solving a Problem in the Doctrine of Chances" published in 1764. The mathematics is somewhat complicated and we will not discuss the details. However, the use of Bayes's procedures is becoming increasingly common among practitioners.

Probability and risk

Once we understand that "personal judgment of how likely" and "what happens in many repetitions" are different ideas, we have a good start toward understanding why the public and the experts disagree so strongly about what is risky and what isn't. The experts use probabilities from data to describe the risk of an unpleasant event. Individuals and society, however, seem to ignore data. We worry about some risks that almost never occur while ignoring others that are much more probable.

EXAMPLE 9 Asbestos in the schools

High exposure to asbestos is dangerous. Low exposure, such as that experienced by teachers and students in schools where asbestos is present in the insulation around pipes, is not very risky. The probability that a teacher who works for 30 years in a school with typical asbestos levels will get cancer from the asbestos is around 15/1,000,000. The risk of dying in a car accident during a lifetime of driving is about 15,000/1,000,000. That is, driving regularly is about 1000 times more risky than teaching in a school where asbestos is present.

Risk does not stop us from driving. Yet the much smaller risk from asbestos launched massive cleanup campaigns and a federal requirement that every school inspect for asbestos and make the findings public.

Why do we take asbestos so much more seriously than driving? Why do we worry about very unlikely threats such as tornadoes and terrorists more than we worry about heart attacks?

- We feel safer when a risk seems under our control than when we cannot control it. We are in control (or so we imagine) when we are driving, but we can't control the risk from asbestos or tornadoes or terrorists.

- It is hard to comprehend very small probabilities. Probabilities of 15 per million and 15,000 per million are both so small that our intuition cannot distinguish between them. Psychologists have shown that we generally overestimate very small risks and underestimate higher risks. Perhaps this is part of the general weakness of our intuition about how probability operates.

- The probabilities for risks like asbestos in the schools are not as certain as probabilities for tossing coins. They must be estimated by experts from complicated statistical studies. Perhaps it is safest to suspect that the experts may have underestimated the level of risk.

What are the odds?
Gamblers often express chance in terms of *odds* rather than probability. Odds of A to B against an outcome means that the probability of that outcome is $B/(A + B)$. So "odds of 5 to 1" is another way of saying "probability 1/6." A probability is always between 0 and 1, but odds range from 0 to infinity. Although odds are mainly used in gambling, they give us a way to make very small probabilities clearer. "Odds of 999 to 1" may be easier to understand than "probability 0.001."

Our reactions to risk depend on more than probability, even if our personal probabilities are higher than the experts' data-based probabilities. We are influenced by our psychological makeup and by social standards. As one writer noted, "Few of us would leave a baby sleeping alone in a house while we drove off on a 10-minute errand, even though car-crash risks are much greater than home risks."

STATISTICS IN SUMMARY

Chapter Specifics

- Some things in the world, both natural and of human design, are **random.** That is, their outcomes have a clear pattern in very many repetitions even though the outcome of any one trial is unpredictable.

- **Probability** describes the long-term regularity of random phenomena. The probability of an outcome is the proportion of very many repetitions on which that outcome occurs. A probability is a number between 0 (the outcome never occurs) and 1 (always occurs). We emphasize this kind of probability because it is based on data.

- Probabilities describe only what happens in the long run. Short runs of random phenomena like tossing coins or shooting a basketball often don't look random to us because they do not show the regularity that in fact emerges only in very many repetitions.

- **Personal probabilities** express an individual's personal judgment of how likely outcomes are. Personal probabilities are also numbers between 0 and 1. Different people can have different personal probabilities, and a personal probability need not agree with a proportion based on data about similar cases.

 This chapter begins our study of the mathematics of chance or "probability." The important fact is that random phenomena are unpredictable in the short run but have a regular and predictable behavior in the long run.

The long-run behavior of random phenomena will help us understand both why and in what way we can trust random samples and randomized comparative experiments, the subjects of Chapters 2 to 6. It is the key to generalizing what we learn from data produced by random samples and randomized comparative experiments to some wider universe or population. We will study how this is done in Part IV. As a first step in this direction, we will look more carefully at the basic rules of probability in the next chapter.

CASE STUDY In the Case Study described at the beginning of this chapter you were EVALUATED told that if birth dates are random and independent, the chance that three children, selected at random, are all born on Leap Day is about 1 in 3 billion.

1. Go to the most recent *Statistical Abstract* (online at **www.census.gov/compendia/ statab/**) and look under *Population, Households and Families, Families by Number of Own Children under 18 Years Old.* How many families in the United States have at least three children under 18 years old?

2. For the time being, assume that the families you found in the previous question all have exactly three children. Explain why the probability that *some family* in the United States has three children all born on Leap Day is much larger than 1 in 3 billion.

3. Next, consider the fact that not all the families have exactly three children, that the number of families in Question 1 does not include those with children over the age of 18, and that parents might intentionally try to conceive children with the same birth date (in fact, the Associated Press news article mentioned that after they had a child born on Leap Day, the couple in Utah tried to have a child on subsequent Leap Days). Write a paragraph discussing whether the "surprising" coincidence described in the Case Study that began this chapter is as surprising as it might first appear. ∎

CHAPTER 17 EXERCISES

For Exercise 17.1, see page 406; for Exercise 17.2, see page 410.

17.3 Nickels spinning. Hold a nickel upright on its edge under your forefinger on a hard surface, then snap it with your other forefinger so that it spins for some time before falling. Based on 40 spins, estimate the probability of heads.

17.4 Nickels falling over. You may feel that it is obvious that the probability of a head in tossing a coin is about 1/2 because the coin has two faces. Such opinions are not always correct. The previous exercise asked you to spin a nickel rather than toss it—that changes the probability of a head. Now try another variation. Stand a nickel on edge on a hard, flat surface. Pound the surface with your hand so that the nickel falls over. What is the probability that it falls with heads upward? Make at least 40 trials to estimate the probability of a head.

17.5 Random digits. The table of random digits (Table A) was produced by a random mechanism that gives each digit probability 0.1 of being a 0. What proportion of the first 400 digits in the table are 0s? This proportion is an estimate, based on 400 repetitions, of the true probability, which in this case is known to be 0.1.

17.6 How many tosses to get a head? When we toss a penny, experience shows that the probability (long-term proportion) of a head is close to 1/2. Suppose now that we toss the penny repeatedly until we get a head. What is the probability that the first head comes up in an odd number of tosses (1, 3, 5, and so on)? To find out, repeat this experiment 40 times, and keep a record of the number of tosses needed to get a head on each of your 40 trials.

(a) From your experiment, estimate the probability of a head on the first toss. What value should we expect this probability to have?

(b) Use your results to estimate the probability that the first head appears on an odd-numbered toss.

17.7 Tossing a thumbtack. Toss a thumbtack on a hard surface 50 times. How many times did it land with the point up? What is the approximate probability of landing point up?

17.8 Three of a kind. You read in a book on poker that the probability of being dealt three of a kind in a five-card poker hand is about 1/47. Explain in simple language what this means.

17.9 From words to probabilities. Probability is a measure of how likely an event is to occur. Match one of the probabilities that follow with each statement of likelihood given. (The probability is usually a more exact measure of likelihood than is the verbal statement.)

$$0 \quad 0.01 \quad 0.3 \quad 0.6 \quad 0.99 \quad 1$$

(a) This event is impossible. It can never occur.

(b) This event is certain. It will occur on every trial.

(c) This event is very unlikely, but it will occur once in a while in a long sequence of trials.

(d) This event will occur more often than not.

17.10 Winning a baseball game. Over the period from 1965 to 2011 the champions of baseball's two major leagues won 63% of their home games during the regular season. At the end of each season, the two league champions meet in the baseball World Series. Would you use the results from the regular season to assign probability 0.63 to the event that the home team wins a World Series game? Explain your answer.

17.11 Will you have an accident? The probability that a randomly chosen driver will be involved in an accident in the next year is about 0.2. This is based on the proportion of millions of drivers who have accidents. "Accident" includes things like crumpling a fender in your own driveway, not just highway accidents.

(a) What do you think is your own probability of being in an accident in the next year? This is a personal probability.

(b) Give some reasons why your personal probability might be a more accurate prediction of your "true chance" of having an accident than the probability for a random driver.

(c) Almost everyone says that their personal probability is lower than the random driver probability. Why do you think this is true?

17.12 Marital status. The probability that a randomly chosen 65-year-old woman is divorced is about 0.14. This probability is a long-run proportion based on all the millions of women aged 65. Let's suppose that the proportion stays at 0.14 for the next 45 years. Bridget is now 20 years old and is not married.

(a) Bridget thinks her own chances of being divorced at age 65 are about 5%. Explain why this is a personal probability.

(b) Give some good reasons why Bridget's personal probability might differ from the proportion of all women aged 65 who are divorced.

(c) You are a government official charged with looking into the impact of the Social Security system on retirement-aged divorced women. You care only about the probability 0.14, not about anyone's personal probability. Why?

17.13 Personal probability versus data. Give an example in which you would rely on a probability found as a long-term proportion from data on many trials. Give an example in which you would rely on your own personal probability.

17.14 Personal probability? When there are few data, we often fall back on personal probability. There had been just 24 space shuttle launches, all successful, before the *Challenger* disaster in January 1986. The shuttle program management thought the chances of such a failure were only 1 in 100,000.

(a) Suppose 1 in 100,000 is a correct estimate of the chance of such a failure. If a shuttle was launched every day, about how many failures would one expect in 300 years?

(b) Give some reasons why such an estimate is likely to be too optimistic.

17.15 Personal random numbers? Ask several of your friends (at least 10 people) to choose a four-digit number "at random." How many of the numbers chosen start with 1 or 2? How many start with 8 or 9? (There is strong evidence that people in general tend to choose numbers starting with low digits.)

17.16 Playing Pick 4. The Pick 4 games in many state lotteries announce a four-digit winning number each day. The winning number is essentially a four-digit group from a table of random digits. You win if your choice matches the winning digits, in exact order. The winnings are divided among all players who matched the winning digits. That suggests a way to get an edge.

(a) The winning number might be, for example, either 2873 or 9999. Explain why these two outcomes have exactly the same probability. (It is 1 in 10,000.)

(b) If you asked many people which outcome is more likely to be the randomly chosen winning number, most would favor one of them. Use the information in this chapter to say which one and to explain why. If you choose a number that people think is unlikely, you have the same chance to win, but you will win a larger amount because few other people will choose your number.

17.17 Surprising? During the broadcast of the October 2, 2010, football game between the Ohio State University and the University of Illinois, the announcers reported that the head coach, the offensive coordinator, and the defensive coordinator of the Ohio State football team all had the same first name, Jim. No other Division I football team had a head coach, offensive coordinator, and defensive coordinator with the same first name. Should this fact surprise you? Explain your answer.

17.18 An eerie coincidence? An October 6, 2002, ABC News article reported that the winning New York State lottery numbers on the one-year anniversary of the attacks on America were 911. Should this fact surprise you? Explain your answer.

17.19 Nash's free throws. The basketball player Steve Nash is the all-time career free-throw shooter among active players. He makes 90.3% of his free throws. In today's game, Nash misses his first two free throws. The TV commentator says, "Nash's technique looks out of rhythm today." Explain why the claim that Nash's technique has deteriorated is not justified.

17.20 In the long run. Probability works not by compensating for imbalances but by overwhelming them. Suppose that the first 10 tosses of a coin give 10 tails and that tosses after that are exactly half heads and half tails. (Exact balance is unlikely, but the example illustrates how the first 10 outcomes are swamped by later outcomes.) What is the proportion of heads after the first 10 tosses? What is the proportion of heads after 100 tosses if half of the last 90 produce heads (45 heads)? What is the proportion of heads after 1000 tosses if half of the last 990 produce heads? What is the proportion of heads after 10,000 tosses if half of the last 9990 produce heads?

17.21 The "law of averages." The baseball player Ichiro Suzuki gets a hit about 33.1% of the time over an entire season. After he has failed to hit safely in nine straight at-bats, the TV commentator says, "Ichiro is due for a hit by the law of averages." Is that right? Why?

17.22 Snow coming. A meteorologist, predicting below-average snowfall this winter, says, "First, in looking at the past few winters, there has been above-average snowfall. Even though we are not supposed to use the law of averages, we are due." Do you think that "due by the law of averages" makes sense in talking about the weather? Explain.

17.23 An unenlightened gambler. (a) A gambler knows that red and black are equally likely to occur on each spin of a roulette wheel. He observes five consecutive reds occur and bets heavily on black at the next spin. Asked why, he explains that black is "due by the law of averages." Explain to the gambler what is wrong with this reasoning.

(b) After listening to you explain why red and black are still equally likely after five reds on the roulette wheel, the gambler moves to a poker game. He is dealt five straight red cards. He remembers what you said and assumes that the next card dealt in the same hand is equally likely to be red or black. Is the gambler right or wrong, and why?

17.24 Reacting to risks. The probability of dying if you play high school football is about 10 per million each year you play. The risk of getting cancer from asbestos if you attend a school in which asbestos is present for 10 years is about 5 per million. If we ban asbestos from schools, should we also ban high school football? Briefly explain your position.

17.25 Reacting to risks. National newspapers such as *USA Today* and the *New York Times* carry many more stories about deaths from airplane crashes than about deaths from motor vehicle crashes. Motor vehicle accidents killed about 44,000 people in the United States in 2007. Crashes of all scheduled air carriers worldwide, including commuter carriers, killed 587 people in 2007, and only 1 of these involved a U.S. air carrier.

(a) Why do the news media give more attention to airplane crashes?

(b) How does news coverage help explain why many people consider flying more dangerous than driving?

17.26 What probability doesn't say. The probability of a head in tossing a coin is 1/2. This means that as we make more tosses, the *proportion* of heads will eventually get close to 0.5. It does not mean that the *count* of heads will get close to 1/2 the number of tosses. To see why, imagine that the proportion of heads is 0.49 in 100 tosses, 0.493 in 1000 tosses, 0.4969 in 10,000 tosses, and 0.49926 in 100,000 tosses of a coin. How many heads came up in each set of tosses? How close is the number of heads to half the number of tosses?

EXPLORING THE WEB

17.27 Web-based exercise. Search the Web to see if you can find an example of a misuse or misstatement of the law of averages. Explain why the statement you find is incorrect. (We found some examples by doing a Google search on the phrase "law of averages.")

17.28 Web-based exercise. One of the best ways to grasp the idea of probability is to watch the proportion of trials on which an outcome occurs gradually settle down at the outcome's probability. Computer simulations can show this. Go to the *Statistics: Concepts and Controversies* Web site, **www.whfreeman.com/scc8e,** and look at the *Probability* applet. Select a probability for heads between 0.2 and 0.8 and run the applet for a total of 400 trials. Describe the pattern you see. How quickly did the proportion of heads settle down to the true probability?

17.29 Web-based exercise. The probabilities of various poker hands refer to the proportion of times these hands will appear in many, many games. You can find a simulation of the results of dealing many poker hands at **www.stat.uiuc.edu/courses/stat100/cuwu/Games.html.** Visit this Web site and simulate the results of several thousand poker hands by repeatedly clicking on the 1000 button under # draws. How quickly does the percentage of hands containing a straight converge to the expected percentage?

NOTES AND DATA SOURCES

Page 404 Example 3: See **www.cdc.gov/ nchs/deaths.htm** for mortality statistics.

Page 404 More historical detail can be found in the opening chapters of F. N. David, *Games, Gods and Gambling,* Courier Dover Publications, 1998. The historical information given here comes from this excellent and entertaining book.

Page 406 Example 5: T. Gilovich, R. Vallone, and A. Tversky, "The hot hand in basketball: on the misperception of random sequences," *Cognitive Psychology,* 17 (1985), pp. 295–314.

Page 408 Example 7: for more on Fallon, see the Web site **www.cdc.gov/nceh/clusters/**

Fallon/default.htm. For Randolph, see R. Day, J. H. Ware, D. Wartenberg, and M. Zelen, "An investigation of a reported cancer cluster in Randolph, Ma.," Harvard School of Public Health Technical Report, June 27, 1988. For Woburn, see S. W. Lagakos, B. J. Wessen, and M. Zelen, "An analysis of contaminated well water and health effects in Woburn, Massachusetts," *Journal of the American Statistical Association,* 81 (1986), pp. 583–596.

Page 409 For a discussion and amusing examples, see A. E. Watkins, "The law of averages," *Chance,* 8, No. 2 (1995), pp. 28–32.

Page 410 The quotation is from Chapter 3 of C. S. Lewis, *Miracles*, Macmillan Co., 1947.

Page 411 The presentation of personal probability and long-term proportion as distinct ideas is influenced by psychological research that appears to show that people judge single-case questions differently from distributional or relative frequency questions. At least some of the biases found in the classic work of Tversky and Kahneman on our perception of chance seem to disappear when it is made clear to subjects which interpretation is intended. This is a complex area and *SCC* is a simple book, but we judged it somewhat behind the times to put too much emphasis on the Tversky-Kahneman findings. See Gerd Gigerenzer, "How to make cognitive illusions disappear: beyond heuristics and biases," in Wolfgang Stroebe and Miles Hewstone (eds.), *European Review of Social Psychology*, Vol. 2, Wiley, 1991, pp. 83–115. Two additional references are G. Gigerenzer and R. Selten, *Bounded Rationality: The Adaptive Toolbox*, MIT Press, 2001; and G. Gigerenzer, P. M. Todd, and the ABC Research Group, *Simple Heuristics That Make Us Smart*, Oxford University Press, 1999.

Page 412 The 1764 essay by Thomas Bayes, "An essay towards solving a problem in the doctrine of chances," was published in the *Philosophical Transactions of the Royal Society of London*, 53, 370–418. A fascimile is available online at **www.stat.ucla.edu/history/essay.pdf.**

Page 413 Example 9: Estimated probabilities are from R. D'Agostino, Jr., and R. Wilson, "Asbestos: the hazard, the risk, and public policy," in K. R. Foster, D. E. Bernstein, and P. W. Huber (eds.), *Phantom Risk: Scientific Inference and the Law*, MIT Press, 1994, pp. 183–210. See also the similar conclusions in B. T. Mossman et al., "Asbestos: scientific developments and implications for public policy," *Science*, 247 (1990), pp. 294–301.

Page 414 The quotation is from R. J. Zeckhauser and W. K. Viscusi, "Risk within reason," *Science*, 248 (1990), pp. 559–564.

Page 417 Exercise 17.15: See T. Hill, "Random-number guessing and the first digit phenomenon," *Psychological Reports*, 62 (1988), pp. 967–971.

Probability Models

CASE STUDY Shortly after the New York Giants won Super Bowl XLVI, Web sites were already posting the probabilities of winning Super Bowl XLVII for the various NFL teams. Table 18.1 lists the probabilities posted on one Web site. These are perhaps best interpreted as personal probabilities. They are likely to change as the 2012 season approaches because the players on teams will change due to trades, the draft, injuries, and retirements.

Because these are personal probabilities, are there any restrictions on the choices for the probabilities? In this chapter we will answer this question. We will learn that probabilities must obey certain rules in order to make sense. By the end of this chapter you will be able to assess whether the probabilities in Table 18.1 make sense. ■

Probability models

Choose a woman aged 25 to 29 years old at random and record her marital status. "At random" means that we give every such woman the same chance to be the one we choose. That is, we choose a random sample of size 1. The probability of any marital status is just the proportion of all women aged 25 to 29 who have that status—if we chose many women, this is the proportion we would get. Here is the set of probabilities:

Marital status:	Never married	Married	Widowed	Divorced
Probability:	0.455	0.494	0.005	0.046

This table gives a *probability model* for drawing a young woman at random and finding out her marital status. It tells us what are the possible outcomes (there are only four) and it assigns probabilities to these outcomes. The probabilities here are the proportions of all women in each

TABLE 18.1 Probabilities of winning Super Bowl XLVII

Team	Probability	Team	Probability
Green Bay Packers	1/6	Cincinnati Bengals	1/41
New Orleans Saints	1/7	Indianapolis Colts	1/41
New England Patriots	1/8	Buffalo Bills	1/51
New York Giants	1/11	Carolina Panthers	1/51
Baltimore Ravens	1/13	Denver Broncos	1/51
Houston Texans	1/13	Kansas City Chiefs	1/51
Philadelphia Eagles	1/13	Miami Dolphins	1/51
Pittsburgh Steelers	1/13	Oakland Raiders	1/51
San Diego Chargers	1/16	Tennessee Titans	1/51
Atlanta Falcons	1/21	Seattle Seahawks	1/61
Dallas Cowboys	1/21	Cleveland Browns	1/101
Detroit Lions	1/21	Jacksonville Jaguars	1/101
New York Jets	1/21	Minnesota Vikings	1/101
San Francisco 49ers	1/21	St. Louis Rams	1/101
Arizona Cardinals	1/31	Tampa Bay Buccaneers	1/101
Chicago Bears	1/31	Washington Redskins	1/101

Source: **www.footballlocks.com/nfl_futures_odds.shtml** (as of February 13, 2012).

Politically correct In 1950, the Russian mathematician B. V. Gnedenko (1912–1995) wrote a book, *The Theory of Probability,* that was popular around the world. The introduction contains a mystifying paragraph that begins "We note that the entire development of probability theory shows evidence of how its concepts and ideas were crystallized in a severe struggle between materialistic and idealistic conceptions." It turns out that "materialistic" is jargon for "Marxist-Leninist." It was good for the health of Russian scientists in the Stalin era to add such statements to their books.

marital class. That makes it clear that the probability that a woman is not married is just the sum of the probabilities of the three classes of unmarried women:

$$P(\text{not married}) = P(\text{never married}) + P(\text{widowed})$$
$$+ P(\text{divorced})$$
$$= 0.455 + 0.005 + 0.046 = 0.506$$

As a shorthand, we often write $P(\text{not married})$ for "the probability that the woman we choose is not married." You see that our model does more than assign a probability to each individual outcome—we can find the probability of any collection of outcomes by adding up individual outcome probabilities.

> ### Probability model
>
> A **probability model** for a random phenomenon describes all the possible outcomes and says how to assign probabilities to any collection of outcomes. We sometimes call a collection of outcomes an **event.**

Probability rules

Because the probabilities in this example are just the proportions of all women who have each marital status, they follow rules that say how proportions behave. Here are some basic rules that any probability model must obey:

A. **Any probability is a number between 0 and 1.** Any proportion is a number between 0 and 1, so any probability is also a number between 0 and 1. An event with probability 0 never occurs, and an event with probability 1 occurs on every trial. An event with probability 0.5 occurs in half the trials in the long run.

B. **All possible outcomes together must have probability 1.** Because some outcome must occur on every trial, the sum of the probabilities for all possible outcomes must be exactly 1.

C. **The probability that an event does not occur is 1 minus the probability that the event does occur.** If an event occurs in (say) 70% of all trials, it fails to occur in the other 30%. The probability that an event occurs and the probability that it does not occur always add to 100%, or 1.

D. **If two events have no outcomes in common, the probability that one or the other occurs is the sum of their individual probabilities.** If one event occurs in 40% of all trials, a different event occurs in 25% of all trials, and the two can never occur together, then one or the other occurs on 65% of all trials because 40% + 25% = 65%.

EXAMPLE 1 Marital status of young women

Look again at the probabilities for the marital status of young women. Each of the four probabilities is a number between 0 and 1. Their sum is

$$0.455 + 0.494 + 0.005 + 0.046 = 1$$

This assignment of probabilities satisfies Rules A and B. **Any assignment of probabilities to all individual outcomes that satisfies Rules A and B is legitimate.** That is, it makes sense as a set of probabilities. Rules C and D are then automatically true. Here is an example of the use of Rule C.

The probability that the woman we draw is not married is, by Rule C,

$$P(\text{not married}) = 1 - P(\text{married})$$
$$= 1 - 0.494 = 0.506$$

That is, if 49.4% are married, then the remaining 50.6% are not married. Rule D says that you can also find the probability that a woman is not married by adding the probabilities of the three distinct ways of being not married, as we did earlier. This gives the same result.

EXAMPLE 2   Rolling two dice

Rolling two dice is a common way to lose money in casinos. There are 36 possible outcomes when we roll two dice and record the up-faces in order (first die, second die). Figure 18.1 displays these outcomes. What probabilities should we assign?

Casino dice are carefully made. Their spots are not hollowed out, which would give the faces different weights, but are filled with white plastic of the same density as the red plastic of the body. For casino dice it is reasonable to assign the same probability to each of the 36 outcomes in Figure 18.1. Because these 36 probabilities must have sum 1 (Rule B), each outcome must have probability 1/36.

We are interested in the sum of the spots on the up-faces of the dice. What is the probability that this sum is 5? The event "roll a 5" contains four outcomes, and

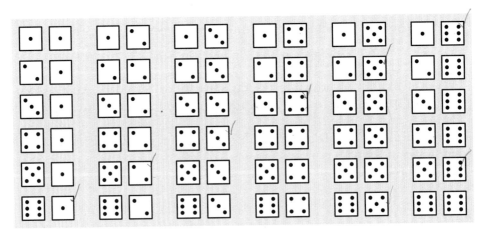

Figure 18.1 The 36 possible outcomes from rolling two dice, for Example 2.

its probability is the sum of the probabilities of these outcomes:

$$P(\text{roll a 5}) = P\left(\boxed{\cdot}\ \boxed{\vdots}\right) + P\left(\boxed{\cdot\cdot}\ \boxed{\vdots}\right) + P\left(\boxed{\vdots}\ \boxed{\cdot\cdot}\right) + P\left(\boxed{\vdots}\ \boxed{\cdot}\right)$$

$$= \frac{1}{36} + \frac{1}{36} + \frac{1}{36} + \frac{1}{36}$$

$$= \frac{4}{36} = 0.111$$

NOW IT'S YOUR TURN | **18.1 Rolling dice.** Suppose you roll two casino dice, as in Example 2. What is the probability that the sum of the spots on the up-faces is 7? 11? 7 or 11?

The rules tell us only what probability models *make sense.* They don't tell us whether the probabilities are *correct,* that is, whether they describe what actually happens in the long run. The probabilities in Example 2 are correct for casino dice. Inexpensive dice with hollowed-out spots are not balanced, and this probability model does not describe their behavior.

What about personal probabilities? Because they are personal, can't they be anything you want them to be? If your personal probabilities don't obey Rules A and B, you are entitled to your opinion, but we can say that your personal probabilities are **incoherent.** That is, they don't go together in a way that makes sense. So we usually insist that personal probabilities for all the outcomes of a random phenomenon obey Rules A and B. That is, the same rules govern both kinds of probability. For example, if you believe that the Green Bay Packers, the New Orleans Saints, and the New England Patriots each have probability 0.4 of winning Super Bowl XLVII, your personal probabilities would not obey Rule B.

Probability models for sampling

Choosing a random sample from a population and calculating a statistic such as the sample proportion is certainly a random phenomenon. The *distribution* of the statistic tells us what values it can take and how often it takes those values. That sounds a lot like a probability model.

EXAMPLE 3 A sampling distribution

Take a simple random sample of 2527 adults. Ask each whether they favor a constitutional amendment that would define marriage as being between a man and a woman. The proportion who say "Yes"

$$\hat{p} = \frac{\text{number who say "Yes"}}{2527}$$

is the sample proportion $\hat{p}$. Do this 1000 times and collect the 1000 sample proportions $\hat{p}$ from the 1000 samples. The histogram in Figure 18.2 shows the distribution of 1000 sample proportions when the truth about the population is that 50% would favor such an amendment. The results of random sampling are of course random: we can't predict the outcome of one sample, but the figure shows that the outcomes of many samples have a regular pattern.

We have seen Figure 18.2 before, in Chapter 13. In fact, we saw the histogram part of this figure earlier, in Chapters 3 and 11. This repetition reminds us that the regular pattern of repeated random samples is one of the big ideas of statistics. The Normal curve in the figure is a good approximation to the histogram. The histogram is the result of these particular 1000 SRSs. Think of the Normal curve as the idealized pattern we would get if we kept on taking SRSs from this population forever. That's exactly the idea of probability—the pattern we would see in the very long run. *The Normal curve assigns probabilities to sample proportions computed from random samples.*

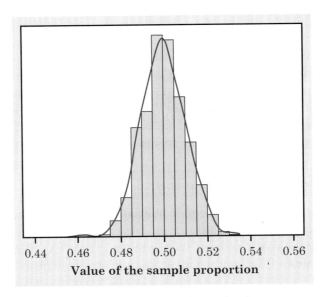

Figure 18.2 The sampling distribution of a sample proportion $\hat{p}$ from SRSs of size 2527 drawn from a population in which 50% of the members would give positive answers, for Example 3. The histogram shows the distribution from 1000 samples. The Normal curve is the ideal pattern that describes the results of a very large number of samples. (This figure was created using the Stata software package.)

This Normal curve has mean 0.5 and standard deviation about 0.010. The "95" part of the 68–95–99.7 rule says that 95% of all samples will give a $\hat{p}$ falling within 2 standard deviations of the mean. That's within 0.02 of 0.50, or between 0.48 and 0.52. We now have more concise language for this fact: the *probability* is 0.95 that between 48% and 52% of the people in a sample will say "Yes." The word "probability" says we are talking about what would happen in the long run, in very many samples.

A statistic from a large sample has a great many possible values. Assigning a probability to each individual outcome worked well for 4 marital classes or 36 outcomes of rolling two dice but is awkward when there are thousands of possible outcomes. Example 3 uses a different approach: assign probabilities to intervals of outcomes by using areas under a Normal density curve. Density curves have area 1 underneath them, which lines up nicely with total probability 1. The total area under the Normal curve in Figure 18.2 is 1, and the area between 0.48 and 0.52 is 0.95, which is the probability that a sample gives a result in that interval. When a Normal curve assigns probabilities, you can calculate probabilities from the 68–95–99.7 rule or from Table B of percentiles of Normal distributions. These probabilities satisfy Rules A to D.

Sampling distribution

The **sampling distribution** of a statistic tells us what values the statistic takes in repeated samples from the same population and how often it takes those values.

We think of a sampling distribution as assigning probabilities to the values the statistic can take. Because there are usually many possible values, sampling distributions are often described by a **density curve** such as a Normal curve.

EXAMPLE 4 Do you approve of gambling?

An opinion poll asks an SRS of 501 teens, "Generally speaking, do you approve or disapprove of legal gambling or betting?" Suppose that in fact exactly 50% of all teens would say "Yes" if asked. (This is close to what polls show to be true.) The poll's statisticians tell us that the sample proportion who say "Yes" will vary in repeated samples according to a Normal distribution with mean 0.5 and standard deviation about 0.022. This is the *sampling distribution* of the sample proportion $\hat{p}$.

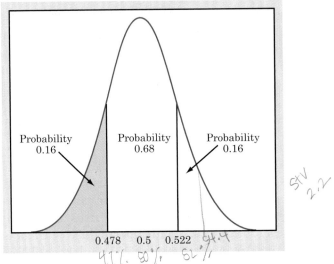

Figure 18.3 The Normal sampling distribution for Example 4. Because 0.478 is one standard deviation below the mean, the area under the curve to the left of 0.478 is 0.16.

The 68–95–99.7 rule says that the probability is 0.16 that the poll gets a sample in which fewer than 47.8% say "Yes." Figure 18.3 shows how to get this result from the Normal curve of the sampling distribution.

> **NOW IT'S YOUR TURN** | **18.2 Teen opinion poll.** Refer to Example 4. Using the 68–95–99.7 rule, what is the probability that fewer than 45.6% say "Yes"?

EXAMPLE 5 Using Normal percentiles*

What is the probability that the opinion poll in Example 4 will get a sample in which 52% or more say "Yes"? Because 0.52 is not 1, 2, or 3 standard deviations away from the mean, we can't use the 68–95–99.7 rule. We will use Table B of percentiles of Normal distributions.

To use Table B, first turn the outcome $\hat{p} = 0.52$ into a standard score by subtracting the mean of the distribution and dividing by its standard deviation:

$$\frac{0.52 - 0.5}{0.022} = 0.9$$

---

*Example 5 is optional.

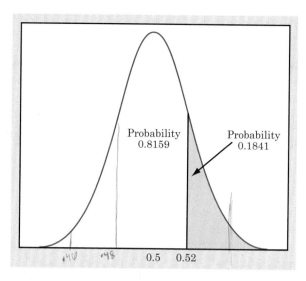

Figure 18.4 The Normal sampling distribution for Example 5. The outcome 0.52 has standard score 0.9, so Table B tells us that the area under the curve to the left of 0.52 is 0.8159.

Now look in Table B. A standard score of 0.9 is the 81.59 percentile of a Normal distribution. This means that the probability is 0.8159 that the poll gets a smaller result. By Rule C (or just the fact that the total area under the curve is 1), this leaves probability 0.1841 for outcomes with 52% or more answering "Yes." Figure 18.4 shows the probabilities as areas under the Normal curve.

STATISTICS IN SUMMARY

Chapter Specifics

- A **probability model** describes a random phenomenon by telling what outcomes are possible and how to assign probabilities to them.

- There are two simple ways to give a probability model. The first assigns a probability to each individual outcome. These probabilities must be numbers between 0 and 1 (Rule A) and they must add to exactly 1 (Rule B). To find the probability of any **event,** add the probabilities of the outcomes that make up the event.

- The second kind of probability model assigns probabilities as areas under a **density curve,** such as a Normal curve. The total probability is 1 because the total area under the curve is 1. This kind of probability model is often used to describe the **sampling distribution** of a statistic. This is the pattern of values of the statistic in many samples from the same population.

- All legitimate assignments of probability, whether data based or personal, obey the same **probability rules.** So the mathematics of probability is always the same.

 This chapter continues the discussion of probability that we began in Chapter 17. Here we examine the formal mathematics of probability, embodied in probability models and probability rules. Probability models and rules provide the tools for describing and predicting the long-run behavior of random phenomena.

In this chapter we also begin to formalize this process, first mentioned in Chapter 3, of using a statistic to estimate an unknown parameter. In particular, the sampling distribution will be the "probabilistic" tool we use to generalize from data produced by random samples and randomized comparative experiments to some wider population. Exactly how we do this will be the subject of Part IV.

CASE STUDY Look again at Table 18.1, discussed in the Case Study that opened the
EVALUATED chapter.

1. Do the probabilities in this table follow the rules given on page 425?

2. On some Web sites, the probabilities are given as betting odds in the form "A to B." In this form the probabilities tell you that a bet of $\$B$ will pay you $\$A$ if the team wins the Super Bowl. Thus, odds of A to B represent a probability of $B/(A + B)$ of winning. If you are a bookie setting betting odds, should you set the odds so the probabilities sum to greater than 1, exactly 1, or less than 1? Discuss. ■

CHAPTER 18 EXERCISES

For Exercise 18.1, see page 427; for Exercise 18.2, see page 430.

18.3 Moving up. A sociologist studying social mobility in Denmark finds that the probability that the son of a lower-class father remains in the lower class is 0.46. What is the probability that the son moves to one of the higher classes?

18.4 Causes of death. Government data assign a single cause for each death that occurs in the United States. The data show that the probability is 0.34 that a randomly chosen death was due to heart disease, and 0.23 that it was due to cancer. What is the probability that a death was due either to heart disease or to cancer? What is the probability that the death was due to some other cause?

18.5 Land in Canada. Choose an acre of land in Canada at random. The probability is 0.45 that it is forest and 0.03 that it is pasture.

(a) What is the probability that the acre chosen is not forested?

(b) What is the probability that it is either forest or pasture?

(c) What is the probability that a randomly chosen acre in Canada is something other than forest or pasture?

18.6 Our next president? A Gallup Poll on Presidents Day 2008 interviewed a random sample of 1007 adult Americans. Those in the sample were asked which former president they would like to bring back as the next president if they could. Here are the results:

Outcome	Probability
John F. Kennedy	0.23
Ronald Reagan	0.22
Abraham Lincoln	0.10
Someone else	?

These proportions are probabilities for the random phenomenon of choosing an adult American at random and asking her or his opinion.

(a) What must be the probability that the person chosen selects someone other than John F. Kennedy, Ronald Reagan, or Abraham Lincoln? Why?

(b) The event "I would select either John F. Kennedy or Ronald Reagan" contains the first two outcomes. What is its probability?

18.7 Rolling a die. Figure 18.5 displays several assignments of probabilities to the six faces of a die. We can learn which assignment is actually *correct* for a

	Probability			
Outcome	Model 1	Model 2	Model 3	Model 4
⚀	1/7	1/3	1/3	1
⚁	1/7	1/6	1/6	1
⚂	1/7	1/6	1/6	2
⚃	1/7	0	1/6	1
⚄	1/7	1/6	1/6	1
⚅	1/7	1/6	1/6	2

Figure 18.5 Four probability models for rolling a die, for Exercise 18.7.

particular die only by rolling the die many times. However, some of the assignments are not *legitimate* assignments of probability. That is, they do not obey the rules. Which are legitimate and which are not? In the case of the illegitimate models, explain what is wrong.

18.8 High school academic rank. Select a first-year college student at random and ask what his or her academic rank was in high school. Here are the probabilities, based on proportions from a large sample survey of first-year students:

Rank:	Top 20%	Second 20%	Third 20%	Fourth 20%	Lowest 20%
Probability:	0.44	0.26	0.23	0.06	0.01

(a) What is the sum of these probabilities? Why do you expect the sum to have this value?

(b) What is the probability that a randomly chosen first-year college student was not in the top 20% of his or her high school class?

(c) What is the probability that a first-year student was in the top 40% in high school?

18.9 Tetrahedral dice. Psychologists sometimes use tetrahedral dice to study our intuition about chance behavior. A tetrahedron (Figure 18.6) is a pyramid with 4 faces, each a triangle with all sides equal in length. Label the 4 faces of a tetrahedral die with 1, 2, 3, and 4 spots. Give a probability model for rolling such a die and recording the number of spots on the down-face. Explain why you think your model is at least close to correct.

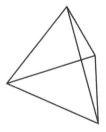

Figure 18.6 A tetrahedron. Exercises 18.9 and 18.11 concern dice with this shape.

18.10 Birth order. A couple plan to have three children. There are 8 possible arrangements of girls and boys. For example, GGB means the first two children are girls and the third child is a boy. All 8 arrangements are (approximately) equally likely.

(a) Write down all 8 arrangements of the sexes of three children. What is the probability of any one of these arrangements?

(b) What is the probability that the couple's children are 2 girls and 1 boy?

18.11 More tetrahedral dice. Tetrahedral dice are described in Exercise 18.9. Give a probability model for rolling two such dice. That is, write down all possible outcomes and give a probability to each. (Example 2 and Figure 18.1 may help you.) What is the probability that the sum of the down-faces is 5?

18.12 Roulette. A roulette wheel has 38 slots, numbered 0, 00, and 1 to 36. The slots 0 and 00 are colored green, 18 of the others are red, and 18 are black. The dealer spins the wheel and at the same time rolls a small ball along the wheel in the opposite direction. The wheel is carefully balanced so that the ball is equally likely to land in any slot when the wheel slows. Gamblers can bet on various combinations of numbers and colors.

(a) What is the probability of any one of the 38 possible outcomes? Explain your answer.

(b) If you bet on "red," you win if the ball lands in a red slot. What is the probability of winning?

(c) The slot numbers are laid out on a board on which gamblers place their bets. One column of numbers on the board contains all multiples of 3, that is, 3, 6, 9, ..., 36. You place a "column bet" that wins if any of these numbers comes up. What is your probability of winning?

18.13 Colors of M&M'S. If you draw an M&M candy at random from a bag of the candies, the candy you draw will have one of six colors. The probability of drawing each color depends on the proportion of each color among all candies made.

(a) Here are the probabilities of each color for a randomly chosen plain M&M:

Color:	Brown	Red	Yellow	Green	Orange	Blue
Probability:	0.30	0.20	0.20	0.10	0.10	?

What must be the probability of drawing a blue candy?

(b) The probabilities for peanut M&M'S are a bit different. Here they are:

Color:	Brown	Red	Yellow	Green	Orange	Blue
Probability:	0.20	0.20	0.20	0.10	0.10	?

What is the probability that a peanut M&M chosen at random is blue?

(c) What is the probability that a plain M&M is any of red, yellow, or orange? What is the probability that a peanut M&M has one of these colors?

18.14 Legitimate probabilities? In each of the following situations, state whether or not the given assignment of probabilities to individual outcomes is legitimate, that is, satisfies the rules of probability. If not, give specific reasons for your answer.

(a) When a coin is spun, $P(H) = 0.55$ and $P(T) = 0.45$.

(b) When two coins are tossed, $P(HH) = 0.4$, $P(HT) = 0.4$, $P(TH) = 0.4$, and $P(TT) = 0.4$.

(c) Plain M&M'S have not always had the mixture of colors given in Exercise 18.13. In the past there were no red candies and no blue candies. Tan had probability 0.10 and the other four colors had the same probabilities that are given in Exercise 18.13.

18.15 Immigration. Suppose that 45% of all adult Americans think that the level of immigration to the United States should be decreased. An opinion poll interviews 1020 randomly chosen Americans and records the sample proportion who feel that the level of immigration to this country should be decreased. This statistic will vary from sample to sample if the poll is repeated. The sampling distribution is approximately Normal with mean 0.45 and standard deviation about 0.016. Sketch this Normal curve and use it to answer the following questions.

(a) The mean of the population is 0.45. In what range will the middle 95% of all sample results fall?

(b) What is the probability that the poll gets a sample in which fewer than 43.4% say the level of immigration to this country should be decreased?

18.16 Airplane safety. Suppose that 68% of all adults think that airplanes would be safer places if airline passengers were banned from carrying on board any luggage, including purses, computers, and briefcases. An opinion poll plans to ask an SRS of 1023 adults about airplane safety. The proportion of the sample who think that airplanes would be safer if passengers were banned from carrying on board any luggage, including purses, computers, and briefcases, will vary if we take many samples from this same population. The sampling distribution of the sample proportion is approximately Normal with mean 0.68 and standard deviation about 0.015. Sketch this Normal curve and use it to answer the following questions.

(a) What is the probability that the poll gets a sample in which more than 71% of the people think that airplanes would be safer if passengers were banned from carrying on board any luggage, including purses, computers, and briefcases?

(b) What is the probability of getting a sample that misses the truth (68%) by 3.0% or more?

18.17 Immigration (optional). In the setting of Exercise 18.15, what is the probability of getting a sample in which more than 50% of those sampled think that the level of immigration to this country should be decreased? (Use Table B.)

18.18 Airplane safety (optional). In the setting of Exercise 18.16, what is the probability of getting a sample in which more than 70% think that airplanes would be safer if passengers were banned from carrying on board any luggage, including purses, computers, and briefcases? (Use Table B.)

18.19 Do you jog? An opinion poll asks an SRS of 1500 adults, "Do you happen to jog?" Suppose (as is approximately correct) that the population proportion who jog is $p = 0.15$. In a large number of samples, the proportion $\hat{p}$ who answer "Yes" will be approximately Normally distributed with mean 0.15 and standard deviation 0.009. Sketch this Normal curve and use it to answer the following questions.

(a) What percentage of many samples will have a sample proportion who jog that is 0.15 or less? Explain clearly why this percentage is the probability that $\hat{p}$ is 0.15 or less.

(b) What is the probability that $\hat{p}$ will take a value between 0.141 and 0.159? (Use the 68–95–99.7 rule.)

(c) Now use Rule C for probability: what is the probability that $\hat{p}$ does not lie between 0.141 and 0.159?

18.20 Applying to college. You ask an SRS of 1500 college students whether they applied for admission to any other college. Suppose that in fact 35% of all college students applied to colleges besides the one they are attending. (That's close to the truth.) The sampling distribution of the proportion $\hat{p}$ of your sample who say "Yes" is approximately Normal with mean 0.35 and standard deviation 0.01. Sketch this Normal curve and use it to answer the following questions.

(a) Explain in simple language what the sampling distribution tells us about the results of our sample.

(b) What percentage of many samples would have a $\hat{p}$ larger than 0.37? (Use the 68–95–99.7 rule.) Explain in simple language why this percentage is the probability of an outcome larger than 0.37.

(c) What is the probability that your sample will have a $\hat{p}$ less than 0.33?

(d) Use Rule D: what is the probability that your sample result will be either less than 0.33 or greater than 0.35?

18.21 Generating a sampling distribution. Let us illustrate the idea of a sampling distribution in the case of a very small sample from a very small population. The population is the scores of 10 students on an exam:

Student:	0	1	2	3	4	5	6	7	8	9
Score:	82	62	80	58	72	73	65	66	74	62

The parameter of interest is the mean score in this population. The sample is an SRS of size $n = 4$ drawn from the population. Because the students are labeled 0 to 9, a single random digit from Table A chooses one student for the sample.

(a) Find the mean of the 10 scores in the population. This is the population mean.

(b) Use Table A to draw an SRS of size 4 from this population. Write the four scores in your sample and calculate the mean $\bar{x}$ of the sample scores. This statistic is an estimate of the population mean.

(c) Repeat this process 10 times using different parts of Table A. Make a histogram of the 10 values of $\bar{x}$. You are constructing the sampling distribution of $\bar{x}$. Is the center of your histogram close to the population mean you found in (a)?

EXPLORING THE WEB

18.22 Web-based exercise. Oddsmakers often list the odds for certain sporting events on the Web. For example, one can find the current odds of winning the next Super Bowl for each NFL team. We found a list of such odds at **www.nsawins.com/super-bowl-odds.shtml.** When an oddsmaker says the odds are A to B of winning, he or she means that the probability of winning is $B/(A + B)$. For example, when we checked the Web site listed above, the odds that the Green Bay Packers would win Super Bowl XLVII were 6 to 1. This corresponds to a probability of winning of $1/(6 + 1) = 1/7$.

On the Web, find the current odds, according to an oddsmaker, of winning the Super Bowl for each NFL team. Convert these odds to probabilities. Do these probabilities satisfy Rules A and B given in this chapter? If they don't, can you think of a reason why?

18.23 Web-based exercise. Tables of areas under a Normal curve, like Table B at the back of this book, are still common but are also giving way to "applets" that let you find areas visually. Go to the *Statistics: Concepts and Controversies* Web site, **www.whfreeman.com/scc8e,** and look at the *Normal Curve* applet. This applet can be used for Normal curve calculations. If you did either Exercise 18.17 or 18.18, check your calculations using the *Normal Curve* applet.

NOTES AND DATA SOURCES

Page 435 Exercise 18.13: We found information about probabilities for the colors in a package of M&M'S at the Web site **www.madehow.com/Volume-3/M-M-Candy.html.**

Simulation

CASE STUDY In a horse race, the starting position can affect the outcome. It is advantageous to have a starting position that is near the inside of the track. To ensure fairness, starting position is determined by a random draw before the race. All positions are equally likely, so no horse has an advantage.

During the summer and autumn of 2007, the members of the Ohio Racing Commission noticed that one trainer appeared to have an unusually good run of luck. In 35 races, the horses of this trainer received one of the three inner positions 30 times. The number of horses in a race ranges from 6 to 10, but most of the time the number of entries is 9. Thus, for most races the chance of getting one of the three inside positions is 1/3.

The Ohio Racing Commission believed that the trainer's run of luck was too good to be true. A mathematician can show that the chance of receiving one of the three inside positions at least 30 times in 35 races

Daikusan/Getty Images

is very small. The Ohio Racing Commission therefore suspected that cheating had occurred. But the trainer had entered horses in nearly 1000 races over several years. Perhaps it was inevitable that at some time over the course of these nearly 1000 races the trainer would have a string of 35 races in which he received one of the three inside positions at least 30 times. It came to the attention of the Ohio Racing Commission only because it was one of those seemingly surprising coincidences discussed in Chapter 17. Perhaps the accusation of cheating was unfounded.

Calculating the probability that in a sequence of 1000 races with varying numbers of horses, there would occur a string of 35 consecutive races in which one would receive one of the three inside positions at least 30 times is very difficult. How one could go about determining this probability is the subject of this chapter. By the end of the chapter you will be able to describe how to find this probability. ■

Where do probabilities come from?

The probabilities of heads and tails in tossing a coin are very close to 1/2. In principle, these probabilities come from *data* on many coin tosses. Joe's personal probabilities for the winner of next year's Super Bowl come from Joe's own *individual judgment*. What about the probability that we get a run of three straight heads somewhere in 10 tosses of a coin? We can find this probability by *calculation from a model* that describes tossing coins. That is, once we have used data to give a probability model for the random fall of coins, we don't have to go back to the beginning every time we want the probability of a new event.

The big advantage of probability models is that they allow us to calculate the probabilities of complicated events starting from an assignment of probabilities to simple events like "heads on one toss." This is true whether the model reflects probabilities from data or personal probabilities. Unfortunately, the math needed to do probability calculations is often tough. Technology rides to the rescue: once we have a probability model, we can use a computer to *simulate* many repetitions. This is easier than math and much faster than actually running many repetitions in the real world. You might compare finding probabilities by simulation to practicing flying in a computer-controlled flight simulator. Both kinds of simulation are in wide use. Both have similar drawbacks: they are only as good as the model you start with. Flight simulators use a software model of how an airplane reacts. Simulations of probabilities use a probability model. We set the model in motion by using our old friends the random digits from Table A.

Simulation

Using random digits from a table or from computer software to imitate chance behavior is called **simulation.**

We look at simulation partly because it is how scientists and engineers really do find probabilities in complex situations. Simulations are used to develop strategies for reducing waiting times in lines to speak to a teller at banks, in lines to check in at airports, and in lines to vote during elections. Simulations are used to study the effects of changes in greenhouse gases on the climate. Simulations are used to study the effects of catastrophic events, such as the failure of a nuclear power plant, the effects on a structure of the explosion of a nuclear device, or the progression of a deadly, infectious disease in a densely populated city.

We also look at simulation because simulation forces us to think clearly about probability models. We'll do the hard part (setting up the model) and leave the easy part (telling a computer to do 10,000 repetitions) to those who really need the right probability at the end.

Simulation basics

Simulation is an effective tool for finding probabilities of complex events once we have a trustworthy probability model. We can use random digits to simulate many repetitions quickly. The proportion of repetitions on which an event occurs will eventually be close to its probability, so simulation can give good estimates of probabilities. The art of simulation is best learned from a series of examples.

EXAMPLE 1 Doing a simulation

Toss a coin 10 times. What is the probability of a run of at least 3 consecutive heads or 3 consecutive tails?

Really random digits
For purists, the RAND Corporation long ago published a book titled *One Million Random Digits*. The book lists 1,000,000 digits that were produced by a very elaborate physical randomization and really are random. An employee of RAND once told one of us that this is not the most boring book that RAND has ever published . . .

Step 1. Give a probability model. Our model for coin tossing has two parts:

- Each toss has probabilities 0.5 for a head and 0.5 for a tail.

- Tosses are *independent* of each other. That is, knowing the outcome of one toss does not change the probabilities for the outcomes of any other toss.

Step 2. Assign digits to represent outcomes. Digits in Table A of random digits will stand for the outcomes, in a way that matches the probabilities from Step 1. We know that each digit in Table A has probability 0.1 of being any one of 0, 1, 2, 3, 4, 5, 6, 7, 8, or 9, and that successive digits in the table are independent. Here is one assignment of digits for coin tossing:

- One digit simulates one toss of the coin.

- Odd digits represent heads; even digits represent tails.

This works because the 5 odd digits give probability 5/10 to heads (but any other assignment where half the digits represent heads is equally good). Successive digits in the table simulate independent tosses.

Step 3. Simulate many repetitions. Ten digits simulate 10 tosses, so looking at 10 consecutive digits in Table A simulates one repetition. Read many groups of 10 digits from the table to simulate many repetitions. Be sure to keep track of whether or not the event we want (a run of 3 heads or 3 tails) occurs on each repetition.

Here are the first three repetitions, starting at line 101 in Table A. We have underlined all runs of 3 or more heads or tails.

	Repetition 1	Repetition 2	Repetition 3
Digits	1 9 2 2 3 9 5 0 3 4	0 5 7 5 6 2 8 7 1 3	9 6 4 0 9 1 2 5 3 1
Heads/tails	H H T T H H H T H T	T H H H T T T H H H	H T T T H H T H H H
Run of 3?	YES	YES	YES

Continuing in Table A, we did 25 repetitions; 23 of them did have a run of 3 or more heads or tails. So we estimate the probability of a run by the proportion

$$\text{estimated probability} = \frac{23}{25} = 0.92$$

Of course, 25 repetitions are not enough to be confident that our estimate is accurate. Now that we understand how to do the simulation, we can tell a computer to do many thousands of repetitions. A long simulation (or hard mathematics) finds that the true probability is about 0.826. Most people think runs are somewhat unlikely, so even our short simulation challenges our intuition by showing that runs of 3 occur most of the time in 10 tosses.

Once you have gained some experience in simulation, setting up the probability model (Step 1) is usually the hardest part of the process. Although coin tossing may not fascinate you, the model in Example 1 is typical of many probability problems because it consists of *independent trials* (the tosses) all having the *same possible outcomes* with the *same probabilities*. Shooting 10 free throws and observing the sexes of 10 children have similar models and are simulated in much the same way. The new part of the model is independence, which simplifies our work because it allows us to simulate each of the 10 tosses in exactly the same way.

> **Independence**
>
> Two random phenomena are **independent** if knowing the outcome of one does not change the probabilities for outcomes of the other.

Independence, like all aspects of probability, can be verified only by observing many repetitions. It is plausible that repeated tosses of a coin are independent (the coin has no memory), and observation shows that they are. It seems less plausible that successive shots by a basketball player are

"I've had it! Simulated wood, simulated leather, simulated coffee, and now simulated probabilities!"

independent, but observation shows that they are at least very close to independent.

Step 2 (assigning digits) rests on the properties of the random digit table. Here are some examples of this step.

EXAMPLE 2 Assigning digits for simulation

(a) Choose a person at random from a group of which 70% are employed. One digit simulates one person:

$$0, \ 1, \ 2, \ 3, \ 4, \ 5, \ 6 = \text{employed}$$
$$7, \ 8, \ 9 = \text{not employed}$$

(b) Choose one person at random from a group of which 73% are employed. Now *two* digits simulate one person:

$$00, \ 01, \ 02, \ \ldots, \ 72 = \text{employed}$$
$$73, \ 74, \ 75, \ \ldots, \ 99 = \text{not employed}$$

We assigned 73 of the 100 two-digit pairs to "employed" to get probability 0.73. Representing "employed" by 01, 02, …, 73 and "not employed" by 74, 75, …, 99, 00 would also be correct.

(c) Choose one person at random from a group of which 50% are employed, 20% are unemployed, and 30% are not in the labor force. There are now three possible outcomes, but the principle is the same. One digit simulates one person:

$$0, \ 1, \ 2, \ 3, \ 4 = \text{employed}$$
$$5, \ 6 = \text{unemployed}$$
$$7, \ 8, \ 9 = \text{not in the labor force}$$

NOW IT'S YOUR TURN

19.1 Selecting cards at random. In a standard deck of 52 cards, there are 13 spades, 13 hearts, 13 diamonds, and 13 clubs. How would you assign digits for a simulation to determine the suit (spades, hearts, diamonds, or clubs) of a card chosen at random from a standard deck of 52 cards?

Thinking about independence

Before discussing more elaborate simulations, it is worth discussing the concept of independence further. We said above that independence can be verified only by observing many repetitions of random phenomena. It is probably more accurate to say that a *lack* of independence can be verified only by observing many repetitions of random phenomena. How does one recognize that two random phenomena are not independent? For example, how can we tell if tosses of a fair coin (that is, one for which the probability of a head is 0.5 and the probability of a tail is 0.5) are not independent?

One approach might be to apply the definition of "independence." For a sequence of tosses of a fair coin, one could compute the proportion of times in the sequence that a toss is followed by the same outcome: in other words, the frequency with which a head is followed by a head or a tail is followed by a tail. This proportion should be close to 0.5 if tosses are independent (knowing the outcome of one toss does not change the probabilities for outcomes of the next) and if many tosses have been observed.

EXAMPLE 3 Investigating independence

Suppose we tossed a fair coin 15 times and obtained the following sequence of outcomes:

Toss:	1	2	3	4	5	6	7	8	9	10	11	12	13	14	15
Outcome:	H	H	H	T	H	H	T	T	T	T	T	H	H	T	T

For the first 14 tosses, the following toss is the same 9 times. The proportion of times a toss is followed by the same outcome is

$$\text{proportion} = \frac{9}{14} = 0.64$$

For so few tosses, this would not be considered a large departure from 0.5.

Unfortunately, if the proportion of times a head is followed by a head or a tail is followed by a tail is close to 0.5, this does not necessarily imply that the tosses are independent. For example, suppose that instead of tossing the coin we simply placed the coin heads up or tails up according to the following pattern:

Trial:	1	2	3	4	5	6	7	8	9	10	11	12	13	14	15
Outcome:	H	H	T	T	H	H	T	T	H	H	T	T	H	H	T

We begin with two heads, followed by two tails, followed by two heads, etc. If we know the previous outcomes, we know exactly what the next outcome will be. Successive outcomes are not independent. However, looking at the first 14 outcomes, the proportion of times in this sequence that a head is followed by a head or a tail is followed by a tail is

$$\text{proportion} = \frac{7}{14} = 0.5$$

Thus, our approach can help us recognize when independence is lacking, but not when independence is present.

Another method for assessing independence is based on the concept of correlation, which we discussed in Chapter 14. If two random phenomena have numerical outcomes, and we observe both phenomena in a sequence of n trials, we can compute the correlation for the resulting data. If the random phenomena are independent, there will be no straight-line relationship between them, and the correlation should be close to 0.

It is not necessarily true that two random phenomena are independent if their correlation is 0. In Exercise 14.19 (page 328), there is a clear curved relationship between speed and mileage, but the correlation is 0. Independence implies no relationship at all, but correlation measures the strength of only a straight-line relationship.

Because independence implies no relationship, we would expect to see no overall pattern in a scatterplot of the data if the variables are independent. Looking at scatterplots is another method for determining if independence is lacking.

Was he good or was he lucky? When a baseball player hits .300, everyone applauds. A .300 hitter gets a hit in 30% of times at bat. Could a .300 year just be luck? Typical major leaguers bat about 500 times a season and hit about .260. A hitter's successive tries seem to be independent. From this model, we can calculate or simulate the probability of hitting .300. It is about 0.025. Out of 100 run-of-the-mill major league hitters, 2 or 3 each year will bat .300 just because they were lucky.

Many methods for assessing independence exist. For example, if trials are not independent and, say, tossing a head increases the probability that the next toss is also a head, then in a sequence of tosses we might expect to see unusually long runs of heads. We mentioned this idea of unusually long runs in Example 5 (page 406) of Chapter 17. Unusually long runs of made free throws would be expected if a basketball player has a "hot hand." However, careful study has shown that runs of baskets made or missed are no more frequent in basketball than would be expected if each shot is independent of the player's previous shots.

More elaborate simulations

The building and simulation of random models constitute a powerful tool of contemporary science, yet a tool that can be understood without advanced mathematics. What is more, doing a few simulations will increase your understanding of probability more than many pages of our prose. Having in mind these two goals of understanding simulation for itself and understanding simulation to understand probability, let us look at two more elaborate examples. The first still has independent trials, but there is no longer a fixed number of trials as there was when we tossed a coin 10 times.

EXAMPLE 4 We want a girl

A couple plan to have children until they have a girl or until they have three children, whichever comes first. What is the probability that they will have a girl among their children?

Step 1. The probability model is like that for coin tossing:

- Each child has probability 0.49 of being a girl and 0.51 of being a boy. (Yes, more boys than girls are born. Boys have higher infant mortality, so the sexes even out soon.)

- The sexes of successive children are independent.

Step 2. Assigning digits is also easy. Two digits simulate the sex of one child. We assign 49 of the 100 pairs to "girl" and the remaining 51 to "boy":

$$00, 01, 02, \ldots, 48 = \text{girl}$$
$$49, 50, 51, \ldots, 99 = \text{boy}$$

Step 3. To simulate one repetition of this childbearing strategy, read pairs of digits from Table A until the couple have either a girl or three children. The number of pairs needed to simulate one repetition depends on how quickly the couple get a girl. Here are 10 repetitions, simulated using line 130 of Table A. To interpret the pairs of digits, we have written G for girl and B for boy under them, have added space to separate repetitions, and under each repetition have written "+" if a girl was born and "−" if not.

6905	16	48	17	8717	40	9517	845340	648987	20
B G	G	G	G	B G	G	B G	B B G	B B B	G
+	+	+	+	+	+	+	+	−	+

In these 10 repetitions, a girl was born 9 times. Our estimate of the probability that this strategy will produce a girl is therefore

$$\text{estimated probability} = \frac{9}{10} = 0.9$$

Some mathematics shows that, if our probability model is correct, the true probability of having a girl is 0.867. Our simulated answer came quite close. Unless the couple are unlucky, they will succeed in having a girl.

Our final example has stages that are *not independent*. That is, the probabilities at one stage depend on the outcome of the preceding stage.

EXAMPLE 5   A kidney transplant

Morris's kidneys have failed and he is awaiting a kidney transplant. His doctor gives him this information for patients in his condition: 90% survive the transplant operation, and 10% die. The transplant succeeds in 60% of those who survive, and the other 40% must return to kidney dialysis. The proportions who survive for at least five years are 70% for those with a new kidney and 50% for those who return to dialysis. Morris wants to know the probability that he will survive for at least five years.

Step 1. The **tree diagram** in Figure 19.1 organizes this information to give a probability model in graphical form. The tree shows the three stages and the possible outcomes and probabilities at each stage. Each path through the tree leads to either survival for five years or to death in less than five years. To simulate Morris's fate, we must simulate each of the three stages. The probabilities at Stage 3 depend on the outcome of Stage 2.

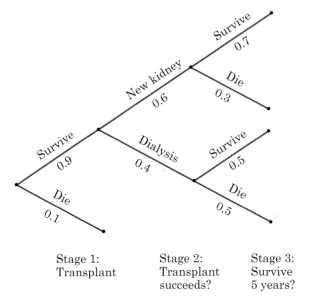

Stage 1:   Stage 2:   Stage 3:
Transplant   Transplant   Survive
                            succeeds?   5 years?

Figure 19.1 A tree diagram for the probability model of Example 5. Each branch point starts a new stage, with outcomes and their probabilities written on the branches. One repetition of the model follows the tree to one of its endpoints.

Step 2. Here is our assignment of digits to outcomes:

Stage 1:

$$0 = \text{die}$$

$$1, \ 2, \ 3, \ 4, \ 5, \ 6, \ 7, \ 8, \ 9 = \text{survive}$$

Stage 2:

$$0, \ 1, \ 2, \ 3, \ 4, \ 5 = \text{transplant succeeds}$$

$$6, \ 7, \ 8, \ 9 = \text{return to dialysis}$$

Stage 3 with kidney:

$$0, \ 1, \ 2, \ 3, \ 4, \ 5, \ 6 = \text{survive five years}$$

$$7, \ 8, \ 9 = \text{die}$$

Stage 3 with dialysis:

$$0, \ 1, \ 2, \ 3, \ 4 = \text{survive five years}$$

$$5, \ 6, \ 7, \ 8, \ 9 = \text{die}$$

The assignment of digits at Stage 3 depends on the outcome of Stage 2. That's lack of independence.

Step 3. Here are simulations of several repetitions, each arranged vertically. We used random digits from line 110 of Table A.

	Repetition 1	Repetition 2	Repetition 3	Repetition 4
Stage 1	3 Survive	4 Survive	8 Survive	9 Survive
Stage 2	8 Dialysis	8 Dialysis	7 Dialysis	1 Kidney
Stage 3	4 Survive	4 Survive	8 Die	8 Die

Morris survives five years in 2 of our 4 repetitions. Now that we understand how to arrange the simulation, we should turn it over to a computer to do many repetitions. From a long simulation or from mathematics, we find that Morris has probability 0.558 of living for at least five years.

NOW IT'S YOUR TURN   **19.2 Selecting cards at random.** In a standard deck of 52 cards, there are 13 spades, 13 hearts, 13 diamonds, and 13 clubs. Carry out a simulation to determine the probability that, when two cards are selected together at random from a standard deck, both have the same suit. Follow the steps given in Example 5 to help set up the simulation. Do your simulation 10 times and use the result to estimate the probability.

STATISTICS IN SUMMARY

Chapter Specifics

- We can use random digits to **simulate** random outcomes if we know the probabilities of the outcomes. Use the fact that each random digit has probability 0.1 of taking any one of the 10 possible digits and that all digits in the random number table are **independent** of each other.

- To simulate more complicated random phenomena, string together simulations of each stage. A common situation is several independent trials with the same possible outcomes and probabilities on each trial. Other simulations may require varying numbers of trials or different probabilities at each stage or may have stages that are not independent so that the probabilities at some stage depend on the outcome of earlier stages.

- The key to successful simulation is thinking carefully about the probability model. A **tree diagram** can be helpful by giving the probability model in graphical form.

In Chapter 18 we discussed probability models and the basic rules of probability. These allow us to compute the probabilities of simple events, but the math needed to find the probabilities of complicated events is often tough. In this chapter we learn how we can use simulations to determine the probabilities of complicated events. The underlying idea, introduced in Chapter 17, is that probability is the long-run proportion of times an event occurs. Simulating such an event many, many times using technology allows us to estimate this long-run proportion.

CASE STUDY Look again at the Case Study that opened the chapter. For simplicity,
EVALUATED assume that there are always 9 horses in a race so that the probability of receiving one of the three inside positions is 1/3 if assignments are random. Use what you have learned in this chapter to describe how you would do a simulation to estimate the probability that somewhere in a sequence of 1000 races a string of 35 consecutive races would occur in which at least 30 times one of the three inside positions was assigned. Begin by describing how you would simulate one case of 1000 races and, for these races, how you would determine whether somewhere in the 1000 races there was a string of 35 consecutive races in which at least 30 times one of the three inside positions was assigned. Do not try to carry out the simulation. This would be very time-consuming and is best left to a computer. (In fact, the Ohio Racing Commission hired a statistician to compute the probability that in a sequence of 1000 races with varying numbers of horses, there would occur a string of 35 consecutive races in which one would receive one of the three inside positions at least 30 times. The statistician used simulation to estimate this probability.) ∎

CHAPTER 19 EXERCISES

For Exercise 19.1, see page 444; for Exercise 19.2, see page 449.

19.3 Which party does it better? An opinion poll selects adult Americans at random and asks them, "Which political party, Democratic or Republican, do you think is better able to manage the economy?" Explain carefully how you would assign digits from Table A to simulate the response of one person in each of the following situations.

(a) Of all adult Americans, 50% would choose the Democrats and 50% the Republicans.

(b) Of all adult Americans, 60% would choose the Democrats and 40% the Republicans.

(c) Of all adult Americans, 40% would choose the Democrats, 40% would choose the Republicans, and 20% are undecided.

(d) Of all adult Americans, 49% would choose the Democrats, 38% the Republicans, and 13% are undecided. (These were the percentages in a September 2010 Gallup Poll.)

19.4 A small opinion poll. Suppose that 90% of a university's students favor abolishing evening exams. You ask 10 students chosen at random. What is the probability that all 10 favor abolishing evening exams?

(a) Give a probability model for asking 10 students independently of each other.

(b) Assign digits to represent the answers "Yes" and "No."

(c) Simulate 25 repetitions, starting at line 129 of Table A. What is your estimate of the probability?

19.5 Basic simulation. Use Table A to simulate the responses of 10 independently chosen adults in each of the four situations of Exercise 19.3.

(a) For situation (a), use line 110.

(b) For situation (b), use line 111.

(c) For situation (c), use line 112.

(d) For situation (d), use line 113.

19.6 Simulating an opinion poll. A Gallup Poll on Presidents Day 2008 interviewed a random sample of 1007 adult Americans. Those in the sample were asked which former president they would like to bring back as the next president if they could. The poll showed that about 10% of adult Americans would bring back Abraham Lincoln. Suppose that this is exactly true. Choosing an adult American at random then has probability 0.1 of getting one who would bring back Abraham Lincoln. If we interview adult Americans separately, we can assume that their responses are independent. We want to know the probability that a simple random sample of 100 adult Americans will contain at least 15 who say that they would bring back Abraham Lincoln. Explain carefully how to do this simulation and simulate *one* repetition of the poll using line 112 of Table A. How many of the 100 adult Americans said they would bring back Abraham Lincoln? Explain how you would estimate the probability by simulating many repetitions.

19.7 Course grades. Choose a student at random from all who took beginning statistics at Upper Wabash Tech in recent years. The probabilities for the student's grade are

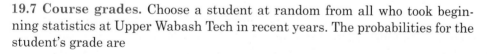

Grade:	A	B	C	D or F
Probability:	0.2	0.4	0.3	?

(a) What must be the probability of getting a D or an F?

(b) To simulate the grades of randomly chosen students, how would you assign digits to represent the four possible outcomes listed?

19.8 Class rank. Choose a college student at random and ask his or her class rank in high school. Probabilities for the outcomes are as follows:

Rank:	Top 20%	Second 20%	Third 20%	Lowest 40%
Probability:	0.40	0.30	0.20	?

(a) What must be the probability that a randomly chosen student was in the bottom 40% of his or her high school class?

(b) To simulate the class standing of randomly chosen students, how would you assign digits to represent the four possible outcomes listed?

19.9 More on course grades. In Exercise 19.7 you explained how to simulate the grade of a randomly chosen student in a statistics course. Five students on the same floor of a dormitory are taking this course. They don't study together, so their grades are independent. Use simulation to estimate the probability that all five get a C or better in the course. (Simulate 20 repetitions.)

19.10 More on class rank. In Exercise 19.8 you explained how to simulate the high school class rank of a randomly chosen college student. The Random Foundation decides to offer 8 randomly chosen students full college scholarships. What is the probability that no more than 3 of the 8 students chosen are in the bottom 40% of their high school class? Simulate 10 repetitions of the foundation's choices to estimate this probability.

19.11 LeBron's three-point shooting. The basketball player LeBron James makes about 30% of his three-point shots over an entire season. Take his probability of a success to be 0.3 on each shot. Using line 122 of Table A, simulate 25 repetitions of his performance in a game in which he shoots 10 three-point shots.

(a) Estimate the probability that LeBron makes at least half of his three-point shots.

(b) Examine the sequence of hits and misses in your 25 repetitions. How long was the longest run of shots made?

19.12 Tonya's free throws. Tonya makes 40% of her free throws in a long season. In a tournament game she shoots 5 shots late in the game and misses all of them. The fans think she was nervous, but the misses may simply be chance. Let's shed some light by estimating a probability.

(a) Describe how to simulate a single shot if the probability of making each shot is 0.4. Then describe how to simulate 5 independent shots.

(b) Simulate 50 repetitions of the 5 shots and record the number missed on each repetition. Use Table A, starting at line 125. What is the approximate probability that Tonya will miss all of the 5 shots?

19.13 Repeating an exam. Elaine is enrolled in a self-paced course that allows three attempts to pass an examination on the material. She does not study and has probability 2/10 of passing on any one attempt by luck. What is Elaine's probability of passing in three attempts? (Assume the attempts are independent because she takes a different examination on each attempt.)

(a) Explain how you would use random digits to simulate one attempt at the exam.

(b) Elaine will stop taking the exam as soon as she passes. (This is much like Example 4.) Simulate 50 repetitions, starting at line 120 of Table A. What is your estimate of Elaine's probability of passing the exam?

(c) Do you think the assumption that Elaine's probability of passing the exam is the same on each trial is realistic? Why?

19.14 A better model for repeating an exam. A more realistic probability model for Elaine's attempts to pass an exam in the previous exercise is as follows. On the first try she has probability 0.2 of passing. If she fails on the first try, her probability on the second try increases to 0.3 because she learned something from her first attempt. If she fails on two attempts, the probability of passing on a third attempt is 0.4. She will stop as soon as she passes. The course rules force her to stop after three attempts in any case.

(a) Make a tree diagram of Elaine's progress. Notice that she has different probabilities of passing on each successive try.

(b) Explain how to simulate one repetition of Elaine's tries at the exam.

(c) Simulate 50 repetitions and estimate the probability that Elaine eventually passes the exam. Use Table A, starting at line 130.

19.15 Gambling in ancient Rome. Tossing four astragali was the most popular game of chance in Roman times. Many throws of a present-day sheep's astragalus show that the approximate probability distribution for the four sides of the bone that can land uppermost are

Outcome	Probability
Narrow flat side of bone	1/10
Broad concave side of bone	4/10
Broad convex side of bone	4/10
Narrow hollow side of bone	1/10

The best throw of four astragali was the "Venus," when all four uppermost sides were different.

(a) Explain how to simulate the throw of a single astragalus. Then explain how to simulate throwing four astragali independently of each other.

(b) Simulate 25 throws of four astragali. Estimate the probability of throwing a Venus. Be sure to say what part of Table A you used.

19.16 The Asian stochastic beetle. We can use simulation to examine the fate of populations of living creatures. Consider the Asian stochastic beetle.

Females of this insect have the following pattern of reproduction:

- 20% of females die without female offspring, 30% have 1 female offspring, and 50% have 2 female offspring.

- Different females reproduce independently.

What will happen to the population of Asian stochastic beetles: will they increase rapidly, barely hold their own, or die out? It's enough to look at the female beetles, as long as there are some males around.

(a) Assign digits to simulate the offspring of one female beetle.

(b) Make a tree diagram for the female descendants of one beetle through three generations. The second generation, for example, can have 0, 1, or 2 females. If it has 0, we stop. Otherwise, we simulate the offspring of each second-generation female. What are the possible numbers of beetles after three generations?

(c) Use line 105 of Table A to simulate the offspring of 5 beetles to the third generation. How many descendants does each have after three generations? Does it appear that the beetle population will grow?

19.17 Two warning systems. An airliner has two independent automatic systems that sound a warning if there is terrain ahead (that means the airplane is about to fly into a mountain). Neither system is perfect. System A signals in time with probability 0.9. System B does so with probability 0.8. The pilots are alerted if either system works.

(a) Explain how to simulate the response of System A to terrain.

(b) Explain how to simulate the response of System B.

(c) Both systems are in operation simultaneously. Draw a tree diagram with System A as the first stage and System B as the second stage. Simulate 100 trials of the reaction to terrain ahead. Estimate the probability that a warning will sound. The probability for the combined system is higher than the probability for either A or B alone.

19.18 Playing craps. The game of craps is played with two dice. The player rolls both dice and wins immediately if the outcome (the sum of the faces) is 7 or 11. If the outcome is 2, 3, or 12, the player loses immediately. If he rolls any other outcome, he continues to throw the dice until he either wins by repeating the first outcome or loses by rolling a 7.

(a) Explain how to simulate the roll of a single fair die. (*Hint:* Just use digits 1 to 6 and ignore the others.) Then explain how to simulate a roll of two fair dice.

(b) Draw a tree diagram for one play of craps. In principle, a player could continue forever, but stop your diagram after four rolls of the dice. Use Table A, beginning at line 114, to simulate plays and estimate the probability that the player wins.

19.19 The airport van. Your company operates a van service from the airport to downtown hotels. Each van carries 7 passengers. Many passengers who reserve seats don't show up—in fact, the probability is 0.2 that a randomly chosen passenger will fail to appear. Passengers are independent. If you allow 9 reservations for each van, what is the probability that more than 7 passengers will appear? Do a simulation to estimate this probability.

19.20 A multiple-choice exam. Matt has lots of experience taking multiple-choice exams without doing much studying. He is about to take a quiz that has 10 multiple-choice questions, each with four possible answers. Here is Matt's personal probability model. He thinks that in 75% of questions he can eliminate one answer as obviously wrong; then he guesses from the remaining three. He then has probability 1/3 of guessing the right answer. For the other 25% of questions, he must guess from all four answers, with probability 1/4 of guessing correctly.

(a) Make a tree diagram for the outcome of a single question. Explain how to simulate Matt's success or failure on one question.

(b) Questions are independent. To simulate the quiz, just simulate 10 questions. Matt needs to get at least 5 questions right to pass the quiz. You could find his probability of passing by simulating many tries at the quiz, but we ask you to simulate just one try. Did Matt pass this quiz?

19.21 More on the airport van. Let's continue the simulation of Exercise 19.19. You have a backup van, but it serves several stations. The probability that it is available to go to the airport at any one time is 0.6. You want to know the probability that some passengers with reservations will be left stranded because the first van is full and the backup van is not available. Draw a tree diagram with the first van (full or not) as the first stage and the backup (available or not) as the second stage. In Exercise 19.19 you simulated a number of repetitions of the first stage. Add simulations of the second stage whenever the first van is full. What is your estimate of the probability of stranded passengers?

19.22 The birthday problem. A famous example in probability theory shows that the probability that at least two people in a room have the same birthday is already greater than 1/2 when 23 people are in the room. The probability model is

- The birth date of a randomly chosen person is equally likely to be any of the 365 dates of the year.

- The birth dates of different people in the room are independent.

To simulate birthdays, let each three-digit group in Table A stand for one person's birth date. That is, 001 is January 1 and 365 is December 31. Ignore leap years and skip groups that don't represent birth dates. Use line 139 of Table A to simulate birthdays of randomly chosen people until you hit the same date a second time. How many people did you look at to find two with the same birthday?

   With a computer, you could easily repeat this simulation many times. You could find the probability that at least 2 out of 23 people have the same birthday, or you could find the expected number of people you must question to find two with the same birthday. These problems are a bit tricky to do by math, so they show the power of simulation.

19.23 The multiplication rule. Here is another basic rule of probability: *if several events are independent, the probability that all of the events happen is the product of their individual probabilities.* We know, for example, that a child has probability 0.49 of being a girl and probability 0.51 of being a boy, and that the sexes of successive children are independent. So the probability that a couple's two children are two girls is $(0.49)(0.49) = 0.2401$. You can use this multiplication rule to calculate the probability that we simulated in Example 4.

(a) Write down all 8 possible arrangements of the sexes of three children, for example, BBB and BBG. Use the multiplication rule to find the probability of each outcome. Check your work by verifying that your 8 probabilities add to 1.

(b) The couple in Example 4 plan to stop when they have a girl or to stop at 3 children even if all are boys. Use your work from (a) to find the probability that they get a girl.

EXPLORING THE WEB

19.24 Web-based exercise. The basketball player LeBron James makes about 30% of his three-point shots over an entire season. At the end of a game, an announcer states that "LeBron had a hot hand tonight when it counted. He made all 3 of his three-point shots in the final two minutes of the game." Take LeBron's probability of making a three-point shot to be 0.3 on each shot. Simulate 100 repetitions of his performance at the end of a game in which he shoots 3 three-point shots, using the *Probability* applet at the *Statistics: Concepts and Controversies* Web site, **www.whfreeman .com/scc8e.** To do this, set the probability of a heads in the applet to be 0.3. What proportion of the time did LeBron make all 3 of his three-point shots?

19.25 Web-based exercise. There are Web sites that will generate random digits for simulations, providing an alternative to using Table A. One example is The Research Randomizer at **www.randomizer.org.** We discussed how to use the Research Randomizer in Example 4 (page 30) of Chapter 2. To use the Research Randomizer for simulations you will need to select "No" for the box that asks "Do you wish each number in a set to remain unique?" This will allow the same digit to appear multiple times. If you did any of Exercises 19.4, 19.5, 19.9, 19.10, 19.11, 19.12, 19.13, 19.14, 19.15, 19.16, 19.17, 19.18, 19.19, 19.20, 19.21, or 19.22, repeat your simulation using the Research Randomizer as a random number generator, in place of Table A.

19.26 Web-based exercise. The Web abounds in applets that simulate various random phenomena. One amusing probability problem is named *Buffon's needle.* Draw lines 1 inch apart on a piece of paper, then drop a 1-inch-long needle on the paper. What is the probability that the needle crosses a line? You can find both a solution by mathematics and a simulation at George Reese's site, **www.mste.uiuc.edu/reese/buffon/buffon .html.** Visit the site and try the simulation 100 times. What do you get as your estimate of the probability?

Personal sites sometimes vanish; a search on "Buffon's needle" will turn up alternative sites. The probability turns out to be $2/\pi$, where any circle's circumference is π times its diameter. So the simulation is also a way to calculate π, one of the most famous numbers in mathematics.

NOTES AND DATA SOURCES

Page 453 Exercise 19.15: F. N. David, *Games, Gods and Gambling,* Courier Dover Publications, 1998.

Page 453 Exercise 19.16: Stochastic beetles are well known in the folklore of simulation, if not in entomology. They are said to be the invention of Arthur Engle of the School Mathematics Study Group.

The House Edge: Expected Values

CASE STUDY If you gamble, you care about how often you will win. The probability of winning tells you what proportion of a large number of bets will be winners. You care even more about *how much* you will win, because winning a lot is better than winning a little.

There are a lot of ways to gamble. You can play games, like some of the multistate lotteries, that have enormous jackpots but very small probabilities of winning. You can play games like roulette for which the probability of winning is much larger than for a multistate lottery, but with smaller jackpots. Which is the better gamble: an enormous jackpot with extremely small odds or a modest jackpot with more reasonable odds?

In this chapter you will learn about expected values. Expected values provide one way to compare games of chance that have huge jackpots but small chances of winning with games with more modest jackpots but more reasonable chances of winning. By the end of this chapter you will be able to determine whether buying a multistate lottery ticket or simply playing red in roulette is a better bet. ▪

© Radius Images/Corbis

Expected values

Gambling on chance outcomes goes back to ancient times and has continued throughout history. Both public and private lotteries were common in the early years of the United States. After disappearing for a century or so, government-run gambling reappeared in 1964, when New Hampshire caused a furor by introducing a lottery to raise public revenue without raising taxes. The furor subsided quickly as larger states adopted the idea. Forty-two states and all Canadian provinces now sponsor lotteries. State lotteries made gambling acceptable as entertainment. Some form of legal

gambling is allowed in 48 of the 50 states. Over half of all adult Americans have gambled legally. They spend more betting than on spectator sports, video games, theme parks, and movie tickets combined. If you are going to bet, you should understand what makes a bet good or bad. As our introductory Case Study says, we care about how much we win as well as about our probability of winning.

EXAMPLE 1 The Tri-State Daily Numbers

Here is a simple lottery wager: the "Straight" from the Pick 3 game of the Tri-State Daily Numbers offered by New Hampshire, Maine, and Vermont. You pay $0.50 and choose a three-digit number. The state chooses a three-digit winning number at random and pays you $250 if your number is chosen. Because there are 1000 three-digit numbers, you have probability 1/1000 of winning. Here is the probability model for your winnings:

Outcome:	$0	$250
Probability:	0.999	0.001

What are your average winnings? The ordinary average of the two possible outcomes $0 and $250 is $125, but that makes no sense as the average winnings because $250 is much less likely than $0. In the long run you win $250 once in every 1000 bets and $0 on the remaining 999 of 1000 bets. (Of course if you play the game regularly, buying one ticket each time you play, after you have bought exactly 1000 Pick 3 tickets, there is no guarantee that you will win exactly once. Probabilities are only *long-run* proportions.) Your long-run average winnings from a ticket are

$$\$250 \frac{1}{1000} + \$0 \frac{999}{1000} = \$0.25$$

or 25 cents. You see that in the long run the state pays out one-half of the money bet and keeps the other half.

Here is a general definition of the kind of "average outcome" we used to evaluate the bets in Example 1.

Expected value

The **expected value** of a random phenomenon that has numerical outcomes is found by multiplying each outcome by its probability and then adding all the products.

In symbols, if the possible outcomes are $a_1, a_2, \ldots, a_k$ and their probabilities are $p_1, p_2, \ldots, p_k$, the expected value is

$$\text{expected value} = a_1 p_1 + a_2 p_2 + \cdots + a_k p_k$$

An expected value is an average of the possible outcomes, but it is not an ordinary average in which all outcomes get the same weight. Instead, each outcome is weighted by its probability, so that outcomes that occur more often get higher weights.

EXAMPLE 2 The Tri-State Daily Numbers, continued

The Straight wager in Example 1 pays off if you match the three-digit winning number exactly. You can choose instead to make a $1 Straight-Box (6-way) wager. You again choose a three-digit number, but you now have two ways to win. You win $292 if you exactly match the winning number, and you win $42 if your number has the same digits as the winning number, in any order. For example, if your number is 123, you win $292 if the winning number is 123 and $42 if the winning number is any of 132, 213, 231, 312, and 321. In the long run, you win $292 once every 1000 bets and $42 five times for every 1000 bets.

The probability model for the amount you win is

Outcome:	$0	$42	$292
Probability:	0.994	0.005	0.001

The expected value is

$$\text{expected value} = (\$0)(0.994) + (\$42)(0.005) + (\$292)(0.001) = \$0.502$$

We see that the StraightBox is a slightly better bet than the Straight bet, because the state pays out slightly more than half the money bet.

Rigging the lottery We have all seen televised lottery drawings in which numbered balls bubble about and are randomly popped out by air pressure. How might we rig such a drawing? In 1980, the Pennsylvania lottery was rigged by the host and several stage hands. They injected paint into all balls bearing 8 of the 10 digits. This weighed them down and guaranteed that all 3 balls for the winning number would have the remaining 2 digits. The perps then bet on all combinations of these digits. When 6-6-6 popped out, they won $1.2 million. Yes, they were caught.

The Tri-State Daily Numbers is unusual among state lottery games in that it pays a fixed amount for each type of bet. Most states pay off on the "pari-mutuel" system. New Jersey's Pick 3 game is typical: the state pools the money bet and pays out half of it, equally divided among the winning tickets. You still have probability 1/1000 of winning a Straight bet, but the amount your number 123 wins depends both on how much was bet on Pick 3 that day and on how many other players chose the number 123. Without fixed amounts, we can't find the expected value of today's bet on 123, but one thing is constant: the state keeps half the money bet.

The idea of expected value as an average applies to random outcomes other than games of chance. It is used, for example, to describe the uncertain return from buying stocks or building a new factory. Here is a different example.

EXAMPLE 3 How many vehicles per household?

What is the average number of motor vehicles in American households? The Census Bureau tells us that the distribution of vehicles per household (based on the 2000 census) is as follows:

Number of vehicles:	0	1	2	3	4	5
Proportion of households:	0.10	0.34	0.39	0.13	0.03	0.01

This is a probability model for choosing a household at random and counting its vehicles. (We ignored the very few households with more than 5 vehicles.) The expected value for this model is the average number of vehicles per household. This average is

$$\text{expected value} = (0)(0.10) + (1)(0.34) + (2)(0.39)$$
$$+ (3)(0.13) + (4)(0.03) + (5)(0.01)$$
$$= 1.68 \text{ vehicles per household}$$

> **NOW IT'S YOUR TURN**
>
> **20.1 Number of children.** The Census Bureau gives this distribution for the number of a family's own children under the age of 18 in American families in 2009:
>
> | Number of children: | 0 | 1 | 2 | 3 | 4 |
> |---|---|---|---|---|---|
> | Proportion: | 0.55 | 0.19 | 0.17 | 0.07 | 0.02 |
>
> In this table, 4 actually represents 4 or more. But for purposes of this exercise, assume that it means only families with exactly 4 children under the age of 18. This is also the probability distribution for the number of children under 18 in a randomly chosen family. The expected value of this distribution is the average number of children under 18 in a family. What is this expected value?

The law of large numbers

The definition of "expected value" says that it is an average of the possible outcomes, but an average in which outcomes with higher probability count more. We argued that the expected value is also the average outcome in another sense—it represents the long-run average we will actually see if we repeat a bet many times or choose many households at random. This is more than intuition. Mathematicians can prove, starting from just the basic rules of probability, that the expected value calculated from a probability model really is the "long-run average." This famous fact is called the *law of large numbers*.

> **The law of large numbers**
>
> According to the **law of large numbers,** if a random phenomenon with numerical outcomes is repeated many times independently, the mean of the actually observed outcomes approaches the expected value.

The law of large numbers is closely related to the idea of probability. In many independent repetitions, the proportion of each possible outcome will be close to its probability, and the average outcome obtained will be close to the expected value. These facts express the long-run regularity of chance events. They are the true version of the "law of averages," as we mentioned in Chapter 17.

The law of large numbers explains why gambling, which is a recreation or an addiction for individuals, is a business for a casino. The "house" in

High-tech gambling There are more than 700,000 slot machines in the United States. Once upon a time, you put in a coin and pulled the lever to spin three wheels, each with 20 symbols. No longer. Now the machines are video games with flashy graphics and outcomes produced by random number generators. Machines can accept many coins at once, can pay off on a bewildering variety of outcomes, and can be networked to allow common jackpots. Gamblers still search for systems, but in the long run the random number generator guarantees the house its 5% profit.

a gambling operation is not gambling at all. The average winnings of a large number of customers will be quite close to the expected value. The house has calculated the expected value ahead of time and knows what its take will be in the long run. There is no need to load the dice or stack the cards to guarantee a profit. Casinos concentrate on inexpensive entertainment and cheap bus trips to keep the customers flowing in. If enough bets are placed, the law of large numbers guarantees the house a profit. Life insurance companies operate much like casinos—they bet that the people who buy insurance will not die. Some do die, of course, but the insurance company knows the probabilities and relies on the law of large numbers to predict the average amount it will have to pay out. Then the company sets its premiums high enough to guarantee a profit.

Thinking about expected values

As with probability, it is worth exploring a few fine points about expected values and the law of large numbers.

How large is a large number?

The law of large numbers says that the actual average outcome of many trials gets closer to the expected value as more trials are made. It doesn't say how many trials are needed to guarantee an average outcome close to the expected value. That depends on the *variability* of the random outcomes.

The more variable the outcomes, the more trials are needed to ensure that the mean outcome is close to the expected value. Games of chance must be quite variable if they are to hold the interest of gamblers. Even a long evening in a casino has an unpredictable outcome. Gambles with extremely variable outcomes, like state lottos with their very large but very improbable jackpots, require impossibly large numbers of trials to ensure that the average outcome is close to the expected value. (The state doesn't rely on the law of large numbers—most lotto payoffs, unlike casino games, use the pari-mutuel system. In a pari-mutuel system, payoffs and payoff odds are determined by the actual amounts bet. In state lottos, for example, the payoffs are determined by the total amount bet after the state removes its share. In horse racing, payoff odds are determined by the relative amounts bet on the different horses.)

Though most forms of gambling are less variable than lotto, the practical answer to the applicability of the law of large numbers is that the expected value of the winnings for the house is positive and the house plays often enough to rely on it. Your problem is that the expected value of your winnings is negative. As a group, gamblers play as often as the house. Because their expected value is negative, as a group they lose money over time. However, this loss is not spread evenly among the many individual gamblers. Some win big, some lose big, and some break even. Much of the psychological allure of gambling is its unpredictability for the player. The business of gambling rests on the fact that the result is not unpredictable for the house.

STATISTICAL CONTROVERSIES

The State of Legalized Gambling

Most voters think that some forms of gambling should be legal, and the majority has its way: lotteries and casinos are common both in the United States and in other nations. The arguments in favor of allowing gambling are straightforward. Many people find betting entertaining and are willing to lose a bit of money in exchange for some excitement. Gambling doesn't harm other people, at least not directly. A democracy should allow entertainments that a majority supports and that don't do harm. State lotteries raise money for good causes such as education and are a kind of voluntary tax that no one is forced to pay.

These are some of the arguments for legalized gambling. What are some of the arguments against legalized gambling? Ask yourself, from which socioeconomic class do people who tend to play the lot-

tery come, and hence who bears the burden for this "voluntary tax"? For more information, see the sources listed in the Notes and Data Sources at the end of this chapter.

"I think the lottery is a great idea. If they raised taxes instead, we'd have to pay them."

Is there a winning system?

Serious gamblers often follow a system of betting in which the amount bet on each play depends on the outcome of previous plays. You might, for example, double your bet on each spin of the roulette wheel until you win—or, of course, until your fortune is exhausted. Such a system tries to take advantage of the fact that you have a memory even though the roulette wheel does not. Can you beat the odds with a system? No. Mathematicians have established a stronger version of the law of large numbers that says that if you do not have an infinite fortune to gamble with, your average winnings (the expected value) remain the same as long as successive trials of the game (such as spins of the roulette wheel) are independent. Sorry.

Finding expected values by simulation

How can we calculate expected values in practice? You know the mathematical recipe, but that requires that you start with the probability of each outcome. Expected values that are too difficult to compute in this way can be found by simulation. The procedure is as before: give a probability model, use random digits to imitate it, and simulate many repetitions. By the law of large numbers, the average outcome of these repetitions will be close to the expected value.

EXAMPLE 4 We want a girl, again

A couple plan to have children until they have a girl or until they have three children, whichever comes first. We simulated 10 repetitions of this scheme in Example 4 of Chapter 19 (page 446). There, we estimated the probability that they will have a girl among their children. Now we ask a different question: how many children, on the average, will couples who follow this plan have? That is, we want the expected number of children.

The simulation is exactly as before. The probability model says that the sexes of successive children are independent and that each child has probability 0.49 of being a girl. Here are our earlier simulation results—but rather than noting whether the couple did have a girl, we now record the number of children they have. Recall that a pair of digits simulates one child, with 00 to 48 (probability 0.49) standing for a girl.

6905	16	48	17	8717	40	9517	845340	648987	20
B G	G	G	G	B G	G	B G	B B G	B B B	G
2	1	1	1	2	1	2	3	3	1

The mean number of children in these 10 repetitions is

$$\overline{x} = \frac{2+1+1+1+2+1+2+3+3+1}{10} = \frac{17}{10} = 1.7$$

We estimate that if many couples follow this plan, they will average 1.7 children each. This simulation is too short to be trustworthy. Math or a long simulation shows that the actual expected value is 1.77 children.

> **NOW IT'S YOUR TURN** | **20.2 Kobe Bryant's field-goal shooting.** Kobe Bryant makes about 45% of the field-goal shots that he attempts. On average, how many field-goal shots must he take in a game before he makes his first shot? In other words, we want the expected number of shots he takes before he makes his first. Estimate this by using 10 simulations of sequences of shots, stopping when he makes his first. Use Example 4 to help you set up your simulation. What is your estimate of the expected value?

STATISTICS IN SUMMARY

Chapter Specifics

- The **expected value** is found as an average of all the possible outcomes, each weighted by its probability.

- When the outcomes are numbers, as in games of chance, we are often interested in the long-run average outcome. The **law of large numbers** says that the mean outcome in many repetitions eventually gets close to the expected value.

- If you don't know the outcome probabilities, you can estimate the expected value (along with the probabilities) by simulation.

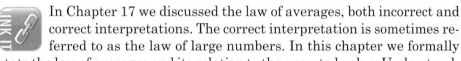 In Chapter 17 we discussed the law of averages, both incorrect and correct interpretations. The correct interpretation is sometimes referred to as the law of large numbers. In this chapter we formally state the law of averages and its relation to the expected value. Understanding the law of large numbers and expected values is helpful in understanding the behavior of games of chance, including state lotteries. Expected values provide a way you can compare games of chance with huge jackpots but small chances of winning with games with more modest jackpots but more reasonable chances of winning.

CASE STUDY EVALUATED Using what you learned in this chapter, answer the following questions.

1. An American roulette wheel has 38 slots, of which 18 are black, 18 are red, and 2 are green. When the wheel is spun, the ball is equally likely to come to rest in any of the slots. A bet of $1 on red will win $2 if the ball lands in a red slot. (When gamblers bet on red or black, the two green slots belong to the house.) Give a probability model for the winnings of a $1 bet on red and find the expected value of this bet.

2. The Web site for the Mega Millions lottery game gives the following table for the various prizes and the probability of winning:

Prize:	Jackpot	$250,000	$10,000	$150	$150
Probability:	1 in 175,711,536	1 in 3,904,701	1 in 689,065	1 in 15,313	1 in 13,781

Prize:	$10	$7	$3	$2
Probability:	1 in 844	1 in 306	1 in 141	1 in 75

(a) The jackpot always starts at $12,000,000 for the cash payout for a $1 ticket and grows each time there is no winner. What is the expected value of a $1 bet when the jackpot is $12,000,000? If there are multiple winners, they share the jackpot, but for purposes of this problem ignore this.

(b) The record jackpot was $656,000,000 on March 30, 2012. For a jackpot of this size, what is the expected value of a $1 bet, assuming the jackpot is not shared?

(c) For what size jackpot are the expected winnings of the Mega Millions the same size as the expected winnings for roulette that you calculated in Question 1?

3. Do you think roulette or the Mega Millions is the better game to play? Discuss. You may want to consider the fact that, as the jackpot grows, ticket sales increase. Thus, the chance that the jackpot is shared increases. The expected values in Question 2 overestimate the actual expected winnings. ∎

CHAPTER 20 EXERCISES

For Exercise 20.1, see page 463; for Exercise 20.2, see page 467.

20.3 The numbers racket. Pick 3 lotteries (Example 1, page 460) copy the numbers racket, an illegal gambling operation common in the poorer areas of large cities. States usually justify their lotteries by donating a portion of the proceeds to education. One version of a numbers racket operation works as follows. You choose any one of the 1000 three-digit numbers 000 to 999 and pay your local numbers runner $1 to enter your bet. Each day, one three-digit number is chosen at random and pays off $600. What is the expected value of a bet on the numbers? Is the numbers racket more or less favorable to gamblers than the Pick 3 game in Example 1?

20.4 Pick 4. The Tri-State Daily Numbers Pick 4 is much like the Pick 3 game of Example 1. Winning numbers for both are reported on television and in local newspapers. You pay $0.50 and pick a four-digit number. The state chooses a four-digit number at random and pays you $2500 if your number is chosen. What are the expected winnings from a $0.50 Pick 4 wager?

20.5 More Pick 4. Just as with Pick 3 (Example 2), you can make more elaborate bets in Pick 4. In the $1 StraightBox (24-way) bet, if you choose 1234 you win $2604 if the randomly chosen winning number is 1234, and you win $104 if the winning number has the digits 1, 2, 3, and 4 in any other order (there are 15 such other orders). What is the expected amount you win?

20.6 More roulette. An American roulette wheel has 38 slots, of which 18 are black, 18 are red, and 2 are green. When the wheel is spun, the ball is equally likely to come to rest in any of the slots. Gamblers bet on roulette by placing chips on a table that lays out the numbers and colors of the 38 slots in the roulette wheel. The red and black slots are arranged on the table in three columns of 12 slots each. A $1 column bet wins $3 if the ball lands in one of the 12 slots in that column. What is the expected amount such a bet wins? If you did the Case Study Evaluated, is a column bet more or less favorable to a gambler than a bet on red or black (see Question 1 in the Case Study Evaluated on page 467)?

20.7 Making decisions. The psychologist Amos Tversky did many studies of our perception of chance behavior. In its obituary of Tversky, the *New York Times* cited the following example.

(a) Tversky asked subjects to choose between two public health programs that affect 600 people. The first has probability 1/2 of saving all 600 and probability 1/2 that all 600 will die. The other is guaranteed to save exactly 400 of the 600 people. Find the expected number of people saved by the first program.

(b) Tversky then offered a different choice. One program has probability 1/2 of saving all 600 and probability 1/2 of losing all 600, while the other will definitely lose exactly 200 lives. What is the difference between this choice and that in (a)?

(c) Given option (a), most subjects choose the second program. Given option (b), most subjects choose the first program. Do the subjects appear to use expected values in their choice? Why do you think the choices differ in the two cases?

20.8 Making decisions. A six-sided die has two green and four red faces and is balanced so that each face is equally likely to come up. You must choose one of the following three sequences of colors:

    RGRRR
    RGRRRG
    GRRRRR

Now start rolling the die. You will win $25 if the first rolls give the sequence you chose.

(a) Which sequence has the highest probability? Why? (You can see which is most probable without actually finding the probabilities.) Because the $25 payoff is fixed, the most probable sequence has the highest expected value.

(b) In a psychological experiment, 63% of 260 students who had not studied probability chose the second sequence. Based on the discussion of "myths about chance behavior" in Chapter 17, explain why most students did not choose the sequence with the best chance of winning.

20.9 Estimating sales. Gain Communications sells aircraft communications units. Next year's sales depend on market conditions that cannot be predicted exactly. Gain follows the modern practice of using probability estimates of sales. The sales manager estimates next year's sales as follows:

Units sold:	7000	8000	9000	10,000
Probability:	0.3	0.4	0.2	0.1

These are personal probabilities that express the informed opinion of the sales manager. What is the sales manager's expected value of next year's sales?

20.10 Keno. Keno is a popular game in casinos. Balls numbered 1 to 80 are tumbled in a machine as the bets are placed, then 20 of the balls are chosen at random. Players select numbers by marking a card. Here are two of the simpler Keno bets. Give the expected winnings for each.

(a) A $1 bet on "Mark 1 number" pays $3 if the single number you mark is one of the 20 chosen; otherwise, you lose your dollar.

(b) A $1 bet on "Mark 2 numbers" pays $12 if both your numbers are among the 20 chosen. The probability of this is about 0.06. Is Mark 2 a more or a less favorable bet than Mark 1?

20.11 Rolling two dice. Example 2 of Chapter 18 (page 426) gives a probability model for rolling two casino dice and recording the number of spots on each of the two up-faces. That example also shows how to find the probability that the total number of spots showing is 5. Follow that method to give a probability model for the *total* number of spots. The possible outcomes are 2, 3, 4, ..., 12. Then use the probabilities to find the expected value of the total.

20.12 The Asian stochastic beetle. We met this insect in Exercise 19.16 (page 453). Females have this probability model for their number of female offspring:

Offspring:	0	1	2
Probability:	0.2	0.3	0.5

(a) What is the expected number of female offspring?

(b) Use the law of large numbers to explain why the population should grow if the expected number of female offspring is greater than 1 and die out if this expected value is less than 1.

20.13 An expected rip-off? A "psychic" runs the following ad in a magazine:

Expecting a baby? Renowned psychic will tell you the sex of the unborn child from any photograph of the mother. Cost, $20. Money-back guarantee.

This may be a profitable con game. Suppose that the psychic simply replies "boy" to all inquiries. In the worst case, everyone who has a girl will ask for her money back. Find the expected value of the psychic's profit by filling in the table below.

Sex of child	Probability	The psychic's profit
Boy	0.51	
Girl	0.49	

20.14 The Asian stochastic beetle again. In Exercise 20.12 you found the expected number of female offspring of the Asian stochastic beetle. Simulate the offspring of 100 beetles and find the mean number of offspring for these 100 beetles. Compare this mean with the expected value from Exercise 20.12. (The law of large numbers says that the mean will be very close to the expected value if we simulate enough beetles.)

20.15 Life insurance. You might sell insurance to a 21-year-old friend. The probability that a man aged 21 will die in the next year is about 0.0015. You decide to charge $2000 for a policy that will pay $1,000,000 if your friend dies.

(a) What is your expected profit on this policy?

(b) Although you expect to make a good profit, you would be foolish to sell a single policy only to your friend. Why?

(c) A life insurance company that sells thousands of policies, on the other hand, would do very well selling policies on exactly these same terms. Explain why.

20.16 Family size. The Census Bureau gives this distribution for the number of people in American families in 2009:

Family size:	2	3	4	5	6	7
Proportion:	0.45	0.22	0.19	0.09	0.03	0.02

(*Note:* In this table, 7 actually represents families of size 7 or greater. But for purposes of this exercise, assume that it means only families of size exactly 7.)

(a) This is also the probability distribution for the size of a randomly chosen family. The expected value of this distribution is the average number of people in a family. What is this expected value?

(b) Suppose you take a random sample of 1000 American families. How many of these families will be of size 2? Sizes 3 to 7?

(c) Based on your calculations in part (b), how many people are represented in your sample of 1000 families? (*Hint:* The number of individuals in your sample who live in families of size 7 is 7 times the number of families of size 7. Repeat this reasoning to determine the number of individuals in families of sizes 2 to 6. Add the results to get the total number of people represented in your sample.)

(d) Calculate the probability distribution for the family size lived in by individual people. Describe the shape of this distribution. What does this shape tell you about family structure?

20.17 Course grades. The distribution of grades in a large statistics course is as follows:

Grade:	A	B	C	D	F
Probability:	0.1	0.4	0.3	0.1	0.1

To calculate student grade point averages, grades are expressed in a numerical scale with A = 4, B = 3, and so on down to F = 0.

(a) Find the expected value. This is the average grade in this course.

(b) Explain how to simulate choosing students at random and recording their grades. Simulate 50 students and find the mean of their 50 grades. Compare this estimate of the expected value with the exact expected value from (a). (The law of large numbers says that the estimate will be very accurate if we simulate a very large number of students.)

20.18 We really want a girl. Example 4 (page 466) estimates the expected number of children a couple will have if they keep going until they get a girl or until they have three children. Suppose that they set no limit on the number of children but just keep going until they get a girl. Their expected number of children must now be higher than in Example 4. How would you simulate such a couple's children? Simulate 25 repetitions. What is your estimate of the expected number of children?

20.19 Play this game, please. OK, friends, we've got a little deal for you. We have a fair coin (heads and tails each have probability 1/2). Toss it twice. If two heads come up, you win right there. If you get any result other than two heads, we'll give you another chance: toss the coin twice more, and if you get two heads, you win. (Of course, if you fail to get two heads on the second try, we win.) Pay us a dollar to play. If you win, we'll give you your dollar back plus another dollar.

(a) Make a tree diagram for this game. Use the diagram to explain how to simulate one play of this game.

(b) Your dollar bet can win one of two amounts: 0 if we win and $2 if you win. Simulate 50 plays, using Table A, starting at line 125. Use your simulation to estimate the expected value of the game.

20.20 A multiple-choice exam. Charlene takes a quiz with 10 multiple-choice questions, each with five answer choices. If she just guesses independently at each question, she has probability 0.20 of guessing right on each. Use simulation to estimate Charlene's expected number of correct answers. (Simulate 20 repetitions.)

20.21 Repeating an exam. Exercise 19.14 (page 453) gives a model for up to three attempts at an exam in a self-paced course. In that exercise, you simulated 50 repetitions to estimate Elaine's probability of passing the exam. Use those simulations (or do 50 new repetitions) to estimate the expected number of tries Elaine will make.

20.22 A common expected value. Here is a common setting that we simulated in Chapter 19: there are a fixed number of independent trials with the same two outcomes and the same probabilities on each trial. Tossing a coin, shooting basketball free throws, and observing the sex of newborn babies are all examples of this setting. Call the outcomes "hit" and "miss." We can see what the expected number of hits should be. If LeBron James shoots 12 three-point shots and has probability 0.3 of making each one, the expected number of hits is 30% of 12, or 3.6. By the same reasoning, if we have n trials with probability p of a hit on each trial, the expected number of hits is np. This fact can be proved mathematically. Can we verify it by simulation?

Simulate 10 tosses of a fair coin 50 times. (To do this quickly, use the first 10 digits in each of the 50 rows of Table A, with odd digits meaning a head and even digits a tail.) What is the expected number of heads by the np formula? What is the mean number of heads in your 50 repetitions?

20.23 Casino winnings. What is a secret, at least to naive gamblers, is that in the real world a casino does much better than expected values suggest. In fact, casinos keep a bit over 20% of the money gamblers spend on roulette chips. That's because players who win keep on playing. Think of a player who gets back exactly 95% of each dollar bet. After one bet, he has 95 cents.

(a) After two bets, how much does he have of his original dollar bet?

(b) After three bets, how much does he have of his original dollar bet?

Notice that the longer he keeps recycling his original dollar, the more of it the casino keeps. Real gamblers don't get a fixed percentage back on each bet, but even the luckiest will lose his stake if he plays long enough. The casino keeps 5.3 cents of every dollar bet but 20 cents of every dollar that walks in the door.

EXPLORING THE WEB

20.24 Web-based exercise. Most states have a lotto game that offers large prizes for choosing (say) 6 out of 51 numbers. If your state has a lotto game, find out what percentage of the money bet is returned to the bettors in the form of prizes. You should be able to find this information on the Web. What percentage of the money bet is used by the state to pay lottery expenses? What percentage is net revenue to the state? For what purposes does the state use lottery revenue?

20.25 Web-based exercise. As mentioned in Exercise 19.25, there are Web sites that will generate random numbers. These Web sites can be used in place of Table A to select random digits for simulations. The Research Randomizer at **www.randomizer.org** that we discussed in Example 4 (page 30) of Chapter 2 is one such site. To use the Research Randomizer for simulations you will need to select "No" for the box that asks "Do you wish each number in a set to remain unique?" This will allow the same digit to appear multiple times. If you did any of Exercises 20.14, 20.17, 20.18, 20.19, 20.20, 20.21, or 20.22, repeat your simulation using the Research Randomizer as a random number generator, in place of Table A.

20.26 Web-based exercise. Information about basketball players can be found at **www.basketball-reference.com.** Go to this Web site and find the percentage of three-point shots that Steve Nash has made in his career. (Nash is among the career leaders in percentage of three-point shots made.) On average, how many three-point shots must he take in a game before he makes his first shot? In other words, we want the expected number of shots he takes before he makes his first. Estimate this by using 10 simulations of sequences of three-point shots, stopping when he makes his first. What is your estimate of the expected value?

NOTES AND DATA SOURCES

Page 465 For the Statistical Controversies feature, visit the Web site of the National Coalition against Legalized Gambling at **www.ncalg.org.** For the defense by the casino industry, visit the Web site of the American Gaming Association at **www.americangaming.org.** State lotteries make their case via the North American Association of State and Provincial Lotteries, **www.naspl.org.** The National Indian Gaming Association, **www.indiangaming.org,** is more assertive; click on *Resources* to go to the media center index, and then click on *Indian Gaming Facts,* for example. The report of a commission established by Congress to study the impact of gambling is at **govinfo .library.unt.edu/ngisc/.** You'll find lots of facts and figures at all these sites. See also "Gambling on the future," *Economist,* June 26, 1999.

Page 468 The Web site for the Mega Millions lottery is **www.megamillions.com.**

Page 469 Exercise 20.7: Obituary by Karen Freeman, *New York Times,* June 6, 1996.

Page 469 Exercise 20.8: Based on A. Tversky and D. Kahneman, "Extensional versus intuitive reasoning: the conjunction fallacy in probability judgment," *Psychological Review,* 90 (1983), pp. 293–315.

Some phenomena are random. That is, although their individual outcomes are unpredictable, there is a regular pattern in the long run. Gambling devices (rolling dice, spinning roulette wheels) and taking an SRS are examples of random phenomena. Probability and expected value give us a language to describe randomness. Random phenomena are not haphazard or chaotic any more than random sampling is haphazard. Randomness is instead a kind of order in the world, a long-run regularity as opposed to either chaos or a determinism that fixes events in advance. Chapter 17 discusses the idea of randomness, Chapter 18 presents basic facts about probability, and Chapter 20 discusses expected values.

When randomness is present, probability answers the question "How often in the long run?" and expected value answers the question "How much on the average in the long run?" The two answers are tied together by the definition of "expected value" in terms of probabilities. Much work with probability starts with a probability model that assigns probabilities to the basic outcomes. Any such model must obey the rules of probability. Another kind of probability model uses a density curve such as a Normal curve to assign probabilities as areas under the curve. Personal probabilities express an individual's judgment of how likely some event is. Personal probabilities for several possible outcomes must also follow the rules of probability if they are to be consistent with each other.

To calculate the probability of a complicated event without using complicated math, we can use random digits to simulate many repetitions. You can also find expected values by simulation. Chapter 19 shows how to do simulations. First give a probability model for the outcomes, then assign random digits to imitate the assignment of probabilities. The table of random digits now imitates repetitions. Keep track of the proportion of repetitions on which an event occurs to estimate its probability. Keep track of the mean outcome to estimate an expected value.

PART III SUMMARY

Here are the most important skills you should have acquired after reading Chapters 17 to 20.

A. RANDOMNESS AND PROBABILITY

1. Recognize that some phenomena are random. Probability describes the long-run regularity of random phenomena.

2. Understand the idea of the probability of an event as the proportion of times the event occurs in very many repetitions of a random phenomenon. Use the idea of probability as long-run proportion to think about probability.

3. Recognize that short runs of random phenomena do not display the regularity described by probability. Accept that randomness is unpredictable in the short run, and avoid seeking causal explanations for random occurrences.

B. PROBABILITY MODELS

1. Use basic probability facts to detect illegitimate assignments of probability: any probability must be a number between 0 and 1, and the total probability assigned to all possible outcomes must be 1.

2. Use basic probability facts to find the probabilities of events that are formed from other events: The probability that an event does not occur is 1 minus its probability. If two events cannot occur at the same time, the probability that one or the other occurs is the sum of their individual probabilities.

3. When probabilities are assigned to individual outcomes, find the probability of an event by adding the probabilities of the outcomes that make it up.

4. When probabilities are assigned by a Normal curve, find the probability of an event by finding an area under the curve.

C. EXPECTED VALUE

1. Understand the idea of expected value as the average of numerical outcomes in very many repetitions of a random phenomenon.

2. Find the expected value from a probability model that lists all outcomes and their probabilities (when the outcomes are numerical).

D. SIMULATION

1. Specify simple probability models that assign probabilities to each of several stages when the stages are independent of each other.

2. Assign random digits to simulate such models.

3. Estimate either a probability or an expected value by repeating a simulation many times.

PART III REVIEW EXERCISES

Review exercises are short and straightforward exercises that help you solidify the basic ideas and skills in each part of this book. We have provided "hints" that indicate where you can find the relevant material for the odd-numbered problems.

III.1 What's the probability? Open your local Yellow Pages telephone directory to any page in the Business White Pages listing. Look at the last four digits of each telephone number, the digits that specify an individual number within an exchange given by the first three digits. Note the first of these four digits in each of the first 100 telephone numbers on the page.

(a) How many of the digits are 1, 2, or 3? What is the approximate probability that the first of the four "individual digits" in a telephone number is 1, 2, or 3? (*Hint:* See page 403.)

(b) If all 10 possible digits had the same probability, what would be the probability of getting a 1, 2, or 3? Based on your work in (a), do you think the first of the four "individual digits" in telephone numbers is equally likely to be any of the 10 possible digits? (*Hint:* See page 425.)

III.2 Course grades. Choose a student at random from all who took Math 101 in recent years. The probabilities for the student's grade are

Grade:	A	B	C	D	F
Probability:	0.2	0.3	0.3	0.1	?

(a) What must be the probability of getting an F?

(b) To simulate the grades of randomly chosen students, how would you assign digits to represent the five possible outcomes listed?

III.3 Blood types. Choose a person at random and record his or her blood type. Here are the probabilities for each blood type:

Blood type:	Type O	Type A	Type B	Type AB
Probability:	0.4	0.3	0.2	?

(a) What must be the probability that a randomly chosen person has Type AB blood? (*Hint:* See page 425.)

(b) To simulate the blood types of randomly chosen people, how would you assign digits to represent the four types? (*Hint:* See page 441.)

III.4 Course grades. If you choose 4 students at random from all those who have taken the course described in Exercise III.2, what is the probability that all the students chosen got a B or better? Simulate 10 repetitions of this random choosing and use your results to estimate the probability. (Your estimate from only 10 repetitions isn't reliable, but if you can do 10, you could do 10,000.)

III.5 Blood types. People with Type B blood can receive blood donations from other people with either Type B or Type O blood. Tyra has Type B blood. What is the probability that 2 or more of Tyra's 6 close friends can donate blood to her? Using your work in Exercise III.3, simulate 10 repetitions and estimate this probability. (Your estimate from just 10 repetitions isn't reliable, but you have shown in principle how to find the probability.) (*Hint:* See page 441.)

III.6 Course grades. Choose a student at random from the course described in Exercise III.2 and observe what grade that student earns (A = 4, B = 3, C = 2, D = 1, F = 0).

(a) What is the expected grade of a randomly chosen student?

(b) The expected grade is not one of the 5 grades possible for one student. Explain why your result nevertheless makes sense as an expected value.

III.7 Dice. What is the expected number of spots observed in rolling a carefully balanced die once? (*Hint:* See page 461.)

III.8 Profit from a risky investment. Rotter Partners is planning a major investment. The amount of profit X is uncertain, but a probabilistic estimate gives the following distribution (in millions of dollars):

Profit:	1	2	3	5	20
Probability:	0.4	0.2	0.2	0.1	0.1

What is the expected value of the profit?

III.9 Poker. Deal a five-card poker hand from a shuffled deck. The probabilities of several types of hand are approximately as follows:

Hand:	Worthless	One pair	Two pairs	Better hands
Probability:	0.50	0.42	0.05	?

(a) What must be the probability of getting a hand better than two pairs? (*Hint:* See page 425.)

(b) What is the expected number of hands a player is dealt before the first hand better than one pair appears? Explain how you would use simulation to answer this question, then simulate just 2 repetitions. (*Hint:* See page 466.)

III.10 How much education? The 2010 *Statistical Abstract* gives this distribution of education for a randomly chosen American over 25 years old:

Education:	Less than high school	High school graduate	College, no bachelor's	Associate's degree	Bachelor's degree	Advanced degree
Probability:	0.134	0.312	0.172	0.088	0.191	0.103

(a) How do you know that this is a legitimate probability model?

(b) What is the probability that a randomly chosen person over age 25 has at least a high school education?

(c) What is the probability that a randomly chosen person over age 25 has at least a bachelor's degree?

III.11 Language study. Choose a student in grades 9 to 12 at random and ask if he or she is studying a language other than English. Here is the distribution of results:

Language:	Spanish	French	German	Latin	All others	None
Probability:	0.302	0.080	0.021	0.013	0.022	0.562

(a) Explain why this is a legitimate probability model. (*Hint:* See page 425.)

(b) What is the probability that a randomly chosen student is studying a language other than English? (*Hint:* See page 425.)

(c) What is the probability that a randomly chosen student is studying French, German, or Spanish? (*Hint:* See page 425.)

III.12 Choosing at random. Abby, Deborah, Mei-Ling, Sam, and Roberto work in a firm's public relations office. Their employer must choose two of them to attend a conference in Paris. To avoid unfairness, the choice will be made by drawing two names from a hat. (This is an SRS of size 2.)

(a) Write down all possible choices of two of the five names. These are the possible outcomes.

(b) The random drawing makes all outcomes equally likely. What is the probability of each outcome?

(c) What is the probability that Mei-Ling is chosen?

(d) What is the probability that neither of the two men (Sam and Roberto) is chosen?

III.13 Anxiety about college tuition. The National Center for Public Policy and Higher Education asked randomly chosen adults their response to the following statement: "Students have to borrow too much money to pay for their college education." Assume that the results of the poll accurately reflect the opinions of all adults. Here is the distribution of responses:

	Agree strongly	Agree somewhat	Disagree somewhat	Disagree strongly	Don't know
Response:					
Probability:	0.65	0.18	0.08	0.06	?

(a) What is the probability that a randomly chosen adult doesn't know enough to say? (*Hint:* See page 425.)

(b) What is the probability that a randomly chosen adult agrees strongly or agrees somewhat that students have to borrow too much to pay for their college tuition? (*Hint:* See page 425.)

III.14 An IQ test. The Wechsler Adult Intelligence Scale (WAIS) is a common IQ test for adults. The distribution of WAIS scores for persons over 16 years of age is approximately Normal with mean 100 and standard deviation 15. Use the 68–95–99.7 rule to answer these questions.

(a) What is the probability that a randomly chosen individual has a WAIS score of 115 or higher?

(b) In what range do the scores of the middle 95% of the adult population lie?

III.15 We like opinion polls. Are Americans interested in opinion polls about the major issues of the day? Suppose that 40% of all adults are very interested in such polls. (According to sample surveys that ask this question, 40% is about right.) A polling firm chooses an SRS of 2400 people. If they do this many times, the percentage of the sample who say they are very interested will vary from sample to sample following a Normal distribution with mean 40% and standard deviation 1.0%. Use the 68–95–99.7 rule to answer these questions.

(a) What is the probability that one such sample gives a result within ±1.0% of the truth about the population? (*Hint:* See pages 427–430.)

(b) What is the probability that one such sample gives a result within ±2% of the truth about the population? (*Hint:* See pages 427–430.)

III.16 An IQ test (optional). Use the information in Exercise III.14 and Table B to find the probability that a randomly chosen person has a WAIS score of 112 or higher.

III.17 We like opinion polls (optional). Use the information in Exercise III.15 and Table B to find the probability that one sample misses the truth about the population by 2.5% or more. (This is the probability that the sample result is either less than 37.5% or greater than 42.5%.) (*Hint:* See pages 430–431.)

III.18 An IQ test (optional). How high must a person score on the WAIS test to be in the top 10% of all scores? Use the information in Exercise III.14 and Table B to answer this question.

III.19 Models, legitimate and not. A bridge deck contains 52 cards, four of each of the 13 face values ace, king, queen, jack, ten, nine, ..., two. You deal a single card from such a deck and record the face value of the card dealt. Give an assignment of probabilities to the possible outcomes that should be correct if the deck is thoroughly shuffled. Give a second assignment of probabilities that is legitimate (that is, obeys the rules of probability) but differs from your first choice. Then give a third assignment of probabilities that is *not* legitimate, and explain what is wrong with this choice. (*Hint:* See page 425.)

III.20 Mendel's peas. Gregor Mendel used garden peas in some of the experiments that revealed that inheritance operates randomly. The seed color of Mendel's peas can be either green or yellow. Suppose we produce seeds by "crossing" two plants, both of which carry the G (green) and Y (yellow) genes. Each parent has probability 1/2 of passing each of its genes to a seed, independently of the other parent. A seed will be yellow unless both parents contribute the G gene. Seeds that get two G genes are green.

What is the probability that a seed from this cross will be green? Set up a simulation to answer this question, and estimate the probability from 25 repetitions.

III.21 Predicting the winner. There are 12 teams in the Big Ten athletic conference. Here's one set of personal probabilities for next year's basketball champion: Michigan State has probability 0.3 of winning. Iowa, Minnesota, Nebraska, Northwestern, and Penn State have no chance. That leaves 6 teams. Purdue, Michigan, Indiana, Illinois, and Wisconsin also all have the same probability of winning, but that probability is one-half that of Ohio State. What probability does each of the 12 teams have? (*Hint:* See page 425.)

III.22 Selling cars. Bill sells new cars in a small town for a living. On a weekday afternoon, he will deal with 1 customer with probability 0.5, 2 customers with probability 0.4, and 3 customers with probability 0.1. Each customer has probability 0.2 of buying a car. Customers buy independently of each other.

Describe how you would simulate the number of cars Bill sells in an afternoon. You must first simulate the number of customers, then simulate the buying decisions of 1, 2, or 3 customers. Simulate one afternoon to demonstrate your procedure.

PART III PROJECTS

Projects are longer exercises that require gathering information or producing data and that emphasize writing a short essay to describe your work. Many are suitable for teams of students.

Project 1. A bit of history. On page 405, we said, "The systematic study of randomness ... began when 17th-century French gamblers asked French mathematicians for help in figuring out the 'fair value' of bets on games of chance." Pierre de Fermat and Blaise Pascal were two of the mathematicians who responded. Both are interesting characters. Choose one of these men. Write a brief essay giving his dates, some anecdotes you find noteworthy from his life, and at least one example of a probability problem he studied. (A Web search on the name will produce abundant information. Remember to use your own words in writing your essay.)

Project 2. Reacting to risks. On page 414, we quoted a writer as saying, "Few of us would leave a baby sleeping alone in a house while we drove off on a

10-minute errand, even though car-crash risks are much greater than home risks." Take it as a fact that the probability that the baby will be injured in the car is very much higher than the probability of any harm occurring at home in the same time period. Would you leave the baby alone? Explain your reasons in a short essay. If you would not leave the baby alone, be sure to explain why you choose to ignore the probabilities.

Project 3. First digits. Here is a remarkable fact: the first digits of the numbers in long tables are usually *not* equally likely to have any of the 10 possible values 0, 1, 2, 3, 4, 5, 6, 7, 8, and 9. The digit 1 tends to occur with probability roughly 0.3, the digit 2 with probability about 0.17, and so on. You can find more information about this fact, called "Benford's law," on the Web or in two articles by Theodore P. Hill, "The Difficulty of Faking Data," *Chance,* 12, No. 3 (1999), pp. 27–31, and "The First Digit Phenomenon," *American Scientist,* 86 (1998), pp. 358–363. You don't have to read these articles for this project.

Locate at least two long tables whose entries could plausibly begin with any digit. You may choose data tables, such as populations of many cities or the number of shares traded on the New York Stock Exchange on many days, or mathematical tables such as logarithms or square roots. We hope it's clear that you can't use the table of random digits. Let's require that your examples each contain at least 300 numbers. Tally the first digits of all entries in each table. Report the distributions (in percentages) and compare them with each other, with Benford's law, and with the "equally likely" distribution.

Project 4. Personal probability. Personal probabilities are personal, so we expect them to vary from person to person. Choose an event that most students at your school should have an opinion about, such as rain next Friday or a victory in your team's next game. Ask many students (at least 50) to tell you what probability they would assign to rain or a victory. Then analyze the data with a graph and numbers—shape, center, spread, and all that. What do your data show about personal probabilities for this future event?

Project 5. Making decisions. Exercise 20.7 (page 469) reported the results of a study by the psychologist Amos Tversky on the effect of wording on people's decisions about chance outcomes. His subjects were college students. Repeat Tversky's study at your school. Prepare two typed cards. One says:

You are responsible for treating 600 people who have been exposed to a fatal virus. Treatment A has probability 1/2 of saving all 600 and probability 1/2 that all 600 will die. Treatment B is guaranteed to save exactly 400 of the 600 people. Which treatment will you give?

The second card says:

You are responsible for treating 600 people who have been exposed to a fatal virus. Treatment A has probability 1/2 of saving all 600 and probability 1/2 that all 600 will die. Treatment B will definitely lose exactly 200 of the lives. Which treatment will you give?

Show each card to at least 25 people (25 different people for each, chosen as randomly as you can conveniently manage, and chosen from people who have not studied probability). Record the choices. Tversky claims that people shown the first card tend to choose B, while those shown the second card tend to choose A. Do your results agree with this claim? Write a brief summary of your findings: Do people use expected values in their decisions? Does the frame in which a decision is presented (the wording, for example) influence choices?

NOTES AND DATA SOURCES

Page 479 Exercise III.13: From a survey commissioned in 2009 by Public Agenda for the National Center for Public Policy and Higher Education, found at **www.highereducation.org/reports/ squeeze_play_10/index.shtml.**

Page 482 Project 5: You can find lots of background on similar quirks of the mind in Thomas Gilovich, *How We Know What Isn't So: The Fallibility of Reason in Everyday Life,* Free Press, 1991.

Inference

To *infer* is to draw a conclusion from evidence. *Statistical inference* draws a conclusion about a population from evidence provided by a sample. Drawing conclusions in mathematics is a matter of starting from a hypothesis and using logical argument to prove without doubt that the conclusion follows. Statistics isn't like that. Statistical conclusions are uncertain, because the sample isn't the entire population. So statistical inference has to not only state conclusions but also say how uncertain they are. We use the language of probability to express uncertainty.

Because inference must both give conclusions and say how uncertain they are, it is the most technical part of statistics. Texts and courses intended to train people to *do* statistics spend most of their time on inference. Our aim in this book is to help you *understand* statistics, which takes less technique but often more thought. We will look only at a few basic techniques of inference. The techniques are simple, but the ideas are subtle, so prepare to think. To start, think about what you already know and don't be too impressed by elaborate statistical techniques: even the fanciest inference cannot remedy basic flaws such as voluntary response samples or uncontrolled experiments.

What Is a Confidence Interval?

CASE STUDY Know someone who is prone to anger? Nature has a way to calm such people: they get heart disease more often. Several observational studies have discovered a link between anger and heart disease. The best study looked at 12,986 people, of all races, chosen at random from four communities. When first examined, all subjects were between the ages of 45 and 64 and were free of heart disease. Let's focus on the 8474 people in this sample who had normal blood pressure.

A short psychological test, the Spielberger Trait Anger Scale, measured how easily each person became angry. There were 633 people in the high range of the anger scale, 4731 in the middle, and 3110 in the low range. Now follow these people forward in time for almost six years and compare the rate of heart disease in the high and low groups. There are some lurking variables: people in the high-anger group are somewhat more likely to be male, to have less than a high school education, and to be smokers and drinkers. After adjusting for these differences, high-anger people were 2.2 times as likely to get heart disease and 2.7 times as likely to have an acute heart attack as low-anger people.

That makes anger sound serious. But only 53 people in the low group and 27 in the high group got heart disease during the period of the study. We also know that the numbers 2.2 and 2.7 won't be exactly right for the population of all people aged 45 to 64 years with normal blood pressure. But are they close?

In this chapter we study confidence intervals. These are intervals that help us see how accurate numbers like 2.2 and 2.7 are. By the end of this chapter you will be able to construct such intervals for proportions and means, and you will be able to interpret what such intervals represent. ∎

Estimating

Statistical inference draws conclusions about a population on the basis of data about a sample. One kind of conclusion answers questions like "What percentage of employed women have a college degree?" or "What is the mean survival time for patients with this type of cancer?" These questions ask about a number (a percentage, a mean) that describes a population. Numbers that describe a population are **parameters.** To estimate a population parameter, choose a sample from the population and use a **statistic,** a number calculated from the sample, as your estimate. Here's an example.

EXAMPLE 1 Binge drinking among college graduates

How common is risky behavior such as binge drinking? The Behavioral Risk Factor Surveillance System (BRFSS) is the world's largest, ongoing telephone health survey system, tracking health conditions and risk behaviors in the United States yearly since 1984. Results from this survey can be found at the Centers for Disease Control and Prevention Web site. Data are collected monthly in all 50 states, the District of Columbia, Puerto Rico, the U.S. Virgin Islands, and Guam. In California in 2010 the BRFSS interviewed a random sample of 6911 college graduates. Of these, 792 said that they had engaged in binge drinking that year. Binge drinking is defined as having five or more drinks on one occasion for males and four or more drinks on one occasion for females. This result may be biased by reluctance to tell the truth about alcoholic consumption (see Exercise 21.14). For now, assume that the people in the sample told the truth. Based on these data, what can we say about the percentage of all California college graduates who engaged in binge drinking in 2010?

Our population is college graduates in California. The parameter is the proportion who engaged in binge drinking in 2010. Call this unknown parameter p, for "proportion." The statistic that estimates the parameter p is the **sample proportion**

$$\hat{p} = \frac{\text{count in the sample}}{\text{size of the sample}}$$

$$= \frac{792}{6911} = 0.115$$

A basic move in statistical inference is to use a sample statistic to estimate a population parameter. Once we have the sample in hand, we estimate that the proportion of all college graduates who engaged in binge drinking in 2010 is "about 11.5%" because the proportion in the sample was exactly 11.5%. We can only estimate that the truth about the population is "about" 11.5% because we know that the sample result is unlikely to be exactly the same as the true population proportion. A confidence interval makes that "about" precise.

95% confidence interval

A **95% confidence interval** is an interval calculated from sample data by a process that is guaranteed to capture the true population parameter in 95% of all samples.

We will first march straight through to the interval for a population proportion, and then we will reflect on what we have done and generalize a bit.

Estimating with confidence

We want to estimate the proportion p of the individuals in a population who have some characteristic—they are employed, or they approve the president's performance, for example. Let's call the characteristic we are looking for a "success." We use the proportion $\hat{p}$ of successes in a simple random sample (SRS) to estimate the proportion p of successes in the population. How good is the statistic $\hat{p}$ as an estimate of the parameter p? To find out, we ask, "What would happen if we took many samples?" Well, we know that $\hat{p}$ would vary from sample to sample. We also know that this sampling variability isn't haphazard. It has a clear pattern in the long run, a pattern that is pretty well described by a Normal curve. Here are the facts.

Sampling distribution of a sample proportion

The **sampling distribution** of a statistic is the distribution of values taken by the statistic in all possible samples of the same size from the same population.

Take an SRS of size n from a large population that contains proportion p of successes. Let $\hat{p}$ be the **sample proportion** of successes,

$$\hat{p} = \frac{\text{count of successes in the sample}}{n}$$

Then, if the sample size is large enough:

- The sampling distribution of $\hat{p}$ is **approximately Normal.**
- The **mean** of the sampling distribution is p.
- The **standard deviation** of the sampling distribution is

$$\sqrt{\frac{p(1-p)}{n}}$$

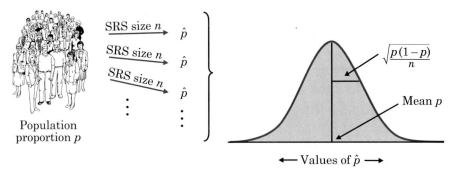

Figure 21.1 Repeat many times the process of selecting an SRS of size n from a population in which the proportion p are successes. The values of the sample proportion of successes $\hat{p}$ have this Normal sampling distribution.

These facts can be proved by mathematics, so they are a solid starting point. Figure 21.1 summarizes them in a form that also reminds us that a sampling distribution describes the results of lots of samples from the same population.

EXAMPLE 2 Binge drinking

Suppose, for example, that the truth is that 12% of college graduates in California engaged in binge drinking in 2010. Then, in the setting of Example 1, $p = 0.12$. The BRFSS sample of size $n = 6911$ would, if repeated many times, produce sample proportions $\hat{p}$ that closely follow the Normal distribution with

$$\text{mean} = p = 0.12$$

and

$$\text{standard deviation} = \sqrt{\frac{p(1-p)}{n}}$$

$$= \sqrt{\frac{(0.12)(0.88)}{6911}}$$

$$= \sqrt{0.0000153} = 0.0039$$

The center of this Normal distribution is at the truth about the population. That's the absence of bias in random sampling once again. The standard deviation is small because the sample is quite large. So almost all samples will produce a statistic $\hat{p}$ that is close to the true p. In fact, the 95 part of

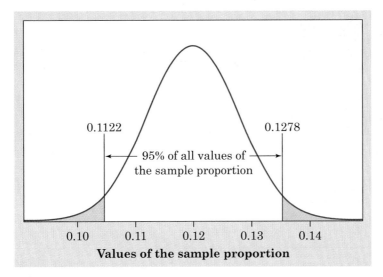

Figure 21.2 Repeat many times the process of selecting an SRS of size 6911 from a population in which the proportion $p = 0.12$ are successes. The middle 95% of the values of the sample proportion $\hat{p}$ will lie between 0.1122 and 0.1278.

the 68–95–99.7 rule says that 95% of all sample outcomes will fall between

$$\text{mean} - 2 \text{ standard deviations} = 0.12 - 0.0078 = 0.1122$$

and

$$\text{mean} + 2 \text{ standard deviations} = 0.12 + 0.0078 = 0.1278$$

Figure 21.2 displays these facts.

So far, we have just put numbers on what we already knew: we can trust the results of large random samples because almost all such samples give results that are close to the truth about the population. The numbers say that in 95% of all samples of size 6911, the statistic $\hat{p}$ and the parameter p are within 0.0078 of each other. We can put this another way: 95% of all samples give an outcome $\hat{p}$ such that the population truth p is captured by the interval from $\hat{p} - 0.0078$ to $\hat{p} + 0.0078$.

The 0.0078 came from substituting $p = 0.12$ into the formula for the standard deviation of $\hat{p}$. For any value of p, the general fact is:

When the population proportion has the value p, 95% of all samples catch p in the interval extending 2 standard deviations on either side of $\hat{p}$.

That's the interval

$$\hat{p} \pm 2\sqrt{\frac{p(1-p)}{n}}$$

Is this the 95% confidence interval we want? Not quite. The interval can't be found just from the data, because the standard deviation involves the population proportion p, and in practice we don't know p. In Example 2, we used $p = 0.12$ in the formula, but this may not be the true p.

What can we do? Well, the standard deviation of the statistic $\hat{p}$ does depend on the parameter p, but it doesn't change a lot when p changes. Go back to Example 2 and redo the calculation for other values of p. Here's the result:

Value of p:	0.10	0.11	0.12	0.13	0.14
Standard deviation:	0.0036	0.0038	0.0039	0.0040	0.0042

We see that, if we guess a value of p reasonably close to the true value, the standard deviation found from the guessed value will be about right. We know that, when we take a large random sample, the statistic $\hat{p}$ is almost always close to the parameter p. So we will use $\hat{p}$ as the guessed value of the unknown p. Now we have an interval that we can calculate from the sample data.

95% confidence interval for a proportion

Choose an SRS of size n from a large population that contains an unknown proportion p of successes. Call the proportion of successes in this sample $\hat{p}$. An **approximate 95% confidence interval** for the parameter p is

$$\hat{p} \pm 2\sqrt{\frac{\hat{p}(1-\hat{p})}{n}}$$

EXAMPLE 3 A confidence interval for binge drinking

The BRFSS random sample of 6911 college graduates in California found that 792 had engaged in binge drinking in 2010, a sample proportion $\hat{p} = 0.115$. The 95% confidence interval for the proportion of all college graduates in California who engaged in binge drinking in 2010 is

$$\hat{p} \pm 2\sqrt{\frac{\hat{p}(1-\hat{p})}{n}} = 0.115 \pm 2\sqrt{\frac{(0.115)(0.885)}{6911}}$$

$$= 0.115 \pm (2)(0.0038)$$

$$= 0.115 \pm 0.0076$$

$$= 0.1074 \text{ to } 0.1226$$

Interpret this result as follows: we got this interval by using a recipe that catches the true unknown population proportion in 95% of all samples. The shorthand is: we are **95% confident** that the true proportion of all college graduates in California who engaged in binge drinking in 2010 lies between 10.74% and 12.26%.

 Who is a smoker? When estimating a proportion p, be sure you know what counts as a "success." The news says that 20% of adolescents smoke. Shocking. It turns out that this is the percentage who smoked at least once in the past month. If we say that a smoker is someone who smoked in at least 20 of the past 30 days and smoked at least half a pack on those days, fewer than 4% of adolescents qualify.

NOW IT'S YOUR TURN | **21.1 Gambling.** A May 2011 Gallup Poll consisting of a random sample of 1018 adult Americans found that 31% believe gambling is morally wrong. Find a 95% confidence interval for the proportion of all adult Americans who believe gambling is morally wrong. How would you interpret this interval?

Understanding confidence intervals

Our 95% confidence interval for a population proportion has the familiar form

$$\text{estimate} \pm \text{margin of error}$$

News reports of sample surveys, for example, usually give the estimate and the margin of error separately: "A new Gallup Poll shows that 65% of women favor new laws restricting guns. The margin of error is plus or minus four percentage points." News reports usually leave out the level of confidence, although it is almost always 95%.

The next time you hear a report about the result of a sample survey, consider the following. If most confidence intervals reported in the media have a 95% level of confidence, then in about 1 in 20 poll results that you hear about, the confidence interval does *not* contain the true proportion.

Not all confidence intervals are expressed in the form "estimate ± margin of error." Here's a complete description of a confidence interval.

Confidence interval

A **level C confidence interval** for a parameter has two parts:

- An **interval** calculated from the data.
- A **confidence level** C, which gives the probability that the interval will capture the true parameter value in repeated samples.

There are many recipes for statistical confidence intervals for use in many situations. Be sure you understand how to interpret a confidence interval. The interpretation is the same for any recipe, and you can't use a calculator or a computer to do the interpretation for you.

Confidence intervals use the central idea of probability: ask what would happen if we repeated the sampling many times. The 95% in a 95% confidence interval is a probability, the probability that the method produces an interval that does capture the true parameter.

EXAMPLE 4 How confidence intervals behave

The BRFSS sample of 6911 college graduates in California in 2010 found that 792 had engaged in binge drinking, so the sample proportion was

$$\hat{p} = \frac{792}{6911} = 0.115$$

and the 95% confidence interval was

$$\hat{p} \pm 2\sqrt{\frac{\hat{p}(1 - \hat{p})}{n}} = 0.115 \pm 0.0076$$

Draw a second sample from the same population. It finds that 826 of its 6911 respondents had engaged in binge drinking. For this sample,

$$\hat{p} = \frac{826}{6911} = 0.120$$

$$\hat{p} \pm 2\sqrt{\frac{\hat{p}(1 - \hat{p})}{n}} = 0.120 \pm 0.0078$$

Draw another sample. Now the count is 752 and the sample proportion and confidence interval are

$$\hat{p} = \frac{752}{6911} = 0.109$$

$$\hat{p} \pm 2\sqrt{\frac{\hat{p}(1 - \hat{p})}{n}} = 0.109 \pm 0.0075$$

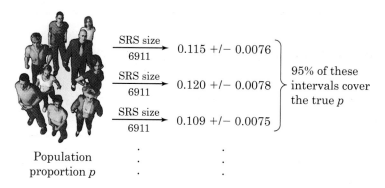

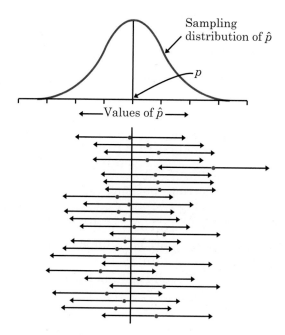

Figure 21.3 Repeated samples from the same population give different 95% confidence intervals, but 95% of these intervals capture the true population proportion p.

Keep sampling. Each sample yields a new estimate $\hat{p}$ and a new confidence interval. *If we sample forever, 95% of these intervals capture the true parameter.* This is true no matter what the true value is. Figure 21.3 summarizes the behavior of the confidence interval in graphical form.

On the assumption that two pictures are better than one, Figure 21.4 gives a different view of how confidence intervals behave. Example 4 and

Figure 21.4 Twenty-five samples from the same population give these 95% confidence intervals. In the long run, 95% of all such intervals cover the true population proportion, marked by the vertical line.

Figure 21.3 remind us that repeated samples give different results and that we are guaranteed only that 95% of the samples give a correct result. Figure 21.4 goes behind the scenes. The vertical line is the true value of the population proportion p. The Normal curve at the top of the figure is the sampling distribution of the sample statistic $\hat{p}$, which is centered at the true p. We are behind the scenes because in real-world statistics we usually don't know p.

The 95% confidence intervals from 25 SRSs appear below, one after the other. The central dots are the values of $\hat{p}$, the centers of the intervals. The arrows on either side span the confidence interval. In the long run, 95% of the intervals will cover the true p and 5% will miss. Of the 25 intervals in Figure 21.4, 24 hit and 1 misses. (Remember that probability describes only what happens in the long run—we don't expect exactly 95% of a fixed set of intervals to capture the true parameter.)

Don't forget that our interval is only *approximately* a 95% confidence interval. It isn't exact for two reasons. The sampling distribution of the sample proportion $\hat{p}$ isn't exactly Normal. And we don't get the standard deviation of $\hat{p}$ exactly right because we used $\hat{p}$ in place of the unknown p. We use a new estimate of the standard deviation of the sampling distribution every time, even though the true standard deviation never changes. Both of these difficulties go away as the sample size n gets larger. So our recipe is good only for large samples. What is more, the recipe assumes that the population is really big—at least 10 times the size of the sample. Professional statisticians use more elaborate methods that take the size of the population into account and work even for small samples. But our method works well enough for many practical uses. More important, it shows how we get a confidence interval from the sampling distribution of a statistic. That's the reasoning behind any confidence interval.

More on confidence intervals for a population proportion*

We used the 95 part of the 68–95–99.7 rule to get a 95% confidence interval for the population proportion. Perhaps you think that a method that works 95% of the time isn't good enough. You want to be 99% confident. For that, we need to mark off the central 99% of a Normal distribution. For any probability C between 0 and 1, there is a number z^* such that any Normal distribution has probability C within z^* standard deviations of the mean. Figure 21.5 shows how the probability C and the number z^* are related.

---

*This section is optional.

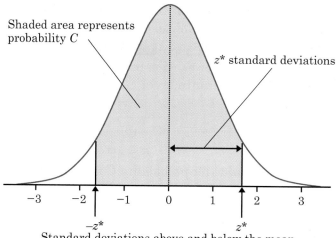

Figure 21.5 Critical values z^* of the Normal distributions. In any Normal distribution, there is area (probability) C under the curve between $-z^*$ and z^* standard deviations away from the mean.

Table 21.1 gives the numbers z^* for various choices of C. For convenience, the table gives C as a confidence level in percent. The numbers z^* are called **critical values** of the Normal distributions. Table 21.1 shows that any Normal distribution has probability 99% within ± 2.58 standard deviations of its mean. The table also shows that any Normal distribution has probability 95% within ± 1.96 standard deviations of its mean. The 68–95–99.7 rule uses 2 in place of the critical value $z^* = 1.96$. That is good enough for practical purposes, but the table gives the more exact value.

From Figure 21.5 we see that, with probability C, the sample proportion $\hat{p}$ takes a value within z^* standard deviations of p. That is just to say that, with probability C, the interval extending z^* standard deviations on either side of the observed $\hat{p}$ captures the unknown p. Using the estimated standard deviation of $\hat{p}$ produces the following recipe.

TABLE 21.1 Critical values of the Normal distributions

Confidence level C	Critical value z^*	Confidence level C	Critical value z^*
50%	0.67	90%	1.64
60%	0.84	95%	1.96
70%	1.04	99%	2.58
80%	1.28	99.9%	3.29

Confidence interval for a population proportion

Choose an SRS of size n from a population of individuals of which proportion p are successes. The proportion of successes in the sample is $\hat{p}$. When n is large, an **approximate level C confidence interval for p** is

$$\hat{p} \pm z^* \sqrt{\frac{\hat{p}(1 - \hat{p})}{n}}$$

where z^* is the critical value for probability C from Table 21.1.

EXAMPLE 5 A 99% confidence interval

The BRFSS random sample of 6911 college graduates in California in 2010 found that 792 had engaged in binge drinking in the past year. We want a 99% confidence interval for the proportion p of all college graduates in California who engaged in binge drinking in the past year. Table 21.1 says that for 99% confidence, we must go out $z^* = 2.58$ standard deviations. Here are our calculations:

$$\hat{p} = \frac{792}{6911} = 0.115$$

$$\hat{p} \pm z^* \sqrt{\frac{\hat{p}(1 - \hat{p})}{n}} = 0.115 \pm 2.58 \sqrt{\frac{(0.115)(0.885)}{6911}}$$

$$= 0.115 \pm (2.58)(0.0038)$$

$$= 0.115 \pm 0.0098$$

$$= 0.1052 \text{ to } 0.1248$$

We are 99% confident that the true population proportion of all college graduates in California who engaged in binge drinking is between 10.52% and 12.48%. That is, we got this range of percentages by using a method that gives a correct answer 99% of the time.

Compare Example 5 with the calculation of the 95% confidence interval in Example 3. The only difference is the use of the critical value 2.58 for 99% confidence in place of 2 for 95% confidence. That makes the margin of error for 99% confidence larger and the confidence interval wider. Higher confidence isn't free—we pay for it with a wider interval. Figure 21.5 reminds us why this is true. To cover a higher percentage of the area under a Normal curve, we must go farther out from the center.

NOW IT'S YOUR TURN
21.2 Gambling. A May 2011 Gallup Poll consisting of a random sample of 1018 adult Americans found that 31% believe that gambling is morally wrong. Find a 99% confidence interval for the proportion of all adult Americans who believe that gambling is morally wrong. How would you interpret this interval?

The sampling distribution of a sample mean*

What is the mean number of hours your college's first-year students study each week? What was their mean grade point average in high school? We often want to estimate the mean of a population. To distinguish the population mean (a parameter) from the sample mean $\bar{x}$, we write the population mean as μ, the Greek letter mu. We use the mean $\bar{x}$ of an SRS to estimate the unknown mean μ of the population.

Like the sample proportion $\hat{p}$, the sample mean $\bar{x}$ from a large SRS has a sampling distribution that is close to Normal. Because the sample mean of an SRS is an unbiased estimator of μ, the sampling distribution of $\bar{x}$ has μ as its mean. The standard deviation of $\bar{x}$ depends on the standard deviation of the population, which is usually written as σ, the Greek letter sigma. By mathematics we can discover the following facts.

Sampling distribution of a sample mean

Choose an SRS of size n from a population in which individuals have mean μ and standard deviation σ. Let $\bar{x}$ be the mean of the sample. Then:

- The sampling distribution of $\bar{x}$ is **approximately Normal** when the sample size n is large.
- The **mean** of the sampling distribution is equal to μ.
- The **standard deviation** of the sampling distribution is $\sigma/\sqrt{n}$.

It isn't surprising that the values that $\bar{x}$ takes in many samples are centered at the true mean μ of the population. That's the lack of bias in random sampling once again. The other two facts about the sampling distribution make precise two very important properties of the sample mean $\bar{x}$:

- The mean of a number of observations is less variable than individual observations.

---

*This section is optional.

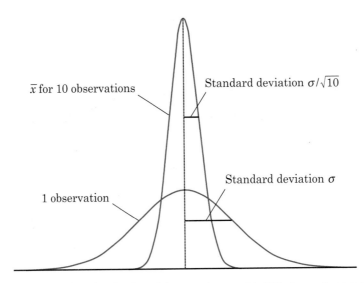

$\bar{x}$ for 10 observations

Standard deviation $\sigma/\sqrt{10}$

Standard deviation σ

1 observation

Figure 21.6 The sampling distribution of the sample mean $\bar{x}$ of 10 observations compared with the distribution of individual observations.

- The distribution of a mean of a number of observations is more Normal than the distribution of individual observations.

Figure 21.6 illustrates the first of these properties. It compares the distribution of a single observation with the distribution of the mean $\bar{x}$ of 10 observations. Both have the same center, but the distribution of $\bar{x}$ is less spread out. In Figure 21.6, the distribution of individual observations is Normal. If that is true, then the sampling distribution of $\bar{x}$ is exactly Normal for any size sample, not just approximately Normal for large samples. A remarkable statistical fact, called the **central limit theorem,** says that as we take more and more observations at random from *any* population, the distribution of the mean of these observations eventually gets close to a Normal distribution. (There are some technical qualifications to this big fact, but in this text we assume these qualifications are satisfied.) The central limit theorem lies behind the use of Normal sampling distributions for sample means.

EXAMPLE 6 The central limit theorem in action

Figure 21.7 shows the central limit theorem in action. The top-left density curve describes individual observations from a population. It is strongly right-skewed. Distributions like this describe the time it takes to repair a household appliance, for example. Most repairs are quickly done, but some are lengthy.

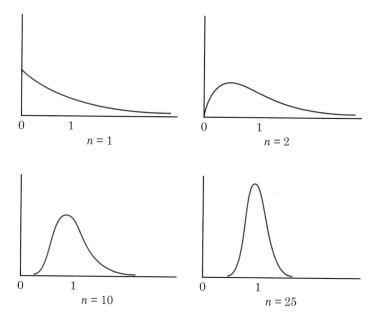

Figure 21.7 The distribution of a sample mean $\bar{x}$ becomes more Normal as the size of the sample increases. The distribution of individual observations ($n = 1$) is far from Normal. The distributions of means of 2, 10, and finally 25 observations move closer to the Normal shape.

The other three density curves in Figure 21.7 show the sampling distributions of the sample means of 2, 10, and 25 observations from this population. As the sample size n increases, the shape becomes more Normal. The mean remains fixed and the standard deviation decreases, following the pattern $\sigma/\sqrt{n}$. The distribution for 10 observations is still somewhat skewed to the right but already resembles a Normal curve. The density curve for $n = 25$ is yet more Normal. The contrast between the shapes of the population distribution and of the distribution of the mean of 10 or 25 observations is striking.

Confidence intervals for a population mean*

The standard deviation of $\bar{x}$ depends on both the sample size n and the standard deviation σ of individuals in the population. We know n but not σ. When n is large, the sample standard deviation s is close to σ and can be used to estimate it, just as we use the sample mean $\bar{x}$ to estimate the population mean μ. The estimated standard deviation of $\bar{x}$ is therefore $s/\sqrt{n}$.

---

*This section is optional.

Now we can find confidence intervals for μ following the same reasoning that led us to confidence intervals for a proportion p. The big idea is that to cover the central area C under a Normal curve, we must go out a distance z^* on either side of the mean. Look again at Figure 21.5 to see how C and z^* are related.

Confidence interval for a population mean

Choose an SRS of size n from a large population of individuals having mean μ. The mean of the sample observations is $\bar{x}$. When n is reasonably large, an **approximate level C confidence interval for μ** is

$$\bar{x} \pm z^* \frac{s}{\sqrt{n}}$$

where z^* is the critical value for confidence level C from Table 21.1 (page 497).

The cautions we noted in estimating p apply here as well. The recipe is valid only when an SRS is drawn and the sample size n is reasonably large. How large is reasonably large? The answer depends upon the true shape of the population distribution. A sample size of $n \geq 15$ is usually adequate unless there are extreme outliers or strong skewness. For clearly skewed distributions, a sample size of $n \geq 40$ often suffices if there are no outliers.

The margin of error again decreases only at a rate proportional to $\sqrt{n}$ as the sample size n increases. And it bears repeating that $\bar{x}$ and s are strongly influenced by outliers. Inference using $\bar{x}$ and s is suspect when outliers are present. Always look at your data.

EXAMPLE 7 NAEP quantitative scores

The National Assessment of Educational Progress (NAEP) includes a mathematics test for high school seniors. Scores on the test range from 0 to 300. Demonstrating the ability to use the Pythagorean theorem to determine the length of a hypotenuse is an example of the skills and knowledge associated with performance at the Basic level. An example of the knowledge and skills associated with the Proficient level is using trigonometric ratios to determine length.

In 2009, 51,000 12th-graders were in the NAEP sample for the mathematics test. The mean mathematics score was $\bar{x} = 153$, and the standard deviation of their scores was $s = 34$. Assume that these 51,000 students were a random sample from the population of all 12th-graders. On the basis of this sample, what can we say about the mean score μ in the population of all 12th-grade students?

The 95% confidence interval for μ uses the critical value $z^* = 1.96$ from Table 21.1. The interval is

$$\bar{x} \pm z^* \frac{s}{\sqrt{n}} = 153 \pm 1.96 \frac{34}{\sqrt{51,000}}$$
$$= 153 \pm (1.96)(0.151) = 153 \pm 0.3$$

We are 95% confident that the mean score for all 12th-grade students lies between 152.7 and 153.3.

NOW IT'S YOUR TURN **21.3 Blood pressure of executives.** The medical director of a large company looks at the medical records of 72 executives between the ages of 35 and 44 years. He finds that the mean systolic blood pressure in this sample is $\bar{x} = 126.1$ and the standard deviation is $s = 15.2$. Assuming the sample is a random sample of all executives in the company, find a 95% confidence interval for μ, the unknown mean blood pressure of all executives in the company.

STATISTICS IN SUMMARY

Chapter Specifics

- **Statistical inference** draws conclusions about a population on the basis of data from a sample. Because we don't have data for the entire population, our conclusions are uncertain.

- A **confidence interval** estimates an unknown parameter in a way that tells us how uncertain the estimate is. The interval itself says how closely we can pin down the unknown parameter. The **confidence level** is a probability that says how often in many samples the method would produce an interval that does catch the parameter. We find confidence intervals starting from the **sampling distribution** of a statistic, which shows how the statistic varies in repeated sampling.

- We estimate a population proportion p using the sample proportion $\hat{p}$ of an SRS from the population. Confidence intervals for p are based on the

sampling distribution of $\hat{p}$. When the sample size n is large, this distribution is approximately Normal.

- We estimate a population mean μ using the sample mean $\overline{x}$ of an SRS from the population. Confidence intervals for μ are based on the **sampling distribution** of $\overline{x}$. When the sample size n is large, the **central limit theorem** says that this distribution is approximately Normal. Although the details of the methods differ, inference about μ is quite similar to inference about a population proportion p because both are based on Normal sampling distributions.

The reason we collect data is not to learn about the individuals that we observed but to infer from the data to some wider population that the individuals represent. Chapters 1 to 6 tell us that the way we produce the data (sampling, experimental design) affects whether we have a good basis for generalizing to some wider population—in particular, whether a sample statistic provides insight into the value of the corresponding population parameter. Chapters 17 to 20 discuss probability, the formal mathematical tool that determines the nature of the inferences we make. Chapter 18 discusses sampling distributions, which tell us how statistics computed from repeated SRSs behave and hence what a statistic (in particular, a sample proportion) computed from our sample is likely to tell us about the corresponding parameter of the population (in particular, a population proportion) from which the sample was selected.

In this chapter we discuss the basic reasoning of statistical estimation of a population parameter, with emphasis on estimating a population proportion and population mean. To an estimate of a population parameter, such as a population proportion, we attach a margin of error and a confidence level. The result is a confidence interval. The sampling distribution, first introduced in Chapter 3 and discussed more fully in Chapter 18, provides the mathematical basis for constructing confidence intervals and understanding their properties. We will provide more advice on interpreting confidence intervals in Chapter 23.

CASE STUDY News reports of the study discussed in the Case Study at the beginning
EVALUATED of the chapter reported only that high-anger people were 2.2 times as likely to get heart disease and 2.7 times as likely to have an acute heart attack as low-anger people, but the full report in the medical journal *Circulation* gave confidence intervals. With 95% confidence, high-anger people are between 1.36 and 3.55 times as likely to get heart disease as low-anger people. They are between 1.48 and 4.90 times as likely to have an acute heart attack. Interpret these intervals in plain language that someone who knows no statistics will understand. ∎

CHAPTER 21 EXERCISES

For Exercise 21.1, see page 493; for Exercise 21.2, see page 499; and for Exercise 21.3, see page 503.

21.4 A student survey. Tonya wants to estimate what proportion of the students in her dormitory like the dorm food. She interviews an SRS of 50 of the 175 students living in the dormitory. She finds that 14 think the dorm food is good.

(a) What population does Tonya want to draw conclusions about?

(b) In your own words, what is the population proportion p in this setting?

(c) What is the numerical value of the sample proportion $\hat{p}$ from Tonya's sample?

21.5 Fire the coach? A college president says, "99% of the alumni support my firing of Coach Boggs." You contact an SRS of 200 of the college's 15,000 living alumni and find that 66 of them support firing the coach.

(a) What population does the inference concern here?

(b) Explain clearly what the population proportion p is in this setting.

(c) What is the numerical value of the sample proportion $\hat{p}$?

21.6 Are students engaged in school? Results from a summer 2010 Gallup Student Survey reveal that 37% of students in the 5th through 12th grades feel either not engaged or actively disengaged in school. The report of this sample survey of 642 students aged 10 to 18 stated that, with 95% confidence, the margin of sampling error was ±4.94%. Explain to someone who knows no statistics what the phrase "95% confidence" means in this report.

21.7 Gun control. An October 2011 Gallup Poll asked a sample of 1005 adults whether they thought there should or should not be a law that would ban the possession of handguns except by the police and other authorized persons. Only 261 of the respondents thought there should be such a law. Although the samples in national polls are not SRSs, they are similar enough that our method gives approximately correct confidence intervals.

(a) Say in words what the population proportion p is for this poll.

(b) Find a 95% confidence interval for p.

(c) Gallup announced a margin of error of plus or minus four percentage points for this poll result. How well does your work in (b) agree with this margin of error?

21.8 Children of illegal immigrants. An October 2011 CBS News Poll on illegal immigrants interviewed 1012 randomly selected American adults. Of those in the sample, 688 said that they "oppose allowing the children of illegal immigrants to attend state college at the lower tuition rate of state residents." Although the samples in national polls are not SRSs, they are similar enough that our method gives approximately correct confidence intervals.

(a) Explain in words what the parameter p is in this setting.

(b) Use the poll results to give a 95% confidence interval for p.

(c) Write a short explanation of your findings in (b) for someone who knows no statistics.

21.9 Computer crime. Adults are spending more and more time on the Internet, and the number experiencing computer- or Internet-based crime is rising. A 2010 Gallup Poll of 1025 adults, aged 18 and over, found that 113 of those in the sample said that they or a member of their household were victims of computer or Internet crime on their home computer in the past year. Although the samples in national polls are not SRSs, they are similar enough that our method gives approximately correct confidence intervals.

(a) Explain in words what the parameter p is in this setting.

(b) Use the poll results to give a 95% confidence interval for p.

21.10 Gun control. In Exercise 21.7, you constructed a 95% confidence interval based on a random sample of $n = 1005$ adults. How large a sample would be needed to get a margin of error half as large as the one in Exercise 21.7? You may find it helpful to refer to the discussion surrounding Example 5 in Chapter 3 (page 47).

21.11 The effect of sample size. An October 2011 CBS News/*New York Times* Poll found that 78% of its sample thought that it was probably a good idea to significantly cut taxes for small businesses in order to try and create jobs. Give a 95% confidence interval for the proportion of all adults who feel this way, assuming that the result $\hat{p} = 0.78$ comes from a sample of size

(a) $n = 750$

(b) $n = 1500$

(c) $n = 3000$

(d) Explain briefly what your results show about the effect of increasing the size of a sample on the width of the confidence interval.

21.12 Random digits. We know that the proportion of 0s among a large set of random digits is $p = 0.1$ because all 10 possible digits are equally probable. The entries in a table of random digits are a random sample from the population of all random digits. To get an SRS of 200 random digits, look at the first digit in each of the 200 five-digit groups in lines 101 to 125 of Table A in the back of the book. How many of these 200 digits are 0s? Give a 95% confidence interval for the proportion of 0s in the population from which these digits are a random sample. Does your interval cover the true parameter value, $p = 0.1$?

21.13 Tossing a thumbtack. If you toss a thumbtack on a hard surface, what is the probability that it will land point up? Estimate this probability p by tossing a thumbtack 100 times. The 100 tosses are an SRS of size 100 from the population of all tosses. The proportion of these 100 tosses that land point up is the sample proportion $\hat{p}$. Use the result of your tosses to give a 95% confidence interval for p. Write a brief explanation of your findings for someone who knows no statistics but wonders how often a thumbtack will land point up.

21.14 Don't forget the basics. The Behavioral Risk Factor Surveillance System survey found that 792 individuals in its 2010 random sample of 6911 college graduates in California said that they had engaged in binge drinking in the past year. We used this finding to calculate confidence intervals for the proportion of all college graduates in California who engaged in binge drinking in the past year. This sample survey may have bias that our confidence intervals do not take into account. Why is some bias likely to be present? Does the sample proportion 11.5% probably overestimate or underestimate the true population proportion?

21.15 Count Buffon's coin. The 18th-century French naturalist Count Buffon tossed a coin 4040 times. He got 2048 heads. Give a 95% confidence interval for the probability that Buffon's coin lands heads up. Are you confident that this probability is not 1/2? Why?

21.16 Share the wealth. The *New York Times* conducted a nationwide poll of 1650 randomly selected American adults. Of these, 1089 felt that the distribution of money and wealth in this country should be more evenly distributed among more people. We can consider the sample to be an SRS.

(a) Give a 95% confidence interval for the proportion of all American adults who, at the time of the poll, felt that the distribution of money and wealth in this country should be more evenly distributed among more people.

(b) The news article says, "In theory, in 19 cases out of 20, the poll results will differ by no more than 3 percentage points in either direction from what would have been obtained by seeking out all American adults." Explain how your results agree with this statement.

21.17 Harley motorcycles. Harley-Davidson motorcycles make up 28% of all the motorcycles registered in the United States. You plan to interview an SRS of 600 motorcycle owners.

(a) What is the sampling distribution of the proportion of your sample who own Harleys?

(b) How likely is your sample to contain 29.8% or more who own Harleys? How likely is it to contain at least 26.2% Harley owners? Use the 68–95–99.7 rule and your answer to (a).

21.18 Do you jog? Suppose that 10% of all adults jog. An opinion poll asks an SRS of 400 adults if they jog.

(a) What is the sampling distribution of the proportion $\hat{p}$ in the sample who jog?

(b) According to the 68–95–99.7 rule, what is the probability that the sample proportion who jog will be 7.3% or greater?

21.19 The quick method. The quick method of Chapter 3 (page 46) uses $\hat{p} \pm 1/\sqrt{n}$ as a rough recipe for a 95% confidence interval for a population proportion. The margin of error from the quick method is a bit larger than needed. It differs most from the more accurate method of this chapter when $\hat{p}$ is close to 0 or 1. An SRS of 500 motorcycle registrations finds that 68 of the motorcycles are Harley-Davidsons. Give a 95% confidence interval for the proportion of all motorcycles that are Harleys by the quick method and then by the method of this chapter. How much larger is the quick-method margin of error?

21.20 68% confidence. We used the 95 part of the 68–95–99.7 rule to give a recipe for a 95% confidence interval for a population proportion p.

(a) Use the 68 part of the rule to give a recipe for a 68% confidence interval.

(b) Explain in simple language what "68% confidence" means.

(c) Use the result of the Behavioral Risk Factor Surveillance System (Example 3, pages 492–493) to give a 68% confidence interval for the proportion of college graduates in California in 2010 who had engaged in binge drinking. How does your interval compare with the 95% interval in Example 3?

21.21 Simulating confidence intervals. In Exercise 21.20, you found the recipe for a 68% confidence interval for a population proportion p. Suppose that (unknown to anyone) 60% of the voters in her district favor the reelection of Congresswoman Caucus.

(a) How would you simulate the votes of an SRS of 25 voters?

(b) Simulate choosing 10 SRSs, using a different row in Table A for each sample. What are the 10 values of the sample proportion $\hat{p}$ who favor Caucus?

(c) Find the 68% confidence interval for p from each of your 10 samples. How many of the intervals capture the true parameter value $p = 0.6$? (Samples of size 25 are not large enough for our recipe to be very accurate, but even a small simulation illustrates how confidence intervals behave in repeated samples.)

The following exercises concern the optional sections of this chapter.

21.22 Gun control. Exercise 21.7 (page 505) reports a Gallup Poll in which 261 of a random sample of 1005 adults thought there should be a law that would ban the possession of handguns except by the police and other authorized persons. Use Table 21.1 to give a 90% confidence interval for the proportion of all adults who feel this way. How does your interval compare with the 95% confidence interval from Exercise 21.7?

21.23 Children of illegal immigrants. Exercise 21.8 (page 505) reports a CBS News Poll that found that 688 in a random sample of 1012 American adults said that they "oppose allowing the children of illegal immigrants to attend state college at the lower tuition rate of state residents." Use Table 21.1 (page 497) to give a 99% confidence interval for the proportion of all American adults who feel this way. How does your interval compare with the 95% confidence interval of Exercise 21.8?

21.24 The effect of confidence level. An October 2011 CBS News/*New York Times* Poll found that 78% of a random sample of 1650 American adults thought that it was probably a good idea to significantly cut taxes for small businesses in order to create jobs. Use this poll result and Table 21.1 (page 497) to give 70%, 80%, 90%, and 99% confidence intervals for the proportion of all adults who feel this way. What do your results show about the effect of changing the confidence level?

21.25 Unhappy HMO patients. How likely are patients who file complaints with a health maintenance organization (HMO) to leave the HMO? In one year, 639 of the more than 400,000 members of a large New England HMO filed complaints. Fifty-four of the complainers left the HMO voluntarily. (That is, they were not forced to leave by a move or a job change.) Consider this year's complainers as an SRS of all patients who will complain in the future. Give a 90% confidence interval for the proportion of complainers who voluntarily leave the HMO.

21.26 Estimating unemployment. The Bureau of Labor Statistics (BLS) uses 90% confidence in presenting unemployment results from the monthly Current Population Survey (CPS). The January 2008 survey interviewed 134,444 people. Of these, 62,409 were employed and 4991 were unemployed. The CPS is not an SRS, but for the purposes of this exercise, we will act as though the BLS took an SRS of 134,444 people. Give a 90% confidence interval for the proportion of those surveyed who were unemployed. (*Note:* Example 3 in Chapter 8 on pages 161–162 explains how unemployment is measured.)

21.27 Safe margin of error. The margin of error $z^*\sqrt{\hat{p}(1-\hat{p})/n}$ is 0 when $\hat{p}$ is 0 or 1 and is largest when $\hat{p}$ is 1/2. To see this, calculate $\hat{p}(1-\hat{p})$ for $\hat{p} = 0, 0.1, 0.2, \ldots, 0.9$, and 1. Plot your results vertically against the values of $\hat{p}$ horizontally. Draw a curve through the points. You have made a graph of $\hat{p}(1-\hat{p})$. Does the graph reach its highest point when $\hat{p} = 1/2$? You see that taking $\hat{p} = 1/2$ gives a margin of error that is always at least as large as needed.

21.28 The idea of a sampling distribution. Figure 21.1 (page 490) shows the idea of the sampling distribution of a sample proportion $\hat{p}$ in picture form. Draw a similar picture that shows the idea of the sampling distribution of a sample mean $\bar{x}$.

21.29 IQ test scores. Here are the IQ test scores of 31 seventh-grade girls in a Midwest school district:

114	100	104	89	102	91	114	114	103	105
108	130	120	132	111	128	118	119	86	72
111	103	74	112	107	103	98	96	112	112	93

(a) We expect the distribution of IQ scores to be close to Normal. Make a histogram of the distribution of these 31 scores. Does your plot show outliers, clear skewness, or other non-Normal features? Using a calculator, find the mean and standard deviation of these scores.

(b) Treat the 31 girls as an SRS of all middle-school girls in the school district. Give a 95% confidence interval for the mean score in the population.

(c) In fact, the scores are those of all seventh-grade girls in one of the several schools in the district. Explain carefully why we cannot trust the confidence interval from (b).

21.30 Averages versus individuals. Scores on the ACT college entrance examination vary Normally with mean $\mu = 18$ and standard deviation $\sigma = 6$. The range of reported scores is 1 to 36.

(a) What range contains the middle 95% of all individual scores?

(b) If the ACT scores of 25 randomly selected students are averaged, what range contains the middle 95% of the averages $\bar{x}$?

21.31 Blood pressure. A randomized comparative experiment studied the effect of diet on blood pressure. Researchers divided 54 healthy white males at random into two groups. One group received a calcium supplement, and the other group received a placebo. At the beginning of the study, the researchers measured many variables on the subjects. The average seated systolic blood pressure of the 27 members of the placebo group was reported to be $\bar{x} = 114.9$ with a standard deviation of $s = 9.3$.

(a) Give a 95% confidence interval for the mean blood pressure of the population from which the subjects were recruited.

(b) The recipe you used in part (a) requires an important assumption about the 27 men who provided the data. What is this assumption?

21.32 Testing a random number generator. Our statistical software has a "random number generator" that is supposed to produce numbers scattered at random between 0 and 1. If this is true, the numbers generated come from a population with $\mu = 0.5$. A command to generate 100 random numbers gives outcomes with mean $\bar{x} = 0.536$ and $s = 0.312$. Give a 90% confidence interval for the mean of all numbers produced by the software.

21.33 Will they charge more? A bank wonders whether omitting the annual credit card fee for customers who charge at least $2500 in a year will increase the amount charged on its credit cards. The bank makes this offer to an SRS of 200 of its credit card customers. It then compares how much these customers charge this year with the amount that they charged last year. The mean increase in the sample is $346, and the standard deviation is $112. Give a 99% confidence interval for the mean amount charges would have increased if this benefit had been extended to all such customers.

21.34 A sampling distribution. Exercise 21.32 concerns the mean of the random numbers generated by a computer program. The mean is supposed to be 0.5 because the numbers are supposed to be spread at random between 0 and 1. We asked the software to generate samples of 100 random numbers repeatedly. Here are the sample means $\bar{x}$ for 50 samples of size 100:

0.532	0.450	0.481	0.508	0.510	0.530	0.499	0.461	0.543	0.490
0.497	0.552	0.473	0.425	0.449	0.507	0.472	0.438	0.527	0.536
0.492	0.484	0.498	0.536	0.492	0.483	0.529	0.490	0.548	0.439
0.473	0.516	0.534	0.540	0.525	0.540	0.464	0.507	0.483	0.436
0.497	0.493	0.458	0.527	0.458	0.510	0.498	0.480	0.479	0.499

The sampling distribution of $\bar{x}$ is the distribution of the means from all possible samples. We actually have the means from 50 samples. Make a histogram of these 50 observations. Does the distribution appear to be roughly Normal, as the central limit theorem says will happen for large enough samples?

21.35 Will they charge more? In Exercise 21.33, you carried out the calculations for a confidence interval based on a bank's experiment in changing the rules for its credit cards. You ought to ask some questions about this study.

(a) The distribution of the amount charged is skewed to the right, but outliers are prevented by the credit limit that the bank enforces on each card. Why can we use a confidence interval based on a Normal sampling distribution for the sample mean $\bar{x}$?

(b) The bank's experiment was not comparative. The increase in amount charged over last year may be explained by lurking variables rather than by the rule change. What are some plausible reasons why charges might go up? Outline the design of a comparative randomized experiment to answer the bank's question.

21.36 A sampling distribution, continued. Exercise 21.34 presents 50 sample means $\bar{x}$ from 50 random samples of size 100. Using a calculator, find the mean and standard deviation of these 50 values. Then answer these questions.

(a) The mean of the population from which the 50 samples were drawn is $\mu = 0.5$ if the random number generator is accurate. What do you expect the mean of the distribution of $\bar{x}$'s from all possible samples to be? Is the mean of these 50 samples close to this value?

(b) The standard deviation of the distribution of $\bar{x}$ from samples of size $n = 100$ is supposed to be $\sigma/10$, where σ is the standard deviation of individuals in the population. Use this fact with the standard deviation you calculated for the 50 $\bar{x}$'s to estimate σ.

EXPLORING THE WEB

21.37 Web-based exercise. If you go to the Centers for Disease Control and Prevention Web site, you will find that the proportion of college graduates in California who engaged in binge drinking in 2010 is reported as 15.8%. Go to the Web site at **apps.nccd.cdc.gov/brfss/.** Why is the percentage reported at the Web site different from the value calculated in Example 1? Was the sample obtained by the Behavioral Risk Factor Surveillance System a simple random sample?

21.38 Web-based exercise. Results from the Behavioral Risk Factor Surveillance System can be found at **apps.nccd.cdc.gov/brfss/.** Find the sample proportion of college graduates in your state who engaged in binge drinking for the most recent year reported at the Web site. Construct a 95% confidence interval for the population proportion (assume that the sample is a random sample). Write a short explanation of your findings for someone who knows no statistics.

21.39 Web-based exercise. Search the Web for a recent poll in which the sample statistic is a proportion, for example, the proportion in the sample responding "Yes" to a question. Calculate a 95% confidence interval for the population proportion (assume that the sample is a random sample). Write a short explanation of your findings for someone who knows no statistics. Your report should include the exact wording of the question asked, how the sample was collected, the sample size, the population of interest, and how the reader should interpret your confidence interval. One possible Web site for recent polls, including the information needed for your report, is the Gallup Poll Web site at **www.gallup.com/.**

NOTES AND DATA SOURCES

Page 487 Janice E. Williams et al., "Anger proneness predicts coronary heart disease risk," *Circulation,* 101 (2000), pp. 2034–2039.

Page 492 For the state of the art in confidence intervals for *p*, see Alan Agresti and Brent Coull, "Approximate is better than 'exact' for interval estimation of binomial proportions," *American Statistician,* 52 (1998), pp. 119–126. You can also consult the book *An Introduction to Categorical Data Analysis,* Wiley, 1996, by Alan Agresti. We note that the accuracy of our confidence interval for *p* can be greatly improved by simply "adding 2 successes and 2 failures." That is, replace $\hat{p}$ by (count of successes + 2)/(n + 4). Texts on sample surveys give confidence intervals that take into account the fact that the population has finite size and also give intervals for sample designs more complex than an SRS.

Page 502 Example 7: Information about the NAEP test can be found online at **nationsreportcard.gov/math_2009/.**

Page 505 Exercise 21.6: The poll can be found in the archives of the Gallup Web site. Go to **www.gallup.com/poll/Topics.aspx** and look under the topic "Education."

Page 505 Exercise 21.7: The poll can be found in the archives of the Gallup Web site. Go to **www.gallup.com/poll/Topics.aspx** and look under the topic "Guns."

Page 505 Exercise 21.8: The poll results can be found at **www.pollingreport.com/ immigration.htm.**

Page 506 Exercise 21.11: The poll results can be found at **www.pollingreport.com/ budget.htm.**

Page 507 Exercise 21.16: Jeff Zeleny and Megan Thee-Brenan, "New poll finds a deep distrust of government," *New York Times,* October 25, 2011.

Page 509 Exercise 21.25: Sara J. Solnick and David Hemenway, "Complaints and disenrollment at a health maintenance organization," *Journal of Consumer Affairs,* 26 (1992), pp. 90–103.

Page 509 Exercise 21.26: The data are from the CPS Web site, **www.bls.gov/cps/**. Data from this Web site can be downloaded using the DataFerrett software that is available at the Web site **dataferrett.census.gov/.** The variable we downloaded had three values: $-1 =$ in universe, met no conditions to assign; $1 =$ employed; and $2 =$ unemployed. 67,044 cases had value -1; 62,409 had value 1; and 4991 had value 2.

Page 509 Exercise 21.29: Data provided by Darlene Gordon, Purdue University.

What Is a Test of Significance?

© Randy Faris/Corbis

CASE STUDY A January 26, 2011, article in the *New York Times* reported record stress levels in college freshmen. What was the basis for this finding?

Every year since 1985, the Higher Education Research Institute at UCLA has conducted a survey of college freshmen. The 2010 survey involved a random sample of 201,818 of the 1.5 million first-time, full-time, first-year students at 279 of the nation's baccalaureate colleges and universities. A disturbing finding of the 2010 survey was that the percentage of students reporting that their emotional health was "above average" reached a record low of 51.9%, down from 55.3% in 2009. Following a similar trend, the percentage who reported being frequently "overwhelmed by all I had to do" was 29.1%, up from 27.1% in 2009.

The *New York Times* article reported that these findings were consistent with the experiences of campus counselors. Every day they see students in their offices who are depressed, under stress, and receiving psychiatric medication. One counselor commented that the state of the economy, student loans, and concerns about employment after graduating contribute to the decline in emotional health. At the same time, questions asking students to rate their own emotional health compared with that of others are hard to assess, since they required students to create their own definition of emotional health and to make judgments of how they compare with others.

The sample size in the 2009 survey was 219,864, so these findings come from two very large samples. On the other hand, the changes in the percentages were small. Could it be that the difference between the two samples is just due to the luck of the draw in randomly choosing the respondents?

In this chapter we discuss methods, called tests of significance, that help us decide whether an observed difference can plausibly be attributed to chance. By the end of this chapter you will know how to interpret such tests and whether the differences in the Higher Education Research Institute's surveys of college freshmen can be plausibly attributed to chance. ∎

The reasoning of statistical tests of significance

The local hot-shot playground basketball player claims to make 80% of his free throws. "Show me," you say. He shoots 20 free throws and makes 8 of them. "Aha," you conclude, "if he makes 80%, he would almost never make as few as 8 of 20. So I don't believe his claim." That's the reasoning of statistical **tests of significance** at the playground level: *an outcome that is very unlikely if a claim is true is good evidence that the claim is not true.*

Statistical inference uses data from a sample to draw conclusions about a population. So once we leave the playground, statistical tests deal with claims about a population. Statistical tests ask if sample data give good evidence *against* a claim. A statistical test says, "If we took many samples and the claim were true, we would rarely get a result like this." To get a numerical measure of how strong the sample evidence is, replace the vague term "rarely" by a probability. Here is an example of this reasoning at work.

> **EXAMPLE 1** Is the coffee fresh?
>
> People of taste are supposed to prefer fresh-brewed coffee to the instant variety. But perhaps many coffee drinkers just need their caffeine fix. A skeptic claims that coffee drinkers can't tell the difference. Let's do an experiment to test this claim.
>
> Each of 50 subjects tastes two unmarked cups of coffee and says which he or she prefers. One cup in each pair contains instant coffee; the other, fresh-brewed coffee. The statistic that records the result of our experiment is the proportion $\hat{p}$ of the sample who say they like the fresh-brewed coffee better. We find that 36 of our 50 subjects choose the fresh coffee. That is,
>
> $$\hat{p} = \frac{36}{50} = 0.72 = 72\%$$
>
> To make a point, let's compare our outcome $\hat{p} = 0.72$ with another possible result. If only 28 of the 50 subjects like the fresh coffee better than instant coffee, the sample proportion is
>
> $$\hat{p} = \frac{28}{50} = 0.56 = 56\%$$

Surely 72% is stronger evidence against the skeptic's claim than 56%. But how much stronger? Is even 72% in favor in a *sample* convincing evidence that a majority of the *population* prefer fresh coffee? Statistical tests answer these questions. Here's the answer in outline form:

- **The claim.** The skeptic claims that coffee drinkers can't tell fresh from instant, so that only half will choose fresh-brewed coffee. That is, he claims that the population proportion p is only 0.5. *Suppose for the sake of argument that this claim is true.*

- **The sampling distribution (from page 489).** If the claim $p = 0.5$ were true and we tested many random samples of 50 coffee drinkers, the sample proportion $\hat{p}$ would vary from sample to sample according to (approximately) the Normal distribution with

$$\text{mean} = p = 0.5$$

and

$$\text{standard deviation} = \sqrt{\frac{p(1-p)}{n}}$$
$$= \sqrt{\frac{(0.5)(0.5)}{50}}$$
$$= 0.0707$$

Figure 22.1 displays this Normal curve.

- **The data.** Place the sample proportion $\hat{p}$ on the sampling distribution. You see in Figure 22.1 that $\hat{p} = 0.56$ isn't an unusual value, but that $\hat{p} = 0.72$ is unusual. We would rarely get 72% of a sample of 50 coffee drinkers preferring fresh-brewed coffee if only 50% of all coffee drinkers felt that way. So the sample data do give evidence against the claim.

- **The probability.** We can measure the strength of the evidence against the claim by a probability. What is the probability that a

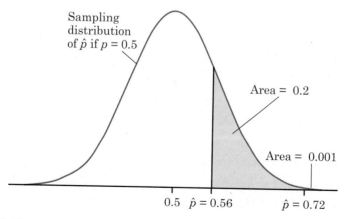

Figure 22.1 The sampling distribution of the proportion of 50 coffee drinkers who prefer fresh-brewed coffee if the truth about all coffee drinkers is that 50% prefer fresh coffee, for Example 1. The shaded area is the probability that the sample proportion is 56% or greater.

sample gives a $\hat{p}$ this large or larger if the truth about the population is that $p = 0.5$? If $\hat{p} = 0.56$, this probability is the shaded area under the Normal curve in Figure 22.1. This area is 0.20. Our sample actually gave $\hat{p} = 0.72$. The probability of getting a sample outcome this large is only 0.001, an area too small to see in Figure 22.1. An outcome that would occur just by chance in 20% of all samples is *not* strong evidence against the claim. But an outcome that would happen only 1 in 1000 times *is* good evidence.

Be sure you understand why this evidence is convincing. There are two possible explanations of the fact that 72% of our subjects prefer fresh to instant coffee:

1. The skeptic is correct ($p = 0.5$), and by bad luck a very unlikely outcome occurred.

2. In fact, the population proportion favoring fresh coffee is greater than 0.5, so the sample outcome is about what would be expected.

We cannot be certain that Explanation 1 is untrue. Our taste test results *could* be due to chance alone. But the probability that such a result would occur by chance is so small (0.001) that we are quite confident that Explanation 2 is right.

Hypotheses and *P*-values

Tests of significance refine (and perhaps hide) this basic reasoning. In most studies, we hope to show that some definite effect is present in the population. In Example 1, we suspect that a majority of coffee drinkers prefer fresh-brewed coffee. A statistical test begins by supposing for the sake of argument that the effect we seek is *not* present. We then look for evidence against this supposition and in favor of the effect we hope to find. The first step in a test of significance is to state a claim that we will try to find evidence *against*.

Null hypothesis H_0

The claim being tested in a statistical test is called the **null hypothesis**. The test is designed to assess the strength of the evidence against the null hypothesis. Usually the null hypothesis is a statement of "no effect" or "no difference."

The term "null hypothesis" is abbreviated H_0 and is read as "H-nought," "H-oh," and sometimes even "H-null." It is a statement about the population and so must be stated in terms of a population parameter. In Example 1, the parameter is the proportion p of all coffee drinkers who prefer fresh to instant coffee. The null hypothesis is

$$H_0:\ p = 0.5$$

The statement we hope or suspect is true instead of H_0 is called the **alternative hypothesis** and is abbreviated H_a. In Example 1, the alternative hypothesis is that a majority of the population favor fresh coffee. In terms of the population parameter, this is

$$H_a:\ p > 0.5$$

A significance test looks for evidence against the null hypothesis and in favor of the alternative hypothesis. The evidence is strong if the outcome we observe would rarely occur if the null hypothesis is true but is more probable if the alternative hypothesis is true. For example, it would be surprising to find 36 of 50 subjects favoring fresh coffee if in fact only half of the population feel this way. How surprising? A significance test answers this question by giving a probability: the probability of getting an outcome at least as far as the actually observed outcome from what we would expect when H_0 is true. What counts as "far from what we would expect" depends on H_a as well as H_0. In the taste test, the probability we want is the probability that 36 or more of 50 subjects favor fresh coffee. If the null hypothesis $p = 0.5$ is true, this probability is very small (0.001). That's good evidence that the null hypothesis is not true.

Gotcha! A tax examiner suspects that Ripoffs, Inc., is issuing phony checks to inflate its expenses and reduce the tax it owes. To learn the truth without examining every check, she boots up her computer. The first digits of real data follow well-known patterns that do *not* give digits 0 to 9 equal probabilities. If the check amounts don't follow this pattern, she will investigate. Down the street, a hacker is probing a company's computer files. He can't read them, because they are encrypted. But he may be able to locate the key to the encryption anyway— if it's the only long string that really *does* give equal probability to all possible characters. Both the tax examiner and the hacker need a method for testing whether the pattern they are looking for is present.

P-value

The probability, computed assuming that H_0 is true, that the sample outcome would be as extreme or more extreme than the actually observed outcome is called the ***P*-value** of the test. The smaller the *P*-value is, the stronger is the evidence against H_0 provided by the data.

In practice, most statistical tests are carried out by computer software that calculates the *P*-value for us. It is usual to report the *P*-value in describing the results of studies in many fields. You should therefore understand what *P*-values say even if you don't do statistical tests yourself, just

The truth about Count Buffon's coin-tossing experiment.

as you should understand what "95% confidence" means even if you don't calculate your own confidence intervals.

EXAMPLE 2   Count Buffon's coin

The French naturalist Count Buffon (1707–1788) considered questions ranging from evolution to estimating the number "pi" and made it his goal to answer them. One question he explored was whether a "balanced" coin would come up heads half of the time when tossed. To investigate, he tossed a coin 4040 times. He got 2048 heads. The sample proportion of heads is

$$\hat{p} = \frac{2048}{4040} = 0.507$$

That's a bit more than one-half. Is this evidence that Buffon's coin was not balanced? This is a job for a significance test.

The hypotheses. The null hypothesis says that the coin is balanced ($p = 0.5$). We did not suspect a bias in a specific direction before we saw the data, so the alternative hypothesis is just "the coin is not balanced." The two hypotheses are

$$H_0: p = 0.5$$
$$H_a: p \neq 0.5$$

The sampling distribution. *If the null hypothesis is true*, the sample proportion of heads has approximately the Normal distribution with

$$\text{mean} = p = 0.5$$
$$\text{standard deviation} = \sqrt{\frac{p(1-p)}{n}}$$
$$= \sqrt{\frac{(0.5)(0.5)}{4040}}$$
$$= 0.00787$$

The data. Figure 22.2 shows this sampling distribution with Buffon's sample outcome $\hat{p} = 0.507$ marked. The picture already suggests that this is not an unlikely outcome that would give strong evidence against the claim that $p = 0.5$.

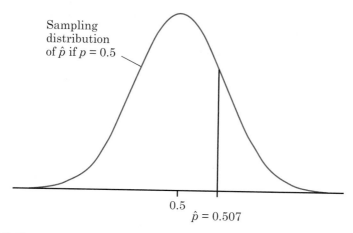

Figure 22.2 The sampling distribution of the proportion of heads in 4040 tosses of a balanced coin, for Example 2. Count Buffon's result, proportion 0.507 heads, is marked.

The *P*-value. How unlikely is an outcome as far from 0.5 as Buffon's $\hat{p} = 0.507$? Because the alternative hypothesis allows p to lie on either side of 0.5, values of $\hat{p}$ far from 0.5 in either direction provide evidence against H_0 and in favor of H_a. The *P*-value is therefore the probability that the observed $\hat{p}$ lies as far from 0.5 *in either direction* as the observed $\hat{p} = 0.507$. Figure 22.3 shows this probability as area under the Normal curve. It is $P = 0.37$.

The conclusion. A truly balanced coin would give a result this far or farther from 0.5 in 37% of all repetitions of Buffon's trial. His result gives no reason to think that his coin was not balanced.

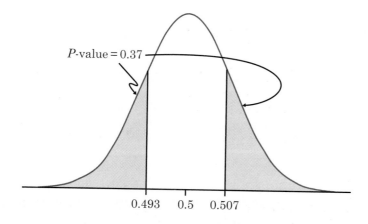

Figure 22.3 The *P*-value for testing whether Count Buffon's coin was balanced, for Example 2. This is the probability, calculated assuming a balanced coin, of a sample proportion as far or farther from 0.5 as Buffon's result of 0.507.

The alternative H_a: $p > 0.5$ in Example 1 is a **one-sided alternative** because the effect we seek evidence for says that the population proportion is greater than one-half. The alternative H_a: $p \neq 0.5$ in Example 2 is a **two-sided alternative** because we ask only whether or not the coin is balanced. Whether the alternative is one-sided or two-sided determines whether sample results that are extreme in one direction or in both directions count as evidence against H_0 in favor of H_a.

> **NOW IT'S YOUR TURN**   **22.1 Coin tossing.** We do not have the patience of Count Buffon, so we tossed a coin only 50 times. We got 21 heads. The proportion of heads is
>
> $$\hat{p} = \frac{21}{50} = 0.42$$
>
> This is less than one-half. Is this evidence that our coin is not balanced? Formulate the hypotheses for an appropriate significance test and determine the sampling distribution of the sample proportion of heads if the null hypothesis is true.

Statistical significance

We can decide in advance how much evidence against H_0 we will insist on. The way to do this is to say, before any data are collected, how small a P-value we require. The decisive value of P is called the **significance level**. It is usual to write it as α, the Greek letter alpha. If we choose $\alpha = 0.05$, we are requiring that the data give evidence against H_0 so strong that it would happen no more than 5% of the time (1 time in 20) when H_0 is true. If we choose $\alpha = 0.01$, we are insisting on stronger evidence against H_0, evidence so strong that it would appear only 1% of the time (1 time in 100) if H_0 is in fact true.

> **Statistical significance**
>
> If the P-value is as small or smaller than α, we say that the data are **statistically significant at level α.**

"Significant" in the statistical sense does not mean "important." It means simply "not likely to happen just by chance." We used these words in Chapter 5 (page 101). Now we have attached a number to statistical significance to say what "not likely" means. You will often see significance at level 0.01 expressed by the statement "The results were significant ($P < 0.01$)." Here P stands for the P-value.

We don't have to make use of traditional levels of significance such as 5% and 1%. The *P*-value is more informative, because it allows us to assess significance at any level we choose. For example, a result with $P = 0.03$ is significant at the $\alpha = 0.05$ level but not significant at the $\alpha = 0.01$ level. Nonetheless, the traditional significance levels are widely accepted guidelines for "how much evidence is enough." We might say that $P < 0.10$ indicates "some evidence" against the null hypothesis, $P < 0.05$ is "moderate evidence," and $P < 0.01$ is "strong evidence." Don't take these guidelines too literally, however. We will say more about interpreting tests in Chapter 23.

Calculating *P*-values*

Finding the *P*-values we gave in Examples 1 and 2 requires doing Normal distribution calculations using Table B of Normal percentiles. That was optional reading in Chapter 13 (pages 299–300). In practice, software does the calculation for us, but here is an example that shows how to use Table B.

EXAMPLE 3   Tasting coffee

The hypotheses. In Example 1, we want to test the hypotheses

$$H_0: p = 0.5$$
$$H_a: p > 0.5$$

Here *p* is the proportion of the population of all coffee drinkers who prefer fresh coffee to instant coffee.

The sampling distribution. If the null hypothesis is true, so that $p = 0.5$, we saw in Example 1 that $\hat{p}$ follows a Normal distribution with mean 0.5 and standard deviation 0.0707.

The data. A sample of 50 people found that 36 preferred fresh coffee. The sample proportion is $\hat{p} = 0.72$.

The *P*-value. The alternative hypothesis is one-sided on the high side. So the *P*-value is the probability of getting an outcome at least as large as 0.72. Figure 22.1 displays this probability as an area under the Normal sampling distribution curve. To find any Normal curve probability, move to the standard scale. The standard score for the outcome $\hat{p} = 0.72$ is

$$\text{standard score} = \frac{\text{observation} - \text{mean}}{\text{standard deviation}}$$
$$= \frac{0.72 - 0.5}{0.0707} = 3.1$$

---

*This section is optional.

Table B says that standard score 3.1 is the 99.9 percentile of a Normal distribution. That is, the area under a Normal curve to the left of 3.1 (in the standard scale) is 0.999. The area to the right is therefore 0.001, and that is our *P*-value.

The conclusion. The small *P*-value means that these data provide very strong evidence that a majority of the population prefers fresh coffee.

NOW IT'S YOUR TURN | **22.2 Coin tossing.** Refer to Exercise 22.1 (page 522). We tossed a coin only 50 times and got 21 heads, so the proportion of heads is

$$\hat{p} = \frac{21}{50} = 0.42$$

This is less than one-half. Is this evidence that our coin is not balanced? For the hypotheses you formulated in Exercise 22.1, find the *P*-value based on the results of our 50 tosses. Are the results significant at the 0.05 level?

Tests for a population mean*

The reasoning that leads to significance tests for hypotheses about a population mean μ follows the reasoning that leads to tests about a population proportion p. The big idea is to use the sampling distribution that the sample mean $\bar{x}$ would have if the null hypothesis were true. Locate the value of $\bar{x}$ from your data on this distribution and see if it is unlikely. A value of $\bar{x}$ that would rarely appear if H_0 were true is evidence that H_0 is not true. The four steps are also similar to those in tests for a proportion. Here are two examples, the first one-sided and the second two-sided.

EXAMPLE 4 Can you balance your checkbook?

In a discussion of the education level of the American workforce, a pessimist says, "The average young person can't even balance a checkbook." The National Assessment of Adult Literacy (NAAL) survey indicates that a score of 290 or higher on its quantitative test reflects skills that include those needed to balance a checkbook. The NAAL administered the test to a random sample of 2168 young men (aged 19 to 24) who had mean score $\bar{x} = 276$, a bit below the checkbook-balancing level. Is this sample result good

---

*This section is optional.

evidence that the mean for *all* young men is less than 290? The standard deviation of the scores in the sample is $s = 112$.

The hypotheses. The pessimist's claim is that the mean NAAL score is less than 290. That's our alternative hypothesis, the statement we seek evidence *for*. The hypotheses are

$$H_0: \mu = 290$$

$$H_a: \mu < 290$$

The sampling distribution. *If the null hypothesis is true,* the sample mean $\bar{x}$ has approximately the Normal distribution with mean $\mu = 290$ and standard deviation

$$\frac{s}{\sqrt{n}} = \frac{112}{\sqrt{2168}} = 2.4$$

We once again use the sample standard deviation s in place of the unknown population standard deviation σ.

The data. The NAAL sample gave $\bar{x} = 276$. The standard score for this outcome is

$$\text{standard score} = \frac{\text{observation} - \text{mean}}{\text{standard deviation}}$$

$$= \frac{276 - 290}{2.4} = -5.83$$

That is, the sample result is about 5.83 standard deviations below the mean we would expect if, on the average, young men had just enough skill to balance a checkbook.

The P-value. Figure 22.4 locates the sample outcome -5.83 (in the standard scale) on the Normal curve that represents the sampling distribution if H_0 is true. This curve has mean 0 and standard deviation 1 because we are using the standard scale. The P-value for our one-sided test is the area to the left of -5.83 under the Normal curve. Figure 22.4 indicates that this area is very small. The smallest value in Table B is -3.4 and Table B says that -3.4 is the 0.03 percentile, so the area to its left is 0.0003. Because -5.83 is smaller than -3.4, we know that the area to its left is smaller than 0.0003. Thus, our P-value is smaller than 0.0003. (The Minitab software package gives $P < 0.0000000$.)

 Catching cheaters Lots of students take a long multiple-choice exam. Can the computer that scores the exam also screen for papers that are suspiciously similar? Clever people have created a measure that takes into account not just identical answers but the popularity of those answers and the total score on the similar papers. The measure has close to a Normal distribution, and the computer flags pairs of papers with a measure outside ±4 standard deviations as significant.

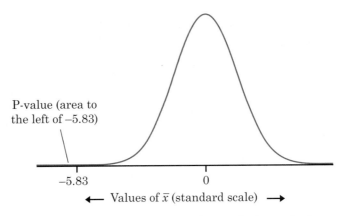

Figure 22.4 The *P*-value for a one-sided test when the standard score for the sample mean is −5.83, for Example 4.

The conclusion. A *P*-value of less than 0.0003 is strong evidence that the mean score for all young men (aged 19 to 24) is below the level that includes the skills needed to balance a checkbook.

EXAMPLE 5 Executives' blood pressures

The National Center for Health Statistics reports that the mean systolic blood pressure for males 35 to 44 years of age is 128. The medical director of a large company looks at the medical records of 72 executives in this age group and finds that the mean systolic blood pressure in this sample is $\bar{x} = 126.1$ and that the standard deviation is $s = 15.2$. Is this evidence that the company's executives have a different mean blood pressure from the general population?

The hypotheses. The null hypothesis is "no difference" from the national mean. The alternative is two-sided because the medical director did not have a particular direction in mind before examining the data. So the hypotheses about the unknown mean μ of the executive population are

$$H_0: \mu = 128$$
$$H_a: \mu \neq 128$$

The sampling distribution. *If the null hypothesis is true,* the sample mean $\bar{x}$ has approximately the Normal distribution with mean $\mu = 128$ and standard deviation

$$\frac{s}{\sqrt{n}} = \frac{15.2}{\sqrt{72}} = 1.79$$

The data. The sample mean is $\bar{x} = 126.1$. The standard score for this outcome is

$$\text{standard score} = \frac{\text{observation} - \text{mean}}{\text{standard deviation}}$$

$$= \frac{126.1 - 128}{1.79} = -1.06$$

We know that an outcome just slightly more than 1 standard deviation away from the mean of a Normal distribution is not very surprising. The last step is to make this formal.

The *P*-value. Figure 22.5 locates the sample outcome -1.06 (in the standard scale) on the Normal curve that represents the sampling distribution if H_0 is true. The two-sided *P*-value is the probability of an outcome at least this far out *in either direction*. This is the shaded area under the curve. To use Table B, round the standard score to -1.1. This is the 13.57 percentile of a Normal distribution. So the area to the left of -1.1 is 0.1357. The area to the left of -1.1 and to the right of 1.1 is double this, or about 0.27. This is our approximate *P*-value. (The exact *P*-value, from software, is $P = 0.289$.)

The conclusion. The large *P*-value gives us no reason to think that the mean blood pressure of the executive population differs from the national average.

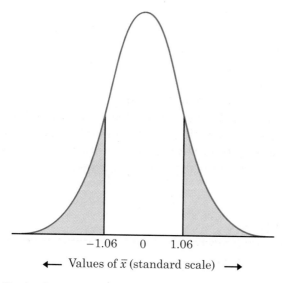

$-1.06 \quad 0 \quad 1.06$

$\longleftarrow$ Values of $\bar{x}$ (standard scale) $\longrightarrow$

Figure 22.5 The *P*-value for a two-sided test when the standard score for the sample mean is -1.06, for Example 5.

The test assumes that the 72 executives in the sample are an SRS from the population of all middle-aged male executives in the company. We should check this assumption by asking how the data were produced. If medical records are available only for executives with recent medical problems, for example, the data are of little value for our purpose. It turns out that all executives are given a free annual medical exam, and that the medical director selected 72 exam results at random.

What if the medical director did not draw a random sample? You should be very cautious about inference based on nonrandom samples, such as convenience samples, or random samples with large nonresponse. Although it is possible for a convenience sample to be representative of the population and hence yield reliable inference, establishing this is not easy. You must be certain that the method for selecting the sample is unrelated to the quantity being measured. You must make a case that individuals are independent. You must use additional information to argue that the sample is representative. This can involve comparing other characteristics of the sample to known characteristics of the population. Is the proportion of men and women in the sample about the same as in the population? Is the racial composition of the sample about the same as in the population? What about the distribution of ages or other demographic characteristics? Even after arguing that your sample appears to be representative, the case is not as compelling as that for a truly random sample. You should proceed with caution.

The data in Example 5 do *not* establish that the mean blood pressure μ for this company's executives is 128. We sought evidence that μ differed from 128 and failed to find convincing evidence. That is all we can say. No doubt the mean blood pressure of the entire executive population is not exactly equal to 128. A large enough sample would give evidence of the difference, even if it is very small.

NOW IT'S YOUR TURN | **22.3 IQ scores.** The mean IQ for the entire population in any age group is supposed to be 100. Suppose that we measure the IQ of 31 seventh-grade girls in a Midwest school district and find a sample mean of 105.8 and sample standard deviation of 14.3. Treat the scores as if they were an SRS from all middle-school girls in this district. Do the scores provide good evidence that the mean IQ of this population is not 100?

STATISTICS IN SUMMARY

Chapter Specifics

- A confidence interval estimates an unknown parameter. A **test of significance** assesses the evidence for some claim about the value of an unknown parameter.

- In practice, the purpose of a statistical test is to answer the question "Could the effect we see in the sample just be an accident due to chance, or is it good evidence that the effect is really there in the population?"

- Significance tests answer this question by giving the probability that a sample effect as large as the one we see in this sample would arise just by chance. This probability is the **P-value.** A small P-value says that our outcome is unlikely to happen just by chance.

- To set up a test, state a **null hypothesis** that says the effect you seek is *not* present in the population. The **alternative hypothesis** says that the effect *is* present.

- The P-value is the probability, calculated taking the null hypothesis to be true, of an outcome as extreme in the direction specified by the alternative hypothesis as the actually observed outcome.

- A sample result is **statistically significant at the 5% level (or at the 0.05 level)** if it would occur just by chance no more than 5% of the time in repeated samples.

In this chapter we discuss tests of significance, another type of statistical inference. The mathematics of probability, in particular the sampling distributions discussed in Chapter 18, provides the formal basis for a test of significance. The sampling distribution allows us to assess "probabilistically" the strength of evidence against a null hypothesis, through either a level of significance or a P-value. The goal of hypothesis testing, which is used to assess the evidence provided by data about some claim concerning a population, is different from the goal of confidence interval estimation, discussed in Chapter 21, which is used to estimate a population parameter.

Although we have applied the reasoning of tests of significance to population proportions and population means, the same reasoning applies to tests of significance for other population parameters, such as the correlation coefficient, in more advanced settings. In the next chapter we provide more discussion of the practical interpretation of statistical tests.

CASE STUDY Look again at the Case Study at the beginning of this chapter. Could
EVALUATED the 2010 and 2009 random samples in the Higher Education Research
Institute's surveys of college freshmen differ by 51.9% versus 55.3% for those reporting
that their emotional health was above average, and by 29.1% versus 27.1% for those
reporting being frequently "overwhelmed by all I had to do," just by chance? Tests of
significance can help answer these questions. In both cases one finds that the P-values
for the tests of whether two such random samples would differ by the amounts reported
was less than 0.001.

1. Using language that can be understood by someone who knows no statistics, write a paragraph explaining what a *P*-value of less than 0.001 means in the context of the Higher Education Research Institute's surveys of college freshmen.

2. Are the results of the study significant at the 0.05 level? At the 0.01 level? Explain. ■

CHAPTER 22 EXERCISES

For Exercise 22.1, see page 522; for Exercise 22.2, see page 524; and for Exercise 22.3, see page 528.

22.4 Ethnocentrism. A social psychologist reports, "In our sample, ethnocentrism was significantly higher (*P* < 0.05) among church attenders than among nonattenders." Explain to someone who knows no statistics what this means.

22.5 Students' earnings. The financial aid office of a university asks a sample of students about their employment and earnings. The report says, "For academic year earnings, a significant difference (*P* = 0.028) was found between the sexes, with men earning more on average than women. No difference (*P* = 0.576) was found between the earnings of black and white students." Explain both of these conclusions, for the effects of sex and of race on mean earnings, in language understandable to someone who knows no statistics.

22.6 Diet and diabetes. Does eating more fiber reduce the blood cholesterol level of patients with diabetes? A randomized clinical trial compared normal and high-fiber diets. Here is part of the researchers' conclusion: "The high-fiber diet reduced plasma total cholesterol concentrations by 6.7 percent (*P* = 0.02), triglyceride concentrations by 10.2 percent (*P* = 0.02), and very-low-density lipoprotein cholesterol concentrations by 12.5 percent (*P* = 0.01)." A doctor who knows no statistics says that a drop of 6.7% in cholesterol isn't a lot—maybe it's just an accident due to the chance assignment of patients to the two diets. Explain in simple language how "*P* = 0.02" answers this objection.

22.7 Diet and bowel cancer. It has long been thought that eating a healthier diet reduces the risk of bowel cancer. A large study cast doubt on this advice. The subjects were 2079 people who had polyps removed from their bowels in the past six months. Such polyps may lead to cancer. The subjects were randomly assigned to a low-fat, high-fiber diet or to a control group in which subjects ate their usual diets. Did polyps reoccur during the next four years?

(a) Outline the design of this experiment.

(b) Surprisingly, the occurrence of new polyps "did not differ significantly between the two groups." Explain clearly what this finding means.

22.8 Pigs and prestige in ancient China. It appears that pigs in Stone Age China were not just a source of food. Owning pigs was also a display of wealth. Evidence for this comes from examining burial sites. If the skulls of sacrificed pigs tend to appear along with expensive ornaments, that suggests that the pigs, like the ornaments, signal the wealth and prestige of the person buried. A study of burials from around 3500 B.C. concluded that "there are striking differences in grave goods between burials with pig skulls and burials without them. . . . A test indicates that the two samples of total artifacts are significantly different at the 0.01 level." Explain clearly why "significantly different at the 0.01 level" gives good reason to think that there really is a systematic difference between burials that contain pig skulls and those that lack them.

22.9 Ancient Egypt. Settlements in Egypt before the time of the pharaohs are dated by measuring the presence of forms of carbon that decay over time. The first datings of settlements in the Nagada region used hair that had been excavated 60 years earlier. Now researchers have used newer methods and more recently excavated material. Do the dates differ? Here is the conclusion about one location: "There are two dates from Site KH6. Statistically, the two dates are not significantly different. They provide a weighted average corrected date of 3715 ± 90 B.C." Explain to someone interested in ancient Egypt but not interested in statistics what "not significantly different" means.

22.10 What's a gift worth? Do people value gifts from others more highly than they value the money it would take to buy the gift? We would like to think so, because we hope that "the thought counts." A survey of 209 adults asked them to list three recent gifts and then asked, "Aside from any sentimental value, if, without the giver ever knowing, you could receive an amount of money instead of the gift, what is the minimum amount of money that would make you equally happy?" It turned out that most people would need more money than the gift cost to be equally happy. The magic words "significant ($P < 0.01$)" appear in the report of this finding.

(a) The sample consisted of students and staff in a graduate program and of "members of the general public at train stations and airports in Boston and Philadelphia." The report says this sample is "not ideal." What's wrong with the sample?

(b) In simple language, what does it mean to say that the sample thought their gifts were worth "significantly more" than their actual cost?

(c) Now be more specific: what does "significant ($P < 0.01$)" mean?

22.11 Attending church. A 2010 Gallup Poll found that 39% of American adults say they attended religious services last week. This is almost certainly not true.

(a) Why might we expect answers to a poll to overstate true church attendance?

(b) You suspect strongly that the true percentage attending church in any given week is less than 39%. You plan to watch a random sample of adults and see whether or not they go to church. What are your null and alternative hypotheses? (Be sure to say in words what the population proportion p is for your study.)

22.12 Body temperature. We have all heard that 98.6 degrees Fahrenheit (or 37 degrees Celsius) is "normal body temperature." In fact, there is evidence that most people have a slightly lower body temperature. You plan to measure the body temperature of a random sample of people very accurately. You hope to show that a majority have temperatures lower than 98.6 degrees.

(a) Say clearly what the population proportion p stands for in this setting.

(b) In terms of p, what are your null and alternative hypotheses?

22.13 Unemployment. The national unemployment rate in a recent month was 9.0%. You think the rate may be different in your city, so you plan a sample survey that will ask the same questions as the Current Population Survey. To see if the local rate differs significantly from 9.0%, what hypotheses will you test?

22.14 First-year students. A UCLA survey of college freshmen in the 2008–2009 academic year found that 31.0% of all first-year college students identify themselves as politically liberal. You wonder if this percentage is different at your school, but you have no idea whether it is higher or lower. You plan a sample survey of first-year students at your school. What hypotheses will you test to see if your school differs significantly from the UCLA survey result?

22.15 Do our athletes graduate? The National Collegiate Athletic Association (NCAA) requires colleges to report the graduation rates of their athletes. At one large university, 78% of all students who entered in 2004 graduated within six years. One hundred thirty-seven of the 190 students who entered with athletic scholarships graduated. Consider these 190 as a sample of all athletes who will be admitted under present policies. Is there evidence that the percentage of athletes who graduate is less than 78%?

(a) Explain in words what the parameter p is in this setting.

(b) What are the null and alternative hypotheses H_0 and H_a?

(c) What is the numerical value of the sample proportion $\hat{p}$? The P-value is the probability of what event?

(d) The P-value is $P = 0.025$. Explain why this P-value indicates there is some reason to think that graduation rates are lower among athletes than among all students.

22.16 Using the Internet. In 2006, 75.9% of first-year college students responding to a national survey said that they used the Internet frequently for research or homework. Administrators at a large state university believe that their current first-year students use the Internet more frequently than students in 2006. They find that 168 of an SRS of 200 of the university's first-year students said that they used the Internet frequently for research or homework. Is the proportion of first-year students at this university who said that they used the Internet frequently for research or homework larger than the 2006 national value of 75.9%?

(a) Explain in words what the parameter p is in this setting.

(b) What are the null and alternative hypotheses H_0 and H_a?

(c) What is the numerical value of the sample proportion $\hat{p}$? The P-value is the probability of what event?

(d) The P-value is $P = 0.0037$. Explain carefully why this is reasonably good evidence that H_0 is not true and that H_a is true.

22.17 Vote for the best face? We often judge other people by their faces. It appears that some people judge candidates for elected office by their faces. Psychologists showed head-and-shoulders photos of the two main candidates in 32 races for the U.S. Senate to many subjects (dropping subjects who recognized one of the candidates) to see which candidate was rated "more competent" based on nothing but the photos. On election day, the candidates whose faces looked more competent won 22 of the 32 contests. If faces don't influence voting, half of all races in the long run should be won by the candidate with the better face. Is there evidence that the proportion of times the candidate with the better face wins is more than 50%?

(a) Explain in words what the parameter p is in this setting.

(b) What are the null and alternative hypotheses H_0 and H_a?

(c) What is the numerical value of the sample proportion $\hat{p}$? The P-value is the probability of what event?

(d) The P-value is $P = 0.017$. Explain carefully why this is reasonably good evidence that H_0 is not true and that H_a is true.

22.18 Do our athletes graduate? Is the result of Exercise 22.15 statistically significant at the 10% level? At the 5% level?

22.19 Using the Internet. Is the result of Exercise 22.16 statistically significant at the 5% level? At the 1% level?

22.20 Vote for the best face? Is the result of Exercise 22.17 statistically significant at the 5% level? At the 1% level?

22.21 Significant at what level? Explain in plain language why a result that is significant at the 1% level must always be significant at the 5% level. If a result is significant at the 5% level, what can you say about its significance at the 1% level?

22.22 Significance means what? Asked to explain the meaning of "statistically significant at the $\alpha = 0.05$ level," a student says: "This means that the probability that the null hypothesis is true is less than 0.05." Is this explanation correct? Why or why not?

22.23 Finding a P-value by simulation. Is a new method of teaching reading to first-graders (Method B) more effective than the method now in use (Method A)? You design a matched pairs experiment to answer this question. You form

20 pairs of first-graders, with the two children in each pair carefully matched by IQ, socioeconomic status, and reading-readiness score. You assign at random one student from each pair to Method A. The other student in the pair is taught by Method B. At the end of first grade, all the children take a test to determine their reading skill. Assume that the higher the score on this test, the more proficient the student is at reading. Let p stand for the proportion of all possible matched pairs of children for which the child taught by Method B has the higher score. Your hypotheses are

$$H_0: p = 0.5 \quad \text{(no difference in effectiveness)}$$
$$H_a: p > 0.5 \quad \text{(Method B is more effective)}$$

The result of your experiment is that Method B gave the higher score in 12 of the 20 pairs, or $\hat{p} = 12/20 = 0.6$.

(a) If H_0 is true, the 20 pairs of students are 20 independent trials with probability 0.5 that Method B "wins" each trial (is the more effective method). Explain how to use Table A to simulate these 20 trials if we assume for the sake of argument that H_0 is true.

(b) Use Table A, starting at line 105, to simulate 10 repetitions of the experiment. Estimate from your simulation the probability that Method B will do better (be the more effective method) in 12 or more of the 20 pairs when H_0 is true. (Of course, 10 repetitions are not enough to estimate the probability reliably. Once you understand the idea, more repetitions are easy.)

(c) Explain why the probability you simulated in (b) is the P-value for your experiment. With enough patience, you could find all the P-values in this chapter by doing simulations similar to this one.

22.24 Finding a P-value by simulation. A classic experiment to detect extrasensory perception (ESP) uses a shuffled deck of cards containing five suits (waves, stars, circles, squares, and crosses). As the experimenter turns over each card and concentrates on it, the subject guesses the suit of the card. A subject who lacks ESP has probability 1/5 of being right by luck on each guess. A subject who has ESP will be right more often. Julie is right in 5 of 10 tries. (Actual experiments use much longer series of guesses so that weak ESP can be spotted. No one has ever been right half the time in a long experiment!)

(a) Give H_0 and H_a for a test to see if this result is significant evidence that Julie has ESP.

(b) Explain how to simulate the experiment if we assume for the sake of argument that H_0 is true.

(c) Simulate 20 repetitions of the experiment; begin at line 121 of Table A.

(d) The actual experimental result was 5 correct in 10 tries. What is the event whose probability is the P-value for this experimental result? Give an estimate of the P-value based on your simulation. How convincing was Julie's performance?

The following exercises concern the optional section on calculating P-values. To carry out a test, complete the steps (hypotheses, sampling distribution, data, P-value, and conclusion) illustrated in Example 3 (page 523).

22.25 Using the Internet. Return to the study in Exercise 22.16 (page 532), which found that 168 of 200 first-year students said that they used the Internet frequently for research or homework. Carry out the hypothesis test described in Exercise 22.16 and compute the P-value. How does your value compare with the value given in Exercise 22.16(d)?

22.26 Share the wealth. The *New York Times* conducted a nationwide poll of 1650 randomly selected American adults. Of these, 1089 felt that the distribution of money and wealth in this country should be more evenly distributed among more people. We can consider the sample to be an SRS. Is there good evidence that more than half of all American adults feel that the distribution of money and wealth in this country should be more evenly distributed among more people?

22.27 Side effects. An experiment on the side effects of pain relievers assigned arthritis patients to one of several over-the-counter pain medications. Of the 420 patients who took one brand of pain reliever, 21 suffered some "adverse symptom."

(a) If 10% of all patients suffer adverse symptoms, what would be the sampling distribution of the proportion with adverse symptoms in a sample of 420 patients?

(b) Does the experiment provide strong evidence that fewer than 10% of patients who take this medication have adverse symptoms?

22.28 Do chemists have more girls? Some people think that chemists are more likely than other parents to have female children. (Perhaps chemists are exposed to something in their laboratories that affects the sex of their children.) The Washington State Department of Health lists the parents' occupations on birth certificates. Between 1980 and 1990, 555 children were born to fathers who were chemists. Of these births, 273 were girls. During this period, 48.8% of all births in Washington State were girls. Is there evidence, at a significance level of 0.05, that the proportion of girls born to chemists is higher than the state proportion?

22.29 Speeding. It often appears that most drivers on the road are driving faster than the posted speed limit. Situations differ, of course, but here is one set of data. Researchers studied the behavior of drivers on a rural interstate highway in Maryland where the speed limit was 55 miles per hour. They measured speed with an electronic device hidden in the pavement and, to eliminate large trucks, considered only vehicles less than 20 feet long. They found that 5690 out of 12,931 vehicles were exceeding the speed limit. Is this good evidence, at a significance level of 0.05, that (at least in this location) fewer than half of all drivers are speeding?

The following exercises concern the optional section on tests for a population mean. To carry out a test, complete the steps illustrated in Example 4 (page 524) or Example 5 (page 526).

22.30 Student attitudes. The Survey of Study Habits and Attitudes (SSHA) is a psychological test that measures students' study habits and attitudes toward school. Scores range from 0 to 200. The mean score for U.S. college students is about 115, and the standard deviation is about 30. A teacher suspects that older students have better attitudes toward school. She gives the SSHA to 25 students who are at least 30 years old. Assume that scores in the population of older students are Normally distributed with standard deviation $\sigma = 30$. The teacher wants to test the hypotheses

$$H_0: \mu = 115$$
$$H_a: \mu > 115$$

(a) What is the sampling distribution of the mean score $\bar{x}$ of a sample of 25 older students if the null hypothesis is true? Sketch the density curve of this distribution. (*Hint:* Sketch a Normal curve first, then mark the axis using what you know about locating μ and σ on a Normal curve.)

(b) Suppose that the sample data give $\bar{x} = 118.6$. Mark this point on the axis of your sketch. In fact, the outcome was $\bar{x} = 125.7$. Mark this point on your sketch. Using your sketch, explain in simple language why one outcome is good evidence that the mean score of all older students is greater than 115 and why the other outcome is not.

(c) Shade the area under the curve that is the *P*-value for the sample result $\bar{x} = 125.7$.

22.31 Mice in a maze. Experiments on learning in animals sometimes measure how long it takes mice to find their way through a maze. The mean time is 19 seconds for one particular maze. A researcher thinks that a loud noise will cause the mice to complete the maze faster. She measures how long each of several mice takes to find its way through a maze with a noise as stimulus. What are the null hypothesis H_0 and alternative hypothesis H_a?

22.32 Response time. Last year, your company's service technicians took an average of 2.5 hours to respond to trouble calls from business customers who had purchased service contracts. Do this year's data show a significantly different average response time? What null and alternative hypotheses should you test to answer this question?

22.33 Testing a random number generator. Our statistical software has a "random number generator" that is supposed to produce numbers scattered at random between 0 and 1. If this is true, the numbers generated come from a population with $\mu = 0.5$. A command to generate 100 random numbers gives outcomes with $\bar{x} = 0.536$ and $s = 0.312$. Is this good evidence that the mean of all numbers produced by this software is not 0.5?

22.34 Will they charge more? A bank wonders whether omitting the annual credit card fee for customers who charge at least $3000 in a year will increase the amount charged on its credit cards. The bank makes this offer to an SRS of 400 of its credit card customers. It then compares how much these customers charge this year with the amount that they charged last year. The mean increase in the sample is $246, and the standard deviation is $112. Is there significant evidence at the 1% level that the mean amount charged increases under the no-fee offer? State H_0 and H_a and carry out a significance test. Use significance level 0.01.

22.35 Bad weather, bad tip? People tend to be more generous after receiving good news. Are they less generous after receiving bad news? The average tip left by adult Americans is 20%. Give 20 patrons of a restaurant a message on their bill warning them that tomorrow's weather will be bad and record the tip percentage they leave. Here are the tips as a percentage of the total bill:

| 18.0 | 19.1 | 19.2 | 18.8 | 18.4 | 19.0 | 18.5 | 16.1 | 16.8 | 18.2 |
| 14.0 | 17.0 | 13.6 | 17.5 | 20.0 | 20.2 | 18.8 | 18.0 | 23.2 | 19.4 |

Suppose that tip percentages are Normal with $\sigma = 2$ and assume that the patrons in this study are a random sample of all patrons of this restaurant. Is there good evidence that the mean tip percentage for all patrons of this restaurant is less than 20 when they receive a message warning them that tomorrow's weather will be bad? State H_0 and H_a and carry out a significance test. Use significance level 0.05.

EXPLORING THE WEB

22.36 Web-based exercise. Choose a major journal in your field of study. A faculty member in your major or a Web search engine may help you identify an appropriate journal. Use a Web search engine to find the journal's Web site—just search on the journal's name. Find a paper in a recent issue that uses a phrase like "significant ($P = 0.01$)." Write a paragraph explaining what the authors of the paper concluded. Your explanation should be understandable to someone who knows no statistics and is not an expert in your major.

NOTES AND DATA SOURCES

Page 515 Case Study: The case study is based on the survey conducted by the Higher Education Research Institute at UCLA, "The American freshman: national norms fall 2010." A copy of the survey can be found at **www.heri.ucla.edu/PDFs/pubs/briefs/ HERI_ResearchBrief_Norms2010.pdf.**

Page 516 Example 1: An important historical version of this example is discussed by

Sir Ronald A. Fisher, *The Design of Experiments,* Hafner Publishing, 1971, pp. 11–26. A reissue of this book is available from Oxford University Press (1990) in *Statistical Methods, Experimental Design, and Statistical Inference,* by R. A. Fisher. Fisher's version discusses a tea-tasting experiment.

Page 524 Example 4: The data in this example are from the 2003 National Assessment of Adult Literacy, which can be found online at **nces.ed.gov/naal/.**

Page 530 Exercise 22.6: Manisha Chandalia et al., "Beneficial effects of high dietary fiber intake in patients with type 2 diabetes mellitus," *New England Journal of Medicine,* 342 (2000), pp. 1392–1398.

Page 530 Exercise 22.7: Arthur Schatzkin et al., "Lack of effect of a low-fat, high-fiber diet on the recurrence of colorectal adenomas," *New England Journal of Medicine,* 342 (2000), pp. 1149–1155.

Page 530 Exercise 22.8: Seung-Ok Kim, "Burials, pigs, and political prestige in Neolithic China," *Current Anthropology,* 35 (1994), pp. 119–141.

Page 531 Exercise 22.9: Fekri A. Hassan, "Radiocarbon chronology of predynastic Nagada settlements, Upper Egypt," *Current Anthropology,* 25 (1984), pp. 681–683.

Page 531 Exercise 22.10: Sara J. Solnick and David Hemenway, "The deadweight loss of Christmas: comment," *American Economic Review,* 86 (1996), pp. 1299–1305.

Page 531 Exercise 22.11: There is some evidence that part of the reason that 39% may be an overstatement of church attendance is that the same people who go to church are likely to agree to participate in national surveys run by the National Opinion Research Center. See Robert D. Woodberry, "When surveys lie and people tell the truth: how surveys oversample church attenders," *American Sociological Review,* 63 (1998), pp. 119–122.

Page 532 Exercise 22.14: See the Web site **newsroom.ucla.edu/portal/ucla/political-engagement-of-college-78404 .aspx** for details about the 2008–2009 academic year survey.

Page 532 Exercise 22.15: See the Web site **fs.ncaa.org/Docs/newmedia/public/rates/index.html** for information on graduation rates of student athletes.

Page 533 Exercise 22.17: Alexander Todorov et al., "Inferences of competence from faces predict election outcomes," *Science,* 308 (2005), pp. 1623–1626.

Page 535 Exercise 22.26: Jeff Zeleny and Megan Thee-Brenan, "New poll finds a deep distrust of government," *New York Times,* October 25, 2011.

Page 535 Exercise 22.28: Eric Ossiander, letter to the editor, *Science,* 257 (1992), p. 1461.

Page 535 Exercise 22.29: N. Teed, K. L. Adrian, and R. Knoblouch, "The duration of speed reductions attributable to radar detectors," *Accident Analysis and Prevention,* 25 (1991), pp. 131–137. This is one of the Electronic Encyclopedia of Statistical Examples and Exercises (EESEE) case studies, available at **whfreeman .com/catalog/static/whf/eesee.**

Page 537 Exercise 22.35: B. Rind and D. Strohmetz, "Effect of beliefs about future weather conditions on restaurant tipping," *Journal of Applied Social Psychology,* 31 (2001), pp. 2160–2164.

Use and Abuse of Statistical Inference

CASE STUDY Suppose we take a look at the more than 10,000 mutual funds on sale to the investing public. Any Internet investment site worth clicking on will tell you which funds produced the highest returns over the past (say) three years. In 2011, one site claimed that the Pro-Funds Internet Inv Fund was among the top 1% over the three preceding years. If we had bought this fund in 2008, we would have gained 50.2% per year. Comparing this return with the average over this period for all funds, we find that the ProFunds Internet Inv Fund return is significantly higher. Doesn't statistical significance suggest that the ProFunds Internet Inv Fund is a good investment?

In this chapter we will take a careful look at what statistical confidence and statistical significance do and do not mean. We will discuss some abuses of statistical inference. By the end of this chapter you will be able to answer the question of whether there was strong evidence that the ProFunds Internet Inv Fund was a sound investment in 2011. ◼

© Ocean/CORBIS

Using inference wisely

In previous chapters we have met the two major types of statistical inference: confidence intervals and significance tests. We have, however, seen only two inference methods of each type, one designed for inference about a population proportion p and the other designed for inference about a population mean μ. There are libraries of both books and software filled with methods for inference about various parameters in various settings. The reasoning of confidence intervals and significance tests remains the same, but the details can seem overwhelming. The first step in using inference wisely is to understand your data and the questions you want to answer and fit the method to its setting. Here are some tips on inference, adapted to the settings we are familiar with.

The design of the data production matters

"Where do the data come from?" remains the first question to ask in any statistical study. Any inference method is intended for use in a specific setting. For our confidence interval and test for a proportion p:

- The data must be a simple random sample (SRS) from the population of interest. When you use these methods, you are acting as if the data are an SRS. In practice, it is often not possible to actually choose an SRS from the population. Your conclusions may then be open to challenge.

- These methods are not correct for sample designs more complex than an SRS, such as stratified samples. There are other methods that fit these settings.

- There is no correct method for inference from data haphazardly collected with bias of unknown size. Fancy formulas cannot rescue badly produced data.

- Other sources of error, such as dropouts and nonresponse, are important. Remember that confidence intervals and tests use the data you collect and ignore these errors.

EXAMPLE 1 The psychologist and the sociologist

A psychologist is interested in how our visual perception can be fooled by optical illusions. Her subjects are students in Psychology 101 at her university. Most psychologists would agree that it's safe to treat the students as an SRS of all people with normal vision. There is nothing special about being a student that changes visual perception.

A sociologist at the same university uses students in Sociology 101 to examine attitudes toward poor people and antipoverty programs. Students as a group are younger than the adult population as a whole. Even among young people, students as a group come from more prosperous and better-educated homes. Even among students, this university isn't typical of all campuses. Even on this campus, students in a sociology course may have opinions that are quite different from those of engineering students. The sociologist can't reasonably act as if these students are a random sample from any interesting population.

Know how confidence intervals behave

A confidence interval estimates the unknown value of a parameter and also tells us how uncertain the estimate is. All confidence intervals share these behaviors:

- The confidence level says how often the *method* catches the true parameter when sampling many times. We never know whether this specific data set gives us an interval that contains the true value of the parameter. All we can say is that "we got this result from a method that works 95% of the time." This data set might be one of the 5% that produce an interval that misses the true value of the parameter. If that risk is too high for you, use a 99% confidence interval.

- High confidence is not free. A 99% confidence interval will be wider than a 95% confidence interval based on the same data. There is a trade-off between how closely we can pin down the true value of the parameter and how confident we are that we have captured its true value.

- Larger samples give narrower intervals. If we want high confidence *and* a narrow interval, we must take a larger sample. The length of our confidence interval for *p* goes down in proportion to the square root of the sample size. To cut the interval in half, we must take four times as many observations. This is typical of many types of confidence interval.

Dropping out An experiment found that weight loss is significantly more effective than exercise for reducing high cholesterol and high blood pressure. The 170 subjects were randomly assigned to a weight-loss program, an exercise program, or a control group. Only 111 of the 170 subjects completed their assigned treatment, and the analysis used data from these 111. Did the dropouts create bias? Always ask about details of the data before trusting inference.

Know what statistical significance says

Many statistical studies hope to show that some claim is true. A clinical trial compares a new drug with a standard drug because the doctors hope that the health of patients given the new drug will improve. A psychologist studying gender differences suspects that women will do better than men (on the average) on a test that measures social-networking skills. The purpose of significance tests is to weigh the evidence that the data give in favor of such claims. That is, a test helps us know if we found what we were looking for.

To do this, we ask what would happen if the claim were *not* true. That's the null hypothesis—no difference between the two drugs, no difference between women and men. A significance test answers only one question: "How strong is the evidence that the null hypothesis is not true?" A test answers this question by giving a *P*-value. The *P*-value tells us how unlikely data as or more extreme than ours (in the sense of providing evidence against the null hypothesis) would be if the null hypothesis were true. Data that are very unlikely are good evidence that the null hypothesis is not true. We usually don't know whether the hypothesis is true for this specific population. All we can say is that "data as or more extreme than these would occur only 5% of the time if the hypothesis were true."

This kind of indirect evidence against the null hypothesis (and for the effect we hope to find) is less straightforward than a confidence interval. We will say more about tests in the next section.

Know what your methods require

Our significance test and confidence interval for a population proportion p require that the population size be much larger than the sample size. They also require that the sample size itself be reasonably large so that the sampling distribution of the sample proportion $\hat{p}$ is close to Normal. We have said little about the specifics of these requirements because the reasoning of inference is more important. Just as there are inference methods that fit stratified samples, there are methods that fit small samples and small populations. If you plan to use statistical inference in practice, you will need help from a statistician (or need to learn lots more statistics) to manage the details.

Most of us read about statistical studies more often than we actually work with data ourselves. Concentrate on the big issues, not on the details of whether the authors used exactly the right inference methods. Does the study ask the right questions? Where did the data come from? Do the results make sense? Does the study report confidence intervals so you can see both the estimated values of important parameters and how uncertain the estimates are? Does it report P-values to help convince you that findings are not just good luck?

The woes of significance tests

The purpose of a significance test is usually to give evidence for the presence of some effect in the population. The effect might be a probability of heads different from one-half for a coin or a longer mean survival time for patients given a new cancer treatment. If the effect is large, it will show up in most samples—the proportion of heads among our tosses will be far from one-half, or the patients who get the new treatment will live much longer than those in the control group. Small effects, such as a probability of heads only slightly different from one-half, will often be hidden behind the chance variation in a sample. This is as it should be: big effects are easier to detect. That is, the P-value will usually be small when the population truth is far from the null hypothesis.

The "woes" of testing start with the fact that a test measures only the strength of evidence against the null hypothesis. It says nothing about how big or how important the effect we seek in the population really is. For example, our hypothesis might be "This coin is balanced." We express this hypothesis in terms of the probability p of getting a head as $H_0: p = 0.5$. No real coin is exactly balanced, so we know that this hypothesis is not exactly true. If this coin has probability $p = 0.502$ of a head, we might say that for practical purposes it is balanced. A statistical test doesn't think about "practical purposes." It just asks if there is evidence that p is not exactly equal to 0.5. The focus of tests on the strength of the evidence against an exact null hypothesis is the source of much confusion in using tests.

Pay particular attention to the size of the sample when you read the result of a significance test. Here's why:

- Larger samples make tests of significance more sensitive. If we toss a coin hundreds of thousands of times, a test of H_0: $p = 0.5$ will often give a very low P-value when the truth for this coin is $p = 0.502$. The test is right—it found good evidence that p really is not exactly equal to 0.5—but it has picked up a difference so small that it is of no practical interest. **A finding can be statistically significant without being practically important.**

- On the other hand, tests of significance based on small samples are often not sensitive. If you toss a coin only 10 times, a test of H_0: $p = 0.5$ will often give a large P-value even if the truth for this coin is $p = 0.7$. Again the test is right—10 tosses are not enough to give good evidence against the null hypothesis. **Lack of significance does not mean that there is no effect, only that we do not have good evidence for an effect. Small samples often miss important effects that are really present in the population.**

EXAMPLE 2 Antidepressants versus a placebo

Through a Freedom of Information Act request, two psychologists obtained 47 studies used by the Food and Drug Administration for approval of the six antidepressants prescribed most widely between 1987 and 1999. Overall, the psychologists found that there was a statistically significant difference in the effects of antidepressants compared with a placebo, with antidepressants being more effective. However, the psychologists went on to report that antidepressant pills worked 18% better than placebos, a statistically significant difference, "but not meaningful for people in clinical settings."

Whatever the truth about the population, whether $p = 0.7$ or $p = 0.502$, more observations allow us to pin down p more closely. If p is not 0.5, more observations will give more evidence of this, that is, a smaller P-value. Because statistical significance depends strongly on the sample size as well as on the truth about the population, statistical significance tells us nothing about how large or how practically important an effect is. Large effects (like $p = 0.7$ when the null hypothesis is $p = 0.5$) often give data that are insignificant if we take only a small sample. Small effects (like $p = 0.502$) often give data that are highly significant if we take a large sample. Let's return to a favorite example to see how significance changes with sample size.

EXAMPLE 3   Count Buffon's coin again

Count Buffon tossed a coin 4040 times and got 2048 heads. His sample proportion of heads was

$$\hat{p} = \frac{2048}{4040} = 0.507$$

Is the count's coin balanced? Suppose we seek statistical significance at level 0.05. The hypotheses are

$$H_0: p = 0.5$$
$$H_a: p \neq 0.5$$

The test of significance works by locating the sample outcome $\hat{p} = 0.507$ on the sampling distribution that describes how $\hat{p}$ would vary if the null hypothesis were true. Figure 23.1 repeats Figure 22.2. It shows that the observed $\hat{p} = 0.507$ is not surprisingly far from 0.5 and therefore is not good evidence against the hypothesis that the true p is 0.5. The P-value, which is 0.37, just makes this precise.

Suppose that Count Buffon got the *same result*, $\hat{p} = 0.507$, from tossing a coin 1000 times and also from tossing a coin 100,000 times. The sampling distribution of $\hat{p}$ when the null hypothesis is true always has

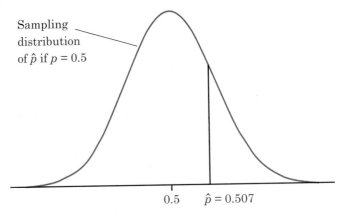

Sampling distribution of $\hat{p}$ if $p = 0.5$

0.5 $\hat{p} = 0.507$

Figure 23.1 The sampling distribution of the proportion of heads in 4040 tosses of a coin if in fact the coin is balanced, for Example 3. Sample proportion 0.507 is not an unusual outcome.

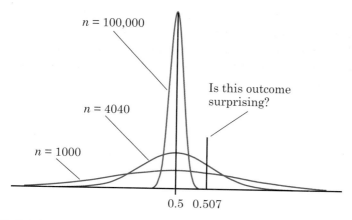

Figure 23.2 The three sampling distributions of the proportion of heads in 1000, 4040, and 100,000 tosses of a balanced coin, for Example 3. Sample proportion 0.507 is not unusual in 1000 or 4040 tosses but is very unusual in 100,000 tosses.

mean 0.5, but its standard deviation gets smaller as the sample size n gets larger. Figure 23.2 displays the three sampling distributions, for $n = 1000$, $n = 4040$, and $n = 100,000$. The middle curve in this figure is the same Normal curve as in Figure 23.1, drawn on a scale that allows us to show the very tall and narrow curve for $n = 100,000$. Locating the sample outcome $\hat{p} = 0.507$ on the three curves, you see that the same outcome is more or less surprising depending on the size of the sample.

The P-values are $P = 0.66$ for $n = 1000$, $P = 0.37$ for $n = 4040$, and $P = 0.000009$ for $n = 100,000$. Imagine tossing a balanced coin 1000 times repeatedly. You will get a proportion of heads at least as far from one-half as Buffon's 0.507 in about two-thirds of your repetitions. If you toss a balanced coin 100,000 times repeatedly, however, you will almost never (9 times in a million repeats) get an outcome as or more unbalanced than this.

The outcome $\hat{p} = 0.507$ is not evidence against the hypothesis that the coin is balanced if it comes up in 1000 tosses or in 4040 tosses. It is completely convincing evidence if it comes up in 100,000 tosses.

Beware the naked P-value

The P-value of a significance test depends strongly on the size of the sample, as well as on the truth about the population.

It is bad practice to report a naked P-value (a P-value by itself) without also giving the sample size and a statistic or statistics that describe the sample outcome.

NOW IT'S YOUR TURN | **23.1 Weight loss.** A company that sells a weight-loss program conducted a randomized experiment to determine whether people lost weight after eight weeks on the program. The company researchers report that, on average, the subjects in the study lost weight and that the weight loss was statistically significant with a *P*-value of 0.013. Do you find the results convincing? If so, why? If not, what additional information would you like to have?

The advantages of confidence intervals

Examples 2 and 3 suggest that we should not rely on significance alone in understanding a statistical study. In Example 3, just knowing that the sample proportion was $\hat{p} = 0.507$ helps a lot. You can decide whether this deviation from one-half is large enough to interest you. Of course, $\hat{p} = 0.507$ isn't the exact truth about the coin, just the chance result of Count Buffon's tosses. So a confidence interval, whose width shows how closely we can pin down the truth about the coin, is even more helpful. Here are the 95% confidence intervals for the true probability of a head p, based on the three sample sizes in Example 3. You can check that the method of Chapter 21 gives these answers.

Number of tosses	95% confidence interval
$n = 1000$	0.507 ± 0.031, or 0.476 to 0.538
$n = 4040$	0.507 ± 0.015, or 0.492 to 0.522
$n = 100{,}000$	0.507 ± 0.003, or 0.504 to 0.510

The confidence intervals make clear what we know (with 95% confidence) about the true p. The intervals for 1000 and 4040 tosses include 0.5, so we are not confident that the coin is unbalanced. For 100,000 tosses, however, we are confident that the true p lies between 0.504 and 0.510. In particular, we are confident that it is not 0.5.

Give a confidence interval

Confidence intervals are more informative than significance tests because they actually estimate a population parameter. They are also easier to interpret. It is good practice to give confidence intervals whenever possible.

Significance at the 5% level isn't magical

The purpose of a test of significance is to describe the degree of evidence provided by the sample against the null hypothesis. The P-value does this. But how small a P-value is convincing evidence against the null hypothesis? This depends mainly on two circumstances:

- *How plausible is H_0?* If H_0 represents an assumption that the people you must convince have believed for years, strong evidence (small P) will be needed to persuade them.

- *What are the consequences of rejecting H_0?* If rejecting H_0 in favor of H_a means making an expensive changeover from one type of product packaging to another, you need strong evidence that the new packaging will boost sales.

STATISTICAL CONTROVERSIES

Should Significance Tests Be Banned?

Research studies in many fields rely on tests of significance. Often habit leads to over-reliance. Robert Rosenthal, an eminent Harvard psychologist well known for statistical work, says, "Many of us were trained that we're not supposed to look too carefully at the data. You come up with a hypothesis, decide on a statistical test, do the test, and if your results are significant at .05, you've supported your hypothesis. If not, you stick it all in a drawer and never look at your data."

What is your reaction to the above quotation? How does this approach compare with the practices that we have emphasized in this book?

How can it be that many psychologists hardly glance at the data? The tyranny of significance tests and of 0.05 as the magical sign that a result is important, say some psychologists. In particular, custom dictates that results should be significant at the 5% level in order to be published,

so researchers fall into the bad habits that Rosenthal describes. Significant at 5%, good. Not significant at 5%, failure. The limitations of tests are so severe, the risks of misinterpretation so high, and bad habits so ingrained, say these critics, that significance tests should be banned from professional journals in psychology.

These criteria are a bit subjective. Different people will often insist on different levels of significance. Giving the P-value allows each of us to decide individually if the evidence is sufficiently strong. But the level of significance that will satisfy us should be decided before calculating the P-value. Computing the P-value and then deciding that we are satisfied with a level of significance that is just slightly larger than this P-value is an abuse of significance testing.

Users of statistics have often emphasized standard levels of significance such as 10%, 5%, and 1%. For example, courts have tended to accept 5% as a standard in discrimination cases. This emphasis reflects the time when tables of critical values rather than computer software dominated statistical practice. The 5% level ($\alpha = 0.05$) is particularly common. **There is no sharp border between "significant" and "insignificant," only increasingly strong evidence as the P-value decreases.** There is no practical distinction between the P-values 0.049 and 0.051. It makes no sense to treat $P \leq 0.05$ as a universal rule for what is significant.

Beware of searching for significance

Statistical significance ought to mean that you have found an effect that you were looking for. The reasoning behind statistical significance works well if you decide what effect you are seeking, design a study to search for it, and use a test of significance to weigh the evidence you get. In other settings, significance may have little meaning. Here is an example.

EXAMPLE 4 Predicting success of trainees

You want to learn what distinguishes managerial trainees who eventually become executives from those who, after expensive training, don't succeed and leave the company. You have abundant data on past trainees—data on their personalities and goals, their college preparation and performance, even their family backgrounds and hobbies. Statistical software makes it easy to perform dozens of significance tests on these dozens of variables to see which ones best predict later success. Aha! You find that future executives are significantly more likely than washouts to have an urban or suburban upbringing and an undergraduate degree in a technical field.

Before you base future recruiting on these findings, recall that results significant at the 5% level occur 5 times in 100 in the long run even when H_0 is true. When you make dozens of tests at the 5% level, you expect a few of them to be significant by chance alone. Running one test and reaching the $\alpha = 0.05$ level is reasonably good evidence that you have found something. Running several dozen tests and reaching that level once or twice is not.

In Example 4, we tested everything and took the most significant. It is bad practice to confuse the roles of exploratory analysis of data (using graphs, tables, and summary statistics, like those discussed in Part II, to find suggestive patterns in data) and formal statistical inference. Finding statistical significance is not surprising if you use exploratory methods to examine many outcomes, choose the largest, and test to see if it is significantly larger than the others.

Searching data for suggestive patterns is certainly legitimate. Exploratory data analysis is an important part of statistics. But the reasoning of formal inference does not apply when your search for a striking effect in the data is successful. The remedy is clear. Once you have a hypothesis, design a study to search specifically for the effect you now think is there. If the result of this study is statistically significant, you have real evidence.

William Manning/CORBIS

NOW IT'S YOUR TURN | **23.2 Take me out to the ball game.** A researcher compared a random sample of recently divorced men in a large city with a random sample of men from the same city who had been married at least 10 years and had never been divorced. The researcher measured 122 variables on each man and compared the two samples using 122 separate tests of significance. Only the variable measuring how often the men attended Major League Baseball games with their spouse was significant at the 1% level, with the married men attending a higher proportion of games with their spouse, on average, than the divorced men did while they were married. Is this strong evidence that attendance at Major League Baseball games improves the chance that a man will remain married? Discuss.

STATISTICS IN SUMMARY

Chapter Specifics

- Statistical inference is less widely applicable than exploratory analysis of data. Any inference method requires the right setting, in particular the right design for a random sample or randomized experiment.

- Understanding the meaning of confidence levels and statistical significance helps prevent improper conclusions.

- Increasing the number of observations has a straightforward effect on confidence intervals: the interval gets shorter for the same level of confidence.

- Taking more observations usually decreases the *P*-value of a test when the truth about the population stays the same, making significance tests harder to interpret than confidence intervals.

- A finding with a small *P*-value may not be practically interesting if the sample is large, and an important truth about the population may fail to be significant if the sample is small. Avoid depending on fixed significance levels such as 5% to make decisions.

 In Chapters 21 and 22 we introduced the basic reasoning behind statistical estimation and tests of significance. We applied this reasoning to the problem of making inferences about a population proportion and a population mean. In this chapter we provided some cautions about confidence intervals and tests of significance. Some of these echo statements made in Chapters 1 to 6 that where the data come from matters. Some cautions are based on the behavior of confidence intervals and significance tests. These cautions will help you evaluate studies that report confidence intervals or the results of a test of significance.

CASE STUDY   Look again at the Case Study at the beginning of this chapter. The Pro-
EVALUATED   Funds Internet Inv Fund was identified as among the top 1% in returns after looking at more than 10,000 mutual funds.

1. Does it make sense to look at past data, take a fund that is one of the best out of over 10,000 funds, and ask if this fund was above average? What would you expect the outcome of a test of significance to show about the performance of the fund compared with the average performance of all funds?

2. In fact, the ProFunds Internet Inv Fund was in the *bottom* 42% out of over 10,000 mutual funds in 2011. Is this surprising given the fact that a test of significance shows that its average return over the previous three years was significantly higher than the average return for all funds? Discuss.

3. Significance tests work when we form a hypothesis, such as "the ProFunds Internet Inv Fund will have higher-than-average returns over the next three years" and then collect data. Suppose you did this for a particular fund and found that its average return was significantly higher than the average return for all funds. Discuss whether this provides good evidence that the fund is a sound investment. You may wish to review the material on understanding prediction in Chapter 15. In particular, read pages 340 to 341. ■

CHAPTER 23 EXERCISES

For Exercise 23.1, see page 546; and for Exercise 23.2, see page 549.

23.3 A television poll. A television news program conducts a call-in poll about the salaries of business executives. Of the 2372 callers, 1921 believe that business executives are vastly overpaid and that their pay should be substantially reduced. The station, following recommended practice, makes a confidence statement: "81% of the Channel 13 Pulse Poll sample believe that business executives are vastly overpaid and that their pay should be substantially reduced. We can be 95% confident that the true proportion of citizens who believe that business executives are vastly overpaid and that their pay should be substantially reduced is within 1.6% of the sample result." The confidence interval calculation is correct, but the conclusion is not justified. Why not?

23.4 Ages of presidents. Joe is writing a report on the backgrounds of American presidents. He looks up the ages of all 44 presidents when they entered office. Because Joe took a statistics course, he uses these 44 numbers to get a 95% confidence interval for the mean age of all men who have been president. This makes no sense. Why not?

23.5 How do we feel? A Gallup Poll taken in late October 2011 found that 58% of the American adults surveyed, reflecting on the day before they were surveyed, said they experienced a lot of happiness and enjoyment without a lot of stress and worry. The poll had a margin of sampling error of plus or minus three percentage points at 95% confidence. A news commentator at the time said the poll is surprising, given the current economic climate, in that it supports the notion that the *majority* of American adults are experiencing a lot of happiness and enjoyment without a lot of stress and worry. Why do you think the news commentator said this?

23.6 How far do rich parents take us? How much education children get is strongly associated with the wealth and social status of their parents. In social science jargon, this is "socioeconomic status," or SES. But the SES of parents has little influence on whether children who have graduated from college go on to yet more education. One study looked at whether college graduates took the graduate admissions tests for business, law, and other graduate programs. The effects of the parents' SES on taking the LSAT test for law school were "both statistically insignificant and small."

(a) What does "statistically insignificant" mean?

(b) Why is it important that the effects were small in size as well as insignificant?

23.7 Searching for ESP. A researcher looking for evidence of extrasensory perception (ESP) tests 200 subjects. Only one of these subjects does significantly better ($P < 0.01$) than random guessing.

(a) Do the results of this study provide strong evidence that this person has ESP? Explain your answer.

(b) What should the researcher now do to test whether the subject has ESP?

23.8 Are the drugs really effective? A March 29, 2012, article in the *Columbus Dispatch* reported that a former researcher at a major pharmaceutical company found that many basic studies on the effectiveness of new cancer drugs appeared to be unreliable. Among the studies the former researcher reviewed was one that had been published in a reputable journal. In this published study, a cancer drug was reported as having a statistically significant positive effect on treating cancer. For purposes of this problem, assume that statistically significant means significant at level 0.05.

(a) Explain in language that is understandable to someone who knows no statistics what "statistically significant at level 0.05" means.

(b) The former researcher interviewed the lead author of the published paper. The newspaper article reported that the lead author admitted that they had repeated their experiment six times and got a significant result only once, but put it in the paper because it made the best story. In light of this admission, do you think that it is accurate to claim in their published study that the findings were significant at the 0.05 level? Explain your answer. (*Note:* A statistician can show that an event that has only probability 0.05 of occurring on any given trial will occur at least once in six trials with probability about 0.26.)

23.9 Comparing bottle designs. A company compares two designs for bottles of an energy drink by placing bottles with both designs on the shelves of several markets in a large city. Checkout scanner data on more than 10,000 bottles bought show that more shoppers bought Design A than Design B. The difference is statistically significant ($P = 0.018$). Can we conclude that consumers strongly prefer Design A? Explain your answer.

23.10 Color blindness in Africa. An anthropologist suspects that color blindness is less common in societies that live by hunting and gathering than in settled agricultural societies. He tests a number of adults in two populations in Africa, one of each type. The proportion of color-blind people is significantly lower ($P < 0.05$) in the hunter-gatherer population. What additional information would you want to help you decide whether you accept the claim about color blindness?

23.11 Blood types in Southeast Asia. One way to assess whether two human groups should be considered separate populations is to compare their distributions of blood types. An anthropologist finds significantly different ($P = 0.01$) proportions of the main human blood types (A, B, AB, O) in different tribes in central Malaysia. What other information would you want before you agree that these tribes are separate populations?

23.12 Why we seek significance. Asked why statistical significance appears so often in research reports, a student says, "Because saying that results are significant tells us that they cannot easily be explained by chance variation alone." Do you think that this statement is essentially correct? Explain your answer.

23.13 What is significance good for? Which of the following questions does a test of significance answer?

(a) Is the sample or experiment properly designed?

(b) Is the observed effect due to chance?

(c) Is the observed effect important?

23.14 What distinguishes schizophrenics? Psychologists once measured 77 variables on a sample of schizophrenic people and a sample of people who were not schizophrenic. They compared the two samples using 77 separate significance tests. Two of these tests were significant at the 5% level. Suppose that there is in fact no difference in any of the 77 variables between people who are and people who are not schizophrenic in the adult population. That is, all 77 null hypotheses are true.

(a) What is the probability that one specific test shows a difference that is significant at the 5% level?

(b) Why is it not surprising that 2 of the 77 tests were significant at the 5% level?

23.15 Why are larger samples better? Statisticians prefer large samples. Describe briefly the effect of increasing the size of a sample (or the number of subjects in an experiment) on each of the following.

(a) The margin of error of a 95% confidence interval.

(b) The P-value of a test, when H_0 is false and all facts about the population remain unchanged as n increases.

23.16 Is this convincing? You are planning to test a vaccine for a virus that now has no vaccine. Since the disease is usually not serious, you will expose 100 volunteers to the virus. After some time, you will record whether or not each volunteer has been infected.

(a) Explain how you would use these 100 volunteers in a designed experiment to test the vaccine. Include all important details of designing the experiment (but don't actually do any random allocation).

(b) You hope to show that the vaccine is more effective than a placebo. State H_0 and H_a. (Notice that this test compares *two* population proportions.)

(c) The experiment gave a P-value of 0.15. Explain carefully what this means.

(d) Your fellow researchers do not consider this evidence strong enough to recommend regular use of the vaccine. Do you agree?

The following exercises require carrying out the methods described in the optional sections of Chapters 21 and 22.

23.17 Do our athletes graduate? Return to the study in Exercise 22.15 (page 532), which found that 137 of 190 athletes admitted to a large university graduated within six years. The proportion of athletes who graduated was significantly lower ($P = 0.025$) than the 78% graduation rate for all students. It may be more informative to give a 95% confidence interval for the graduation rate of athletes. Do this.

23.18 Using the Internet. Return to the study in Exercise 22.16 (page 532), which found that 168 of an SRS of 200 entering students at a large state university said that they used the Internet frequently for research or homework. This differed significantly ($P = 0.0037$) from the 75.9% of all first-year students in the country who claim this. It may be more informative to give a 95% confidence interval for the proportion of this university's entering students who claim to use the Internet frequently for research or homework. Do this.

23.19 What's your weight? A Gallup Poll asked a national random sample of 501 adult women to state their current weight. The mean weight in the sample was $\bar{x} = 159$. We will treat these data as an SRS from a Normally distributed population with standard deviation $\sigma = 35$.

(a) Give a 95% confidence interval for the mean weight of adult women based on these data.

(b) Do you trust the interval you computed in part (a) as a 95% confidence interval for the mean weight of all U.S. adult women? Why or why not?

23.20 Is it significant? Over several years and many thousands of students, 85% of the high school students in a large city have passed the competency test that is one of the requirements for a diploma. Now reformers claim that a new mathematics curriculum will increase the percentage who pass. A random sample of 1000 students follow the new curriculum. The school board wants to see an improvement that is statistically significant at the 5% level before it will adopt the new program for all students. If p is the proportion of all students who would pass the exam if they followed the new curriculum, we must test

$$H_0: p = 0.85$$
$$H_a: p > 0.85$$

(a) Suppose that 868 of the 1000 students in the sample pass the test. Show that this is *not* significant at the 5% level. (Follow the method of Example 3, page 523, in Chapter 22.)

(b) Suppose that 869 of the 1000 students pass. Show that this *is* significant at the 5% level.

(c) Is there a practical difference between 868 successes in 1000 tries and 869 successes? What can you conclude about the importance of a fixed significance level?

23.21 We like confidence intervals. The previous exercise compared significance tests about the proportion p of all students who would pass a competency test, based on data showing that either 868 or 869 of an SRS of 1000 students passed. Give the 95% confidence interval for p in both cases. The intervals make clear how uncertain we are about the true value of p and how little difference there is between the two sample outcomes.

EXPLORING THE WEB

23.22 Web-based exercise. Find an example of a study in which a statistically significant result may not be practically important. Write a paragraph summarizing the study and its conclusions in your own words. Be sure to explain why the results are not practically important. The CHANCE Web site at **www.causeweb.org/wiki/chance/index.php/Main_Page** is a good place to look for examples. You can also try entering a phrase such as "statistically significant but not practically significant" or "statistically significant but not meaningful" into a Web search engine.

23.23 Web-based exercise. The report of the American Psychological Association's Task Force on Statistical Inference is an excellent brief introduction to wise use of inference. The report appeared in the journal *American Psychologist* in 1999. You can find a copy on the Web at **www.apa.org/science/leadership/bsa/statistical/tfsi-followup-report.pdf.** Read the report. Are the authors opposed to the use of hypothesis testing? Describe one abuse of hypothesis testing that is cited in this report.

NOTES AND DATA SOURCES

Page 539 We looked at the list of top funds at the *Smart Money* Web site, **www.smartmoney.com,** in November 2011. Here's part of the abstract of a typical study (Mark Carhart, "On persistence in mutual fund performance," *Journal of Finance,* 52 [1997], pp. 57–82): "Using a sample free of survivor bias, I demonstrate that common factors in stock returns and investment expenses almost completely explain persistence in equity mutual funds' mean and risk-adjusted returns.... The only significant persistence not explained is concentrated in strong underperformance by the worst-return mutual funds. The results do not support the existence of skilled or informed mutual fund portfolio managers."

Page 543 Example 2: Marilyn Elias, "Study: antidepressant barely better than placebo," *USA Today,* July 7, 2002.

Page 547 Robert Rosenthal is quoted in B. Azar, "APA statistics task force prepares to release recommendations for public comment," *APA Monitor Online,* 30 (May 1999), available online at **www.apa.org/monitor.** The task force report, Leland Wilkinson et al., "Statistical methods in psychology journals: guidelines and explanations," *American Psychologist,* 54 (August 1999), offers a summary of the elements of good statistical practice.

Page 551 Exercise 23.5: The Gallup Poll results can be found online at **www.gallup.com/poll/106915/Gallup-Daily-US-Mood.aspx.**

Page 551 Exercise 23.6: Ross M. Stolzenberg, "Educational continuation by college graduates," *American Journal of Sociology,* 99 (1994), pp. 1042–1077.

Two-Way Tables and the Chi-Square Test*

CASE STUDY Purdue is a Big Ten university that emphasizes engineering, scientific, and technical fields. In the 2010–2011 academic year, Purdue had 1755 professors, of whom 479 were women. That's just over 27%, or slightly more than one out of every four professors. These numbers don't tell us much about the fields of expertise of women on the faculty. As usual, we must look at relationships among several variables, not just at sex alone. For example, female faculty members are more common in the humanities than in agriculture.

Let's look at the relationship between sex and a variable particularly important to faculty members, academic rank. Professors typically start as assistant professors, are promoted to associate professor (and gain tenure then), and finally reach the rank of full professor. Universities tend to be run by full professors. Here is a *two-way table* that categorizes Purdue's 1755 faculty members by both sex and academic rank:

	Female	Male	Total
Assistant professors	178	244	422
Associate professors	169	361	530
Professors	132	671	803
Total	479	1276	1755

What does this table tell us about the rank of women on the faculty? In this chapter we will learn how to interpret such tables. By the end of the chapter you will be able to interpret this table. ■

---

*This more advanced chapter is optional.

Two-way tables

EXAMPLE 1 Discrimination in admissions?

A university offers only two degree programs, one in electrical engineering and one in English. Admission to these programs is competitive, and the women's caucus suspects discrimination against women in the admissions process. The caucus obtains the following data from the university, a two-way table of all applicants classified by sex and admission decision:

	Male	Female	Total
Admit	35	20	55
Deny	45	40	85
Total	80	60	140

How should we evaluate the information in this table?

The admission status and sex of applicants are both *categorical variables*. That is, they place individuals into categories but do not have numerical values that allow us to describe relationships by scatterplots, correlation, or regression lines. To display relationships between two categorical variables, use a **two-way table** like the table of admission status and sex of applicants. Admission status is the **row variable** because each row in the table describes one of the possible admission decisions for an applicant. Sex is the **column variable** because each column describes one sex. The entries in the table are the counts of applicants in each admission status–by–sex class.

How can we best grasp the information contained in this table? First, *look at the distribution of each variable separately*. The distribution of a categorical variable says how often each outcome occurred. The "Total" column at the right of the table contains the totals for each of the rows. These row totals give the distribution of admission status for all applicants, men and women combined. The "Total" row at the bottom of the table gives the distribution of sex for applicants, both categories of admission status combined. It is often clearer to present these distributions using percentages. We might report the distribution of sex as

$$\text{percentage male} = \frac{80}{140} = 0.57 = 57\%$$

$$\text{percentage female} = \frac{60}{140} = 0.43 = 43\%$$

The two-way table contains more information than the two distributions of admission status alone and sex alone. The nature of the relationship between admission status and sex cannot be deduced from the separate distributions but requires the full table. **To describe relationships among categorical variables, calculate appropriate percentages from the counts given.**

EXAMPLE 2 Discrimination in admissions?

Because there are only two categories of admission status, we can see the relationship between sex and admission status by comparing the percentages of male and female applicants admitted:

$$\text{percentage of male applicants admitted} = \frac{35}{80} = 44\%$$

$$\text{percentage of female applicants admitted} = \frac{20}{60} = 33\%$$

Almost half of the males but only one-third of the females who applied were admitted.

In working with two-way tables, you must calculate lots of percentages. Here's a tip to help decide what fraction gives the percentage you want. Ask, "What group represents the total that I want a percentage of?" The count for that group is the denominator of the fraction that leads to the percentage. In Example 2, we wanted the percentage *of each sex* who were admitted, so the counts of each sex form the denominators.

Inference for a two-way table

We often gather data and arrange them in a two-way table to see if two categorical variables are related to each other. The sample data are easy to investigate: turn them into percentages and look for an association between the row and column variables. Is the association in the sample evidence of an association between these variables in the entire population? Or could the sample association easily arise just from the luck of random sampling? This is a question for a significance test.

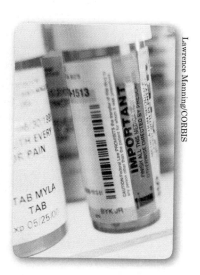

EXAMPLE 3 Treating cocaine addiction

Cocaine addicts need the drug to feel pleasure. Perhaps giving them a medication that fights depression will help them resist cocaine. A three-year study compared an antidepressant called desipramine, lithium (a standard treatment for cocaine addiction) and a placebo. The subjects were 72 chronic users of cocaine who wanted to break their drug habit. An equal number of the subjects were randomly assigned to each treatment. Here are the counts and percentages of the subjects who succeeded in not using cocaine during the study:

Group	Treatment	Subjects	Successes	Percent
1	Desipramine	24	14	58.3
2	Lithium	24	6	25.0
3	Placebo	24	4	16.7

The sample proportions of subjects who did not use cocaine are quite different. In particular, the percentage of subjects in the desipramine group who did not use cocaine was much higher than for the lithium or placebo group. The bar graph in Figure 24.1 compares the results visually. Are these data good evidence that there is a relationship between treatment and outcome in the population of all cocaine addicts?

The test that answers this question starts with a two-way table. Here's the table for the data of Example 3:

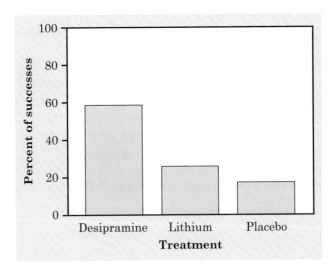

Figure 24.1 Bar graph comparing the success rates of three treatments for cocaine addiction, for Example 3.

	Success	Failure	Total
Desipramine	14	10	24
Lithium	6	18	24
Placebo	4	20	24
Total	24	48	72

Our null hypothesis, as usual, says that the treatments have no effect. That is, addicts do equally well on any of the three treatments. The differences in the sample are just the play of chance. Our null hypothesis is

H_0: There is no association between the treatment an addict receives and whether or not there is success in not using cocaine in the population of all cocaine addicts.

Expressing this hypothesis in terms of population parameters can be a bit complicated, so we will be content with the verbal statement. The alternative hypothesis just says, "Yes, there is some association between the treatment an addict receives and whether or not he succeeds in staying off cocaine." The alternative doesn't specify the nature of the relationship. It doesn't say, for example, "Addicts who take desipramine are more likely to succeed than addicts given lithium or a placebo."

To test H_0, we compare the observed counts in a two-way table with the *expected counts,* the counts we would expect—except for random variation—if H_0 were true. If the observed counts are far from the expected counts, that is evidence against H_0. We can guess the expected counts for the cocaine study. In all, 24 of the 72 subjects succeeded. That's an overall success rate of one-third, because 24/72 is one-third. If the null hypothesis is true, there is no difference among the treatments. So we expect one-third of the subjects in each group to succeed. There were 24 subjects in each group, so we expect 8 successes and 16 failures in each group. If the treatment groups differ in size, the expected counts will differ also, even though we still expect the same proportion in each group to succeed. Fortunately, there is a rule that makes it easy to find expected counts. Here it is.

Expected counts

The **expected count** in any cell of a two-way table when H_0 is true is

$$\text{expected count} = \frac{\text{row total} \times \text{column total}}{\text{table total}}$$

Try it. For example, the expected count of successes in the desipramine group is

$$\text{expected count} = \frac{\text{row 1 total} \times \text{column 1 total}}{\text{table total}}$$

$$= \frac{(24)(24)}{72} = 8$$

If the null hypothesis of no treatment differences is true, we expect 8 of the 24 desipramine subjects to succeed. That's just what we guessed.

NOW IT'S YOUR TURN

24.1 Video-gaming and grades. The popularity of computer, video, online, and virtual reality games has raised concerns about their ability to negatively impact youth. Based on a recent survey, 1808 fourteen- to eighteen-year-olds in Connecticut high schools were classified by their average grades and by whether they had or had not played such games. The following table summarizes the findings.

	Grade Average		
	A's and B's	**C's**	**D's and F's**
Played games	736	450	193
Never played games	205	144	80

Find the expected count of students with average grades of A's and B's who have played computer, video, online, or virtual reality games under the null hypothesis that there is no association between grades and game playing in the population of students. Assume that the sample was a random sample of students in Connecticut high schools.

statistics in Your World

More chi-square tests
There are also chi-square tests for hypotheses more specific than "no relationship." Place people in classes by social status, wait 10 years, then classify the same people again. The row and column variables are the classes at the two times. We might test the hypothesis that there has been no change in the overall distribution of social status. Or we might ask if moves up in status are balanced by matching moves down. These hypotheses can be tested by variations of the chi-square test.

The chi-square test

To see if the data give evidence against the null hypothesis of "no relationship," compare the counts in the two-way table with the counts we would expect if there really were no relationship. If the observed counts are far from the expected counts, that's the evidence we were seeking. The significance test uses a statistic that measures how far apart the observed and expected counts are.

Chi-square statistic

The **chi-square statistic**, denoted χ^2, is a measure of how far the observed counts in a two-way table are from the expected counts. The formula for the statistic is

$$\chi^2 = \sum \frac{(\text{observed count} - \text{expected count})^2}{\text{expected count}}$$

The symbol $\sum$ means "sum over all cells in the table."

The chi-square statistic is a sum of terms, one for each cell in the table. In the cocaine example, 14 of the desipramine group succeeded. The expected count for this cell is 8. So the term in the chi-square statistic from this cell is

$$\frac{(\text{observed count} - \text{expected count})^2}{\text{expected count}} = \frac{(14 - 8)^2}{8}$$

$$= \frac{36}{8} = 4.5$$

EXAMPLE 4 The cocaine study

Here are the observed and expected counts for the cocaine study side by side:

	Observed		Expected	
	Success	**Failure**	**Success**	**Failure**
Desipramine	14	10	8	16
Lithium	6	18	8	16
Placebo	4	20	8	16

We can now find the chi-square statistic, adding 6 terms for the 6 cells in the two-way table:

$$\chi^2 = \frac{(14 - 8)^2}{8} + \frac{(10 - 16)^2}{16} + \frac{(6 - 8)^2}{8}$$

$$+ \frac{(18 - 16)^2}{16} + \frac{(4 - 8)^2}{8} + \frac{(20 - 16)^2}{16}$$

$$= 4.50 + 2.25 + 0.50 + 0.25 + 2.00 + 1.00 = 10.50$$

NOW IT'S YOUR TURN **24.2 Video-gaming and grades.** The popularity of computer, video, online, and virtual reality games has raised concerns about their ability to negatively impact youth. Based on a recent survey, 1808 fourteen- to eighteen-year-olds in Connecticut high schools were classified by their average grades and by whether they had or had not played such games. The following table summarizes the findings. The observed and expected counts are given side by side.

	Observed			Expected		
	A's and B's	C's	D's and F's	A's and B's	C's	D's and F's
Played games	736	450	193	717.7	453.1	208.2
Never played games	205	144	80	223.3	140.9	64.8

Find the chi-square statistic.

Because χ^2 measures how far the observed counts are from what would be expected if H_0 were true, large values are evidence against H_0. Is $\chi^2 = 10.5$ a large value? You know the drill: compare the observed value 10.5 against the *sampling distribution* that shows how χ^2 would vary if the null hypothesis were true. This sampling distribution is *not* a Normal distribution. It is a right-skewed distribution that allows only nonnegative values because χ^2 can never be negative. Moreover, the sampling distribution is different for two-way tables of different sizes. Here are the facts.

The chi-square distributions

The sampling distribution of the chi-square statistic χ^2 when the null hypothesis of no association is true is called a **chi-square distribution.**

The chi-square distributions are a family of distributions that take only nonnegative values and are skewed to the right. A specific chi-square distribution is specified by giving its **degrees of freedom.**

The **chi-square test** for a two-way table with r rows and c columns uses critical values from the chi-square distribution with $(r - 1)(c - 1)$ degrees of freedom.

Figure 24.2 shows the density curves for three members of the chi-square family of distributions. As the degrees of freedom (df) increase, the

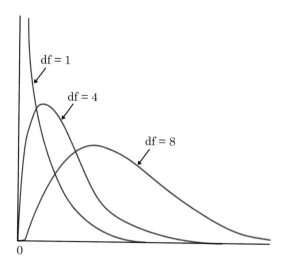

Figure 24.2 The density curves for three members of the chi-square family of distributions. The sampling distributions of chi-square statistics belong to this family.

density curves become less skewed and larger values become more probable. We can't find P-values as areas under a chi-square curve by hand, though software can do it for us. Table 24.1 is a shortcut. It shows how large the chi-square statistic χ^2 must be in order to be significant at various levels. This isn't as good as an actual P-value, but it is often good enough. Each number of degrees of freedom has a separate row in the table. We see, for example, that a chi-square statistic with 3 degrees of freedom is significant at the 5% level if it is greater than 7.81 and is significant at the 1% level if it is greater than 11.34.

TABLE 24.1 To be significant at level α, a chi-square statistic must be larger than the table entry for α

df	\multicolumn{7}{c}{Significance Level α}
	0.25	0.20	0.15	0.10	0.05	0.01	0.001
1	1.32	1.64	2.07	2.71	3.84	6.63	10.83
2	2.77	3.22	3.79	4.61	5.99	9.21	13.82
3	4.11	4.64	5.32	6.25	7.81	11.34	16.27
4	5.39	5.99	6.74	7.78	9.49	13.28	18.47
5	6.63	7.29	8.12	9.24	11.07	15.09	20.51
6	7.84	8.56	9.45	10.64	12.59	16.81	22.46
7	9.04	9.80	10.75	12.02	14.07	18.48	24.32
8	10.22	11.03	12.03	13.36	15.51	20.09	26.12
9	11.39	12.24	13.29	14.68	16.92	21.67	27.88

EXAMPLE 5 The cocaine study, conclusion

We have seen that desipramine produced markedly more successes and fewer failures than lithium or a placebo. Comparing observed and expected counts gave the chi-square statistic $\chi^2 = 10.5$. The last step is to assess significance.

The two-way table of 3 treatments by 2 outcomes for the cocaine study has 3 rows and 2 columns. That is, $r = 3$ and $c = 2$. The chi-square statistic therefore has degrees of freedom

$$(r - 1)(c - 1) = (3 - 1)(2 - 1) = (2)(1) = 2$$

Look in the df = 2 row of Table 24.1. We see that $\chi^2 = 10.5$ is larger than the critical value 9.21 required for significance at the $\alpha = 0.01$ level but smaller than the critical value 13.82 for $\alpha = 0.001$. The cocaine study shows a significant relationship ($P < 0.01$) between treatment and success.

The significance test says only that we have strong evidence of *some* association between treatment and success. We must look at the two-way table to see the nature of the relationship: desipramine works better than the other treatments.

NOW IT'S YOUR TURN | **24.3 Video-gaming and grades.** The popularity of computer, video, online, and virtual reality games has raised concerns about their ability to negatively impact youth. Based on a recent survey, 1808 fourteen- to eighteen-year-olds in Connecticut high schools were classified by their average grades and by whether they had or had not played such games. The following table summarizes the findings. The observed and expected counts are given side by side.

	Observed			Expected		
	A's and B's	C's	D's and F's	A's and B's	C's	D's and F's
Played games	736	450	193	717.7	453.1	208.2
Never played games	205	144	80	223.3	140.9	64.8

From these counts, we find that the chi-square statistic is 6.74. Does the study show that there is a statistically significant relationship between playing games and average grades? Use a significance level of 0.05.

Using the chi-square test

Like our test for a population proportion, the chi-square test uses some approximations that become more accurate as we take more observations. Here is a rough rule for when it is safe to use this test.

Cell counts required for the chi-square test

You can safely use the chi-square test when no more than 20% of the expected counts are less than 5 and all individual expected counts are 1 or greater.

The cocaine study easily passes this test: all the expected cell counts are either 8 or 16. Here is a concluding example that outlines the examination of a two-way table.

EXAMPLE 6 Do angry people have a greater incidence of heart disease?

People who get angry easily tend to have more heart disease. That's the conclusion of a study that followed a random sample of 12,986 people from three locations for about four years. All subjects were free of heart disease at the beginning of the study. The subjects took the Spielberger Trait Anger Scale test, which measures how prone a person is to sudden anger. Here are data for the 8474 people in the sample who had normal blood pressure. CHD stands for "coronary heart disease." This includes people who had heart attacks and those who needed medical treatment for heart disease.

	Anger Score		
	Low	**Moderate**	**High**
Sample size	3110	4731	633
CHD count	53	110	27
CHD percent	1.7%	2.3%	4.3%

There is a clear trend: as the anger score increases, so does the percentage who suffer heart disease. Is this relationship between anger and heart disease statistically significant?

The first step is to write the data as a two-way table by adding the counts of subjects who did not suffer from heart disease. We also add the row and column totals, which we need to find the expected counts.

	Low anger	Moderate anger	High anger	Total
CHD	53	110	27	190
No CHD	3057	4621	606	8284
Total	3110	4731	633	8474

We can now follow the steps for a significance test, familiar from Chapter 22.

The hypotheses

The chi-square method tests these hypotheses:

H_0: no association between anger and CHD

H_a: some association between anger and CHD

The sampling distribution

We will see that all the expected cell counts are larger than 5, so we can safely apply the chi-square test. The two-way table of anger versus CHD has 2 rows and 3 columns. We will use critical values from the chi-square distribution with degrees of freedom df $= (2 - 1)(3 - 1) = 2$.

The data

First find the expected cell counts. For example, the expected count of high-anger people with CHD is

$$\text{expected count} = \frac{\text{row 1 total} \times \text{column 3 total}}{\text{table total}}$$

$$= \frac{(190)(633)}{8474} = 14.19$$

Here is the complete table of observed and expected counts side by side:

	Observed			Expected		
	Low	Moderate	High	Low	Moderate	High
CHD	53	110	27	69.73	106.08	14.19
No CHD	3057	4621	606	3040.27	4624.92	618.81

Looking at these counts, we see that the high-anger group has more CHD than expected and the low-anger group has less CHD than expected. This is consistent with what the percentages in Example 6 show. The chi-square statistic is

$$\chi^2 = \frac{(53 - 69.73)^2}{69.73} + \frac{(110 - 106.08)^2}{106.08} + \frac{(27 - 14.19)^2}{14.19}$$
$$+ \frac{(3057 - 3040.27)^2}{3040.27} + \frac{(4621 - 4624.92)^2}{4624.92} + \frac{(606 - 618.81)^2}{618.81}$$
$$= 4.014 + 0.145 + 11.564 + 0.092 + 0.003 + 0.265 = 16.083$$

In practice, statistical software can do all this arithmetic for you. Look at the 6 terms that we sum to get χ^2. Most of the total comes from just one cell: high-anger people have more CHD than expected.

Significance?

Look at the df $= 2$ line of Table 24.1 on page 563. The observed chi-square $\chi^2 = 16.083$ is larger than the critical value 13.82 for $\alpha = 0.001$. We have highly significant evidence ($P < 0.001$) that anger and heart disease are related. Statistical software can give the actual P-value. It is $P = 0.0003$.

The conclusion

Can we conclude that proneness to anger *causes* heart disease? This is an observational study, not an experiment. It isn't surprising to find that some lurking variables are confounded with anger. For example, people prone to anger are more likely than others to be those who drink and smoke. The study report used advanced statistics to adjust for many differences among the three anger groups. The adjustments raised the P-value from $P = 0.0003$ to $P = 0.02$ because the lurking variables explain some of the heart disease. This is still good evidence for a relationship if a significance level of 0.05 is used. Because the study started with a random sample of people who had no CHD and followed them forward in time, and because many lurking variables were measured and accounted for, it does give some evidence for causation. The next step might be an experiment that shows anger-prone people how to change. Will this reduce their risk of heart disease?

Simpson's paradox

As is the case with quantitative variables, the effects of lurking variables can change or even reverse relationships between two categorical variables. Let's continue studying the issue of sex and higher education presented in Examples 1 and 2. The numbers here are artificial for simplicity, but they illustrate a phenomenon that often appears in real data.

EXAMPLE 7 Discrimination in admissions?

A university offers only two degree programs, one in electrical engineering and one in English. Admission to these programs is competitive, and the women's caucus suspects discrimination against women in the admissions process. The caucus obtains the following data from the university, a two-way table of all applicants by sex and admission decision:

	Male	Female	Total
Admit	35	20	55
Deny	45	40	85
Total	80	60	140

We studied these data in Examples 1 and 2 and found that they do show an association between the sex of applicants and their success in obtaining admission. In particular we found:

$$\text{percentage of male applicants admitted} = \frac{35}{80} = 44\%$$

$$\text{percentage of female applicants admitted} = \frac{20}{60} = 33\%$$

Aha! Almost half of the males but only one-third of the females who applied were admitted. Isn't this proof of discrimination against women?

© Dennis Hallinan/Alamy

The university replies that although the observed association is correct, it is not due to discrimination. In its defense, the university produces a **three-way table** that classifies applicants by sex, admission decision, and the program to which they applied. We will present a three-way table as two or more two-way tables side by side, one for each value of the third variable. In this case there are two two-way tables, one for each program:

Engineering				English		
	Male	Female			Male	Female
Admit	30	10		Admit	5	10
Deny	30	10		Deny	15	30
Total	60	20		Total	20	40

Check that these entries add to the entries in the two-way table. The university has simply broken down that table by department. We now see that engineering admitted exactly half of all applicants, both male and female, and that English admitted one-fourth of both males and females. There is *no association* between sex and admission decision in either program.

How can no association in either program produce strong association when the two are combined? Look at the data: English is hard to get into, and mainly females apply to that program. Electrical engineering is easier to get into and attracts mainly male applicants. English had 40 female and 20 male applicants, while engineering had 60 male and only 20 female applicants. The original two-way table, which did not take account of the difference between programs, was misleading. This is an example of *Simpson's paradox*.

The Inventor of Simpson's Paradox at Work.

Professor Simpson

"No, I am **NOT** Homer Simpson."

Simpson's paradox

An association or comparison that holds for all of several groups can disappear or even reverse direction when the data are combined to form a single group. This situation is called **Simpson's paradox.**

Simpson's paradox is just an extreme form of the fact that observed associations can be misleading when there are lurking variables. Remember the caution from Chapter 15: *beware the lurking variable.*

EXAMPLE 8 Discrimination in mortgage lending?

Studies of applications for home mortgage loans from banks show a strong racial pattern: banks reject a higher percentage of black applicants than white applicants. One lawsuit against a bank for discrimination in lending in the Washington, DC, area contends that the bank rejected 17.5% of blacks but only 3.3% of whites.

The bank replies that lurking variables explain the difference in rejection rates. Blacks have (on the average) lower incomes, poorer credit records, and less secure jobs than whites. Unlike race, these are legitimate reasons to turn down a mortgage application. It is because these lurking variables are confounded with race, the bank says, that it rejects a higher percentage of black applicants. It is even possible, thinking of Simpson's paradox, that the bank accepts a *higher* percentage of black applicants than of white applicants if we look at people with the same income and credit record.

Who is right? Both sides will hire statisticians to examine the effects of the lurking variables. Both sides will present statistical arguments supporting or refuting a charge of discrimination in lending. Unfortunately, there are no formal guidelines for how juries and judges are to assess statistical arguments. And juries and judges need not have any statistical expertise. Lurking variables and seeming paradoxes such as Simpson's paradox make it difficult for even experts to determine the cause of the disparity in the rejection of mortgage applications. The court will eventually decide as best it can, but the decision may not be based on the statistical arguments.

STATISTICS IN SUMMARY

Chapter Specifics

- Categorical variables group individuals into classes. To display the relationship between two categorical variables, make a **two-way table** of counts for the classes. We describe the nature of an association between categorical variables by comparing selected percentages.

- As always, lurking variables can make an observed association misleading. In some cases, an association that holds for every level of a lurking variable disappears or changes direction when we lump all levels together. This is **Simpson's paradox.**

- The **chi-square test** tells us whether an observed association in a two-way table is statistically significant. The **chi-square statistic** compares the counts in the table with the counts we would expect if there were no association between the row and column variables. The sampling distribution is not Normal. It is a new distribution, the **chi-square distribution.**

 In Chapters 14 and 15 we considered relationships between two quantitative variables. In this chapter we use two-way tables to describe relationships between two *categorical* variables. To explore relationships between two categorical variables, make a two-way table.

Examine the distribution of each row (or each column). Differences in the patterns of the distributions suggest a relationship between the two variables. No change in these patterns suggests that there is no relationship.

As in Chapters 21, 22, and 23, we use formal statistical inference to decide if any difference in the observed patterns of the distributions is simply due to chance. We compare what we would expect cell counts to be based on the distribution of each variable separately with the cell counts actually observed. The chi-square test answers the question of whether differences in these cell counts could be due to chance.

As in Chapters 14 and 15, we must be careful not to assume that the patterns we observe would continue to hold for additional data or in a broader setting. Simpson's paradox is an example of how such an assumption could mislead us. Simpson's paradox occurs when the association or comparison that holds for all of several groups reverses direction when these groups are combined into a single group.

CASE STUDY EVALUATED   Here is the table that was presented in the Case Study at the beginning of this chapter:

	Female	Male	Total
Assistant professors	178	244	422
Associate professors	169	361	530
Professors	132	671	803
Total	479	1276	1755

Use what you have learned in this chapter to answer the following questions.

1. What percentage of the assistant professors are women?
2. What percentage of the associate professors are women?
3. What percentage of the full professors are women?
4. As rank increases from assistant to full professor, how does the percentage of women change? Are women overrepresented, underrepresented, or appropriately represented in the highest rank?
5. Do these data show that women have a harder time gaining promotion? Discuss. ∎

CHAPTER 24 EXERCISES

For Exercise 24.1, see page 562; for Exercise 24.2, see page 564; and for Exercise 24.3 see page 566.

24.4 Is astrology scientific? The University of Chicago's General Social Survey (GSS) is the nation's most important social science sample survey. The GSS asked a random sample of adults their opinion about whether astrology is very or sort of scientific or not at all scientific. Is belief that astrology is scientific related to amount of higher education? Here is a two-way table of counts for people in the sample who had three levels of higher education degrees:

	Degree Held		
Opinion	Junior college	Bachelor	Graduate
Not at all scientific	87	198	111
Very or sort of scientific	43	57	28

Calculate percentages that describe the nature of the relationship between amount of higher education and opinion about whether astrology is very scientific or sort of scientific or not at all scientific. Give a brief summary in words.

24.5 Smoking by students and their families. How are the smoking habits of students related to the smoking habits of their close family members? Here is a two-way table from a survey of male students in six secondary schools in Malaysia:

	Student smokes	Student does not smoke
At least one close family member smokes	115	207
No close family member smokes	25	75

Write a brief answer to the question posed, including a comparison of selected percentages.

24.6 Smoking by parents of preschoolers. How are the smoking habits of parents of preschoolers related to the education of the father? Here is a two-way table from a survey of the parents of preschoolers in Greece:

	Both parents smoke	One parent smokes	Neither parent smokes
University education	42	68	90
Intermediate education	47	69	75
High school education	183	281	273
Primary education or none	69	73	62

Write a brief answer to the question posed, including a comparison of selected percentages.

24.7 Python eggs. How is the hatching of water python eggs influenced by the temperature of the snake's nest? Researchers assigned newly laid eggs to one of three temperatures: hot, neutral (room temperature), or cold. Hot duplicates the extra warmth provided by the mother python, and cold duplicates the absence of the mother. Here are the data on the number of eggs and the number that hatched:

	Eggs	Hatched
Cold	27	16
Neutral	56	38
Hot	104	75

(a) Make a two-way table of temperature by outcome (hatched or not).

(b) Calculate the percentage of eggs in each group that hatched. The researchers anticipated that eggs would not hatch in the cold environment. Do the data support that anticipation?

24.8 Firearm deaths. Here are counts from a study of all firearm-related deaths in Wisconsin between 2000 and 2002 for children and youths under the age of 25 where the type of firearm used was known. We want to compare the types of firearms used in homicides and in suicides. We suspect that long guns (shotguns and rifles) will more often be used in suicides because many people keep them at home for hunting. Make a bar graph to compare homicides and suicides. What does the graph suggest about deaths involving long guns versus handguns?

	Handgun	Long gun	Total
Homicides	106	17	123
Suicides	59	107	166

24.9 Who earns academic degrees? How do women and men compare in the pursuit of academic degrees? The table below presents counts (in thousands), as projected by the National Center for Education Statistics, of degrees earned in 2012–2013 categorized by the level of the degree and the sex of the recipient.

	Bachelor's	Master's	Professional	Doctorate
Female	989	418	49	40
Male	731	266	50	35
Total	1721	684	99	75

(a) How many people earned bachelor's degrees?

(b) What percentage of each level of degree is earned by women? Write a brief description of what the data show about the relationship between sex and degree level.

24.10 Animal testing. "It is right to use animals for medical testing if it might save human lives." The General Social Survey asked 1152 adults to react to this statement. Here is the two-way table of their responses:

Response	Male	Female
Strongly agree	76	59
Agree	270	247
Neither agree nor disagree	87	139
Disagree	61	123
Strongly disagree	22	68

Describe the differences between the distributions of opinions for women and men with percentages, with a graph, and in words.

24.11 Totals aren't enough. Here are the row and column totals for a two-way table with two rows and two columns:

$$
\begin{array}{cc|c}
a & b & 40 \\
c & d & 60 \\
\hline
60 & 40 & 100
\end{array}
$$

Find *two different* sets of counts a, b, c, and d for the body of the table that give these same totals. This shows that the relationship between two variables cannot be obtained from the two individual distributions of the variables.

24.12 Airline flight delays. Here are the numbers of flights on time and delayed for two airlines at five airports during a one-month period. Overall on-time percentages for each airline are often reported in the news. The airport that flights serve is a lurking variable that can make such reports misleading.

	Alaska Airlines		America West	
	On time	Delayed	On time	Delayed
Los Angeles	497	62	694	117
Phoenix	221	12	4840	415
San Diego	212	20	383	65
San Francisco	503	102	320	129
Seattle	1841	305	201	61

(a) What percentage of all Alaska Airlines flights were delayed? What percentage of all America West flights were delayed? These are the numbers usually reported.

(b) Now find the percentage of delayed flights for Alaska Airlines at each of the five airports. Do the same for America West.

(c) America West does worse at *every one* of the five airports, yet does better overall. That sounds impossible. Explain carefully, referring to the data, how this can happen. (The weather in Phoenix and Seattle lies behind this example of Simpson's paradox.)

24.13 Bias in the jury pool? The New Zealand Department of Justice did a study of the composition of juries in court cases. Of interest was whether Maori, the indigenous people of New Zealand, were adequately represented in jury pools. Here are the results for two districts, Rotura and Nelson, in New Zealand (similar results were found in all districts):

Rotura	Maori	Non-Maori
In jury pool	79	258
Not in jury pool	8810	23,751
Total	8889	24,009

Nelson	Maori	Non-Maori
In jury pool	1	56
Not in jury pool	1328	32,602
Total	1329	32,658

(a) Use these data to make a two-way table of race (Maori or non-Maori) versus jury pool status (In or Not in).

(b) Show that Simpson's paradox holds: a higher percentage of Maori are in the jury pool overall, but for both districts a higher percentage of non-Maori are in the jury pool.

c) Use the data to explain the paradox in language that a judge could understand.

24.14 Field goal shooting. Here are data on field goal shooting for two members of the Kent State University 2002–2003 women's basketball team:

	Jamie Rubis		Lindsay Shearer	
	Made	Missed	Made	Missed
Two-pointers	119	115	86	84
Three-pointers	36	61	5	16

(a) What percent of all field goal attempts did Jamie Rubis make? What percent of all field goal attempts did Lindsay Shearer make?

(b) Now find the percent of all two-point field goals and all three-point field goals that Jamie made. Do the same for Lindsay.

(c) Lindsay had a lower percent for *both* types of field goals but had a better overall percent. That sounds impossible. Explain carefully, referring to the data, how this can happen.

24.15 Animal testing. Exercise 24.10 (page 575) gives the responses of a survey of 1152 adults to a statement about animal testing.

(a) Do these data satisfy our guidelines for safe use of the chi-square test?

(b) Is there a statistically significant relationship between the sex and opinion of adults?

24.16 Is astrology scientific? In Exercise 24.4 (page 573), you described the relationship between belief that astrology is scientific and amount of higher education. Is the observed association between these variables statistically significant? To find out, proceed as follows.

(a) Add the row and column totals to the two-way table in Exercise 24.4 (page 573) and find the expected cell counts. Which observed counts differ most from the expected counts?

(b) Find the chi-square statistic. Which cells contribute most to this statistic?

(c) What are the degrees of freedom? Use Table 24.1 (page 565) to say how significant the chi-square test is. Write a brief conclusion for your study.

24.17 Smoking by students and their families. In Exercise 24.5 (page 574), you saw that there is an association between smoking by close family members and smoking by high school students. The students are more likely to smoke if a close family member smokes. We want to know whether this association is statistically significant.

(a) State the hypotheses for the chi-square test. What do you think the population is?

(b) Find the expected cell counts. Write a sentence that explains in simple language what "expected counts" are.

(c) Find the chi-square statistic and its degrees of freedom. What is your conclusion about significance?

24.18 Python eggs. Exercise 24.7 (page 574) presents data on the hatching of python eggs at three different temperatures. Does temperature have a significant effect on hatching? Write a clear summary of your work and your conclusion.

24.19 Stress and heart attacks. You read a newspaper article that describes a study of whether stress management can help reduce heart attacks. The 107 subjects all had reduced blood flow to the heart and so were at risk of a heart attack. They were assigned at random to three groups. The article goes on to say:

One group took a four-month stress management program, another underwent a four-month exercise program and the third received usual heart care from their personal physicians.

In the next three years, only three of the 33 people in the stress management group suffered "cardiac events," defined as a fatal or non-fatal heart attack or a surgical procedure such as a bypass or angioplasty. In the same period, seven of the 34 people in the exercise group and 12 out of the 40 patients in usual care suffered such events.

(a) Use the information in the news article to make a two-way table that describes the study results.

(b) What are the success rates of the three treatments in preventing cardiac events?

(c) Find the expected cell counts under the null hypothesis that there is no difference among the treatments. Verify that the expected counts meet our guideline for use of the chi-square test.

(d) Is there a significant difference among the success rates for the three treatments?

24.20 Standards for child care. Do unregulated providers of child care in their homes follow different health and safety practices in different cities? A study looked at people who regularly provided care for someone else's children in poor areas of three cities. The numbers who required medical releases from parents to allow medical care in an emergency were 42 of 73 providers in Newark, N.J., 29 of 101 in Camden, N.J., and 48 of 107 in South Chicago, Ill.

(a) Use the chi-square test to see if there are significant differences among the proportions of child care providers who require medical releases in the three cities. What do you conclude?

(b) How should the data be produced in order for your test to be valid? (In fact, the samples came in part from asking parents who were subjects in another study who provided their child care. The author of the study wisely did not use a statistical test. He wrote: "Application of conventional statistical procedures appropriate for random samples may produce biased and misleading results.")

EXPLORING THE WEB

24.21 Web-based exercise. The chapter-opening Case Study gave information on faculty by sex and rank at Purdue University for the 2010–2011 academic year. Similar data can be found for other universities. Search the Web to find the number of faculty by rank and sex at another university. Do you see the same patterns as were noted for the Purdue University data? We found several sources by doing a Google search on "faculty headcount by rank and sex."

24.22 Web-based exercise. Find an example of Simpson's paradox and discuss how your example illustrates the paradox. Two examples that we found (thanks to Patricia Humphrey at Georgia Southern University) are **www.nytimes.com/2006/07/15/education/15report.html** and **online.wsj.com/article/SB125970744553071829.html.**

NOTES AND DATA SOURCES

Page 557 These data were obtained online at **www.purdue.edu/datadigest/.**

Page 558 Example 1: An example of real data similar in spirit is P. J. Bickel and J. W. O'Connell, "Is there a sex bias in graduate admissions?" *Science,* 187 (1975), pp. 398–404.

Page 559 Example 3: D. M. Barnes, "Breaking the cycle of addiction," *Science,* 241 (1988), pp. 1029–1030.

Page 567 There are many computer studies of the accuracy of critical values for χ^2. For a brief discussion and some references, see Section 3.2.5 of David S. Moore, "Tests of chi-squared type," in Ralph B. D'Agostino and Michael A. Stephens (eds.), *Goodness-of-Fit Techniques*, Marcel Dekker, 1986, pp. 63–95.

Page 567 Example 6: Janice E. Williams et al., "Anger proneness predicts coronary heart disease risk," *Circulation*, 101 (2000), pp. 2034–2039.

Page 571 Example 8: Data for Nationsbank, from S. A. Holmes, "All a matter of perspective," *New York Times*, October 11, 1995.

Page 573 Exercise 24.4: This General Social Survey exercise presents a table constructed using the search function at the GSS archive, **sda.berkeley.edu/archive.htm.** These data are from the 2008 GSS.

Page 574 Exercise 24.5: K. Shamsuddin and M. Abdul Haris, "Family influence on current smoking habits among secondary school children in Kota Bharu, Kelantan," *Singapore Medical Journal*, 41 (2000), pp. 167–171.

Page 574 Exercise 24.6: C. I. Vardavas et al., "Smoking habits of Greek preschool children's parents," *BMC Public Health*, 7, No. 112 (2007).

Page 574 Exercise 24.7: R. Shine, T. R. L. Madsen, M. J. Elphick, and P. S. Harlow, "The influence of nest temperatures and maternal brooding on hatchling phenotypes in water pythons," *Ecology*, 78 (1997), pp. 1713–1721.

Page 575 Exercise 24.8: T. Shiffler et al., "The burden of suicide and homicide of Wisconsin's children and youth," *Wisconsin Medical Journal*, 104 (2005), pp. 62–67.

Page 575 Exercise 24.9: The data are from the 2009 *Digest of Education Statistics* at the Web site of the National Center for Education Statistics, **nces.ed.gov.**

Page 575 Exercise 24.10: This General Social Survey exercise presents a table constructed using the search function at the GSS archive, **sda.berkeley.edu/archive .htm.** These data are from the 2008 GSS.

Page 576 Exercise 24.12: From reports submitted by airlines to the Department of Transportation, found in A. Barnett, "How numbers can trick you," *Technology Review*, October 1994, pp. 38–45.

Page 576 Exercise 24.13: I. Westbrooke, "Simpson's paradox: An example in a New Zealand survey of jury composition," *Chance*, 11 (1998), pp. 40–42.

Page 577 Exercise 24.14: Found online at **www.math.kent.edu/~darci/simpson/ bballexamples.html,** a Web site maintained by Darci L. Kracht at Kent State University. Thanks to Patricia Humphrey at Georgia Southern University for bringing this example to our attention.

Page 578 Exercise 24.19: Brenda C. Coleman, "Study: heart attack risk cut 74% by stress management," Associated Press dispatch appearing in the *Lafayette (Ind.) Journal and Courier*, October 20, 1997.

Page 579 Exercise 24.20: David M. Blau, "The child care labor market," *Journal of Human Resources*, 27 (1992), pp. 9–39.

Statistical inference draws conclusions about a population on the basis of sample data and uses probability to indicate how reliable the conclusions are. A confidence interval estimates the true value of an unknown parameter. A significance test shows how strong the evidence is for some claim about a parameter. Chapters 21 and 22 present the reasoning of confidence intervals and tests and give optional details for inference about population proportions p and population means μ.

The probabilities in both confidence intervals and significance tests tell us what would happen if we used the formula for the interval or test very many times. A confidence level is the probability that the formula for a confidence interval actually produces an interval that contains the unknown parameter. A 95% confidence interval gives a correct result 95% of the time when we use it repeatedly. Figure IV.1 illustrates the reasoning using the 95% confidence interval for a population proportion p.

A P-value is the probability that the test would produce a result at least as extreme as the observed result if the null hypothesis really were true. Figure IV.2 illustrates the reasoning, placing the sample proportion $\hat{p}$ from our one sample on the Normal curve that shows how $\hat{p}$ would vary in all possible samples *if the null hypothesis were true*. A P-value tells us how surprising the observed outcome is. Very surprising outcomes (small P-values) are good evidence that the null hypothesis is not true.

To detect sound and unsound uses of inference, you must know the basic reasoning and also be aware of some fine points and pitfalls. Chapter 23 will

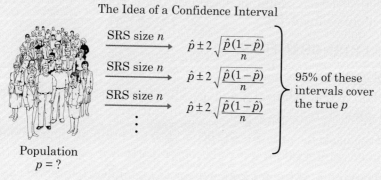

The Idea of a Confidence Interval

SRS size n $\longrightarrow$ $\hat{p} \pm 2 \sqrt{\dfrac{\hat{p}(1-\hat{p})}{n}}$

SRS size n $\longrightarrow$ $\hat{p} \pm 2 \sqrt{\dfrac{\hat{p}(1-\hat{p})}{n}}$

SRS size n $\longrightarrow$ $\hat{p} \pm 2 \sqrt{\dfrac{\hat{p}(1-\hat{p})}{n}}$

95% of these intervals cover the true p

Population $p = ?$

Figure IV.1

The Idea of a Significance Test

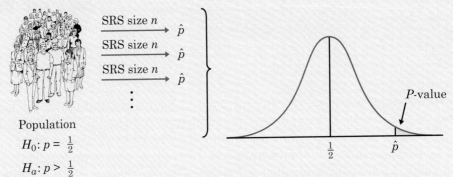

Population
$H_0: p = \frac{1}{2}$
$H_a: p > \frac{1}{2}$

Figure IV.2

help. The optional sections in Chapters 21 and 22 and the optional Chapter 24 discuss a few more specific inference procedures. Chapter 24 deals with two-way tables, both for description and for inference. The descriptive part of this chapter completes Part II's discussion of relationships among variables by describing relationships among categorical variables.

PART IV SUMMARY

Here are the most important skills you should have acquired after reading Chapters 21 to 24. Asterisks mark skills that appear in optional sections of the text.

A. SAMPLING DISTRIBUTIONS

1. Explain the idea of a sampling distribution. See Figure 21.1 (page 490).
2. Use the Normal sampling distribution of a sample proportion $\hat{p}$ and the 68–95–99.7 rule to find probabilities involving $\hat{p}$.
3. *Use the Normal sampling distribution of a sample mean $\bar{x}$ to find probabilities involving $\bar{x}$.

B. CONFIDENCE INTERVALS

1. Explain the idea of a confidence interval. See Figure IV.1.
2. Explain in nontechnical language what is meant by "95% confidence" and other statements of confidence in statistical reports.
3. Use the basic formula $\hat{p} \pm 2\sqrt{\hat{p}(1 - \hat{p})/n}$ to obtain an approximate 95% confidence interval for a population proportion p.
4. Understand how the margin of error of a confidence interval changes with the sample size and the level of confidence.

5. Detect major mistakes in applying inference, such as improper data production, selecting the best of many outcomes, ignoring high nonresponse, and ignoring outliers.

6. *Use the detailed formula $\hat{p} \pm z^* \sqrt{\hat{p}(1 - \hat{p})/n}$ and critical values z^* for Normal distributions to obtain confidence intervals for a population proportion p.

7. *Use the formula $\bar{x} \pm z^* s/\sqrt{n}$ to obtain confidence intervals for a population mean μ.

C. SIGNIFICANCE TESTS

1. Explain the idea of a significance test. See Figure IV.2.

2. State the null and alternative hypotheses in a testing situation when the parameter in question is a population proportion p.

3. Explain in nontechnical language the meaning of the P-value when you are given its numerical value for a test.

4. Explain the meaning of "statistically significant at the 5% level" and other statements of significance. Explain why significance at a specific level such as 5% is less informative than a P-value.

5. Recognize that significance testing does not measure the size or importance of an effect.

6. Recognize and explain the effect of small and large samples on the statistical significance of an outcome.

7. *Use Table B of percentiles of Normal distributions to find the P-value for a test about a proportion p.

8. *Carry out one-sided and two-sided tests about a mean μ using the sample mean $\bar{x}$ and Table B.

D. *TWO-WAY TABLES

1. Create two-way tables for data classified by two categorical variables.

2. Use percentages to describe the relationship between any two categorical variables based on the counts in a two-way table.

3. Explain what null hypothesis the chi-square statistic tests in a specific two-way table.

4. Calculate expected cell counts, the chi-square statistic, and its degrees of freedom from a two-way table.

5. Use Table 24.1 (page 565) for chi-square distributions to assess significance. Interpret the test result in the setting of a specific two-way table.

PART IV REVIEW EXERCISES

Review exercises are short and straightforward exercises that help you solidify the basic ideas and skills in each part of this book. We have provided "hints" that indicate where you can find the relevant material for the odd-numbered problems.

IV.1 Computer crime. An October 2010 Gallup Poll asked a random sample of 1025 adults if they had been a victim of computer or Internet crime in the previous 12 months. Of these adults, 133 said "Yes." We can act as if the sample were an SRS. Give a 95% confidence interval for the proportion of all adults who were victims of computer or Internet crime in the 12 months prior to October 2010. (*Hint:* See pages 489–493.)

IV.2 Do you drink? In a July 2010 Gallup Poll a random sample of 1020 adults were asked whether they drank alcoholic beverages (liquor, wine, or beer) or completely abstained from drinking any alcohol. Among respondents, 337 said that they completely abstain from drinking alcohol. Assume that the sample was an SRS. Give a 95% confidence interval for the proportion of all adults who completely abstain from drinking alcoholic beverages.

IV.3 Computer crime. Exercise IV.1 concerns a random sample of 1025 adults. Suppose that (unknown to the pollsters) exactly 15% of all adults were victims of computer or Internet crime in the 12 months prior to October 2010. Imagine that we take very many SRSs of size 1025 from this population and record the percentage in each sample who claim to have been victims of computer or Internet crime in the 12 months prior to October 2010. Where would the middle 95% of all values of this percentage lie? (*Hint:* See pages 489–490.)

IV.4 Do you drink? Exercise IV.2 concerns a random sample of 1020 adults. Suppose that, in the population of all adults, exactly 30% would say that they completely abstain from drinking alcoholic beverages. Imagine that we take a very large number of SRSs of size 1020. For each sample, we record the proportion $\hat{p}$ of the sample who would say that they totally abstain from alcoholic beverages.

(a) What is the sampling distribution that describes the values $\hat{p}$ would take in our samples?

(b) Use this distribution and the 68–95–99.7 rule to find the approximate percentage of all samples in which more than 34.3% of the respondents would say that they totally abstain from alcoholic beverages.

IV.5 Honesty in the media. A Gallup Poll conducted from November 19 to 21, 2010, asked a random sample of 1037 adults to rate the honesty and ethical standards of people in a variety of professions. Among the respondents, 239 rated the honesty and ethical standards of TV reporters as very high or high. Assume that the sample was an SRS. Give a 95% confidence interval for the proportion of all adults who would rate the honesty and ethical standards of TV reporters as very high or high. (*Hint:* See pages 492–493.)

IV.6 Roulette. A roulette wheel has 18 red slots among its 38 slots. You observe many spins and record the number of times the ball falls in a red slot. Now you want to use these data to test whether the probability p of the ball falling in a red slot has the value that is correct for a fair roulette wheel. State the hypotheses H_0 and H_a that you will test.

IV.7 Why not? Table 11.1 (page 240) records the percentage of residents aged 65 or older in each of the 50 states. You can check that this percentage is 14% or higher in 12 of the states. So the sample proportion of states with at least 14% of elderly residents is $\hat{p} = 12/50 = 0.24$. Explain why it does *not* make sense to go on to calculate a 95% confidence interval for the population proportion p. (*Hint:* See pages 488–489, 506.)

IV.8 Helping welfare mothers. A study compares two groups of mothers with young children who were on welfare two years ago. One group attended a voluntary training program that was offered free of charge at a local vocational school and was advertised in the local news media. The other group did not choose to attend the training program. The study finds a significant difference ($P < 0.01$) between the proportions of the mothers in the two groups who are still on welfare. The difference is not only significant but quite large. The report says that with 95% confidence the percentage of the nonattending group still on welfare is 21% ± 4% higher than that of the group who attended the program. You are on the staff of a member of Congress who is interested in the plight of welfare mothers and who asks you about the report.

(a) Explain in simple language what "a significant difference ($P < 0.01$)" means.

(b) Explain clearly and briefly what "95% confidence" means.

(c) This study is not good evidence that requiring job training of all welfare mothers would greatly reduce the percentage who remain on welfare. Explain this to the member of Congress.

IV.9 Beating the system. Some doctors think that health plan rules restrict their ability to treat their patients effectively, so they bend the rules to help patients get reimbursed by their health plans. Here's a sentence from a study on this topic: "Physicians who agree with the statement 'Today it is necessary to game the system to provide high-quality care' reported manipulating reimbursement systems more often than those who did not agree with the statement (64.3% vs 35.7%; $P < .001$)."

(a) Explain to a doctor what "$P < .001$" means in the context of this specific study. (*Hint:* See pages 519–520.)

(b) A result that is statistically significant can still be too small to be of practical interest. How do you know this is not true here? (*Hint:* See pages 542–543.)

IV.10 Smoking in the United States. A July 2011 nationwide random survey of 1016 adults asked whether they had smoked cigarettes in the last week. Among the respondents, 223 said they had. Assume that the sample was an SRS.

(a) Give a 95% confidence interval for the proportion of all American adults who smoked in the week preceding the survey.

(b) Write a short paragraph for a news report based on the survey results.

IV.11 When shall we call you? As you might guess, telephone sample surveys get better response rates during the evening than during the weekday daytime. One study called 2304 randomly chosen telephone numbers on weekday mornings. Of these, 1313 calls were answered and only 207 resulted in interviews. Of 2454 calls on weekday evenings, 1840 were answered and 712 interviews resulted. Give two 95% confidence intervals, for the proportions of all calls that are answered on weekday mornings and on weekday evenings. Are you confident that the proportion is higher in the evening? (*Hint:* See pages 491–493, 546.)

IV.12 Smoking in the United States. Does the survey of Exercise IV.10 provide good evidence that fewer than 1/4 of all American adults smoked in the week prior to the survey?

(a) State the hypotheses to be tested.

(b) If your null hypothesis is true, what is the sampling distribution of the sample proportion $\hat{p}$? Sketch this distribution.

(c) Mark the actual value of $\hat{p}$ on the curve. In your opinion, does it appear surprising enough to give good evidence against the null hypothesis?

IV.13 When shall we call you? Suppose that we know that 57% of all calls made by sample surveys on weekday mornings are answered. We make 2454 calls to randomly chosen numbers during weekday evenings. Of these, 1840 are answered. Is this good evidence that the proportion of answered calls is higher in the evening?

(a) State the hypotheses to be tested. (*Hint:* See pages 516–519.)

(b) If your null hypothesis is true, what is the sampling distribution of the sample proportion $\hat{p}$? Sketch this distribution. (*Hint:* See pages 516–522.)

(c) Mark the actual value of $\hat{p}$ on the curve. In your opinion, does it appear surprising enough to give good evidence against the null hypothesis? (*Hint:* See pages 516–522.)

IV.14 Not significant. The study cited in Exercise IV.9 looked at the factors that may affect whether doctors bend medical plan rules. Perhaps doctors who fear being prosecuted will bend the rules less often. The study report said, "Notably, greater worry about prosecution for fraud did not affect physicians' use of these tactics ($P = .34$)." Explain why the result $P = 0.34$ supports the conclusion that doctors' fears about potential prosecution did not affect behavior.

IV.15 Going to church. Opinion polls show that about 40% of Americans say they attended religious services in the last week. This result has stayed stable for decades. Studies of what people actually *do,* as opposed to what they *say* they do, suggest that actual church attendance is much lower. One study calculated 95% confidence intervals based on what a sample of Catholics said and then based on a sample of actual behavior. In Chicago, for example, the 95% confidence interval from the opinion poll said that between 45.7% and 51.3% of Catholics attended mass weekly. The 95% confidence interval from actual counts said that between 25.7% and 28.9% attended mass weekly.

(a) Why might we expect opinion polls on church attendance to be biased in the direction of overestimating true attendance? (*Hint:* See pages 539–542.)

(b) The poll in Chicago found that 48.5% of Catholics claimed to attend mass weekly. Why don't we just say that "48.5% of all Catholics in Chicago claim to attend mass" instead of giving the interval 45.7% to 51.3%? (*Hint:* See page 546.)

(c) The two results, from reported and observed behavior, are quite different. What does it mean to say that we are "95% confident" in each of the two intervals given? (*Hint:* See pages 493–496, 539–542.)

The following exercises are based on the optional sections of Chapters 21 and 22.

IV.16 Computer crime. An October 2010 Gallup Poll asked a random sample of 1025 adults if they had been a victim of computer or Internet crime in the previous 12 months. Of these adults, 133 said "Yes." Assume that the sample was an SRS. Give 90% and 99% confidence intervals for the proportion of all adults who were victims of computer or Internet crime in the 12 months prior to October 2010. Explain briefly what important fact about confidence intervals is illustrated by comparing these two intervals and the 95% confidence interval from Exercise IV.1.

IV.17 Do you drink? In a July 2010 Gallup Poll a random sample of 1020 adults were asked whether they drank alcoholic beverages (liquor, wine, or beer) or completely abstained from drinking any alcohol. Among respondents, 337 said that they completely abstain from drinking alcohol. Assume that the sample was an SRS. Is this good evidence that more than 1/3 of American adults completely abstain from drinking alcoholic beverages? Show the five steps of the test (hypotheses, sampling distribution, data, *P*-value, conclusion) clearly. Use a significance level of 0.05. (*Hint:* See pages 516–524.)

IV.18 Honesty in the media. A Gallup Poll conducted from November 19 to 21, 2010, asked a random sample of 1037 adults to rate the honesty and ethical standards of people in a variety of professions. Among the respondents, 239 rated the honesty and ethical standards of TV reporters as very high or high. Assume that the sample was an SRS. Give a 90% confidence interval for the proportion of all adults who rate the honesty and ethical standards of TV reporters as very high or high. For what purpose might a 90% confidence interval be less useful than a 95% confidence interval? For what purpose might a 90% interval be more useful?

IV.19 A poll of voters. You are the polling consultant to a member of Congress. An SRS of 500 registered voters finds that 32% name "environmental problems" as the most important issue facing the nation. Give a 90% confidence interval for the proportion of all voters who hold this opinion. Then explain carefully to the member of Congress what your conclusion reveals about voters' opinions. (*Hint:* See pages 496–499.)

IV.20 Smoking in the United States. Carry out the significance test called for in Exercise IV.12 in all detail. Show the five steps of the test (hypotheses, sampling distribution, data, *P*-value, conclusion) clearly.

IV.21 When shall we call you? Carry out the significance test called for in Exercise IV.13 in all detail. Show the five steps of the test (hypotheses, sampling distribution, data, *P*-value, conclusion) clearly. (*Hint:* See pages 523–524.)

The following exercises are based on the optional material in Chapters 21, 22, and 24.

IV.22 CEO pay. A study of 104 corporations found that the pay of their chief executive officers had increased an average of $\bar{x} = 6.9\%$ per year in real terms. The standard deviation of the percentage increases was $s = 17.4\%$.

(a) The 104 individual percentage increases have a right-skewed distribution. Explain why the central limit theorem says that we can nonetheless act as if the mean increase has a Normal distribution.

(b) Give a 95% confidence interval for the mean percentage increase in pay for all corporate CEOs.

(c) What must we know about the 104 corporations studied to justify the inference you did in (b)?

IV.23 Water quality. An environmentalist group collects a liter of water from each of 45 random locations along a stream and measures the amount of dissolved oxygen in each specimen. The mean is 4.62 milligrams (mg) and the standard deviation is 0.92 mg. Is this strong evidence that the stream has a mean oxygen content of less than 5 mg per liter? (*Hint:* See pages 524–528.)

IV.24 Pleasant smells. Do pleasant odors help work go faster? Twenty-one subjects worked a paper-and-pencil maze wearing a mask that was either unscented or carried the smell of flowers. Each subject worked the maze three times with each mask, in random order. (This is a matched pairs design.) Here are the differences in their average times (in seconds), unscented minus scented. If the floral smell speeds work, the difference will be positive because the time with the scent will be lower.

−7.37	−3.14	4.10	−4.40	19.47	−10.80	−0.87
8.70	2.94	−17.24	14.30	−24.57	16.17	−7.84
8.60	−10.77	24.97	−4.47	11.90	−6.26	6.67

(a) We hope to show that work is faster on the average with the scented mask. State null and alternative hypotheses in terms of the mean difference in times μ for the population of all adults.

(b) Using a calculator, find the mean and standard deviation of the 21 observations. Did the subjects work faster with the scented mask? Is the mean improvement big enough to be important?

(c) Make a stemplot of the data (round to the nearest whole second). Are there outliers or other problems that might hinder inference?

(d) Test the hypotheses you stated in (a). Is the improvement statistically significant?

IV.25 Sharks. Great white sharks are big and hungry. Here are the lengths in feet of 44 great whites:

18.7	12.3	18.6	16.4	15.7	18.3	14.6	15.8	14.9	17.6	12.1
16.4	16.7	17.8	16.2	12.6	17.8	13.8	12.2	15.2	14.7	12.4
13.2	15.8	14.3	16.6	9.4	18.2	13.2	13.6	15.3	16.1	13.5
19.1	16.2	22.8	16.8	13.6	13.2	15.7	19.7	18.7	13.2	16.8

(a) Make a stemplot with feet as the stems and 10ths of feet as the leaves. There are 2 outliers, one in each direction. These won't change $\bar{x}$ much but will increase the standard deviation s. (*Hint:* See pages 248–251.)

(b) Give a 90% confidence interval for the mean length of all great white sharks. (The interval may be too wide due to the influence of the outliers on s.) (*Hint:* See pages 501–503.)

(c) What do we need to know about these sharks in order to interpret your result in (b)? (*Hint:* See pages 501–502.)

IV.26 Pleasant smells. Return to the data in Exercise IV.24. Give a 95% confidence interval for the mean improvement in time to solve a maze when wearing a mask with a floral scent. Are you confident that the scent does improve mean working time?

IV.27 Sharks. Return to the data in Exercise IV.25. Is there good evidence that the mean length of sharks in the population that these sharks represent is greater than 15 feet? (*Hint:* See pages 524–528.)

IV.28 Simpson's paradox. If we compare average 2003 National Assessment of Educational Progress mathematics scores, we find that eighth-grade students in Nebraska do better than eighth-grade students in New York. But if we look only at white students, New York does better. If we look only at minority students, New York again does better. That's Simpson's paradox: the comparison reverses when we lump all students together. Explain carefully why this makes sense, using the fact that a much higher percentage of Nebraska eighth-graders are white.

IV.29 Unhappy HMO patients. A study of complaints by HMO members compared those who filed complaints about medical treatment and those who filed nonmedical complaints with an SRS of members who did not complain that year. Here are the data on the number who stayed and the number who voluntarily left the HMO:

	No complaint	Medical complaint	Nonmedical complaint
Stayed	721	173	412
Left	22	26	28

(a) Find the row and column totals. (*Hint:* See pages 558–559.)

(b) Find the percentage of each group who left. (*Hint:* See pages 558–559.)

(c) Find the expected counts and check that you can safely use the chi-square test. (*Hint:* See pages 561, 567–568.)

(d) The chi-square statistic for this table is $X^2 = 31.765$. What null and alternative hypotheses does this statistic test? What are its degrees of freedom? How significant is it? What do you conclude about the relationship between complaints and leaving the HMO? (*Hint:* See pages 559–568.)

IV.30 Treating ulcers. Gastric freezing was once a recommended treatment for stomach ulcers. Use of gastric freezing stopped after experiments showed it had no effect. One randomized comparative experiment found that 28 of the 82 gastric-freezing patients improved, while 30 of the 78 patients in the placebo group improved.

(a) Outline the design of this experiment.

(b) Make a two-way table of treatment versus outcome (subject improved or not). Is there a significant relationship between treatment and outcome?

(c) Write a brief summary that includes the test result and also percentages that compare the success of the two treatments.

IV.31 When shall we call you? In Exercise IV.11, we learned of a study that dialed telephone numbers at random during two periods of the day. Of 2304 numbers called on weekday mornings, 1313 answered. Of 2454 calls on weekday evenings, 1840 were answered.

(a) Make a two-way table of time of day versus answered or not. What percentage of calls were answered in each time period? (*Hint:* See pages 558–559.)

(b) It should be obvious that there is a highly significant relationship between time of day and answering. Why? (*Hint:* See pages 558–560.)

(c) Nonetheless, carry out the chi-square test. What do you conclude? (*Hint:* See pages 562–566.)

PART IV PROJECTS

Projects are longer exercises that require gathering information or producing data and that emphasize writing a short essay to describe your work. Many are suitable for teams of students.

Project 1. Reporting a medical study. Many of the major articles in medical journals concern statistically designed studies and report the results of inference, usually either *P*-values or 95% confidence intervals. You can find summaries of current articles on the Web sites of the *Journal of the American Medical Association* (**jama.ama-assn.org**) and the *New England Journal of Medicine* (**www.nejm.org**). A full copy may require paying a fee or visiting the library. Choose an article that describes a medical experiment on a topic that is understandable to those of us who lack medical training—anger and heart attacks and fiber in the diet to reduce cholesterol are two examples used in Chapters 21 and 22. Write a two-paragraph news article explaining the results of the study.

Then write a brief discussion of how you decided what to put in the news article and what to leave out. For example, if you omitted details of statistical significance or of confidence intervals, explain why. What did you say about the design of the study, and why? News writers must regularly make decisions like these.

Project 2. Use and abuse of inference. Few accounts of really complex statistical methods are readable without extensive training. One that is, and that is also an excellent essay on the abuse of statistical inference, is "The Real Error of Cyril Burt," a chapter in Stephen Jay Gould's *The Mismeasure of Man* (W. W. Norton, 1981). We met Cyril Burt under suspicious circumstances in Exercise 9.22 (page 197). Gould's long chapter shows that Burt and others engaged in discovering dubious patterns by using complex statistics. Read it, and write a brief explanation of why "factor analysis" failed to give a firm picture of the structure of mental ability.

Project 3. Roll your own statistical study. Collect your own data on two categorical variables whose relationship seems interesting. A simple example is the gender of a student and his or her political party preference. A more elaborate example is the year in school of a college undergraduate and his or her plans following graduation (immediate employment, further study, take some time off, . . .). We won't insist on a proper SRS.

Collect your data and make a two-way table. Do an analysis that includes comparing percentages to describe the relationship between your two variables and using the chi-square statistic to assess its significance. Write a description of your study and its findings. Was your sample so small that lack of significance may not be surprising?

Project 4. Car colors. We have heard that more white cars are sold in the United States than any other color. What percentage of the cars driven by students at your school are white? Answer this question by collecting data and giving a confidence interval for the proportion of white cars. You might collect data by questioning a sample of students or by looking at cars in student parking areas. In your discussion, explain how you attempted to get data that are close to an SRS of student cars.

NOTES AND DATA SOURCES

Page 584 Exercises IV.1, IV.2, IV.5, and IV.10 cite results from Gallup polls, found at **www.gallup.com.**

Page 585 Exercise IV.9: Matthew K. Wynia et al., "Physician manipulation of reimbursement rules for patients," *Journal of the American Medical Association,* 283 (2000), pp. 1858–1865.

Page 586 Exercise IV.11: Michael F. Weeks, Richard A. Kulka, and Stephanie A. Pierson, "Optimal call scheduling for a telephone survey," *Public Opinion Quarterly,* 51 (1987), pp. 540–549.

Page 587 Exercise IV.15: C. Kirk Hadaway, Penny Long Marler, and Mark Chaves, "What the polls don't show: a closer look at U.S. church attendance," *American Sociological Review,* 58 (1993), pp. 741–752. There are also many comments and rebuttals about measuring religious attendance in *American Sociological Review,* 63 (1998). Recent estimates of church attendance can be found at the Gallup Poll Web site, **www.gallup.com.** At the time of this writing, the most recent results were for 2010.

Page 588 Exercise IV.22: Charles W. L. Hill and Phillip Phan, "CEO tenure as a determinant of CEO pay," *Academy of Management Journal,* 34 (1991), pp. 707–717.

Page 588 Exercise IV.24: A. R. Hirsch and L. H. Johnston, "Odors and learning," *Journal of Neurological and Orthopedic Medicine and Surgery,* 17 (1996), pp. 119–126. We found the data in a case study in the Electronic Encyclopedia of Statistical Examples and Exercises (EESEE), available at **whfreeman.com/catalog/static/whf/eesee.**

Page 589 The shark data in Exercise IV.25 were provided by Chris Olsen, who found the information in scuba-diving magazines.

Page 589 We found the example in Exercise IV.28 in Howard Wainer's "Visual revelations" column, *Chance,* 12, No. 2 (1999), pp. 43–44. The paradox continued to be true in 2011. NAEP scores can be found at **nces.ed.gov/nationsreportcard/states/.**

Page 590 Exercise IV.29: Sara J. Solnick and David Hemenway, "Complaints and disenrollment at a health maintenance organization," *Journal of Consumer Affairs,* 26 (1992), pp. 90–103.

Page 590 Exercise IV.30: L. L. Miao, "Gastric freezing: an example of the evaluation of medical therapy by randomized clinical trials," in J. P. Bunker, B. A. Barnes, and F. Mosteller (eds.), *Costs, Risks and Benefits of Surgery,* Oxford University Press, 1977, pp. 198–211.

Note to students: These are controversies, and not everyone will agree how best to resolve them. Our answers include the major issues that should be a part of any discussion.

Chapter 3: Should Election Polls Be Banned?

Arguments *against* public preelection polls charge that they influence voter behavior. Voters may decide to stay home if the polls predict a landslide—why bother to vote if the result is a foregone conclusion? Exit polls are particularly worrisome, because they in effect report actual election results before the election is complete. The U.S. television networks agree not to release the results of their exit surveys in any state until the polls close in that state. If a presidential election is not close, the networks may know (or think they know) the winner by midafternoon, but they forecast the vote only one state at a time as the polls close across the country. Even so, a presidential election result may be known (or thought to be known) before voting ends in the western states. Some countries have laws restricting election forecasts. In France, no poll results can be published in the week before a presidential election. Canada forbids poll results in the 72 hours before federal elections. In all, some 30 countries restrict publication of election surveys.

The argument *for* preelection polls is simple: democracies should not forbid publication of information. Voters can decide for themselves how to use the information. After all, supporters of a candidate who is far behind know that fact even without polls. Restricting publication of polls just invites abuses. In France, candidates continue to take private polls (less reliable than the public polls) in the week before the election. They then leak the results to reporters in the hope of influencing press reports.

One argument *for* exit polls is that they provide a means for checking election outcomes. Discrepancies between exit polls and reported election outcomes invite investigation into the reasons for the differences. Such was the case in the 2004 presidential election. Were the exit polls flawed or were the reported election results in error?

Chapter 4: The Harris Poll Online

The Harris Poll Online uses probability sampling and statistical methods to weight responses, and uses recruitment to attempt to create a panel (sampling frame) that is as representative as possible. But the panel also consists of volunteers and hence suffers to some extent from voluntary response. In addition, panel members are Internet users, and it is not clear that such a panel can be representative of a larger population that includes those who do not use the Internet.

As Crouper points out in his *Public Opinion Quarterly* paper, "it is not the fact that a very large panel of volunteers is being used to collect systematic information on a variety of topics that is of concern, but the fact that the proponents of this approach are making claims that these panels are equal to or better than other forms of survey data collection based on probability sampling methods (especially RDD [random digit dialing] surveys). The claim goes beyond saying that these panels are representative of the Internet population to claiming that they are representative of the general population of the United States. These assertions rest on the efficacy of weighting methods to correct deficiencies in sampling frames constituted by volunteers. We need thorough, open, empirical evaluation of these methods to establish their validity."

Thus, the verdict is out on whether the Harris Poll Online provides accurate information about well-defined populations such as all American adults.

Chapter 6: Is It or Isn't It a Placebo?

Should the FDA require natural remedies to meet the same standards as prescription drugs? That's hard to do in practice, because natural substances can't be patented. Drug companies spend millions of dollars on clinical trials because they can patent the drugs that prove effective. Nobody can patent an herb, so nobody has a financial incentive to pay for a clinical trial. Don't look for big changes in the regulations.

Meanwhile, it's easy to find claims that ginkgo biloba is good for (as one Web site says) "hearing and vision problems as well as impotence, edema, varicose veins, leg ulcers, and strokes." Common sense says you should be suspicious of claims that a substance is good for lots of possibly unrelated conditions. Statistical sense says you should be suspicious of claims not backed by comparative experiments. Many untested remedies are no doubt just placebos. Yet they may have real effects in many people—the placebo effect is strong. Just remember that the safety of these concoctions is also untested.

Chapter 7: Hope for Sale?

One issue to consider is whether BMT really keeps patients alive longer than standard treatments. We don't know, but the answer appears to be "probably not." The patients naturally want to try anything that might keep them alive, and some doctors are willing to offer hope not backed by good evidence. One problem was that patients would not join controlled trials that might assign them to standard treatments rather than to BMT. Results from such trials were delayed for years by the difficulty in recruiting subjects. Of the first five trials reported, four found no significant difference between BMT and standard treatments. The fifth favored BMT—but the researcher soon admitted "a serious breach of scientific honesty and integrity." The *New York Times* put it more bluntly: "he falsified data."

Another issue is "smart" compassion. Compassion seems to support making untested treatments available to dying patients. Reason responds that this opens the door to sellers of hope and delays development of treatments that really work. Compare children's cancer, where doctors agree not to offer experimental treatments outside controlled trials. Result: 60% of all children with cancer are in clinical trials, and progress in saving lives has been much faster for children than for adults. BMT for a rare cancer in children was tested immediately and found to be effective. In contrast, one of the pioneers in using BMT for breast cancer, in the light of better evidence, now says, "We deceived ourselves and we deceived our patients."

Chapter 8: SAT Exams in College Admissions

We see that *SAT scores predict college grades about as well as high school grades do.* Combining SAT scores and college grades does a better job than either by itself. The predictions are actually a bit better for private institutions than for public institutions. We also see that *neither SAT scores nor high school grades predict college grades very well.* Students with the same grades and SAT scores often perform quite differently in college. Motivation and study habits matter a lot.

Selective colleges are justified in paying some attention to SAT scores, but they are also justified in looking beyond SAT scores for the motivation that can bring success to students with weaker academic preparation. The SAT debate is not really about the numbers. It is about how colleges should use all the information they have in deciding whom to admit, and also about the goals colleges should have in forming their entering classes.

	All institutions	Private institutions	Public institutions
SAT	28%	32%	27%
School grades	29%	30%	28%
Both together	38%	42%	37%

Chapter 12: Income Inequality

These are complicated issues, with much room for conflicting data and hidden agendas. The political left wants to reduce inequality, and the political right says the rich earn their high incomes. We want to point to just one important statistical twist. Figures 12.4 and 12.5 report "cross-sectional" data that give a snapshot of households in each year. "Longitudinal" data that follow households over time might paint a different picture. Consider a young married couple, Jamal and Tonya. As students they work part-time, then borrow to go to graduate school. They are down in the bottom fifth. When they get out of school, their income grows quickly. By age 40, they are happily in the top fifth. In other words, many poor households are only temporarily poor.

Longitudinal studies are expensive because they must follow the same households for years. They are prone to bias because some households drop out over time. One study of income tax returns found that only 14% of the bottom fifth were still in the bottom fifth 10 years later. But really poor people don't file tax returns. Another study looked at children under 5 years old. Starting in both 1971 and 1981, it found that 60% of children who lived in households in the bottom fifth still lived in bottom-fifth households 10 years later. Many people do move from poor to rich as they grow older, but there are also many households that stay poor for years. Unfortunately, many children live in these households.

Chapter 15: Gun Control and Crime

Is Lott right? We don't know. His work is more sophisticated than most older studies cited to support gun control. Yet large observational studies have many potential weaknesses, especially when they look for trends over time. Lots of things happen in 18 years, not all of which are in Lott's model—for example, police have become more aggressive in seizing illegal guns. Good data are hard to come by—for example, the number of people carrying guns legally is lower than the number of permits issued and is hard to estimate accurately. It would take very detailed study to reach an informed opinion on Lott's statistics.

The best reason to question Lott's findings combines awareness of the weaknesses of observational studies with the fact that statistical studies with stronger designs support reducing the presence of guns. Temporary bans on carrying guns in several cities in Colombia—highly publicized, and enforced by police checkpoints and searches—reduced murder rates. The Kansas City Gun Experiment compared two high-crime areas. In one, police seized guns by searches during traffic stops and after minor offenses. Gun crimes dropped by half in the treatment area and were unchanged in the control area. There seems good reason to think that reducing *illegal* carrying of guns reduces gun crime. Lott, of course, argues for *legal* carrying. This distinction between legal and illegal carrying of guns makes it possible for Lott and some of his critics to both be right. Lower illegal gun carrying may reduce crime. Higher legal gun carrying could also reduce crime. Like many questions of causation, this one remains open.

Chapter 16: Does the CPI Overstate Inflation?

The CPI has an upward bias because it can't track shifts from beef to tofu and back as consumers try to get the same quality of life from whatever products are cheaper this month. This was the basis of the outside experts' criticisms of the CPI: the CPI does not track the "cost of living." Their first recommendation was that "the BLS should establish a cost of living index as its objective in measuring consumer prices." The BLS said it agreed in principle but that neither it nor anyone else knows how to do this in practice. It also said, "Measurement of changes in 'quality of life' may require too many subjective judgments to furnish an acceptable basis for adjusting the CPI." Nonetheless, a new kind of index that in principle comes closer to measuring changes in the cost of living was created in 2002. This new index is called the Chained CPI-U (C-CPI-U). It more closely approximates a cost-of-living index by reflecting substitution among item categories. This new index may be an improvement, but it is unlikely that the difficult problems of defining living standards and measuring changes in the cost of their attainment over time will ever be resolved completely.

Chapter 20: The State of Legalized Gambling

Opponents of gambling have good arguments against legalized gambling. Some people find betting addictive. A study by the National Opinion Research Center estimated that pathological gamblers account for 15% of gambling revenue, and that each such person costs the rest of us $12,000 over his lifetime for social and police work. Gambling does ruin some lives, and it does indirectly harm others.

State-run lotteries involve governments in trying to persuade their citizens to gamble. In the early days of the New York lottery, we recall billboards that said, "Support education—play the lottery." That didn't work, and the ads quickly changed to "Get rich—play the lottery." Lotteries typically pay out only about half the money bet, so they are a lousy way to get rich even when compared with the slots at the local casino. Professional gamblers and statisticians avoid them, not wanting to waste money on so bad a bargain. Poor people spend a larger proportion of their income on lotteries than do the rich and are the main players of daily numbers games. The lottery may be a voluntary tax, but it hits the poor hardest, and states spend hundreds of millions on advertising to persuade the poor to lose yet more money. Some modest suggestions from those who are concerned about state-run lotteries: states should cut out the advertising and pay out more of what is bet.

States license casinos because they pay taxes and attract tourists—and of course because many citizens want them. In fact, most casinos outside Las Vegas draw gamblers mainly from nearby areas. Crime is higher in counties with casinos—but lots of lurking variables may explain this association. Pathological gamblers do have high rates of arrest, but again the causal link is not clear.

The debate continues. Meanwhile, technology in the form of Internet gambling is bypassing governments and creating a new gambling economy that makes many of the old arguments outdated.

Chapter 23: Should Significance Tests Be Banned?

The quotation from Robert Rosenthal should shock you. "Always plot your data" has been one of our mottoes, along with "Always ask where the data come from." Psychologists generally think carefully about how their data are produced.

In response, the American Psychological Association appointed a Task Force on Statistical Inference. Robert Rosenthal was one of the cochairs of this group. The Task Force did not want to ban tests. Its report was in fact a summary of good statistical practice. Define your population clearly. Describe your data production and prefer randomized methods whenever possible. Describe your variables and how they were measured. Give your sample size and explain how you decided on the sample size. If there were dropouts or other practical problems, mention them. "As soon as you have collected your data, before you compute *any* statistics, *look at your data*." Ask whether the results of computations make sense to you. Recognize that "inferring causality from nonrandomized designs is a risky enterprise."

There is more, but here is the punch line about statistical tests: "It is hard to imagine a situation in which a dichotomous accept-reject decision is better than reporting an actual p value or, better still, a confidence interval.... Always provide some effect-size estimate when reporting a p value." Ban tests? "Although this might eliminate some abuses, the committee thought there were enough counterexamples to justify forbearance."

Solutions to "Now It's Your Turn" Exercises

Chapter 1

1.1 *Population:* The population is not explicitly defined. From the context of the problem, we assume it to be all adult Americans. However, as a phone survey, the population might be more appropriately defined as adult Americans with a telephone.

Sample: The sample is the 1024 randomly selected adults who were called in this research poll.

1.2 This is an observational study. The researcher examined student comments, but no treatment was applied to the students.

Chapter 2

2.1 This is not a simple random sample. Not every possible group of 4 students can be selected. For example, 4 students sitting in the same row can never be selected.

2.2 Step 1: Label. For the 20 teaching assistants (TAs), we use labels

$$01, 02, 03, \ldots, 18, 19, 20$$

Specifically, the list of TAs with labels attached is

01	Dobmeier	11	Pflugheisen
02	Joseph	12	Sanders
03	Kil	13	Snyder
04	Kohlschmidt	14	Sonksen
05	Koster	15	Spade
06	Landgraf	16	Springer
07	Leatherman	17	Stagner
08	Martin	18	Stettler
09	Mazzeo	19	Tam
10	Pearl	20	Tirmenstein

Step 2: Software or table. We used the Research Randomizer and requested that it generate one set of numbers with three numbers per set. We specified the number range as 1 to 20. We requested that each number remain unique and that the numbers be sorted least to greatest. We asked to view the outputted numbers with the markers off. After clicking on the "Randomize Now!" button, we obtained the digits 1, 5, and 14. (Of course, when you use the Research Randomizer, you will very likely get a different set of three numbers.) The sample is the TAs labeled 01, 05, and 14. These are Dobmeier, Koster, and Sonksen.

To use the table of random digits, we might enter Table A at line 116 (any line may be used), which is

$$14459 \quad 26056 \quad 31424 \quad 80371 \quad 65103 \quad 62253 \quad 50490 \quad 61181$$

The first 13 two-digit groups in this line are

$$14 \quad 45 \quad 92 \quad 60 \quad 56 \quad 31 \quad 42 \quad 48 \quad 03 \quad 71 \quad 65 \quad 10 \quad 36$$

We used only labels 01 to 20, so we ignore all other two-digit groups. The first 3 labels between 01 and 20 that we encounter in the table choose our sample. Of the first 13 labels in line 116, we ignore 10 of them because they are too high (over 20). The others are 14, 03, and 10. The TAs labeled 14, 03, and 10 go into the sample. The sample is the TAs labeled 03, 10, and 14. These are Kil, Pearl, and Sonksen.

Chapter 3

3.1 Recall our quick method for finding the margin of error for 95% confidence: $1/\sqrt{n}$. Here with $n = 1077$, the margin of error is

$$\frac{1}{\sqrt{1077}} = \frac{1}{32.82} = 0.030$$

3.2 Now, $n = 4000$, so the margin of error for 95% confidence is

$$\frac{1}{\sqrt{4000}} = \frac{1}{63.24} = 0.016$$

which is smaller than for the smaller sample ($n = 1077$).

Chapter 4

4.1 The question is clearly slanted toward a positive (Yes) response because the question asks the respondent to consider "escalating environmental degradation and incipient resource depletion."

4.2 Label faculty 0, 1, 2, 3, 4. Label students 0, 1, ..., 9. Starting at line 111 in Table A, we choose the person labeled 1 as the faculty member and the person labeled 4 as the student.

Chapter 5

5.1

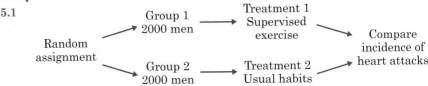

Chapter 6

6.1 There are two explanatory variables. These are baking temperature (300° F, 320° F, and 340° F) and baking time (1 hour and 1 hour and 15 minutes). The response variables are the scores of the panel of tasters for texture and taste. For each cake there is a score for texture and a score for taste from each taster.

There are three baking temperatures and two baking times, so there are a total of six treatments (the six baking temperature and baking time combinations). Ten cakes are baked at each of these six treatments, so 60 cakes are needed. Here is a diagram describing the treatments.

	Baking Temperature		
	300°	320°	340°
Baking Time 1 hour	Treatment 1	Treatment 2	Treatment 3
1 hour, 15 minutes	Treatment 4	Treatment 5	Treatment 6

6.2 In this experiment, the instructors are the blocks, for which the 75 students in each class are split randomly into three groups of 25, each receiving a different version of the test. Exam scores would then be compared as the response variable. Here is a sample diagram:

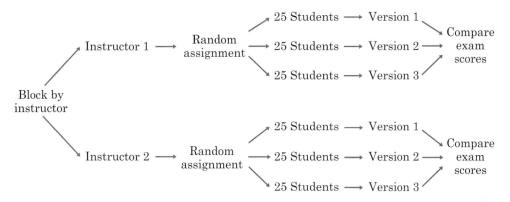

Chapter 7

7.1 This is a complicated situation. This patient's underlying disease appears to be impairing his decision-making capacity. If his wishes are consistent during his lucid periods, this choice may be considered his real preference and followed accordingly. However, as his decision-making capacity is questionable, family members should be contacted about the procedure. Getting a surrogate decision maker involved might help determine what his real wishes are. This example is based on one from the Ethics in Medicine Web site of the University of Washington School of Medicine: **depts.washington.edu/ bioethx.**

7.2 Recall the definition of "anonymity": subjects' names are not known even to the director of the study. Here, the clinic makes no promises about keeping test results secure (and hence does not offer confidentiality), although the procedure offers anonymity (test subjects receive a code to match with their results).

Chapter 8

8.1 The number of drivers is usually much greater between 5 and 6 P.M. (rush hour) than between 1 and 2 P.M. Thus, we would expect the number of accidents to be greater between 5 and 6 P.M. than between 1 and 2 P.M. It is therefore not surprising that the number of traffic accidents attributed to driver fatigue was greater between 5 and 6 P.M. than between 1 and 2 P.M. This is an example where the proportion of accidents attributed to driver fatigue is a more valid measure than the actual count of accidents.

8.2 The results are unreliable because the same restaurants receive drastically different ratings between 2009 and 2010.

Chapter 9

9.1 This is not plausible. From the information given, we can determine how many melons are produced per square foot:

$$\text{melons per sq. foot} = \frac{750{,}000 \text{ melons}}{\text{acre}} \times \frac{1 \text{ acre}}{43{,}560 \text{ sq. feet}} \approx 17.2 \text{ melons per sq. foot}$$

So, as stated, the field would need to produce more than 17 melons per square foot, which is quite unreasonable.

9.2 The percentage increase from the first quiz to the second quiz is

$$\text{percentage change} = \frac{\text{amount of change}}{\text{starting value}} \times 100$$

$$= \frac{10-5}{5} \times 100$$

$$= \frac{5}{5} \times 100 = 1.0 \times 100 = 100\%$$

However, the percentage decrease from the second to the third quiz is

$$\text{percentage change} = \frac{\text{amount of change}}{\text{starting value}} \times 100$$

$$= \frac{10-5}{10} \times 100$$

$$= \frac{5}{10} \times 100 = 0.5 \times 100 = 50\%$$

Chapter 10

10.1 The "state" variable is a categorical variable; for categorical variables, we should use either a bar graph or a pie chart. However, because the percentages do not add up to 100%, a bar graph is more appropriate.

10.2 The line graph is as follows:

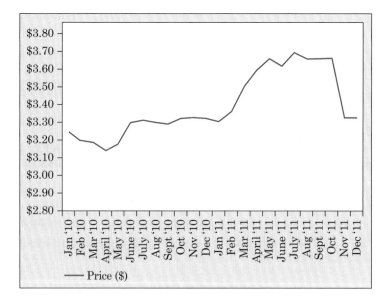

The price of milk was fairly consistent from January 2010 to January 2011 and then increased gradually from January 2011 to May 2011. The price stayed fairly constant again through October 2011 before dropping sharply to the 2010 prices in November 2011.

Chapter 11

11.1 Step 1: Divide the range of the data into classes of equal width. The data in the table range from 20.8 to 31.1, so we choose as our classes

$$20.0 \leq \text{percentage under } 18 < 21.0$$
$$21.0 \leq \text{percentage under } 18 < 22.0$$
$$\vdots$$
$$31.0 \leq \text{percentage under } 18 < 32.0$$

Be sure to specify the classes precisely so that each individual falls into exactly one class. A state with 20.9% of its residents under age 18 would fall into the first class, but 21.0% falls into the second.

Step 2: Count the number of individuals in each class. For example, there are 2 members in the first class, 3 members in the second class, and so on, up to 1 member in the final class.

Step 3: Draw the histogram. Mark on the horizontal axis the scale for the variable whose distribution you are displaying. That's "percentage of residents under age 18" here. The scale runs from 20 to 32 because that range spans the classes we chose. The vertical axis contains the scale of counts. Each bar represents a class. The base of the bar covers the class, and the bar height is the class count. There is no horizontal space between the bars unless a class is empty, so that its bar has height zero. The following figure is our histogram.

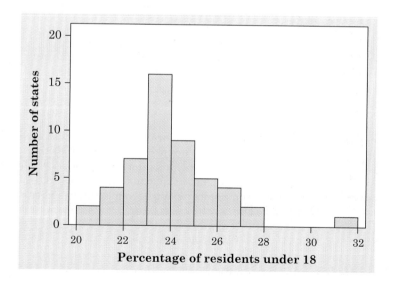

11.2 The distribution is mostly symmetric (perhaps right-skewed), with a middle near 2%. The data (ignoring the outlier) are spread between about 1.2% and 3.0%. There is one outlier (the state represented by the bar to the far left of the graph).

Chapter 12

12.1 There are 22 observations, and the median lies halfway between the middle pair (the 11th and 12th largest). The middle two values are 35 and 41, so the median is

$$M = \frac{35 + 41}{2} = \frac{76}{2} = 38$$

There are 11 observations to the left of the location of the median. The first quartile is the median of these 11 numbers and so is the 6th largest. That is,

$$Q_1 = 11$$

The third quartile is the median of the 11 observations to the right of the median's location:

$$Q_3 = 47$$

12.2 Here is the boxplot for Ruth:

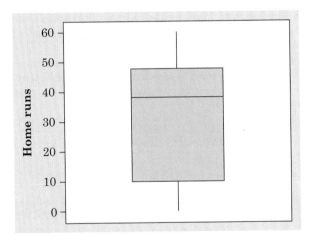

The median (38) and third quartile (47) for Ruth are slightly larger than for Bonds and Aaron. The distribution for Ruth appears more skewed (left-skewed) than for Bonds and Aaron. If one examines Ruth's career, one finds that he was a pitcher for his first six seasons and during those seasons did not have many plate appearances. Hence, he has six seasons of very low home run counts, resulting in a left-skewed distribution.

12.3 To find the mean,

$$\bar{x} = \frac{\text{sum of observations}}{n}$$
$$= \frac{13 + 27 + \cdots + 10}{23}$$
$$= \frac{755}{23} = 32.83$$

To find the standard deviation, use a table layout:

Observation	Squared distance from mean	
13	$(13 - 32.83)^2 = (-19.83)^2 =$	393.2289
27	$(27 - 32.83)^2 = (-5.83)^2 =$	33.9889
$\vdots$		
10	$(10 - 32.83)^2 = (-22.83)^2 =$	521.2089
	sum $=$	2751.3200

The variance is the average

$$\frac{2751.32}{22} = 125.06$$

and the standard deviation is

$$s = \sqrt{125.06} = 11.18$$

The mean is less than the median of 34. This is consistent with the fact that the distribution of Aaron's home runs is slightly left-skewed.

Chapter 13

13.1 The central 95% of any Normal distribution lies within two standard deviations of the mean. Two standard deviations is 5 inches here, so the middle 95% of young men's heights is between 65 inches (that's $70 - 5$) and 75 inches (that's $70 + 5$).

13.2 The standard score of a height of 72 inches is

$$\frac{72 - \text{(mean height)}}{\text{standard deviation of heights}} = \frac{72 - 70}{2.5} = \frac{2}{2.5} = 0.8$$

13.3 To fall in the top 25% of all scores requires a score at or above the 75th percentile. Look in the body of Table B for the percentiles closest to 75. You see that standard score 0.7 is the 75.80 percentile, which is the percentile in the table closest to 75. So, we conclude that a standard score of 0.7 is approximately the 75th percentile *of any Normal distribution*.

To go from the standard score back to the scale of SAT scores, "undo" the standard score calculation as follows:

$$\text{observation} = \text{mean} + \text{standard score} \times \text{standard deviation}$$
$$= 500 + (0.7)(100) = 570$$

A score of 570 or higher will be in the top 25%.

Chapter 14

14.1 The researchers are seeking to predict IQ from brain size. Thus, there is an explanatory variable. Brain size is the explanatory variable. The response variable is IQ. The following figure is a scatterplot of the data.

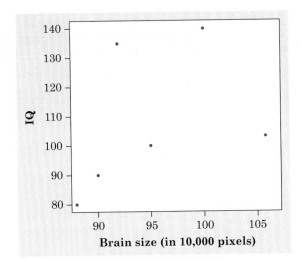

14.2 There is a weak positive association. There is no pronounced form other than evidence of a weak positive association. There are no outliers.

14.3 One might estimate the correlation to be about 0.3 or 0.4. The actual correlation is 0.37.

Chapter 15

15.1 The predicted humerus length for a fossil with a femur 70 cm long is

$$\text{humerus length} = -3.66 + (1.197)(70)$$
$$= 80.1\,\text{cm}$$

15.2 The proportion of variation in hot dog prices explained by the least-squares regression of hot dog prices on 16-ounce soda prices is $r^2 = (0.45)^2 = 0.2025$.

15.3 The observed relationship is certainly not direct causation. Some confounding is possible (food prices may be somewhat standardized at a baseball stadium), but the correlation is most likely due to common response: prices for food and drink probably depend on general economic trends.

Chapter 16

16.1 The CPI for 1984 is 103.9, and the CPI for 2011 is 224.9. So the 1984 median salary in 2011 dollars is

$$2011\text{ dollars} = \$229{,}750 \times \frac{224.9}{103.9} = \$497{,}313$$

16.2 The CPI for 1984 is 103.9, and the CPI for 2010 is 218.1. So the 1984 earnings in 2010 dollars is

$$2010\text{ dollars} = \$17{,}281 \times \frac{218.1}{103.9} = \$36{,}273$$

In terms of real earnings, the change between 1984 and 2010 is

$$\frac{\text{current earnings} - \text{past earnings}}{\text{past earnings}} = \frac{\$28{,}701 - \$36{,}273}{\$36{,}273} = -20.9\%$$

That is, the real earnings have decreased by 20.9%.

Chapter 17

17.1 All the outcomes are equally probable if the coin is fair because heads and tails are equally likely. Thus, all sequences of 10 particular outcomes are equally likely.

17.2 A correct statement might be "If you tossed a coin a billion times, you could predict a nearly equal *proportion* of heads and tails."

Chapter 18

18.1 Let a pair of numbers represent the number of spots on the up-faces of the first and second die, respectively. The probability of rolling a 7 is

$$P(\text{roll a 7}) = P(1, 6) + P(2, 5) + P(3, 4) + P(4, 3) + P(5, 2) + P(6, 1)$$
$$= \frac{1}{36} + \frac{1}{36} + \frac{1}{36} + \frac{1}{36} + \frac{1}{36} + \frac{1}{36}$$
$$= \frac{6}{36} = 0.167$$

The probability of rolling an 11 is

$$P(\text{roll an 11}) = P(5, 6) + P(6, 5)$$
$$= \frac{1}{36} + \frac{1}{36}$$
$$= \frac{2}{36} = 0.056$$

By Rule D, the probability of rolling a 7 or an 11 is

$$P(\text{roll a 7 or an 11}) = P(\text{roll a 7}) + P(\text{roll an 11})$$
$$= \frac{6}{36} + \frac{2}{36}$$
$$= \frac{8}{36} = 0.222$$

18.2 54.4% is two standard deviations above the mean of 50%. The 68–95–99.7 rule tells us that 5% will be more than two standard deviations away from the mean. Half of 5%, namely 2.5%, will be more than two standard deviations above the mean. Thus, the probability that more than 54.4% say "Yes" is 0.025.

Chapter 19

19.1 In a standard deck 25% of the cards are spades, 25% are hearts, 25% are diamonds, and 25% are clubs. We need two digits to simulate one draw:

$$00, 01, 02, \ldots, 24 = \text{spades}$$
$$25, 26, 27, \ldots, 49 = \text{hearts}$$
$$50, 51, 52, \ldots, 74 = \text{diamonds}$$
$$75, 76, 77, \ldots, 99 = \text{clubs}$$

19.2 Step 1. The first card selected can be either a spade, heart, diamond, or club. For each possibility for the first card, the second card can be either a spade, heart, diamond, or club, but the number of spades, hearts, diamonds, and clubs left depends on the suit of the first card selected (there are only 12 of that suit and 13 of the other suits left).

Step 2. *The assignment of probabilities to the first card selected:* 01 to 13 = spade, 14 to 26 = heart, 27 to 39 = diamond, 40 to 52 = club. Skip any other digits.
  The assignment of probabilities to the second card selected: If the first card selected is a spade, then use 01 to 12 = spade, 13 to 25 = heart, 26 to 38 = diamond, 39 to 51 = club. Skip any other digits. If the first card selected is a heart, then use 01 to 13 = spade, 14 to 25 = heart, 26 to 38 = diamond, 39 to 51 = club. Skip any other digits. If the first card selected is a diamond, then use 01 to 13 = spade, 14 to 26 = heart, 27 to 38 = diamond, 39 to 51 = club. Skip any other digits. If the first card selected is a club, then use 01 to 13 = spade, 14 to 26 = heart, 27 to 39 = diamond, 40 to 51 = club. Skip any other digits.

Step 3. The 10 repetitions, starting at line 115 in Table A, gave spade, heart; diamond, heart; club, spade; club, heart; diamond, diamond; heart, heart; club, heart; club, club; spade, spade; and diamond, heart. We got the same suit 4 out of 10 times, so we estimate the probability to be 4/10.

Chapter 20

20.1 The expected value is

$$\text{expected value} = 0.55(0) + 0.19(1) + 0.17(2) + 0.07(3) + 0.02(4)$$

$$= 0.82 \text{ children}$$

20.2 If we expect Kobe to make 45% of his field goals, he must shoot three shots before we expect him to make one field goal (because $0.45(2) = 0.90 < 1$ and $0.45(3) = 1.35 > 1$). To simulate Kobe's field-goal shooting, assign numbers 00 to 44 as makes and 45 to 99 as misses. Starting at line 101 of Table A, Kobe makes his first shot on shot 1, 1, 1, 2, 1, 4, 2, 2, 1, 1. The expected value is then $0.6(1) + 0.3(2) + 0.1(4) = 1.6$.

Chapter 21

21.1 The 95% confidence interval for the proportion of all adult Americans who believe gambling is morally wrong is

$$\hat{p} \pm 2\sqrt{\frac{\hat{p}(1 - \hat{p})}{n}} = 0.31 \pm 2\sqrt{\frac{0.31 \times 0.69}{1018}}$$

$$= 0.31 \pm 2(0.0145)$$

$$= 0.31 \pm 0.029$$

$$= 0.281 \text{ to } 0.339$$

Interpret this result as follows: We are 95% confident that the true proportion of adult Americans who believe gambling is morally wrong is between 28.1% and 33.9%.

21.2 For a 99% confidence interval, we use $z^* = 2.58$. The 99% confidence interval for the proportion of all adult Americans who believe gambling is morally wrong is

$$\hat{p} \pm 2.58\sqrt{\frac{\hat{p}(1 - \hat{p})}{n}} = 0.31 \pm 2.58\sqrt{\frac{0.31 \times 0.69}{1018}}$$

$$= 0.31 \pm 2.58(0.0145)$$

$$= 0.31 \pm 0.037$$
$$= 0.273 \text{ to } 0.347$$

Interpret this result as follows: We are 99% confident that the true proportion of adult Americans who believe gambling is morally wrong is between 27.3% and 34.7%. Note that this interval is wider than the 95% confidence interval in the previous exercise.

21.3 The 95% confidence interval for μ uses the critical value $z^* = 1.96$ from Table 21.1. The interval is

$$\bar{x} \pm z^* \frac{s}{\sqrt{n}} = 126.1 \pm 1.96 \frac{15.2}{\sqrt{72}}$$

$$= 126.1 \pm (1.96)(1.79) = 126.1 \pm 3.5$$

We are 95% confident that the mean for all executives in the company between the ages of 35 and 44 lies between 122.6 and 129.6.

Chapter 22

22.1 The hypotheses. The null hypothesis says that the coin is balanced ($p = 0.5$). We do not suspect a bias in a specific direction before we see the data, so the alternative hypothesis is just "the coin is not balanced." The two hypotheses are

$$H_0: p = 0.5$$
$$H_a: p \neq 0.5$$

The sampling distribution. *If the null hypothesis is true,* the sample proportion of heads has approximately the Normal distribution with

$$\text{mean} = p = 0.5$$

$$\text{standard deviation} = \sqrt{\frac{p(1-p)}{n}}$$

$$= \sqrt{\frac{0.5 \times 0.5}{50}}$$

$$= 0.0707$$

22.2 The data. The sample proportion is $\hat{p} = 0.42$. The standard score for this outcome is

$$\text{standard score} = \frac{\text{observation} - \text{mean}}{\text{standard deviation}}$$

$$= \frac{0.42 - 0.5}{0.0707}$$

$$= -1.13$$

The P-value. To use Table B, round the standard score to -1.1. This is approximately the 26th percentile of a Normal distribution. So the area to the left of -1.1 is approximately 0.26. The area to the left of -1.1 and to the right of 1.1 is double this, or 0.52. This is our approximate P-value.

Conclusion. The large P-value gives us no reason to think that the true proportion of heads differs from 0.5. The results are not significant at the 0.05 level.

22.3 The hypotheses. The null hypothesis is "no difference" from the population mean of 100. The alternative is two-sided, because we did not have a particular direction in mind before examining the data. So the hypotheses about the unknown mean μ of the middle school girls in this district are

$$H_0: \mu = 100$$
$$H_a: \mu \neq 100$$

The sampling distribution. *If the null hypothesis is true*, the sample mean $\bar{x}$ has approximately the Normal distribution with mean $\mu = 100$ and standard deviation

$$\frac{s}{\sqrt{n}} = \frac{14.3}{\sqrt{31}} = 2.57$$

The data. The sample mean is $\bar{x} = 105.8$. The standard score for this outcome is

$$\text{standard score} = \frac{\text{observation} - \text{mean}}{\text{standard deviation}}$$
$$= \frac{105.8 - 100}{2.57} = 2.26$$

The P-value. To use Table B, round the standard score to 2.3. This is the 98.93 percentile of a Normal distribution. So the area to the right of 2.3 is 0.0107. The area to the left of -2.3 and to the right of 2.3 is double this, or 0.0214. This is our approximate P-value.

Conclusion. The small P-value gives some reason to think that the mean IQ score for middle school girls in this district differs from 100.

Chapter 23

23.1 If the P-value is accurate, the results are indeed convincing. However, we would like to know both the sample size and the actual mean weight loss before deciding whether we find the results convincing. Better yet, we would like to know exactly how the study was conducted and to have the actual data. Unfortunately, in many research studies, it is not possible to get the actual data from researchers.

23.2 No. If all 122 null hypotheses of no difference are true, we would expect 1% of 122 (about 1) of these null hypotheses to be significant at the 1% level by chance. Because this is consistent with what was observed, it is not clear if chance explains the result of this study.

Chapter 24

24.1 The expected cell counts for the "Played games" row are 717.7, 453.1, and 208.2; for the "Never played games" row, 223.3, 140.9, and 64.8. For example, for the "Played games" and "A's and B's" cell, the expected count is

$$\text{expected count} = \frac{\text{row 1 total} \times \text{column 1 total}}{\text{table total}}$$
$$= \frac{1379 \times 941}{1808} = 717.7$$

24.2 The chi-square statistic is

$$\chi^2 = \sum \frac{(\text{observed count} - \text{expected count})^2}{\text{expected count}}$$

$$= \frac{(736 - 717.7)^2}{717.7} + \frac{(450 - 453.1)^2}{453.1} + \frac{(193 - 208.3)^2}{208.3}$$

$$+ \frac{(205 - 223.3)^2}{223.3} + \frac{(144 - 140.9)^2}{140.9} + \frac{(80 - 64.8)^2}{64.8} = 6.73$$

24.3 To assess the statistical significance we begin by noting that the two-way table has 2 rows and 3 columns. That is, $r = 2$ and $c = 3$. The chi-square statistic therefore has degrees of freedom

$$(r - 1)(c - 1) = (2 - 1)(3 - 1) = 1 \times 2 = 2$$

Look in the df $= 2$ row of Table 24.1. We see that $\chi^2 = 6.73$ is larger than the critical value 5.99 required for significance at the $\alpha = 0.05$ level but smaller than the critical value 9.21 for $\alpha = 0.01$. The study shows a significant relationship (P-value < 0.05) between playing games and grades.

The significance test says only that we have strong evidence of *some* association between playing games and grades. We must look at the two-way table to see the nature of the relationship: teenagers who play games appear to be more likely to get high grades. However, this is an observational study, and one must be careful about drawing cause-and-effect conclusions from it.

Answers to Odd-Numbered Exercises

Note to students: These answers should be considered guides or checkpoints, not complete solutions. Although explanations and details of computations are not given here, they should be part of a complete answer to an exercise. Solutions to Now It's Your Turn exercises are found on pages 597–609.

Chapter 1

1.3 **(a)** Motor vehicles. **(b)** Make and model; vehicle type; transmission type; number of cylinders; city mpg; highway mpg. The last three variables are numerical.

1.5 For example, whether a household uses its recycling bin.

1.7 The poll measured whether a person favored the death penalty for a person convicted of murder. Adult Americans are the population of interest. The sample consists of the 1005 adults interviewed.

1.9 **(a)** All adults, presumably all adult Americans. **(b)** Video adapter cables in each lot. **(c)** All U.S. households.

1.11 This is a sample survey. A sample (the volunteers) is used to get information about all teenagers.

1.13 **(a)** Doctors (rather than the researchers) chose the treatment for each patient. **(b)** Each patient should be randomly assigned to a treatment.

1.15 **(a)** This is not an experiment. Treatments (level of alcohol consumption) were not randomly assigned to the subjects (women aged 50 to 64). **(b)** Given the large size of the sample, we might assume that the population of interest is all women aged 50 to 64. Variables measured include level of alcohol consumption at two different times and other "detailed information," presumably age, cancer status, etc.

1.17 **(a)** An observational study. **(b)** A sample survey. **(c)** An experiment.

Chapter 2

2.3 Voluntary response samples generally misrepresent the population.

2.5 **(a)** The sample size for this poll is 740. **(b)** Voluntary response samples are often biased.

2.7 **(a)** The newspaper is likely interested in residents of West Lafayette. **(b)** The true proportion is almost certainly larger than the sample proportion of 14/98; the poll relied on voluntary response, and residents who dislike one-way streets are more likely to call in.

2.9 Possible answers: **(a)** A call-in poll. **(b)** Interviewing students as they enter the student center.

2.11 Voluntary response samples, such as phone-in polls, generally misrepresent the population. Random samples are more likely to be representative of the population.

2.13 The first four two-digit labels not exceeding 27 (the total number of names) are 04, 10, 17, and 19. So we choose Darneider, Katzfuss, Schlessman, and Sgambellone.

2.15 (a) The proportion of undergraduates and graduate students selected is the same. **(b)** Some samples of 400 students—for example, all 400 students selected are graduate students—are not possible.

2.17 The first three two-digit lables not exceeding 36 (the total number of names) and not repeating are 04, 11, and 19. So we choose Brooks Edge, Forest Creek, and Jefferson Commons.

2.19 Voluntary response samples, such as texting polls, generally misrepresent the population. Do you think every attendee is equally likely to text in their vote?

2.21 (a) Population: Hispanic residents of Denver. Sample: the adults interviewed at the 200 mailing addresses. **(b)** Respondents may not give truthful answers to a police officer.

Chapter 3

3.3 Statistic.

3.5 Parameter. Statistic.

3.7 (a) Choose 19, 22, 39, 50; $\hat{p} = 0.50$. **(b)** 0.5, 0.5, 0.75, 0.75, 1, 0.5, 0.75, 0, 0.75, 0.5. **(d)** Four samples have $\hat{p} = 0.50$.

3.9 (a) Population: Ontario residents. Sample: the 61,239 people interviewed. **(b)** Such a large sample should represent men and women fairly well.

3.11 51% would not be surprising, but 37% would be surprising.

3.15 It doesn't reduce bias. The margin of error is half as big with the larger sample.

3.17 (a) No. Since the population is at least 100 times larger than the sample, the variability of the sample proportion from a random sample will not depend on the population size, only on the sample size. **(b)** Yes. As in part (a), the variability depends only on sample size, not population size. Here, the sample sizes vary from state to state.

3.19 We used a sampling method that gives results this close to the truth 95% of the time.

3.21 No. 95% refers to how often the sampling method will give a result this close to the truth.

3.23 0.0428 (about 4.3%).

3.25 0.004 (about 0.4%).

3.27 We are 95% confident that between 45% and 49% of all adults favor permitting abortion under "some [but not all] circumstances."

3.29 4020.

3.31 The margin of error for the 95% interval would be larger: a statement with higher confidence must include more values.

3.33 The margin of error of ±9 percentage points belongs to the poll of business executives.

Chapter 4

4.3 (a) Nonsampling error. **(b)** Nonsampling error. **(c)** Sampling error.

4.5 Nonsampling error, such as wrong recording of poll responses.

4.7 This is (random) sampling error, which is accounted for in the margin of error.

4.9 The second question uses more extreme language ("catastrophic") when referring to the effects of ignoring the environment.

4.11 The first question likely elicited the 57% response.

4.15 The survey was carried out by the *New York Times*/CBS News. The sample was randomly chosen from the population of adult U.S. residents and included 1650 people, contacted by phone between December 5 and 10 (the year is not provided). Neither the response rate nor the exact questions asked are given.

4.17 (a) The final result is most accurate, because anonymity is likely to promote honesty. **(b)** Sexual behavior, medical history.

4.19 Each student has a 1-in-10 chance of being interviewed, but each sample must contain exactly 2 students over 21 years of age and 4 students under 21.

4.21 (a) Assign 0000, 0001, ..., 1999 for men and 000, 001, ..., 799 for women, then use the table twice. The first five men are 1387, 0529, 0908, 1369, and 0815. The first five women are 072, 710, 256, 027, 558. **(b)** Men: 1/10. Women: 1/4.

4.23 The margin of error depends on the sample size.

4.25 (a) The sample was selected by a procedure involving several steps, perhaps similar to the Current Population Survey sample. **(b)** The strata are the 17 countries. **(c)** At least one of the stages in choosing the sample involved random selection.

4.27 Choose rooms 35, 75, 115, 155, and 195.

4.29 (a) Assign labels 00000 to 19999 and use random digits. **(b)** Randomly select one of the first 100 and then select the students who are 100, 200, 300, ... places down the list. **(c)** Choose SRSs of size 50 and 150 from each group.

4.31 Mall interviews do not sample randomly from any (useful) population; they only represent people who shop at the mall.

Chapter 5

5.3 This is not the result of an experiment, because the number of students taking the exam was not determined by the researchers. Explanatory variable: number of students taking the exam. Response variable: average SAT Math score.

5.5 Shared activities that might be associated with weight gain are possible lurking variables.

5.7 Designing the experiment in this way will likely confound many lurking variables with conceptual understanding of the moon's phases (for example, if the students at the two universities differ in some systematic way, such as educational background).

5.9 (a) A better design would be to randomize the students from each university into two treatment groups. The response variable might be the score on some unbiased assessment or test. **(b)** *Hint:* Write down the names of all the students from each university that are eligible for participation in the study, then use the table of random digits to select the sample of 20 students (see Chapter 4 methods).

5.11 (a) Explanatory variables: the group of vitamins taken. Response variable: whether the subject developed colon cancer after four years. **(b)** Randomly assign subjects to the four treatments, with 216 people in each group. Compare the rates of colon cancer in the groups after four years. **(c)** Label subjects $000, 001, \ldots, 863$. The first five subjects selected are 731, 253, 304, 470, 296. **(d)** The observed effect could be plausibly attributed to chance. **(e)** One lurking variable might be that people who eat fruits and vegetables have healthier lifestyles.

5.13 This is not a comparative study and does not control for the effects of lurking variables.

5.15 (a) Randomly assign subjects to the four insurance plans. Compare the amount of medical care received and the health of the groups after some time period. **(b)** A practical (and ethical) issue is what to do with low-income subjects who cannot afford out-of-pocket costs. An ethical issue is requiring subjects to be in a treatment group that could have a negative effect on health.

5.17 Possible treatments: mandatory participation in a substance abuse program, suspending the driver's license, and mandatory jail time. Randomly assign those convicted of drunk driving to one of the three treatments. One possible response variable is to measure whether or not a subject is convicted of drunk driving again after a set time period. Compare the rates of second convictions for the three groups.

5.19 The observed effects could plausibly be explained by chance.

5.21 (a) One possible lurking variable is the reason the physicians chose the treatment they did for each patient. For example, did physicians tend to assign the sickest patients to a certain treatment?

5.23 Random assignment of subjects to treatments controls for the effects of lurking variables.

5.25 (a) For one of the assignments, start at line 123 in Table A. The following results are obtained. Diet A infants are numbered 08, 15, 07, 10, 18, 03, 01, 19, 06, and 11. **(b)** Four baby boys were assigned to Diet A in our randomization.

Chapter 6

6.3 "Double-blind" means that neither the subjects nor those who evaluate the subjects know who is getting which treatment; this method prevents the researchers' expectations from affecting the way in which the subjects' conditions are evaluated.

6.5 In this experiment, it would be impossible to blind patients because they would know which side of their face was treated. However, it is possible to blind the researchers responsible for evaluating the improvement in acne. This is what "single-blinded" means here.

6.7 For example, do the benefits arising from large doses of resveratrol given to the mice indicate that similar benefits would occur in humans receiving much lower doses, such as would occur by drinking one glass of wine per day?

6.9 (a) The placebo effect. **(b)** Use a three-treatment, completely randomized design. **(c)** No. To control the placebo effect, patients must not know which treatment they are receiving. **(d)** Since the patient, not the researchers, assesses the degree of pain, the experiment does not need to be double-blind.

6.11 A true placebo can provide genuine pain relief.

6.13 (a) Individuals: children aged 3 to 15 in 49 remote Hutterite communities. Response variable: percentage of colony members with the flu in the following winter. Explanatory variable: receipt of a flu shot or placebo. **(b)** The investigators may feel ethically obligated to compensate for the fact that one group of people received lesser health care. **(c)** The observed difference in the percentage of colony members who had the flu between the two groups is most likely due to the effect of the treatment, not to chance variation.

6.15 (a) Explanatory variables: the change in the price of the rug and source of information about the change. Response variable: subject's rating on the fairness of the change. **(b)** There are four treatments, with five subjects in each group. **(c)** Failing to randomize the order may result in the confounding of variables.

6.17 Completely randomized design: randomly assign 10 students to each treatment group; that is, give 10 students software that highlights pricing trends. Matched pairs design: instruct students to complete the experiment twice, once with the trending software, once without.

6.19 (b) If the progress of the cancer truly differs between the two groups, blocking subjects by sex can remove the presence of this effect when the four therapies are compared.

6.23 Randomly select a sample of taste testers from a group of willing consumers. Administer samples of a double cheeseburger from each restaurant to each subject, randomly assigning which sample is received first.

6.25 (a) Block variable. **(b)** Treatment variable.

Chapter 7

7.3 Answers will vary. Choice (a) seems to qualify as minimal risk; choices (c) and (d) are definitely greater than minimal risk.

7.7 Informed consent should be collected for (b) and (c).

7.11 This practice provides anonymity: names are neither recorded nor revealed.

7.13 Respondents should be notified if anonymity will not be maintained.

7.15 Answers vary; the experiment may be permissible in that it does not prevent participants from receiving medical treatment.

7.17 Answers vary; new drugs may be tested on volunteer subjects, as long as it can be verified that participants are aware of and can understand the risks involved.

7.25 It might be determined that the experiment places participants under a greater-than-normal level of distress.

Chapter 8

8.3 When the total population under consideration (people in the labor force) experiences such drastic changes, using the rate is a better measurement: a count does not account for the growing size of the possible number of people who could be employed.

8.5 (a) The rate of return is a much better measure of customer satisfaction when the population sizes are so different. **(b)** The rate for Sears is 36/1200, the rate for La Boutique Classique is 12/200.

8.7 If the 1:30 P.M. section is much larger than the 7:30 A.M. section, the rate of students who failed the course might actually be smaller for the afternoon course. The total number of students in each section would be required to compare rates. However, the failure rate may not be the best measure of difficulty; we know nothing about the students' abilities in each section.

8.9 The number of executions per million people for these states is as follows:

State	Executions per million people
Oklahoma	25.060
Texas	18.452
Delaware	15.590
Alabama	10.251
Arkansas	9.259
Nevada	4.443
Florida	3.670
Indiana	3.085

We can now see that Texas is still relatively high; Florida is relatively low.

8.11 One possibility is to collect data on the distance of each person's daily commute and the length of time it requires. A low distance-per-time rate might be considered better.

8.13 (1) Cancer deaths increase as the population becomes older. (2) Cancer death rates rise as the health of the population improves and fewer people die of other causes. (3) Survival times could increase if the disease is being discovered earlier; that is, if diagnosis (not treatment) becomes more effective.

8.15 The measurement will be unbiased and should be reliable, but may or may not be valid (we don't know any information about the accuracy of the machine).

8.17 **(a)** No bias means that the measurement process neither systematically overstates nor understates the true value of the property it measures. This implies that the errors will be neither systematically above nor systematically below zero if we measure many pieces of string. **(b)** Perfect reliability means identical guesses for repeated measurements on the same piece of string.

8.19 **(a)** The measurement errors are 0.1, 0, 0.2, and 0.1. **(b)** The variance is 0.0067.

8.21 For example, take $x = 1,000,000$ and $y = 1,500,000$, so that the difference is $y - x = 500,000$. Now, note that $1.03y - 0.97x = 575,000$, such that the difference changes by $\frac{575,000}{500,000} = 1.15$. Thus, the difference between x and y increases by 15% even though each number changed by only 3%.

8.23 The presence of other people would likely lead to underreporting of sexual assaults.

8.25 The reason for the policy change involves the validity of the IQ test as a measure of future job performance.

8.27 These facts are evidence that the *Forbes* ratings are unreliable.

Chapter 9

9.3 Friday through Sunday is 42.8% of the week.

9.5 Both Anacin and Bufferin are brands of aspirin. Bufferin also contains ingredients to prevent upset stomach and is likely specified for that reason.

9.7 For example, over 20 years, 150,000 suicides means an average of 20 suicides per day, which would be hard to ignore.

9.9 57% equals 11.4 studies reporting significant results, and 42% of these 11.4 would be 4.8 studies that agreed on one conclusion.

9.11 (a) 28 miles per driver per year. **(b)** 19,619 miles per driver per year. This seems very low.

9.13 You cannot have more than 100% of a quantity.

9.15 The percentage increase is 74.5%. The population also increased substantially between 1976 and 2009.

9.17 The number of sexual partners men have is highly related to the number of sexual partners women have.

9.19 The number of separate highways across the continent needed to account for 3.9 million miles is $3,900,000/3000 = 1300$.

9.21 Investigate the total number of men and women in the United States in their 40s.

9.23 Household income increased by 4.5% between 1990 and 2009; household income decreased by about 4.8% between 2000 and 2009.

9.25 The count of injuries is probably more accurate. The count of deaths may not increase due to improved health care or increased emergency services.

Part I Review

I.3 It is a convenience sample.

I.5 Label the names from 01 to 25 and then select 07, 20, 24, and 17.

I.7 Sampling error. No.

I.9 Only the error in Exercise I.8 would be reduced.

I.11 (a) Adult Americans. **(b)** The rule of thumb for margin of error is $1/\sqrt{n}$. Here, with $n = 1160$, the margin of error is $1/\sqrt{1160}$, or 0.03. We are 95% confident that between 31% and 37% of adult Americans give Tiger Woods a "favorable" rating.

I.13 Each student has a 1-in-10 chance of being interviewed, but each sample must contain exactly three over-21 students and two under-21 students. This is a stratified sample.

I.15 Randomly select 10 subjects to try each method and compare success rates.

I.17 *Subjects:* 12,064 survivors of myocardial infarction in secondary-care hospitals in the United Kingdom. *Explanatory variable:* the treatment (folic acid/B12 versus placebo). *Response variables:* incidence of major vascular events, incidence of deaths.

I.19 Institutional review board, informed consent, confidentiality.

I.21 Reliability means similar results in repeated measurements. Reliability can be improved by taking many measurements and averaging them.

I.23 (a) This is an observational study because treatments were not assigned. **(b)** The high-fitness group might be younger or healthier; alternatively, confident people may be more likely to exercise.

I.25 The index dropped by $115/1695$ (or 0.0678), a 6.8% drop.

Chapter 10

10.3 A pie chart could be used because we are displaying the distribution of a categorical variable and we have the total for all categories.

10.5 (a) The number of unmarried women is $34,963 + 11,306 + 13,762 = 60,031$. **(c)** A pie chart could be used because we are displaying the distribution of a categorical variable.

10.7 This is a pictogram and is misleadingly scaled.

10.9 (a) Within each year, there is both a high point and a low point. This is expected due to the growing season of oranges. **(b)** After accounting for the seasonal variation, there is a gradual positive trend in the price of oranges.

10.11 (a) This is a pictogram and is misleadingly scaled.

10.13 Adjust the scale and the maximum and minimum values on the vertical axis.

10.15 (a) A downward trend. **(b)** An upward trend. **(c)** No clear trend.

10.17 This is likely just seasonal variation (holiday sales).

10.19 The cycle is about 10 to 11 years. There might be a trend, in that the peaks appear to be higher at the middle of the 1900s and at the end of the 1900s.

10.21 This is a correct graph.

10.23 Use a line graph. The graph is most helpful in highlighting periods of time in which the number of deportable aliens caught by the border patrol is relatively high and relatively low.

10.25 There is overlap between the groups: adolescents could have two or more of these habits and would therefore be represented twice.

10.27 (a) For example,

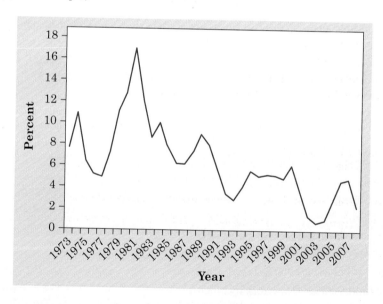

(b) From the line graph, we can see that the interest rate cycle reached temporary high points in 1974, 1981, 1985, 1989, 1995, 2000, and 2007. **(c)** The interest rates had an overall peak in 1981 for the years 1973 to 2008. Since the overall peak, the interest rates have gradually decreased (ignoring the cycles).

Chapter 11

11.3 Roughly symmetric, centered at or about noon, spread from 6:30 A.M. to 5:30 P.M. No outliers.

11.5 Strongly right-skewed; trailing off rapidly from the peak at 1 through 5. Spread is 1 to 50, with few universities awarding more than 15 doctorates to minorities.

11.7 A stemplot would have more information (too many digits) than can be easily absorbed.

11.9 The distribution is approximately symmetric and spread from 18.6% to 34.4% with a center around 26% to 27%.

11.11 **(a)** Strongly right-skewed (many short words, a few quite long words). The center is about 4 letters. The spread is from 1 to 15 letters. **(b)** Shakespeare uses more short words (especially 3 and 4 letters) and fewer very long words.

11.13 It will likely be roughly symmetric but perhaps slightly right-skewed (because some team owners have more money to spend).

11.15 The distribution is irregular in shape: there are two distinct groups, plus a low outlier (the veal brand, with 107 calories).

11.17 Roughly symmetric; 46 in a typical year; 60 is not an outlier.

11.19 Use a histogram or a stemplot. Distribution is right-skewed, with a peak between 10 and 15.

Chapter 12

12.5 Mean incomes are higher than median incomes because the distribution of incomes is skewed to the right. Family incomes are higher than household incomes because households will include persons living alone.

12.7 **(a)** The five-number summary is Minimum = $2000, Q_1 = $2600, Median = $9950, Q_3 = $22,100, Maximum = $38,600. **(b)** The plot is right-skewed; therefore, the mean will be larger than the median.

12.9 **(b)** The five-number summary is Minimum = 15, Q_1 = 24, Median = 26, Q_3 = 27, Maximum = 35. The cars in the bottom fourth of gas mileage are those less than Q_1: Acura RL, Bentley Azure, Jaguar XJ, Lexus GS460, Lincoln Town Car, Maybach 57, Mercedes-Benz E550, Rolls Royce Phantom. **(c)** Cars with gas mileages below 22 are outliers.

12.11 $1,137,680 is the mean and $558,726 is the median. Most income distributions will be right-skewed, and therefore, the mean will be larger than the median.

12.13 Because the distribution is roughly symmetric with no outliers, the mean and standard deviation are suitable.

12.15 Five-number summary: 1, 2, 4, 6, 15.

12.17 With Florida and New York included: mean = 211.8, median = 118.5. With Florida and New York removed: mean = 146.5, median = 98.7. The mean is changed more (as expected).

12.19 $2.36 million must be the mean. 205 players must make more than $2.36 million for it to be the median (unless there are 67 players who make exactly $2.36 million, which is unlikely).

12.21 The mean is more useful because from it we can compute the total number of cans we should have on hand.

12.23 **(a)** Five-number summary: 17, 19, 21, 23, 27. The shape is right-skewed. **(b)** SUVs generally have lower gas mileages than sedans, but there is more variability in the latter.

12.25 **(a)** $\bar{x} = 5.4$. **(b)** $s = 0.6419$.

12.27 **(a)** The new mean is $\bar{x} = 5$ (increased by 2), while $s = 2.19$ is unchanged. **(b)** Adding 10 to each observation would increase the mean by 10 but leave the standard deviation unchanged.

12.29 **(b)** 0, 0, 9, 9 (greatest spread). **(a)** 1, 1, 1, 1 (no spread) is one possible answer. **(c)** There is only 1 correct choice for (b). Any collection of equal numbers has no spread, so there are 10 correct choices for (a).

12.31 **(a)** The mean and median will increase by $1000. **(b)** Both quartiles will increase by $1000, so the distance between them is unchanged. **(c)** There is no change in the spread of the salaries, so the standard deviation is unchanged.

12.33 The relatively small number of top students pull the mean above the median. The median score is not affected much by the inclusion or exclusion of a few students with very high scores.

Chapter 13

13.5 **(a)** B is both the mean and median. **(b)** B is the mean and A the median. **(c)** A is both the mean and median. **(d)** A is the mean and B the median.

13.7 Between 100 and 122.

13.9 0.15%. This is not surprising, because 0.15% of 74 students is only 0.111 students.

13.11 The mean is 10, and the standard deviation is about 2.

13.13 The standard scores for Cobb, Williams, and Brett are 4.15, 4.26, and 4.07, respectively.

13.17 2.5%.

13.19 **(a)** About 50% of all $\hat{p}$-values will be above the mean of 0.47. About 0.15% of all $\hat{p}$-values will be above 0.50. **(b)** 0.45 to 0.49.

13.21 About 18.41%.

13.23 About 1.07%.

13.25 **(a)** About 2.5%. **(b)** About 24.2%.

13.27 **(a)** -23.1% to 48.1%. **(b)** About 0.242. **(c)** About 0.242.

13.29 About 68.25 inches or more.

Chapter 14

14.5 **(a)** Above-average values of length tend to go with above-average values of weight, and below-average values of length tend to go with below-average values of weight. **(b)** No change.

14.7 **(a)** The relationship is positive and roughly linear. **(b)** About 103 and 0.5. **(c)** A, B: low GPA, moderate IQ. C: low IQ, moderate GPA.

14.9 Clearly positive but not near 1.

14.11 Figure 14.11, which shows a stronger linear relationship.

14.13 **(a)** Time is the explanatory variable. **(b)** Negative; lower time requires greater exertion. **(c)** Linear and moderately strong.

14.15 $r = 1$.

14.17 **(a)** $r = -0.746$. **(b)** The correlation would not change.

14.19 The relationship is nonlinear, and correlation describes only *linear* relationships.

14.21 **(a)** Pounds. **(b)** Pounds. **(c)** No units. **(d)** Pounds.

14.23 **(a)** Car manufacturer is not a quantitative variable, so the correlation between it and anything else cannot be computed. **(b)** Correlation can't exceed 1. **(c)** Correlation has no units.

14.25 **(a)** Substantial positive. **(b)** Substantial negative. **(c)** Substantial negative. **(d)** Small.

14.27 Because the instructions specify that soda price should be viewed as explanatory, it should be on the horizontal axis. The relationship is weakly positive. There are no outliers.

14.29 **(a)** Planting rate. **(c)** The pattern is curved rather than linear, and there is neither a positive nor a negative association. Yield increases as you add more plants until they start to become too crowded.

14.31 Plot 1 will have less scatter in the vertical direction and hence a stronger correlation.

Chapter 15

15.5 **(a)** There is a negative association, and if the proportion taking the exam increases by 1 percentage point, we expect the mean SAT Math score to decrease by 109.7 points. **(b)** Approximately 482 points. **(c)** It is reasonable to use the least-squares regression line for proportion-taking values between about 0.05 and 0.95.

15.7 **(a)** When one variable is above average, the other is below average and vice versa. **(b)** Proportion taking the exam explains about $r^2 = 0.71$, or 71%, of the variation in average SAT Math scores. Proportion taking predicts average score moderately well.

15.9 We predict his pulse rate to be about 147.4 beats per minute, about 4.6 beats per minute lower than the actual value.

15.11 **(a)** There is a very strong positive, linear relationship and there is one outlier. **(c)** The regression line explains 94.7% of the variation in the 2004 percentages.

15.13 For 1 liter each year we predict 107.81 deaths per 100,000 people. For 8 liters each year we predict 51.46 deaths per 100,000 people.

15.15 For example, diets and genetic background.

15.17 **(a)** The association is negative; as time goes by, pH decreases. **(b)** Week 1: about 5.425. Week 150: about 4.635. **(c)** The slope is -0.0053; on average, the pH decreased by 0.0053 units per week. **(d)** No. Extrapolation is risky.

15.19 **(a)** Weight $y = 110 + 39x$; the slope is 39 g per week. **(c)** No. At 2 years (104 weeks) we predict a weight of 4166 g (about 9.2 pounds). This is unreasonable.

15.21 **(b)** Set A only.

15.23 (b) In 2020 we predict a population of about -11.2 million people. This is not reasonable.

15.25 The correlation is due to common response (the seriousness of the accident), not cause and effect.

15.27 It is not appropriate to conclude that the relationship is due to cause and effect. There are likely to be lurking variables responsible for the association, for example, houses closer to the freeway are likely to be less expensive which, in turn, may be associated with the resident's lifestyle.

15.29 For example, intelligence and the fact that stronger students (those more likely to succeed in college) are more likely to take these courses. It is incorrect to conclude that the relationship is due to cause and effect.

15.31 For example, children with parents who do not spend much time at home are more likely to spend more time on the Internet and less time on homework because their parents are not present to supervise their activities.

15.33 It is not appropriate to conclude that the relationship is due to cause and effect. Lurking variables might explain the relationship. Students with music experience might have other advantages (wealthier parents, better school systems, etc.).

15.35 The slope is $b = 0.994 \left(\frac{15.89}{13.20} \right) = 1.1966$, which rounds to 1.197, and the intercept is $a = 66 - 58.2\,(1.1966) = -3.642$, which differs slightly from -3.66. The difference is due to round-off in the values for the means, standard deviations, and correlation reported in Example 4.

Chapter 16

16.3 Index number (1991) = 77.46; index number (2001) = 100; index number (2011) = 247.89.

16.5 (a) 170.43 points and 220%. **(b)** 147.89 points and 147.89%.

16.7 (b) The overall trend is increasing, and there were no years in which this trend was reversed. **(c)** Prices were rising fastest between 1972 and 1982; rising slowest between 1992 and 2010.

16.9 Index number (2011) = 133.3.

16.11 In terms of 2004 dollars, $1,180,625.

16.13 In terms of 2011 dollars, $47,418.67.

16.15 In terms of 2009 dollars, Ben Hogan won $163,250 and Jack Nicklaus won $1,004,410.

16.17 In terms of December 2011 dollars, a call in 1976 cost $47.60. The percentage decrease in real cost is 70.4%.

16.19 The adjusted minimum wage figures are $1.00, $1.17, $1.22, $1.16, $1.11, $0.92, $0.86, $0.825, $0.78, $0.835, $0.90, $1.00. In real dollars, the minimum wage increased from 1960 to 1970, then decreased from 1970 to 2008, and increased from 2008 to 2009. In 2009 it is now equivalent to the minimum wage in 1960.

16.21 The percent increase in median income is $\left(\frac{49{,}445 - 48{,}519}{48{,}519} \right) \times 100\% = 1.9\%$, and the percent increase in real income of top earners is $\left(\frac{180{,}810 - 148{,}274}{148{,}274} \right) \times 100\% = 21.9\%$.

16.23 The weight reflects the difference in the cost of owning versus renting and also the fact that more people own than rent.

16.25 Those receiving the wages or payments might be affected by seasonal variation, for example, higher heating costs in the winter. Using the unadjusted CPI means that the payments they receive will be "escalated" to keep up with these seasonal variations.

16.27 The poverty level is going up even after adjusting for inflation.

16.29 Larger sample sizes give more information; more information means less uncertainty, hence less sampling variation and more accuracy.

16.31 (a) Number of crimes reported by type. Arrests and convictions by type of crime. **(b)** Questions such as: Have you been a victim of a crime in the past year? What type of crime? **(c)** Questions such as: Have you changed your habits because of fear of crime? Do you feel safe walking near your home at night?

Part II Review

II.1 The distribution is roughly symmetric, with Mississippi as an outlier.

II.3 7.1, 10.05, 12.05, 14.475, 20.6.

II.5 The mean for all states in Table II.1 is 12.05. Mississippi is a high outlier so we would expect the mean to decrease. The mean without Mississippi is 11.71.

II.7 (a) 500. **(b)** 68%.

II.9 (a) Inches. **(b)** Inches. **(c)** Inches. **(d)** No units.

II.11 Dolphin: 180 kg, 1600 g. Hippo: 1400 kg, 600 g.

II.13 (a) Decrease. **(b)** Increase.

II.15 800 g.

II.17 Roughly a straight line with r close to -1. From use, soap shrinks in size.

II.19 -56.1 g; prediction outside the range of the available data is risky.

II.21 $59,379.54.

II.23 First put all monetary amounts in terms of 2011 dollars; then create a line graph.

Year	Gold price	CPI	2011 price
1985	$327	107.6	$683.60
1987	$484	113.6	$958.37
1989	$399	124.0	$723.80
1991	$353	136.2	$582.99
1993	$392	144.5	$610.22
1995	$387	152.4	$571.21
1997	$290	160.5	$406.43
1999	$290	166.6	$391.55
2001	$277	177.1	$351.83
2003	$416	184.0	$508.56
2005	$513	195.3	$590.86
2007	$834	207.3	$904.97
2009	$1,083	214.5	$1,135.52
2011	$1,567	224.9	$1,567.01

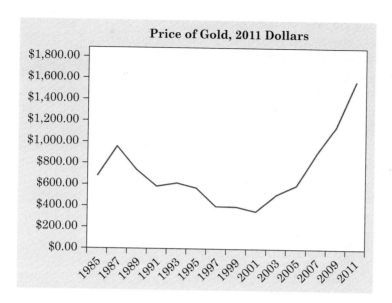

Price of Gold, 2011 Dollars

Although the value of gold dropped from 1987 to 2001 (in terms of 2011 dollars), an investment in gold most certainly has held its value today.

II.25 **(a)** Roughly symmetric; the two highest and one lowest time might be considered outliers.

II.27 The mean would be higher because housing prices have a right-skewed distribution.

II.29 **(a)** Fidelity Technology Fund because of the higher correlation. **(b)** No.

Chapter 17

17.3 Answers will vary, but we got 40% when we did the experiment.

17.5 The proportion from Table A is 41/400, or 0.1025.

17.7 Results will vary with the type of thumbtack used.

17.9 **(a)** 0. **(b)** 1. **(c)** 0.01. **(d)** 0.6.

17.11 **(b)** A personal probability might take into account information about your driving habits. **(c)** Most people believe they are better-than-average drivers.

17.17 This is a surprising coincidence; the exact probability would depend on the probability of a single individual male having the name Jim.

17.19 In a long sequence of free throws (say, 100), Nash will miss about 10% of the time (10 times out of the 100).

17.21 The "law of averages" is no more reliable for predicting whether an outcome will happen (such as getting a hit) in the short run than it is for other predictions. Ichiro is no more "due" for a hit than he was on his previous at-bat.

17.23 **(a)** Because each spin of the roulette wheel is independent, the law of averages says that the probability of a red (or black) on the next spin is the same regardless of what happened on previous rolls. **(b)** The gambler is wrong; now, events are not independent (if all 26 red cards are drawn from the deck, the next card drawn is guaranteed to be black), so the probability of a certain color being drawn is no longer the same for each draw.

17.25 (a) Answers vary; for example, an airplane crash is a rarer occurrence and involves more people. **(b)** The frequency with which people see stories on the news impacts their perceptions of the likelihood of the event.

Chapter 18

18.3 0.54.

18.5 (a) 0.55. **(b)** 0.48. **(c)** 0.52.

18.7 In Models 1, 3, and 4, the probabilities do not sum to 1; Model 4 has probabilities greater than 1. Model 2 is legitimate.

18.9 Each possible value (1, 2, 3, 4) has probability 1/4.

18.11 Possible totals: 2 through 8; probabilities 1/16, 2/16, 3/16, 4/16, 3/16, 2/16, 1/16. The probability is 4/16.

18.13 (a) 0.10. **(b)** 0.20. **(c)** 0.50; 0.50.

18.15 (a) 0.418 to 0.482. **(b)** 0.16.

18.17 0.001, or 0.10%.

18.19 (a) About 50%. Probability is the proportion of times the outcome would occur in a very long series of repetitions. **(b)** About 68%. **(c)** About 32%.

18.21 (a) 69.4. **(b)** Answers will vary. **(c)** Answers will vary.

Chapter 19

19.3 (a) 0 to 4 for Democrats, 5 to 9 for Republicans. **(b)** 0 to 5 for Democrats, 6 to 9 for Republicans. **(c)** 0 to 3 for Democrats, 4 to 7 for Republicans, and 8 or 9 for undecided. **(d)** 00 to 48 for Democrats, 49 to 86 for Republicans, and 87 to 99 for undecided.

19.5 (a) 4 chose Democrats, 6 chose Republicans. **(b)** 3 chose Democrats, 7 chose Republicans. **(c)** 2 chose Democrats, 4 chose Republicans, and 4 were undecided. **(d)** 5 chose Democrats, 3 chose Republicans, and 2 were undecided.

19.7 (a) 0.1. **(b)** 0 to 1 for A, 2 to 5 for B, 6 to 8 for C, 9 for D or F.

19.9 Results will vary depending on the starting point in Table A.

19.11 (a) With 0 to 2 meaning a made three-pointer, he makes at least 5 three-pointers in 2 of the 25 simulations. Thus, we estimate the probability of making at least 5 three-pointers to be 2/25 = 0.08. **(b)** The longest run of shots made was 3.

19.13 (a) 0 or 1 is a pass, 2 to 9 a failure. **(b)** 0.5. **(c)** No. The probability of passing probably increases on each trial because of learning from previous attempts.

19.15 (a) 0 is a narrow flat side, 1 to 4 is a broad concave side, 5 to 8 is a broad convex side, and 9 is a narrow hollow side. **(b)** Results will vary with the starting line in Table A.

19.17 (a) 0 to 8 means System A works. **(b)** 0 to 7 means System B works. **(c)** Results will vary with the starting line in Table A.

19.19 0 to 7 means a passenger shows up. Results will vary with the starting line in Table A.

19.21 0 to 5 means the van is available. Results will vary with the starting line in Table A.

19.23 (a) BBB, BBG, BGB, GBB, GGB, GBG, BGG, GGG. The first has probability 0.132651; the next three, 0.127449; the next three, 0.122451; and the last, 0.117649. (In practice, these should be rounded to 2 or 3 decimal places.) **(b)** 0.867349.

Chapter 20

20.3 $0.60.

20.5 $0.4996.

20.7 (a) 300. **(b)** There is no difference, except in phrasing. Saving 400 is the same as losing 200. **(c)** No. The choice seems to be based on how the options "sound."

20.9 8100 units.

20.11 The probabilities for the outcomes 2, 3, 4, ..., 12 are 1/36, 2/36, 3/36, 4/36, 5/36, 6/36, 5/36, 4/36, 3/36, 2/36, and 1/36, respectively. The expected value is 7.

20.13 The expected profit is $10.20 per customer.

20.15 (a) $500. **(b)** Although the probability is very small, having to pay $1,000,000 would probably be financially disastrous for you. **(c)** The insurance company will earn about $500 per policy by the law of large numbers, and because they sell lots of policies, they can afford the rare case of paying $1,000,000.

20.17 (a) 2.3. **(b)** 0 for A, 1 to 4 for B, 5 to 7 for C, 8 for D, and 9 for F. Results will vary with the starting line in Table A.

20.19 (b) Results will vary depending on the approach to the simulation. Using 0 to 4 for heads and 5 to 9 for tails, the estimate of the expected value is $0.84.

20.21 Results will vary depending on the approach to the simulation.

20.23 (a) $0.9025. **(b)** $0.8574.

Part III Review

III.1 (b) If all are equally likely, this would be 30%.

III.3 (a) 0.1. **(b)** Use 0 to 3 for type O, 4 to 6 for A, 7 and 8 for B, and 9 for AB.

III.5 Results will vary with the starting point in Table A.

III.7 3.5.

III.9 (a) 0.03. **(b)** Use 00 to 91 for one pair or worse and 92 to 99 for two pairs or better. Results will vary with the starting point in Table A.

III.11 (a) All probabilities are between 0 and 1, and the sum of all the probabilities is 1. **(b)** 0.438. **(c)** 0.403.

III.13 (a) 0.03. **(b)** 0.83.

III.15 (a) 68%. **(b)** 95%.

III.17 0.0124, or 1.24%.

III.19 Correct assignment: each face value has probability 1/13.

III.21 Ohio State: 0.2. Purdue, Michigan, Indiana, Illinois, and Wisconsin: 0.1.

Chapter 21

21.5 (a) The 15,000 alumni. **(b)** p is the proportion of all alumni who support the president's decision. **(c)** 0.33.

21.7 (a) p is the true proportion of all U.S. adults who think that there should be a law to ban the possession of handguns. **(b)** Using $z^* = 2$, the 95% confidence interval is 0.26 ± 0.03, or 0.23 to 0.29. **(c)** The margin of error we found here $(2\sqrt{\frac{(0.26)(0.74)}{1005}} \approx 0.03)$ is slightly smaller than what Gallup provides.

21.9 (a) Here, p is the true proportion of adults aged 18 and over who say that they or a member of their household was a victim of computer or Internet crime on their home computer in the past year. **(b)** Using $z^* = 2$, the 95% confidence interval is 0.11 ± 0.02, or (0.09, 0.13).

21.11 (a) 0.75 to 0.81. **(b)** 0.76 to 0.80. **(c)** 0.765 to 0.795. **(d)** As the sample size increases, the width of the confidence interval decreases.

21.13 Answers will vary based on the thumbtack used.

21.15 Using $z^* = 2$, the 95% confidence interval is 0.5069 ± 0.0157, or 0.491 to 0.523. From the confidence interval, we would conclude that the probability is quite likely to be 1/2, since 0.50 is inside the interval.

21.17 (a) The sampling distribution is approximately Normal with mean 0.28 and standard deviation 0.018. **(b)** The likelihood is 16% and 84%, respectively.

21.19 The sample proportion is $\hat{p} = \frac{68}{500} = 0.136$. A 95% confidence interval using the quick method is 0.091 to 0.181; a more accurate confidence interval (using the methods from this chapter) is 0.105 to 0.167.

21.21 (a) In Table A, let the numbers 0 to 5 represent voters who favor the reelection of the congresswoman and the numbers 6 to 9 represent the voters who do not favor her reelection. Starting from a line of your choice, read off the first 25 numbers, and calculate the proportion that are 0, 1, 2, 3, 4, or 5. **(b), (c)** Answers vary based on starting row. For example, starting with row 101, the sample proportion of voters who favor the reelection is 16/25, giving a 68% confidence interval of 0.544 to 0.736.

21.23 For a 99% confidence interval, we want to use $z^* = 2.58$, so the confidence interval is 0.68 ± 0.04, or 0.64 to 0.72. With a higher level of confidence, the interval is wider than the one calculated in Exercise 21.8.

21.25 The 90% confidence interval is 0.085 ± 0.018.

21.29 (a) The mean is 105.8 and the standard deviation is 14.3. **(b)** The 95% confidence interval is 105.8 ± 5.03. **(c)** If the scores come from a single school, we do not have a random sample, and the data are not representative of the population (all middle school girls in the district).

21.31 (a) The 95% confidence interval is 114.9 ± 3.5. **(b)** We assumed that the 27 men from the placebo group are representative of the population from which the subjects were recruited.

21.33 The 99% confidence interval is $\$346.00 \pm \$20.43$.

21.35 (a) The sample size ($n = 200$) is large enough. **(b)** For example, cost-of-living increases or increased unemployment.

Chapter 22

22.5 Conclusion 1: The difference in earnings in our sample was so large that it would rarely occur (probability 0.028) in samples drawn from a population in which men's and women's earnings are equal. Conclusion 2: A difference this large would not be unexpected (probability 0.576) in samples drawn from a population in which black and white earnings are equal.

22.7 (b) The difference could be plausibly attributed to chance.

22.9 The difference could be plausibly attributed to chance.

22.11 (a) People might be embarrassed to admit they did not attend religious services. **(b)** p is the proportion of American adults who attended religious services last week. Use H_0: $p = 0.39$ and H_a: $p < 0.39$.

22.13 H_0: $p = 0.09$ and H_a: $p \neq 0.09$.

22.15 (a) p is the graduation rate for all athletes at this university. **(b)** H_0: $p = 0.78$ and H_a: $p < 0.78$. **(c)** 0.721. The P-value is the probability that $\hat{p} < 0.721$ when we assume a 78% graduation rate. **(d)** P-values as extreme as 0.721 would be unlikely if the true proportion were 0.78.

22.17 (a) Here, the parameter p is the true proportion of times the candidate with the better face wins the contest. **(b)** H_0: $p = 0.50$ and H_a: $p > 0.50$. **(c)** $\hat{p} = 0.6875$ and the P-value is the probability, computed assuming $p = 0.50$, that we would observe a proportion > 0.6875. **(d)** The P-value tells us that it is very unlikely (probability $= 0.017$) that we would obtain a result as extreme as we observed just by chance if H_0 is true.

22.19 From Exercise 22.16, the P-value is 0.0036. This result is statistically significant at both the 5% and 1% levels.

22.21 A test is significant at the 1% level if outcomes as or more extreme than observed occur less than once in 100 times. A test is significant at the 5% level if outcomes as or more extreme than observed occur less than 5 in 100 times. Something that occurs less than once in 100 times also occurs less than 5 in 100 times, but the opposite is not necessarily true.

22.23 (a) Use digits 0 to 4 for "Method A wins trial" and 5 to 9 for "Method B wins trial," and take 20 digits from Table A. **(b)** The numbers of times B won in our 10 simulations were 12, 13, 11, 12, 11, 10, 8, 14, 9, 12. The estimated P-value is 0.5. **(c)** We simulated the probability of observing results at least as extreme as those in our sample.

22.25 The P-value is less than 0.0006 using Table B. It is consistent with part (d) of Exercise 22.16.

22.27 (a) Approximately Normal with mean 0.1 and standard deviation 0.01464. **(b)** The P-value is less than 0.0003, so the evidence is very strong.

22.29 The P-value is less than 0.0003, so the evidence is very strong that fewer than half of all drivers are speeding.

22.31 H_0: $\mu = 19$ and H_a: $\mu < 19$.

22.33 The P-value is about 0.25, so we have little reason to believe that the mean of all numbers produced by this software is not 0.5.

22.35 Here, p is the tip percentage left by customers at the restaurant; H_0: $p = 20$ and H_a: $p < 20$. The standard score for these data is -4.05, which gives a P-value of less than 0.0003; therefore, we conclude that there *is* strong evidence to suggest that the mean tip percentage for all patrons of this restaurant is less than 20 when they receive a message warning them of bad weather.

Chapter 23

23.3 Our confidence method can be applied only to an SRS.

23.5 The entire confidence interval contains values of p larger than 50%.

23.7 (a) No. In a sample of 200 people we would expect to see 2 people with $P < 0.01$. **(b)** Test this person again.

23.9 Yes. The P-value is quite small; this means we would reject the null hypothesis of no preference between designs at a significance level of 0.02.

23.11 For example, we would want to know about the sampling methods—is this a random sample?

23.13 Only question (b) could be answered by a test of significance.

23.15 (a) The width of the confidence interval (or, equivalently, the margin of error) will decrease. **(b)** The P-value will decrease.

23.17 Using $z^* = 1.96$, the confidence interval is 0.72 ± 0.065.

23.19 (a) The confidence interval is 159 ± 3. **(b)** We might suspect the validity of the confidence interval since the data were self-reported.

23.21 For the case in which 868 passed, the confidence interval is 0.868 ± 0.021; for the case in which 869 passed, the confidence interval is 0.869 ± 0.021.

Chapter 24

24.5 At least one close family member smokes: 35.7% of the students smoke. No close family member smokes: 25% of the students smoke. The smoking habits of students are associated with the presence or absence of close family members who smoke.

24.7 (a) Hatched: 16, 38, 75. Did not hatch: 11, 18, 29. **(b)** Cold: 59.3%. Neutral: 67.9%. Hot: 72.1%. Cold did not prevent hatching, but it did make it less likely.

24.9 (a) About 1,721,000 earned bachelor's degrees. **(b)** 57.5%, 61.1%, 49.5%, 53.3%. Women earned the majority of bachelor's, master's, and doctorate degrees and about half of the professional degrees.

24.11 Start by setting a equal to any number between 0 and 40.

24.13 (a) Maori, in jury pool: 80. Non-Maori, in jury pool: 314. Maori, not in jury pool: 10,138. Non-Maori, not in jury pool: 56,353.

24.15 (a) Yes. **(b)** The chi-square statistic is $\chi^2 = 47.55$ with 4 degrees of freedom. The P-value is less than 0.001, so we conclude that there is a statistically significant relationship between gender and opinion of adults on this matter.

24.17 (a) H_0: no relationship between the smoking habits of students and their families. H_a: some relationship between the smoking habits of students and their families.

According to the members in the sample, the population appears to be male students in secondary school in Malaysia. **(b)** The expected cell counts are 106.8, 215.2, 33.2, and 66.8. **(c)** The chi-square statistic is 3.95 with 1 degree of freedom. The result is significant at the $\alpha = 0.05$ level, but not significant at $\alpha = 0.01$.

24.19 (a) The two-way table is as follows:

Treatment	Cardiac event	No cardiac event
Stress management	3	30
Exercise	7	27
Usual care	12	28

(b) If we define success as "no cardiac event," the success rates are 30/33 (90.9%), 27/34 (79.4%), and 28/40 (70%). **(c)** The expected counts are 6.8, 26.2, 7.0, 27.0, 8.2, and 31.8. **(d)** The chi-square statistic is $\chi^2 = 4.84$ with 2 degrees of freedom, which is significant only at the $\alpha = 0.15$ level (that is, not significant by usual standards).

Part IV Review

IV.1 Using $z^* = 1.96$, the confidence interval is between 10.9% and 15.1%.

IV.3 Between about 12.8% and 17.2%.

IV.5 Using $z^* = 1.96$, the confidence interval is between 20.5% and 25.6%.

IV.7 You have information about all states, which is not just a sample.

IV.9 (a) If there were no difference between the two groups of doctors, results like these would almost never happen by chance. **(b)** The two proportions are given for comparison.

IV.11 Mornings: 0.550 to 0.590. Evenings: 0.733 to 0.767. The evening proportion is quite a bit higher and the two intervals do not overlap.

IV.13 (a) H_0: $p = 0.57$ and H_a: $p > 0.57$. **(b)** Normal with mean 0.57 and standard deviation 0.01. **(c)** Yes.

IV.15 (a) Either people are reluctant to admit that they don't attend regularly, or they believe they are more regular than they really are. **(b)** Sample results vary from the population truth. **(c)** Both intervals are based on methods that work 95% of the time.

IV.17 H_0: $p = 1/3$ and H_a: $p > 1/3$; the sampling distribution (assuming H_0 is true) is Normal with mean 1/3 and standard deviation 0.015; the sample proportion is 337/1020, or 0.3304; the P-value is approximately 0.58; the result is nonsignificant at $\alpha = 0.05$, so we conclude that the proportion of American adults who completely abstain from drinking alcoholic beverages is *not* more than 33%.

IV.19 0.286 to 0.354.

IV.21 H_0: $p = 0.57$ and H_a: $p > 0.57$; standard score $= 17.99$; $P < 0.0003$; significant for any reasonable choice of α.

IV.23 Standard score $= -2.77$; $P = 0.0028$.

IV.25 (b) $\bar{x} = 15.59$ ft and $s = 2.550$ ft; 15.0 to 16.2. **(c)** What population are we examining: for example, full-grown sharks, male sharks?

IV.27 Standard score $= 1.53$; $P = 0.063$.

IV.29 **(a)** No complaint: 743. Medical complaint: 199. Nonmedical complaint: 440. Stayed: 1306. Left: 76. **(b)** 2.96%, 13.07%, 6.36%. **(c)** Expected counts: 702.14, 188.06, 415.80; 40.86, 10.94, 24.20. All counts are greater than 5. **(d)** H_0: There is no relationship between a member's complaining and leaving the HMO. H_a: There is some relationship. df $= 2$; this is very significant. Complaining and leaving are associated.

IV.31 **(a)** Yes: 1313, 1840; No: 991, 614. Morning: 57.0%. Evening: 75.0%. **(b)** We have very large samples with very different proportions. **(c)** $\chi^2 = 172$ with df $= 1$; very significant.

Index

Note: Page numbers in **boldface** type indicate pages where key terms are defined.

TABLE A Random digits

101	19223	95034	05756	28713	96409	12531	42544	82853
102	73676	47150	99400	01927	27754	42648	82425	36290
103	45467	71709	77558	00095	32863	29485	82226	90056
104	52711	38889	93074	60227	40011	85848	48767	52573
105	95592	94007	69971	91481	60779	53791	17297	59335
106	68417	35013	15529	72765	85089	57067	50211	47487
107	82739	57890	20807	47511	81676	55300	94383	14893
108	60940	72024	17868	24943	61790	90656	87964	18883
109	36009	19365	15412	39638	85453	46816	83485	41979
110	38448	48789	18338	24697	39364	42006	76688	08708
111	81486	69487	60513	09297	00412	71238	27649	39950
112	59636	88804	04634	71197	19352	73089	84898	45785
113	62568	70206	40325	03699	71080	22553	11486	11776
114	45149	32992	75730	66280	03819	56202	02938	70915
115	61041	77684	94322	24709	73698	14526	31893	32592
116	14459	26056	31424	80371	65103	62253	50490	61181
117	38167	98532	62183	70632	23417	26185	41448	75532
118	73190	32533	04470	29669	84407	90785	65956	86382
119	95857	07118	87664	92099	58806	66979	98624	84826
120	35476	55972	39421	65850	04266	35435	43742	11937
121	71487	09984	29077	14863	61683	47052	62224	51025
122	13873	81598	95052	90908	73592	75186	87136	95761
123	54580	81507	27102	56027	55892	33063	41842	81868
124	71035	09001	43367	49497	72719	96758	27611	91596
125	96746	12149	37823	71868	18442	35119	62103	39244
126	96927	19931	36809	74192	77567	88741	48409	41903
127	43909	99477	25330	64359	40085	16925	85117	36071
128	15689	14227	06565	14374	13352	49367	81982	87209
129	36759	58984	68288	22913	18638	54303	00795	08727
130	69051	64817	87174	09517	84534	06489	87201	97245
131	05007	16632	81194	14873	04197	85576	45195	96565
132	68732	55259	84292	08796	43165	93739	31685	97150
133	45740	41807	65561	33302	07051	93623	18132	09547
134	27816	78416	18329	21337	35213	37741	04312	68508
135	66925	55658	39100	78458	11206	19876	87151	31260
136	08421	44753	77377	28744	75592	08563	79140	92454
137	53645	66812	61421	47836	12609	15373	98481	14592
138	66831	68908	40772	21558	47781	33586	79177	06928
139	55588	99404	70708	41098	43563	56934	48394	51719
140	12975	13258	13048	45144	72321	81940	00360	02428
141	96767	35964	23822	96012	94591	65194	50842	53372
142	72829	50232	97892	63408	77919	44575	24870	04178
143	88565	42628	17797	49376	61762	16953	88604	12724
144	62964	88145	83083	69453	46109	59505	69680	00900
145	19687	12633	57857	95806	09931	02150	43163	58636
146	37609	59057	66967	83401	60705	02384	90597	93600
147	54973	86278	88737	74351	47500	84552	19909	67181
148	00694	05977	19664	65441	20903	62371	22725	53340
149	71546	05233	53946	68743	72460	27601	45403	88692
150	07511	88915	41267	16853	84569	79367	32337	03316

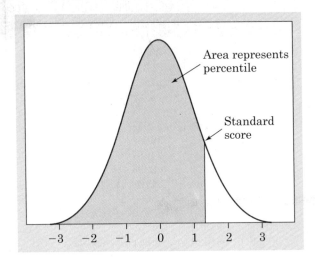

TABLE B Percentiles of the Normal distributions

Standard score	Percentile	Standard score	Percentile	Standard score	Percentile
−3.4	0.03	−1.1	13.57	1.2	88.49
−3.3	0.05	−1.0	15.87	1.3	90.32
−3.2	0.07	−0.9	18.41	1.4	91.92
−3.1	0.10	−0.8	21.19	1.5	93.32
−3.0	0.13	−0.7	24.20	1.6	94.52
−2.9	0.19	−0.6	27.42	1.7	95.54
−2.8	0.26	−0.5	30.85	1.8	96.41
−2.7	0.35	−0.4	34.46	1.9	97.13
−2.6	0.47	−0.3	38.21	2.0	97.73
−2.5	0.62	−0.2	42.07	2.1	98.21
−2.4	0.82	−0.1	46.02	2.2	98.61
−2.3	1.07	0.0	50.00	2.3	98.93
−2.2	1.39	0.1	53.98	2.4	99.18
−2.1	1.79	0.2	57.93	2.5	99.38
−2.0	2.27	0.3	61.79	2.6	99.53
−1.9	2.87	0.4	65.54	2.7	99.65
−1.8	3.59	0.5	69.15	2.8	99.74
−1.7	4.46	0.6	72.58	2.9	99.81
−1.6	5.48	0.7	75.80	3.0	99.87
−1.5	6.68	0.8	78.81	3.1	99.90
−1.4	8.08	0.9	81.59	3.2	99.93
−1.3	9.68	1.0	84.13	3.3	99.95
−1.2	11.51	1.1	86.43	3.4	99.97